Water Wave Scattering

Water Wave Scattering

B.N. Mandal
Physics and Applied Mathematics Unit
Indian Statistical Institute
Kolkata, India

Soumen De
Department of Applied Mathematics
University of Calcutta
Kolkata, India

CRC Press
Taylor & Francis Group
Boca Raton London New York

CRC Press is an imprint of the
Taylor & Francis Group, an **informa** business
A SCIENCE PUBLISHERS BOOK

CRC Press
Taylor & Francis Group
6000 Broken Sound Parkway NW, Suite 300
Boca Raton, FL 33487-2742

Printed on acid-free paper
Version Date: 20150330

International Standard Book Number-13: 978-1-4987-0552-3 (Hardback)

Visit the Taylor & Francis Web site at
http://www.taylorandfrancis.com

and the CRC Press Web site at
http://www.crcpress.com

Preface

The theory of water waves is most varied and a fascinating topic. It includes a wide range of natural phenomena in oceans, rivers and lakes. It is mostly concerned with elucidation of some general aspects of wave motion including the prediction of behaviour of waves in the presence of obstacles of some special configurations that are of interest to ocean engineers. Unfortunately, even the apparently simple problems appear to be difficult to tackle mathematically unless some simplified assumptions are made. Fortunately for water, one can assume it to be an incompressible, inviscid and homogeneous fluid. The linearized theory of water waves is based on the assumption that the amplitude of the motion is small compared to the wave length. If irrotational motion is assumed, then the linearized theory of water waves is essentially concerned with solving the Laplace equation in the water region together with linearized boundary condition. There are varied classes of problems which have been/are being studied mathematically in the literature within the frame work of linearized theory of water waves for last many years. Scattering by obstacles of various geometrical configurations is one such class of water wave problems. This book is devoted to advanced mathematical work related to water wave scattering. Emphasis is given on the mathematical and computational techniques required to study these problems mathematically.

The book contains nine chapters. The first chapter is introductory in nature. It includes the basic equations of linearized theory for a single layer fluid, a two-layer fluid, solution of dispersion equations and a general idea on scattering problems and the energy identity in water with a free surface. Chapter 2 is concerned with wave scattering involving thin rigid plates of various geometrical configurations, namely, plane vertical barriers, or curved barriers, inclined barriers, horizontal barriers and also thin elastic vertical plates. For the horizontal case, the barrier is submerged below an ice-cover modelled as a thin elastic plate floating on water. Chapter 3 discusses wave scattering by a rectangular trench by using the Galerkin technique. Chapter 4 involves wave scattering by a dock by using the Carleman singular integral equation followed by reduction to Riemann-Hilbert problems. Chapter 5 involves several wave scattering problems involving discontinuities at the upper surface of water by using the Wiener-Hopf technique, by reduction to the Carleman singular integral equations. Chapter 6 considers scattering by a long horizontal circular cylinder either half immersed or completely submerged. In Chapter 7, some important energy identities are derived for scattering problems in a single-layer and also in a two-layer fluid. Chapter 8 is concerned with wave scattering in a two-layer fluid by a thin vertical plate and by a long horizontal circular cylinder submerged in either of the two layers. Chapter 9 considers a number of wave scattering problems in a single-layer or a two-layer fluid with variable bottom topography by using a simplified perturbation analysis.

It is hoped that this book will be useful to the researchers on water waves. The several wave scattering problems presented in the book are based mostly on the research work carried out by the authors and their associates. The authors thank their young colleagues Harpreet, Rumpa, Dilip and Paramita for their help in the preparation of the manuscript, Prof. S. Banerjea and Prof. U. Basu for their encouragement. They also thank all their associates who worked on problems on water waves.

January 2015

B.N. Mandal
Soumen De

Contents

CHAPTER I

Introduction

When a particle in a continuous medium is slightly disturbed from its position of rest then the disturbance is transferred to the neighboring particles and these particles disturb further particles. In this manner the disturbance thus created travels with a definite velocity throughout the medium. This is generally referred to as wave motion. Thus a wave may be regarded as a progressive disturbance propagating from point to point in a continuous medium without displacement of the points. Waves are encountered in almost all branches of mathematical physics such as continuum mechanics, quantum mechanics, acoustics, electromagnetic theory, etc. A wave may be intuitively defined as a recognizable signal that is transferred from one part of a medium to another part, and the signal may be any feature of disturbance which is clearly recognizable and its location at any time can be determined. Ever science waves were studied water waves have served the scientists as models since these can be viewed by the naked eye. Waves are generated due to the existence of some kind of restoring force that tends to bring the system back to its undisturbed state and some kind of inertia that causes the system to overshoot after the system returned to the undisturbed state. One of the common wave motions with which we are most familiar is that of waves occurring at free surface of the liquid with gravity playing the role of the restoring force. These waves are called surface gravity waves. Water waves are such waves. The various wave phenomena in water with a free surface and under gravity have attracted the attention of many famous physicists and mathematician from the eighteenth century. An incomplete list of them includes A.L. Cauchy (1758–1857), S.D. Poisson (1781–1840), J.L. Lagrange (1736–1813), G.B. Airy (1801–1892), G.G. Stokes (1819–1903), Lord Kelvin (Sir William Thompson) (1824–1907), J.H. Michell (1936–1921), H. Lamb (1849–1934), J.J. Stoker (1905–1992), F. Ursell (1923–2012), M.J. Lighthill (1924–1998) and many others. This resulted in a systematic development of the theory of water waves from the latter half of the eighteenth century. This theory has provided a background for somewhat rich development of some important mathematical concepts and techniques and consequently, it has become an important branch of applied mathematics and mathematical physics.

The theory of water waves is the most varied and fascinating subject. It includes a wide range of natural phenomena in oceans, rivers and lakes. The research activities in this area accelerated after the Second World War due to the explosive growth in ocean related industrial activities such as offshore drilling for oil production, construction of offshore structures, extraction of wave energy from ocean waves, design of breakwaters to protect ports, sea resorts and marinas from the rough sea, construction of very large floating structures (floating airports), etc. Interest in the mathematical study of problems on water wave scattering by floating and submerged bodies by applied mathematicians got momentum after the Second World War in UK due to unsuccessful attempts to use portable and floating breakwaters in the surprising amphibious landing by the allied army at Normandy.

In the mathematical study of water wave problems, two types of theories are employed, one is the linearized theory of water waves and the other is the nonlinear shallow water theory. If the wave-length is assumed to be much less than the depth of water, then the effect of disturbance diminishes gradually as one moves downwards away from the surface. Waves in this category are termed as surface waves. In this case if it is assumed that the wave amplitude is small compared to the wavelength, then the corresponding theory is called linearized theory of water waves. Under this theory, it is assumed that the velocity components of the wave potential and the deviation of the upper surface from its mean horizontal position together with their partial derivatives, are small quantities so that their products and powers of higher order can be neglected. Also, various simple assumptions regarding the fluid medium (i.e., water) are made, viz. it is homogeneous, incompressible and inviscid. The motion in water is under the action of gravity only and starts from rest so that it is irrotational. Based on linear theory, any water wave problem can be formulated as a boundary value problem or an initial value problem in which the governing partial differential equation (Laplace equation) is linear and the boundary conditions are linear. Such a formulation is based on the linearized theory of water waves.

A natural question arises about the validity of the linearized theory of water waves from the practical point of view. Ursell et al. (1960) experimented with the height of water waves generated by a flat vertical piston wave maker and obtained results which are in very good agreement with theoretical results predicted under the assumption of linearized theory. Dean and Ursell (1959) and Yu and Ursell (1961) also experimented with a horizontal cylinder in deep water as well as in uniform finite depth water. In both the experiments the experimental results on wave amplitude almost coincide with the theoretical results. These experimental evidences confirm and establish the validity of the linearized theory of water waves. In the present book water wave problems based on only the linearized theory have been considered. Since the last century many ocean technologists have been using the linearized theory to study mathematically various problems related to wave phenomena in ocean and successfully utilize various results predicted under linear theory to ocean related industrial problems.

A different kind of approximation from the foresaid linear theory of waves of small amplitude results when one assumes that the depth of water is sufficiently small compared to some significant length, say the wave length. Then it is not necessary to assume that the deviation of the upper surface of water (free surface) and its slope are small and the resulting theory is no longer a linear theory. There are many circumstances in nature under which such a theory leads to good approximations to various natural phenomena. Among such phenomena are the tides in the oceans, the solitary waves in sufficiently shallow water and the breaking of waves on shallow beaches. The assumption for the shallow water theory leads to a set of nonlinear equations even for the first order approximation and constitutes the theory for the study of long waves. The higher order approximations yield solutions corresponding to continuous permanent finite amplitude waves that can propagate without a change in forms and shapes, known as solitary waves. Again, if the wave amplitude is small so that the velocity components and the free surface elevation (or depression) are small, the nonlinear equations arising in the shallow water theory yield linear hyperbolic partial differential equations. These equations constitute the basis for the theory of tides in the ocean. However, these types of problems will not be considered in this book.

When a train of surface water waves travelling from a large distance is incident on an obstacle submerged or partially immersed in water, some part of the wave is reflected back by the obstacle and some part is transmitted over or below it. This type of problem is known as scattering problem. The reflected and transmitted wave fields involve two constant factors known as the reflection and transmission coefficients respectively. It is supposed that the incident wave field is completely known. However, the resulting wave field, after interacting with the obstacle, is unknown. Determination of the unknown scattered wave field together with the unknown reflection and transmission coefficients constitutes a scattering problem. The two physical quantities, namely the reflection and transmission coefficients are very important since they provide a measure for the amount of reflected and transmitted waves. This information is useful in the construction of offshore structures. Water wave scattering problems have practical importance for various engineering applications. Modeling of breakwaters constructed to protect offshore areas from the impact of a rough sea is one of them. Also, these have applications in designing ships, submarines, offshore structures, etc.

Again, if some part of water surface is covered by some materials such as broken ice (inertial surface) or a rigid plate or an elastic plate, then discontinuities arise in the surface boundary conditions at the regions where the materials meet the surface. Such discontinuities also cause hindrance to an incoming train of surface waves and the phenomena of reflection and transmission of the incident wave train occur in this situation also. For the last few decades there is a considerable interest in the study of various types of water wave problems involving water with a floating ice-cover modeled as a thin elastic plate. The study of these problems has gained considerable importance for quite some time due to two reasons, one is to understand the mechanism and effects of wave propagation through Marginal Ice Zone (MIZ) in polar regions while another is due to their applications in the construction of Very Large Floating Structures (VLFS) like floating oil storage bases, offshore pleasure cities, floating airport, artificial harbors, etc.

In a two-layer fluid with a free surface, waves propagate at two different modes, one along the free surface and the other along the interface between the two layers. The study of wave motion in a two-layer fluid has gained importance due to plans to construct an underwater pipe bridge across Norwegian fjords. A fjord consists of a layer of fresh water on the top of a deep layer of salt water. During winter the fjords are covered by ice, and thus a two-layer fluid where the upper layer has an ice-cover becomes a reality.

We now derive the basic equations of linearized theory of water waves for a single-layer and a two-layer fluid.

1 Basic equations in the linearized theory of water waves

The general hydrodynamic theory is based on the two natural laws governing the fluid motion. These are the laws of conservation of mass and momentum. The basic equations in the linearized theory of water waves are derived from these two laws. It is assumed that water is an inviscid, incompressible and homogeneous fluid and the motion in it is under the only the action of gravity and is irrotational. The assumption on the smallness of motion means that the velocity components together with their derivatives are quantities of the first order of smallness so that their squares, products and higher powers can be neglected. Also the deviation of the upper surface from its mean horizontal position together with its derivatives is assumed to be small. These constitute the basis for the linearized theory. The basic equations in a single-layer fluid (water) and also a two-layer fluid with free surface as well as with ice-cover modeled as thin elastic floating plate are derived below.

1.1 Water with free surface

We consider the motion in water which is assumed to be inviscid, incompressible and homogeneous with constant volume density ρ under the action of gravity g only and bounded above by a free surface. A rectangular Cartesian co-ordinate system is chosen in which the y-axis is taken vertically downwards and the plane $y = 0$ is the position of the undisturbed free surface.

We assume that the motion starts from rest so that it is irrotational and thus can be described by a velocity potential $\Phi(x, y, z; t)$. Hence the fluid velocity $\boldsymbol{q}(u, v, w)$ can be expressed as

$$\boldsymbol{q} = \nabla\Phi. \tag{1.1}$$

Now equation of continuity is

$$\nabla.\boldsymbol{q} = \boldsymbol{0} \text{ in the fluid region} \tag{1.2}$$

and the Euler's equation of motion in the fluid region is

$$\frac{\partial \boldsymbol{q}}{\partial t} + (\boldsymbol{q}.\nabla)\boldsymbol{q} = \nabla\left(gy - \frac{p}{\rho}\right) \tag{1.3}$$

where p is the fluid pressure.

Using (1.1) the equation of continuity becomes

$$\nabla^2\Phi = 0 \text{ in the fluid region.} \tag{1.4}$$

Integration of (1.3) produces after linearization the linearized Bernoulli's equation

$$\frac{\partial \Phi}{\partial t} = gy - \frac{p}{\rho} \text{ in the fluid region.} \tag{1.5}$$

The pressure p must be equal to the atmospheric pressure at the free surface $y = \eta(x, z, t)$, where η is the free surface depression below the mean horizontal level is

$y = 0$. Now the atmospheric pressure is constant and by a suitable choice of scale it may be taken as zero so that from (1.5) we obtain

$$\frac{\partial \Phi}{\partial t} = g\eta \text{ on } y = \eta(x, z, t). \tag{1.6}$$

This is the dynamical boundary condition at the free surface. Expanding $\frac{\partial \Phi}{\partial t}$ by Taylor's series about $y = 0$ and neglecting higher order terms, the above condition reduces to

$$\frac{\partial \Phi}{\partial t} = g\eta \text{ on } y = 0. \tag{1.7}$$

Again, the linearized kinematic condition at the free surface is

$$\frac{\partial \eta}{\partial t} = \frac{\partial \Phi}{\partial y} \text{ on } y = 0 \tag{1.8}$$

Eliminating η between (1.7) and (1.8) gives the linearized free surface condition as

$$\frac{\partial^2 \Phi}{\partial t^2} = g\frac{\partial \Phi}{\partial y} \text{ on } y = 0. \tag{1.9}$$

The condition of no motion at the bottom gives

$$\nabla \Phi \to 0 \text{ as } y \to \infty \tag{1.10}$$

for deep water and

$$\frac{\partial \Phi}{\partial y} = 0 \text{ on } y = h \tag{1.11}$$

for water of uniform finite depth h.

If the motion is simple harmonic in time with angular frequency ω, then the velocity potential can be expressed as

$$\Phi(x, y, z, t) = \text{Re}\{\phi(x, y, z, t)e^{-i\omega t}\}, \tag{1.12}$$

so that equation (1.4) and the conditions (1.9) to (1.11) become

$$\nabla^2 \phi = 0 \text{ in the fluid region,} \tag{1.13}$$

$$K\phi + \phi_y = 0 \text{ on } y = 0 \tag{1.14}$$

where $K = \frac{\omega^2}{g}$, is called the wave number,

$$\nabla \phi \to 0 \text{ as } y \to \infty \tag{1.15}$$

for deep water,

$$\phi_y = 0 \text{ on } y = h \tag{1.16}$$

for water of uniform finite depth h.

For the two-dimensional case when ϕ is independent of z, the solution of the Laplace equation representing progressive waves is given by

$$\phi(x,y) = \begin{cases} e^{-Ky \pm iKx}, \text{ for deep water,} & (1.17) \\ \dfrac{\cosh k_0 (h-y)}{\cosh k_0 h} e^{\pm ik_0 x}, \text{ for water of depth } h & (1.18) \end{cases}$$

where k_0 is the unique real positive root of the transcendental equation

$$k \tanh kh = K. \tag{1.19}$$

The local solutions are given by

$$\phi = \begin{cases} (k \cos ky - K \sin ky) e^{-k|x|} \ (k > 0), \text{ for deep water,} & (1.20) \\ \dfrac{\cosh k_n (h-y)}{\cosh k_n h} e^{-k_n |x|}, \text{ for water of depth } h, & (1.21) \end{cases}$$

where $\pm ik_n$'s ($n = 1, 2, \cdots$) are the purely imaginary roots of the transcendental equation (1.19). The equation (1.19) is in fact is a relation between the wave number and the angular frequency of a train of surface gravity waves and is known as the dispersion equation. The term is due to the fact that waves whose velocity depends on wave number disperse or separate. Wave dispersion is a fundamental process in many physical phenomena. The dispersion relation (1.9) has roots $\pm k_0$ and $\pm ik_n$ ($n = 1, 2, \cdots, k_1 < k_2 < \cdots$) and there is no other root. These can be easily computed numerically for given K (i.e., given the angular frequency).

1.2 Water with an ice-cover

In this case we consider the motion in water covered by a thin sheet of ice modeled as a thin elastic plate of uniform surface density $\epsilon\rho$, Young's modulus E and Poisson's ratio γ, ϵ being constant having dimension of length, ρ being the density of water. As before we choose the rectangular Cartesian co-ordinate system so that the water (at rest) occupies the region $y \geq 0$, the plane $y = 0$ is the position of the thin ice-cover at rest.

We assume that the motion starts from rest so that it is irrotational and thus can be described by a velocity potential $\Phi(x, y, z, t)$. Then the equation of continuity gives

$$\nabla^2 \Phi = 0 \tag{2.1}$$

in the fluid region and as before the linearized Bernoulli equation is

$$\Phi_t = gy - \frac{p}{\rho}. \tag{2.2}$$

Let $y = \zeta(x, z, t)$ denote the depression of the ice-covered surface below the mean horizontal level. Then Newton's equation of motion for a small element of the ice-covered surface produces Landau and Lifshitz (1959)

$$\epsilon\rho \frac{\partial^2 \zeta}{\partial t^2} = \epsilon\rho g + \Pi - p - L\Delta^4_{(x,z)}\zeta \text{ on } y = 0 \tag{2.3}$$

where Π is the atmospheric pressure and $L = \frac{Eh_0^3}{12(1-\gamma^2)}$ is the flexural rigidity of the thin ice sheet, h_0 being the very small thickness of ice of which still a smaller part is immersed into water and $\Delta^4_{(x,z)}$ representing the two-dimensional bi-harmonic operator. Using (2.2) in (2.3) we get

$$\epsilon\rho \frac{\partial^2\zeta}{\partial t^2} = \epsilon\rho g + \Pi - \rho\left(g\eta - \frac{\partial\Phi}{\partial t}\right) - L\Delta^4_{(x,z)}\zeta \text{ on } y = 0 \tag{2.4}$$

after linearization. The kinematic condition at the ice-cover is

$$\frac{\partial\zeta}{\partial t} = \frac{\partial\Phi}{\partial y} \text{ on } y = 0. \tag{2.5}$$

Eliminating η between (2.4) and (2.5) we get

$$(\Phi - \epsilon\Phi_y)_{tt} = (1 + D\Delta^4_{(x,z)})g\Phi_y \text{ on } y = 0 \tag{2.6}$$

where $D = \frac{L}{\rho g}$. Also Φ satisfies the bottom condition

$$\nabla\Phi \to 0 \text{ as } y \to \infty \tag{2.7}$$

for deep water and

$$\frac{\partial\Phi}{\partial y} = 0 \text{ on } y = h \tag{2.8}$$

for water of uniform depth h.

Assuming as before the motion to be simple harmonic in time with angular frequency ω, then Φ can be written as

$$\Phi(x, y, z, t) = \text{Re}\{\phi(x, y, z)e^{-i\omega t}\}. \tag{2.9}$$

In this case ice-cover condition (2.6) becomes

$$K\phi + (1 - \epsilon K + D\nabla^4_{(x,y)})\phi_y = 0 \tag{2.10}$$

If the ice-cover is modeled as inertial surface (i.e., non-interacting floating material having no elastic property, e.g., broken ice, floating mat) then $L = 0$ so that $D = 0$. Then the condition (2.10) reduces to

$$K\phi + (1 - \epsilon K)\phi_y = 0 \text{ on } y = 0. \tag{2.11}$$

This takes the form

$$K^*\phi + \phi_y = 0 \text{ on } y = 0, \tag{2.12}$$

if $(1 - \epsilon K) > 0$, where $K^* = \frac{K}{1-\epsilon K} > 0$ and the form

$$K_0\phi - \phi_y = 0 \text{ on } y = 0, \tag{2.13}$$

if $(1 - \epsilon K) < 0$, where $K_0 = \frac{K}{\epsilon K - 1} > 0$. Comparing (2.12) and (2.13) with the usual free surface condition it may be noted that progressive wave along an inertial surface exists if and only if $(1 - \epsilon K) > 0$, i.e., $\omega < \sqrt{g/\epsilon}$. For $(1 - \epsilon K) < 0$, i.e., $\omega > \sqrt{g/\epsilon}$, the condition

(2.13) does not allow existence of any progressive wave at the inertial surface. In this case, the inertial surface is regarded as heavy.

For simplicity let us consider two dimensional motions so that ϕ is a function of x, y only. If we choose $\phi(x,y) = e^{-ky \pm ikx}$, then from the ice-cover condition (2.10) we find that k satisfies the fifty degree polynomial equation

$$D^* k^5 + k - K^* = 0 \tag{2.14}$$

if $(1 - \epsilon K) > 0$, where $D^* = \frac{D}{1-\epsilon K} > 0$,

$$D_0 k^5 + k + K_0 = 0 \tag{2.15}$$

if $(1 - \epsilon K) < 0$, where $D_0 = \frac{D}{\epsilon K - 1} > 0$. It is obvious that the nature of the roots of the polynomial equations (2.14) and (2.15) is the same. Both of them possess unique real positive root. This shows the existence of progressive waves at the ice-cover for any frequency.

For water of uniform finite depth we choose

$$\phi(x, y) = \cosh k(h - y) e^{\pm ikx}$$

then k satisfies the transcendental equation

$$k(D^* k^4 + 1) \sinh kh - K^* \cosh kh = 0 \tag{2.16}$$

for $(1 - \epsilon K) > 0$ and

$$k(D_0 k^4 - 1) \sinh kh - K_0 \cosh kh = 0 \tag{2.17}$$

for $(1 - \epsilon K) < 0$. Each of these two equations (2.16) and (2.17) has a unique real positive root, thus confirming the existence of time-harmonic progressive waves on the ice-cover.

The polynomial equations (2.14) or (2.15) and the transcendental equations (2.16) or (2.17) are the dispersion equations. In section 1.6 we will discuss about the solutions of various dispersion equations in some detail.

For the two-dimensional case, the progressive wave solutions are given by

$$\phi(x, y) = e^{-\lambda_0 y \pm i\lambda_0 x} \tag{2.18}$$

for deep water with ice-cover, where λ_0 is the unique positive root of the polynomial equation

$$Dk^5 + (1 - \epsilon K)k - K = 0 \tag{2.19}$$

and

$$\phi(x, y) = \frac{\cosh \mu_0 (h - y)}{\cosh \mu_0 h} e^{\pm i\mu_0 x} \tag{2.20}$$

for uniform finite depth with an ice-cover, where μ_0 is the unique positive root of the transcendental equation

$$k(Dk^4 + 1 - \epsilon K) \sinh kh - K \cosh kh = 0. \tag{2.21}$$

The local solutions are given by

$$\phi(x,y) = e^{-\lambda_1 y \pm i\lambda_1 x},\ e^{-\bar{\lambda}_1 y \pm i\bar{\lambda}_1 x},\ \{k(Dk^4 + 1 - \epsilon K)\sinh kh - K\cosh kh\}e^{-k|x|} \quad (2.22)$$

for deep water with an ice-cover, λ_1, $\bar{\lambda}_1$ are roots of (2.19) with Re $\lambda_1 > 0$, Im $\lambda_1 > 0$.

The other pair of complex roots of (2.19) is λ_2, $\bar{\lambda}_2$ with Re $\lambda_2 < 0$, Im $\lambda_2 > 0$. An elementary proof of the nature of the roots of the polynomial equation (2.19) is given by Chakrabarti et al. (2003).

For water of uniform finite depth h, the local solutions are given by

$$\phi(x,y) = \frac{\cosh \mu_1 (h-y)}{\cosh \mu_1 h} e^{\pm i\mu_1 x},\ \frac{\cosh \bar{\mu}_1 (h-y)}{\cosh \bar{\mu}_1 h} e^{\pm i\bar{\mu}_1 x},\ \frac{\cosh k_n (h-y)}{\cosh k_n h} e^{-k_n|x|}, \quad (2.23)$$

where $\pm\mu_1$, $\pm\bar{\mu}_1$ are the two pairs of complex conjugate roots of the transcendental equation (2.21) with Re $\mu_1 > 0$, Im $\mu_1 > 0$, (Im $\mu_1 >$ Re μ_1) and $\pm ik_n (n = 1,2,\ldots)$ are the purely imaginary roots of (2.21) where $\left(n-\frac{1}{2}\right)\pi < k_n < n\pi$ and $k_n h \to n\pi$ as $n \to \infty$ cf. Chung and Fox (2002).

1.3 Two-layer fluid with a free surface

We consider the motion under gravity in two superposed, incompressible and inviscid fluids of densities ρ_1 and $\rho_2 (\rho_1 < \rho_2)$ of the upper and lower layers respectively separated by a common interface. The upper fluid is of uniform finite depth h below the mean free surface. The motion is assumed to be irrotational and for simplicity it is further assumed to be two-dimensional. Thus it can be described by the velocity potentials $\Phi(x, y, t)$ and $\Psi(x, y, t)$ in the upper and lower layers respectively. Also the motion is assumed to be simple harmonic in time with angular frequency ω. Then the velocity potentials Φ and Ψ can be written as $\Phi = Re\{\phi(x,y)e^{-i\omega t}\}$ and $\Psi = Re\{\psi(x,y)e^{-i\omega t}\}$, where ϕ and ψ are complex valued potential functions in the upper and lower layers respectively. Then ψ and ϕ satisfy

$$\nabla^2\phi = 0 \text{ in the upper layer,} \quad (3.1)$$

$$\nabla^2\psi = 0 \text{ in the lower layer,} \quad (3.2)$$

the linearized boundary conditions at the interface are obtained from the condition of continuity of the normal velocity and the pressure across the interface and are given by

$$\phi_y = \psi_y \text{ on } y = h, \quad (3.3)$$

$$K\psi + \psi_y = s(K\phi + \phi_y) \text{ on } y = h \quad (3.4)$$

where $s = \frac{\rho_1}{\rho_2} (< 1)$, $K = \frac{\omega^2}{g}$. The condition at the free surface is

$$K\phi + \phi_y = 0 \text{ on } y = h \quad (3.5)$$

and if the lower fluid extends infinitely downwards then condition at infinite depth is

$$\nabla\psi \to 0 \text{ as } y \to h. \quad (3.6)$$

The two-dimensional progressive wave solutions for upper and lower layers are given by

$$\phi(x, y) = e^{\pm ikx} g(y), \text{ for the upper layer,} \tag{3.7}$$

$$\psi(x, y) = e^{\pm ikx - ky}, \text{ for the lower layer,} \tag{3.8}$$

with

$$g(y) = \frac{K-k}{K(\sigma-1)} e^{-ky} + \frac{K\sigma - k}{K(\sigma-1)} e^{ky} \tag{3.9}$$

where k satisfies the dispersion equation

$$(k-K)\{K(\sigma + e^{-2kh}) - k(1-e^{-2kh})\} = 0 \tag{3.10}$$

with $\sigma = \frac{1+s}{1-s}$. The dispersion equation (3.10) has exactly two real roots, one is $k = K$ and the other is $k = v$, say, where v satisfies the equation

$$K(\sigma + e^{-2vh}) = \upsilon(1 - e^{-2\upsilon h}), \tag{3.11}$$

so that

$$K\sigma < \upsilon < K\frac{\sigma+1}{1-e^{-2K\sigma h}}. \tag{3.12}$$

The waves of wave number K propagate along the free surface and the waves of wave number v propagate along the interface between the two layers.

If the lower fluid is of uniform finite depth H below the mean position of interface, then the bottom condition (3.6) changes to

$$\psi_y = 0, \text{ on } y = h + H \tag{3.13}$$

and the dispersion relation (3.10) modifies to

$$k^2(1-s) - kK(\coth kh + \coth kH) + K^2(s + \coth kh \coth kH) = 0. \tag{3.14}$$

The equation (3.14) can be shown to have exactly two real positive roots and thus there exists two types of progressive waves having two different wave numbers. The two dimensional progressive wave solutions for the velocity potentials in the upper and lower layers in this case are given by

$$\phi(x, y) = e^{\pm ikx} f(k, y),\ 0 < y < h, \tag{3.15}$$

$$\psi(x,y) = e^{\pm ikx} \cos k(H-y),\ h < y < h + H \tag{3.16}$$

where

$$f(k, y) = \frac{\sin kH\,\{k \cos k(h+y) - K \sinh k(h+y)\}}{K \cosh kh - k \sinh kh} \tag{3.17}$$

and k is real and satisfies the dispersion relation (3.14).

1.4 Two-layer fluid with an ice-cover

We consider irrotational motion in two superposed fluids under the action of gravity, neglecting any effect due to surface tension at the interface of the two fluids, the upper being of uniform finite depth h and covered by a thin uniform ice sheet modeled as a thin elastic plate. The upper and lower fluids have densities ρ_1 and $\rho_2 (>\rho_1)$ respectively. Rectangular Cartesian co-ordinates are chosen such that xz-plane coincides with the undisturbed interface between the two fluids. The y-axis points vertically downwards with $y = 0$ as the mean position of the interface and $y = -h$ is the mean position of the thin ice-cover. Under the usual assumptions of irrotational motion and linear theory, there exist velocity potentials $\Phi(\text{x, y, t}) = \text{Re}\{\phi(x, y)e^{-i\omega t}\}$, $\Psi(\text{x, y, t}) = \text{Re}\{\psi(x, y)e^{-i\omega t}\}$, in the upper and lower layers respectively, for two dimensional motion, where ω is the circular frequency.

The complex valued potential functions $\phi(x, y)$ and $\psi(x, y)$ satisfy

$$\nabla^2\phi = 0 \text{ in the upper layer,} \tag{4.1}$$

$$\nabla^2\psi = 0 \text{ in the lower layer,} \tag{4.2}$$

The linearized boundary conditions at the interface are

$$\phi_y = \psi_y \text{ on } y = 0, \tag{4.3}$$

$$K\psi + \psi_y = s(K\phi + \phi_y) \text{ on } y = 0 \tag{4.4}$$

where, as before, $s = \frac{\rho_1}{\rho_2} (<1)$, and the linearized boundary condition at the ice-cover is

$$k\phi + \left(D\frac{\partial^4}{\partial x^4} + 1 - \epsilon K\right)\phi_y = 0 \text{ on } y = -h \tag{4.5}$$

where $D = \frac{L}{\rho_2 g}$, L being the flexural rigidity of the elastic ice-cover and $\epsilon = \frac{\rho_0}{\rho_1} h_0$, ρ_0 being the density of ice and h_0 is the very small thickness of the ice-cover. If the lower layer extends infinitely downwards, then the condition at infinite depth is

$$\nabla\psi \to 0 \text{ as } y \to \infty. \tag{4.6}$$

Two dimensional time harmonic progressive wave solutions for the upper and lower layer are given by

$$\phi(x, y) = e^{\pm ikx}[\{k(Dk^4 + 1 - \epsilon K) + K\}e^{-k(y+h)} + \{k(Dk^4 + 1 - \epsilon K) - K\}e^{k(y+h)}], -h < y < 0, \tag{4.7}$$

$$\psi(x, y) = e^{\pm ikx}[\{k(Dk^4 + 1 - \epsilon K) + K\}e^{-kh} + \{k(Dk^4 + 1 - \epsilon K) - K\}e^{kh}]e^{-ky}, -h < y < 0 \tag{4.8}$$

where k satisfies the dispersion equation

$$\begin{aligned}\{k(1 - s) - k\}\{k(Dk^4 + 1 - \epsilon K) \sinh kh - K \cosh kh\}\\ - sk\{k(Dk^4 + 1 - \epsilon K) \cosh kh - K \sinh kh\} = 0.\end{aligned} \tag{4.9}$$

The dispersion equation has exactly two real positive roots λ_1 and $\lambda_2 (\lambda_1 < \lambda_2)$, say. Wave of wave number λ_1 propagates along the ice-cover and the wave of wave number λ_2 propagates along the interface between two layers. It can be shown that the equation (4.9) has one negative real root, two real positive roots and four complex roots situated in the four quadrants of the complex k-plane.

The progressive waves in the two layers are thus of the form

$$\phi(x, y) = e^{\pm i\lambda_j x} g_j(y), j = 1,2, \tag{4.10}$$

$$\psi(x, y) = e^{\pm i\lambda_j x - \lambda_j y}, j = 1,2, \tag{4.11}$$

where

$$g_j(y) = \frac{(\lambda_j(1-s) - K)}{Ks\{\lambda_j(D\lambda_j^4 + 1 - \epsilon K)\cosh\lambda_j h - K\sinh\lambda_j h\}}[\{\lambda_j(D\lambda_j^4 + 1 - \epsilon k) + k\}e^{-\lambda_j(y+h)} + \{\lambda_j(D\lambda_j^4 + 1 - \epsilon k) - k\}e^{\lambda_j(y+h)}], j = 1,2. \tag{4.12}$$

If the lower layer is of uniform finite depth H below the mean interface, then the condition (4.6) changes to

$$\psi_y = 0 \text{ on } y = H. \tag{4.13}$$

In this case, the two-dimensional progressive wave solutions for the upper and lower layers are given by

$$\phi(x, y) = \frac{\sinh kH\, e^{\pm ikx}}{K\cosh kh - (Dk^4 + 1 - \epsilon K)k\sinh kh} \times \{Dk^5\cosh k(h+y) + k(1-\epsilon K)\cosh k(y+h) - K\sinh k(h+y)\}, -h < y < 0, \tag{4.14}$$

$$\psi(x, y) = e^{\pm ikx}\cosh k(H - y), 0 < y < H \tag{4.15}$$

where k satisfies the dispersion relation

$$k^2(Dk^4 + 1 - \epsilon K)(1 - s) - kK\{(Dk^4 + 1 - \epsilon K)(s\coth kh - \coth kH) + (1-s)\coth kh\} + K^2(s + \coth kh\coth kH) = 0. \tag{4.16}$$

It can be shown that this dispersion equation has exactly two real positive roots. Two real negative roots, four complex roots in the four quadrants of the complex k-plane and an infinite number of purely imaginary roots.

1.5 Free surface boundary condition with higher order derivatives

A generalization of the ice-cover condition given in section 1.2 has been introduced by Manam et al. (2006) given by

$$\mathcal{L}\,\phi_y - K\phi = 0 \text{ on } y = 0 \tag{5.1}$$

where $\mathcal{L}$ is a linear differential operator of the form

$$\mathcal{L} = \sum_{m=0}^{m_0} c_m \frac{\partial^{2m}}{\partial x^{2m}}. \tag{5.2}$$

In (5.2), $c_m (m = 0, 1, \ldots, m_0)$ are known constants. Keeping in mind various physical problems involving fluid structure interaction, only the even order partial derivatives in x are considered in the differential operator $\mathcal{L}$. The bottom condition is given by

$$\nabla\phi \to 0 \text{ as } y \to \infty \tag{5.3}$$

for infinitely deep water, or by

$$\phi_y = 0 \text{ on } y = h \tag{5.4}$$

for water of uniform finite depth h.

In this case the two dimensional progressive wave solutions are given by

$$\phi(x, y) = e^{-p_0 y \pm i p_0 x} \tag{5.5}$$

for deep water,

$$\phi(x, y) = \frac{\cosh p_0(h - y)}{\cosh p_0 h} e^{-p_0 x} \tag{5.6}$$

for uniform finite depth h, where p_0 satisfies the dispersion equation

$$\sum_{m=0}^{m_0} (-1)^m c_m p^{2m} = K \tag{5.7}$$

for deep water, and

$$\left(\sum_{m=0}^{m_0} (-1)^m c_m p^{2m} \right) \tanh ph = K \tag{5.8}$$

for water of uniform finite depth h. Under specific assumptions involving the constants $c_m (m = 0, 1, \ldots, m_0)$, the dispersion equations (5.7) or (5.8) is assumed to possess only one real positive root. This is also physically realistic since progressive waves of only one wave number can propagate on the upper surface.

In a two-layer fluid, both the upper and lower layers are assumed to be homogeneous, incompressible and inviscid. Let ρ_1 be the density of the upper fluid and $\rho_2 (> \rho_1)$ be the same for the lower fluid. Let the lower fluid extend infinitely downwards while the upper one has a finite height h above the mean interface. Let y-axis point vertically downwards from the undisturbed interface $y = 0$. Thus the upper layer occupies the region $-h < y < 0$ while the lower layer occupies the region $y > 0$. Under the usual assumption of linear theory and irrotational two-dimensional motion, velocity potentials $\{\phi(x, y)e^{-i\omega t}\}$, $\text{Re}\{\psi(x, y)e^{-i\omega t}\}$ describing the fluid motion in the upper and lower layer, exist. ϕ, ψ satisfy

$$\nabla^2\phi = 0, -h < y < 0, \tag{5.9}$$

$$\nabla^2\psi = 0, y > 0. \tag{5.10}$$

The linearized boundary conditions at the interface $y = 0$ are

$$\phi_y = \psi_y \text{ on } y = 0, \tag{5.11}$$

$$s(\phi_y + K\phi) = \psi_y + K\psi \text{ on } y = 0 \tag{5.12}$$

where $s = \frac{\rho_1}{\rho_2}$ (< 1), while the condition at the upper surface (with higher-order derivatives) at $y = -h$ are

$$\mathcal{L}\phi_y - K\phi = 0 \text{ on } y = -h \tag{5.13}$$

where the differential operator $\mathcal{L}$ has the same form as given in (5.2). The bottom condition is given by

$$\nabla\psi \to 0 \text{ as } y \to \infty \tag{5.14}$$

if the lower layer extends infinitely downwards. In this case the dispersion equation has the form

$$\left(\sum_{m=0}^{m_0}(-1)^m c_m \lambda^{2m+1} \sinh \lambda h - K \cosh \lambda h\right)(\lambda - s\lambda - K)$$
$$-sK\left(\sum_{m=0}^{m_0}(-1)^m c_m \lambda^{2m+1} \cosh \lambda h - K \sinh \lambda h\right) = 0. \tag{5.15}$$

The constants $c_m(m = 0,1,\ldots\ldots, m_0)$ are assumed to be such that the equation (5.15) possesses only two real positive roots, which correspond to the two different wave numbers (modes) at which time-harmonic progressive wave propagate at the upper surface and the interface of the two-layer fluid. It is emphasized that the physical constants $c_m(m = 0,1,\ldots\ldots, m_0)$ are such that the equation (5.15) possesses only two real positive roots. It may be a good idea to find out what should be the forms of these constants for such a situation.

In this case, the two-dimensional progressive wave solution are given by

$$\phi(x, y) = e^{\pm\lambda_1 x} g_1(y), e^{\pm\lambda_2 x} g_2(y), -h < y < 0, \tag{5.16}$$

$$\psi(x, y) = e^{\pm\lambda_1 x - \lambda_1 y}, e^{\pm\lambda_2 x - \lambda_2 y}, y > 0, \tag{5.17}$$

where λ_1, λ_2 are the two real positive roots of the equation (5.15) and

$$g_j(y) = \frac{(1 - \mathrm{s})\lambda_j - K}{sK[\sum_{m=0}^{m_0}(-1)^m c_m \lambda_j^{2m+1} \sinh \lambda_j h - K \cosh \lambda_j h]} \times$$
$$\left[\left\{\sum_{m=0}^{m_0}(-1)^m c_m \lambda_j^{2m+1} + K\right\} e^{-\lambda_j(y+h)} + \left\{\sum_{m=0}^{m_0}(-1)^m c_m \lambda_j^{2m+1} - K\right\} e^{-\lambda_j(y+h)}\right]. \tag{5.18}$$

If the lower fluid is of uniform finite depth H, then the corresponding dispersion equation and the two-dimensional progressive wave solutions can be obtained in a similar manner assuming of course that the dispersion equation has only two real positive roots.

1.6 Solution of some dispersion equations

As mentioned earlier, the relation between the wave number and the angular frequency of a surface gravity wave is known as the dispersion equation. This terminology is derived from the fact that waves for which speed of wave propagation is a function of wave number are called dispersive because waves of different lengths propagating at different speeds disperse or separate. Since the speed of propagation of a wave component depends on its wave number and angular frequency, wave dispersion is a fundamental process in many physical phenomena.

For infinitely deep water as well as uniform finite depth water, it is known that time-harmonic waves with a given frequency can propagate with a wave number which is the only real positive root of the corresponding dispersion equation. The same phenomenon also holds good when water has an ice-cover instead of a free surface. However, for two immiscible superposed fluids, it is known that the time harmonic waves with a given frequency can propagate with two different wave numbers. The wave of smaller wave number propagates along the free surface and the wave of greater wave number propagates along the interface between the two fluids. Das and Mandal (2005) demonstrated how all the roots of different dispersion equations can be determined and that there exists no other root. This is described below.

For infinitely deep water, if k is the wave number and ω is the angular frequency of a surface gravity wave, then the dispersion equation is given by Wehausen and Laitone (1960), p472

$$k = \frac{\omega^2}{g} \ (\equiv K). \tag{6.1}$$

It is trivial to see that the equation (6.1) possesses the solution $k = K$, and this corresponds to the time-harmonic progressive surface waves represented by $\phi(x, y) = e^{-Ky \pm iKx}$ where $\mathrm{Re}\{\phi(x, y)e^{-i\omega t}\}$ is the velocity potential describing the two-dimensional motion in deep water occupying the position $y \geq 0$, y-axis being chosen vertically downwards, the plane $z = 0$ being the mean free surface, and x-direction being the direction of wave propagation.

For water of uniform finite depth h, the equation (6.1) modifies to the transcendental equation (cf. Wehausen and Laitone (1960), p474)

$$k \tanh kh = K. \tag{6.2}$$

It is well known that the transcendental equation (6.2) has real roots $\pm k_0 (k_0 > 0)$ and countably infinite number of purely imaginary roots $\pm ik_n$ $(n = 1,2, \ldots)$ where $0 < k_1 < k_2 < \cdots$ and $k_n \to \frac{n\pi}{h}$ as $n \to \infty$. The positive real root corresponds to progressive surface waves with wave number k_0 while the purely imaginary roots correspond to evanescent modes or local solutions as mentioned in section 1.1 That the equation (6.2) has no other roots except $\pm k_0$ and $\pm ik_n$ $(n = 1,2,3, \ldots)$, can be proved by employing Rouche's theorem of complex variable theory cf. Churchill et al. (1966) to the functions $f(k) = k \sinh kh - K \cosh kh$, $g(k) = k \sinh kh$ within a square with vertices $k = \frac{(2m-1)\pi}{2h}$ $(\pm 1 \pm i)$, m being a large positive integer, in the complex k-plane as was demonstrated by Rhodes-Robinson (1971) for the dispersion equation in which the effect of surface tension at the free surface was included.

For two superposed immiscible fluids separated by a common interface, the upper fluid extending infinitely upwards and the lower fluid extending infinitely downwards, the wave number k of small amplitude interface gravity waves or internal waves is related to the angular frequency ω by the dispersion relation cf. Wehausen and Laitone (1960), p647

$$k = \sigma K \tag{6.3}$$

where $\sigma = \frac{1+s}{1-s}$ with $s = \frac{\rho_2}{\rho_1}$ $(\rho_2 < \rho_1), \rho_1, \rho_2$ being the densities of the lower and upper fluids respectively. If the upper fluid is of uniform finite height h above the mean interface and has a free surface while the lower fluid extends infinitely downwards, then the corresponding dispersion equation is cf. Linton and McIver (1995)

$$(k - K)\ \{K(\sigma + e^{-2kh}) - k(1 - e^{-2kh})\} = 0. \tag{6.4}$$

This equation has two real roots, one is K and the other is v say, where v satisfies

$$K(\sigma + e^{-2vh}) = v(1 - e^{-2vh}) \tag{6.5}$$

so that $K\sigma < v < K\frac{\sigma+1}{1-e^{-2K\sigma h}}$.

Thus there exist time-harmonic progressive waves with two different wave numbers K, v. This is also mentioned earlier in section 1.3. An equivalent form of the equation (6.5) was given in Art.231 of the treatise by Lamb (1932) wherein a description of some of the types of wave motion which can occur in a two-layer fluid with both a free surface and an interface, was also mentioned.

As mentioned in section 1.3, study of wave motion in a two-layer fluid has gained importance due to the plan to construct underwater pipe bridge across Norwegian fjords. A fjord consists of a layer of fresh water over a layer of salt water. If the lower layer is very deep, then the two-layer fluid mentioned above models a fjord. For this type of two-layer fluid, problems on interaction of small amplitude waves on the free surface as well as on the interface have been investigated by Linton and McIver (1995), Linton and Cadby (2002). If the aforesaid fjord is not very deep, then it can be modeled as a two-layer fluid wherein the lower fluid is of uniform finite depth H, say, below the mean interface, and as before, the upper fluid is of height h above the interface and has a free surface. In this case, the dispersion equation (6.4) modifies to

$$(1 - s)k^2 - kK(\coth kh + \coth kH) + K^2(s + \coth kh \coth kH) = 0. \tag{6.6}$$

This equation is given by Sherief et al. (2003, 2004) while investigating forced gravity waves due to a plane and a cylindrical vertical porous wave maker in a two-layer fluid. They simply stated without proof that the equation (6.6) has two real positive roots, two real negative roots and an infinite number of purely imaginary roots of the form $\pm i\lambda_n (n = 1,2, \ldots)$. The proof of this is given below.

Let $\mu = \frac{H}{h}$ and $kh = z$, then the equation (6.6) reduces to

$$f(z) = \coth \mu z \tag{6.7}$$

where

$$f(z) = \frac{Khz \coth z - (1 - s)z^2 - s(Kh)^2}{Kh(Kh \coth z - z)}. \tag{6.8}$$

The real roots of in equation (6.7) can be obtained graphically from plots of $y = \coth \mu x$ and $y = f(x)$. These plots are given in Figs. 1.1 to 1.3 for three values of μ, viz. $\mu = 1(h = H)$, $\mu = 0.7(h > H)$ and $\mu = 1.7(h < H)$. We note that the asymptotes of the curve $y = f(x)$ are $x = \pm k_0 h$, where k_0 is the only two real root of the equation (6.2) (i.e., the real root of the dispersion equation for a single fluid of depth h below its mean free surface).

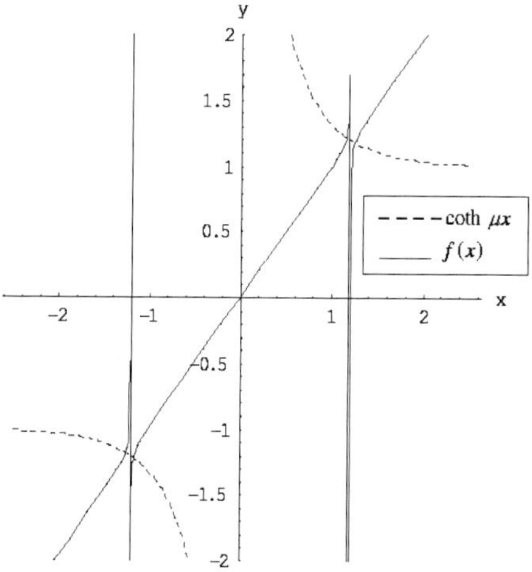

Fig. 1.1 ($\mu = 1$, $s = 0.5$, $Kh = 1$)

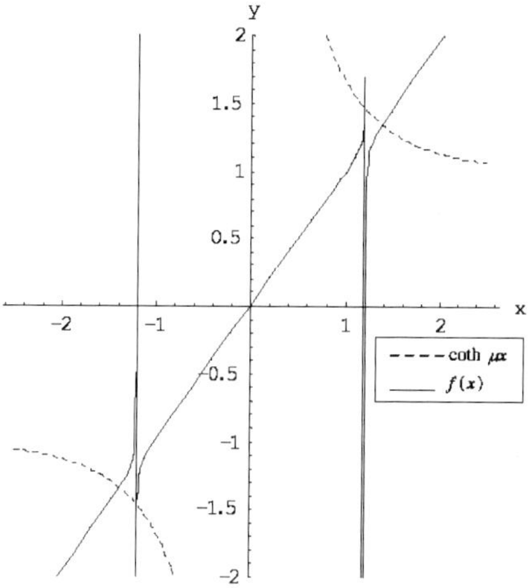

Fig. 1.2 ($\mu = 0.7$, $s = 0.5$, $Kh = 1$)

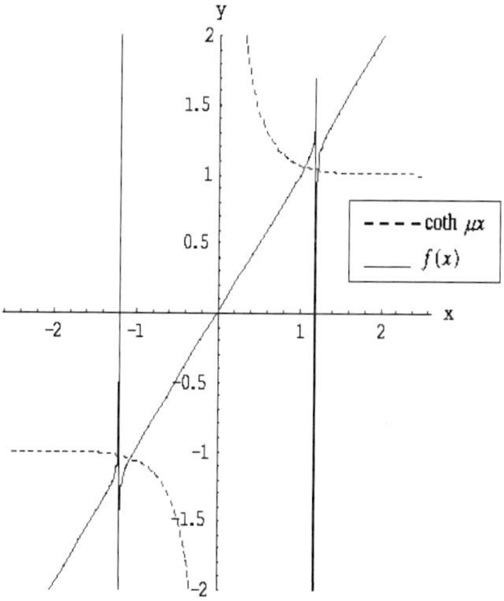

Fig. 1.3 ($\mu = 1.7$, $s = 0.5$, $Kh = 1$)

For the Figs. 1.1 to 1.3, it is obvious that the equation (6.6) has four real roots, two positive and two negative. If the two positive roots are denoted by m_1 and m_2 then the negative roots are $-m_1$, $-m_2$, since the equation (6.6) remains unchanged if k is replaced by $-k$. Also, if $m_1 < m_2$ then $m_1 < k_0 < m_2$.

If we replace k by ik in (6.6), then it becomes, after writing $z = kh$,

$$g(z) = \cot \mu z \tag{6.9}$$

where

$$g(z) = \frac{Khz \cot z + (1-s)z^2 - s(Kh)^2}{Kh(Kh \cot z + z)}. \tag{6.10}$$

Thus the purely imaginary roots of the equation (6.6) can be obtained graphically from the plots of the curves $y = \cot \mu x$ and $y = g(x)$.

Let μ be expressed as $\mu = \frac{p}{q}$ where p and q are integers prime to each other but $p = q$ when $\mu = 1$. In Figs. 1.4 to 1.6, plots of $y = \cot \mu x$ and $y = g(x)$ are given for $\mu = 1, 0.7 (p = 7, q = 10), 1.7 (p = 17, q = 10)$.

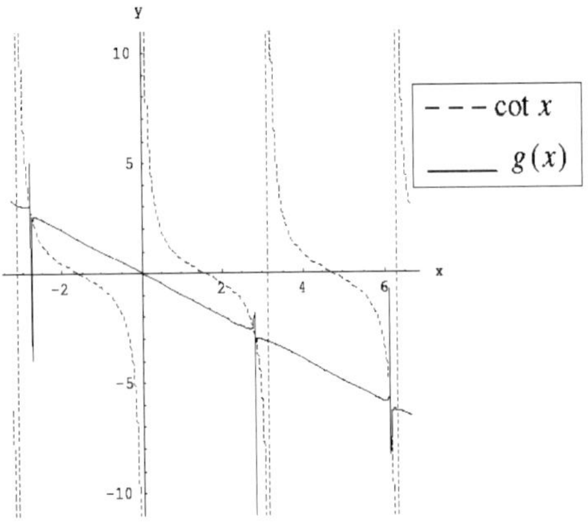

Fig. 1.4 ($\mu = 1, s = 0.5, Kh = 1$)

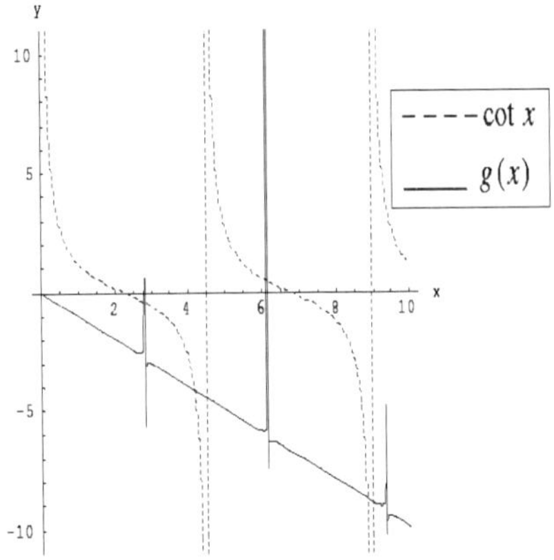

Fig. 1.5 ($\mu = 0.7, s = 0.5, Kh = 1$)

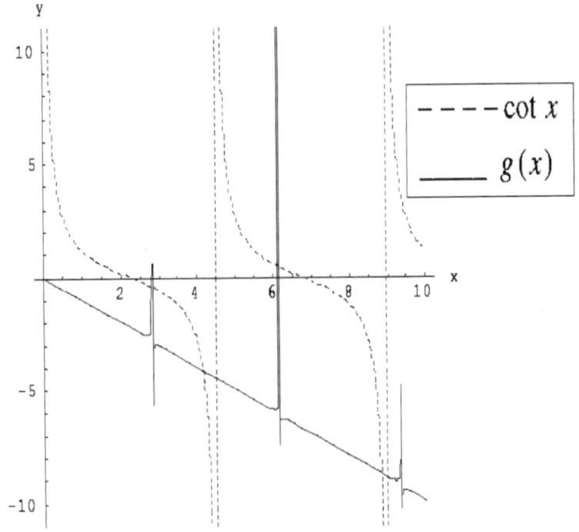

Fig. 1.6 ($\mu = 1.7, s = 0.5, Kh = 1$)

From these figures it is obvious that there exists an infinite number of purely imaginary roots of the equation (6.6) given by $\pm i\lambda_m (m = 1,2, \ldots)$. If m is a multiple of $(p + q)$, i.e., $m = n(p + q)$, say, then it is easy to see that $k_{nq} < \lambda_{n(p+q)} < nq\pi (n = 1,2,\ldots)$ and as n becomes large, $\lambda_{n(p+q)} \to nq\pi$ for any p.

Next it will be shown by using Rouche's theorem that apart from the aforesaid roots, there exists no other root of the dispersion equation (6.6) in the complex k-plane.

Let us define

$$F(z) = (z - Kh \coth \mu z)(z - Kh \coth z), \tag{6.11}$$

$$G(z) = -s(z^2 - (kh)^2) \tag{6.12}$$

where z is complex $(z = kh)$. We consider the contour C of the square with vertices at $(\pm(nq\pi + \epsilon), \pm i(nq\pi + \epsilon))$ in the complex z-plane where n is a large integer and ϵ (> 0) is sufficiently small. The contour C is chosen in such a way that it does not pass through any of the zeros of function $F(z)$. The equation $F(z) = 0$ has four real roots and $2n(p + q)$ purely imaginary roots inside C. Therefore total number of roots of $F(z) = 0$ inside C is $4 + 2n(p + q)$.

Now on the upper side of the square C,

$$z = x + i(nq\pi + \epsilon)$$

where $-(nq\pi + \epsilon) < x < (nq\pi + \epsilon)$. Then on this side

$$\left|\frac{F(z)}{G(z)}\right| = \left|\frac{x + i(nq\pi + \epsilon) - Kh \coth(x + i(nq\pi + \epsilon))}{s\{(x + i(nq\pi + \epsilon))^2 - K^2\}}\{x + i(nq\pi + \epsilon) - Kh \coth(x + i(nq\pi + \epsilon))\}\right|$$

$$= \left|\frac{\left\{1 - \dfrac{Kh \coth\left(x + i(nq\pi + \epsilon)\right)}{n\left(\frac{x}{n} + i(q\pi + \frac{\epsilon}{n})\right)}\right\}\left\{1 - \dfrac{Kh \coth\left(x + i(nq\pi + \epsilon)\right)}{n\left(\frac{x}{n} + i(q\pi + \frac{\epsilon}{n})\right)}\right\}}{s\left\{1 - \dfrac{K^2}{n^2\left(\frac{x}{n} + i(q\pi + \frac{\epsilon}{n})\right)^2}\right\}}\right|.$$

Thus as n is large, the value of $\frac{F(z)}{G(z)}$ on the upper side of C becomes $\frac{1}{s} > 1$. Hence $|F(z)| > |G(z)|$ on the upper side of C. Similarly we can prove that $|F(z)| > |G(z)|$ on the other sides of the contour C. Therefore $|F(z)| > |G(z)|$ uniformly on the contour C. Therefore by Rouche's theorem we find that $F(z)$ and $F(z) + G(z)$ have the same number of zeros inside C. This shows that the dispersion equation (6.6) has four real roots and an infinite number of purely imaginary roots and no other root.

In a similar way the roots of the dispersion equation (4.12) for internal waves in a two-layer fluid in which the upper layer has an ice-cover modeled as thin elastic plate and the lower layer of uniform finite depth below the mean ice-cover, can be obtained. Das et al. (2008) has shown that the dispersion equation (4.12) has four real roots (two real positive, two real negative), four complex roots and an infinite number of purely imaginary roots.

1.7 Scattering problems

As briefly mentioned earlier, when an incoming train of surface water waves encounters an obstacle present in water, it is partially reflected by the obstacle and partially transmitted over (for submerged obstacle) or below the obstacle. Finding the reflection and transmission coefficients either exactly or approximately, is of great challenge to the workers in the area of water waves since these quantities have important physical significance. A scattering problem is generally formulated as a boundary value problem for the purpose of its mathematical study. For two-dimensional motion, let $\phi^{inc}(x, y)$ denote the velocity potential for a surface wave train propagating from the direction of $x = -\infty$ and let it be incident on a body B present in water. If $\phi(x, y)$ denotes the velocity potential for ensuring fluid motion, then $\phi(x, y)$ satisfies the boundary value problem described by

$$\nabla^2\phi = 0 \text{ in the fluid region,} \tag{7.1}$$

$$K\phi + \phi_y = 0 \text{ on the free surface,} \tag{7.2}$$

$$\phi_n = 0 \text{ on } B, \tag{7.3}$$

$$\nabla\phi \to 0 \text{ as } y \to \infty \text{ for deep water,} \tag{7.4a}$$

or

$$\frac{\partial\phi}{\partial y} = 0 \text{ on } y = h \text{ for water of uniform finite depth } h, \tag{7.4b}$$

and

$$\phi(x, y) = \begin{cases} \phi^{inc}(x, y) + R\phi^{inc}(-x, y) \text{ as } x \to \infty, \\ T\phi^{inc}(x, y) \text{ as } x \to \infty \end{cases} \tag{7.5}$$

where R and T are respectively the reflection and transmission coefficients respectively and are unknown.

For an obstacle of an arbitrary geometrical shape, the related water wave scattering problem cannot be solved in closed form in the sense that the scattered potential function as well as the reflection and transmission coefficients cannot be obtained exactly. Only for the case of obstacles in the form of thin vertical barriers present in deep water, it is possible to obtain closed form solutions when the incoming waves are normally incident on the barriers. For other situations or obstacles of other geometrical configurations some approximate methods can be employed for the purpose of obtaining approximate numerical estimates for the reflection and transmission coefficients.

Water wave scattering problems involving thin vertical barriers in the form of four basic configurations, namely a partially immersed plate, a submerged plate extending infinitely downwards, a submerged finite plate or a thin vertical wall with one or more submerged gaps, have been investigated by various researchers from time to time using different mathematical techniques. From an engineering point of view, the thin barrier type obstacles provide simple models for breakwaters which can be implemented easily. Also, as these configurations provide closed form solutions, these solutions can be employed to obtain approximate solutions to some related scattering problems. There exists a wide range of mathematical techniques in the literature to approximately tackle the problems of water wave interaction with obstacles of various geometrical shapes. An account of various techniques is available in the hand book of Linton and McIver (2001), the monograph by Kuznetsov et al. (2002) and for barrier problems in the monograph by Mandal and Chakrabarti (2000).

1.8 Energy identities

There are some very general identities relating to some hydrodynamic quantities in the linearized theory of water waves. These provide some checks that can be used on analytical or numerical results. The reflection and transmission coefficients in any wave scattering problem involving a finite number of bodies present in a single-layer fluid with a free surface satisfy the energy identity. For a two-layer fluid with a free surface there exist two energy identities involving reflection and transmission coefficients of two different modes corresponding to incident waves also of two different modes. Also there are various hydrodynamic relations such as the energy conservation principle and Haskind-Hanaoka relation in water wave problems. These can be determined by a judicious application of the standard Green's integral theorem for harmonic functions in the form

$$\int_S (\phi\psi_n - \psi\phi_n)ds = 0 \tag{8.1}$$

where S denotes the boundaries of the fluid region and ϕ_n and ψ_n denote the partial derivatives along the normal to S.

Let the far field behavior of the potential function $\phi(x, y)$ be given by

$$\phi(x, y) \sim (A^{\pm}e^{\pm ikx} + B^{\pm}e^{\mp ikx})e^{ky} \text{ as } x \to \pm\infty \tag{8.2}$$

for deep water, and

$$\phi(x, y) \sim (A^{\pm}e^{\pm ik_0x} + B^{\pm}e^{\mp ik_0x})\frac{\cosh k_0(h-y)}{\cosh k_0h} \text{ as } x \to \pm\infty \tag{8.3}$$

for water of uniform finite depth for a single layer.

Let $\phi^I(x, y)$ and $\phi^{II}(x, y)$ denote the velocity potentials in the upper and lower layers respectively for a two-layer fluid. The far-field behaviors of $\phi^I(x, y)$ and $\phi^{II}(x, y)$ are given by

$$\phi^I(x, y) \sim (A^{\pm}e^{\pm ikx} + C^{\pm}e^{\mp ikx})e^{-ky} + (B^{\pm}e^{\pm i\upsilon x} + D^{\pm}e^{\mp i\upsilon x})g(y) \text{ as } x \to \pm\infty \tag{8.4}$$

and

$$\phi^{II}(x, y) \sim (A^{\pm}e^{\pm ikx} + C^{\pm}e^{\mp ikx})e^{-ky} + (B^{\pm}e^{\pm i\upsilon x} + D^{\pm}e^{\mp i\upsilon x})e^{-\upsilon y} \text{ as } x \to \pm\infty \tag{8.5}$$

where

$$g(y) = \frac{k\sigma - \upsilon}{K(\sigma - 1)} e^{-\upsilon y} + \frac{K - \upsilon}{K(\sigma - 1)} e^{\upsilon y}, \tag{8.6}$$

and v satisfies

$$K(\sigma + e^{-2\upsilon h}) = \upsilon(1 - e^{-2\upsilon h}).$$

A convenient short hand for (8.2), (8.3) and (8.4), (8.5) is

$$\phi \sim (A^-, B^-; A^+, B^+) \tag{8.7}$$

for single-layer fluid and

$$\phi \sim (A^-, B^-, C^-, D^-; A^+, B^+, C^+, D^+) \tag{8.8}$$

for two-layer fluid.

Single-layer fluid:

Let the boundary of a finite number of bodies present in the fluid be denoted by B. Let ϕ be the solution of a scattering problem with $\phi_n = 0$ on the boundaries B. Let ψ be the solution to the boundary value problem with $\psi_n = 0$ on the boundaries B, with the far-field form of ψ given by

$$\psi \sim (P^-, Q^-; P^+, Q^+). \tag{8.9}$$

We now apply Green's integral theorem (8.1) to φ, ψ, where S is the boundary of the fluid region internally bounded by $x = \pm X$, $0 \le y \le Y$; $y = Y$, $|x| \le X$; $y = 0$, $|x| \le X$ (but outside the portion of $y = 0$ enclosed by bodies if there be any) and bounded externally by the body boundary B and ultimately make X, $Y \to \pm\infty$ for deep water, and the region internally bounded by $x = \pm X$, $0 \le y \le h$; $y = h$, $|x| \le X$; $y = 0$, $|x| \le X$ (but outside the portion of $y = 0$, $y = h$ enclosed by the bodies if there be any) and externally by the body boundary B for uniform finite depth water and ultimately make $X \to \infty$.

Then (8.1) produce

$$\left(\int_{y=Y,|x|\le X} + \int_{x=-X,0\le y\le Y} + \int_{y=0,|x|\le X} + \int_{x=X,0\le y\le Y} + \int_{B} \right) (\phi\psi_n - \psi\phi_n)\, ds = 0 \tag{8.10}$$

for deep water and

$$\left(\int_{y=h,|x|\le X} + \int_{x=-X,\,0\le y\le h} + \int_{y=0,|x|\le X} + \int_{x=X,0\le y\le h} + \int_{B} \right) (\phi\psi_n - \psi\phi_n)\, ds = 0 \tag{8.11}$$

for uniform finite depth water. Calculating all the terms in (8.10) and (8.11) we obtain the relation

$$A^- Q^- - P^- B^- = P^+ B^+ - Q^+ A^+. \tag{8.12}$$

For a scattering problem, the far-field form of ϕ is given by

$$\phi \sim (R, 1; T, 0), \tag{8.13}$$

where R and T are the reflection and transmission coefficients respectively, due to an incident field propagating from the direction of $x = -\infty$. Let $\psi = \phi$ denote the complex conjugate of ϕ and the far-field form of ψ is

$$\psi \sim (1, \bar{R}; 0, \bar{T}). \tag{8.14}$$

Then from (8.12) we obtain

$$|R|^2 + |T|^2 = 1 \tag{8.15}$$

which is the energy identity.

Two-layer fluid:

Let the boundaries of a finite number of bodies lying in the upper layer be denoted by B_I and those in the lower layer by B_{II}. Let ϕ be the solution of a scattering problem with $\phi_n = 0$ on the boundaries B_I, B_{II}. The far-field form of ϕ is then given by

$$\phi \sim \{R_k, r_k, 1{,}0; T_k, t_k, 0{,}0\} \tag{8.16}$$

for the incident wave of wave number K and R_k, and r_k are the reflection coefficients of the waves of wave numbers K and v, respectively, due to an incident wave of wave number K and similarly T_k, t_k for the transmission coefficients. Let $\psi = \bar{\varphi}$, the complex conjugate of ϕ, then

$$\psi \sim \{1{,}0, \bar{R}_k, \bar{r}_k; 0{,}0, \bar{T}_k, \bar{t}_k\}. \tag{8.17}$$

To obtain the energy identity, we use the Green's integral theorem (8.1), where S is the boundary of either the upper or the lower layer cf. Linton and McIver (1995). Taking s to be first the boundary of the upper layer and then the boundary of the lower layer and using the result that on the interface

$$s\left(\phi_I \frac{\partial \psi_I}{\partial y} - \psi_I \frac{\partial \phi_I}{\partial y}\right) = \phi_{II} \frac{\partial \psi_{II}}{\partial y} - \psi_{II} \frac{\partial \phi_{II}}{\partial y} \tag{8.18}$$

we obtain after some algebra, which includes the result

$$s\int_0^h g(y)e^{ky}dy + \int_{-\infty}^0 e^{(k+\upsilon)y}\,dy = 0, \tag{8.19}$$

that

$$J_k(|R_k|^2 + |T_k|^2 - 1) + J_\upsilon(|r_k|^2 + |t_k|^2) = 1 \tag{8.20}$$

where

$$J_k = i\left(1{+}2sk\int_0^h e^{2ky}\,dy\right), \tag{8.21}$$

$$J_v = i\left(1{+}2s\upsilon\int_0^h (g(y))^2\,dy\right). \tag{8.22}$$

Thus we obtain the identity

$$|R_k|^2 + |T_k|^2 + J(|r_k|^2 + |t_k|^2 - 1) = 1 \tag{8.23}$$

where

$$J = \frac{J_v}{J_k}. \tag{8.24}$$

Similarly for the scattering of an incident wave of wave number v, we obtain the identity

$$|R_v|^2 + |T_v|^2 + J(|r_v|^2 + |t_v|^2 - 1) = 1. \tag{8.25}$$

The relations (8.23) and (8.25) are the energy identities. Also Green's integral theorem for harmonic function can be employed to derive the various hydrodynamic relations such as Haskind-Hanaoka relation for scattering and radiation problems and energy conservation principle for any radiation problem involving finite number of bodies present in water having a free surface (cf. Mei (1982); Linton and McIver (2001)). For a two-layer fluid having a free-surface, the hydrodynamic relations for any scattering or radiation problem can also be derived by employing this Green's integral theorem cf. Cadby and Linton (2002).

CHAPTER II

Scattering by Thin Barriers

This chapter is concerned with water wave scattering by thin barriers. As mentioned earlier when a train of surface waves travelling from a large distance is incident on an obstacle, it experiences partial reflection and partial transmission over or below the obstacle. Evaluation of reflection and transmission coefficients is of somewhat mathematical and physical importance. If the obstacle is of arbitrary geometrical shape, the problem of determining the reflection and transmission coefficients is in general a difficult task. The problems of water wave scattering by fixed thin plane vertical barriers are being investigated in the literature for the last seven decades by using various mathematical techniques. Mandal and Chakrabarti (2000) discussed in some detail various methods to solve scattering problems involving a thin vertical barrier in infinitely deep water or in water of uniform finite depth.

In this chapter wave scattering by two thin vertical barriers, by two thin nearly vertical barriers, by thin inclined barriers, by circular arc-shaped thin barriers, by thin vertical elastic barrier will be considered. Instead of a free surface, if the water is covered by a thin ice-cover modeled as a thin elastic floating plate, and a thin vertical rigid plate is submerged beneath the ice-cover, the corresponding wave scattering problem is also considered here.

2.1 Scattering by two thin vertical barriers

The problem of water wave scattering by two thin vertical barriers which are either partially immersed up to the same depth or completely submerged from the same depth and extending infinitely downwards admits of explicit solution. The problem of two partially immersed barriers was solved by Levine and Rodemich (1958) who obtained the expressions for the reflection and transmission coefficients in terms of six definite integrals involving complicated functions of elliptic integrals. They employed the Schwartz-Christoffel transformation of complex variable theory in a somewhat complicated manner to solve the problem. The complementary problem of two submerged plane vertical barriers was investigated by Jarvis (1971) using a procedure similar to that of Levine and Rodemich (1958). Both the problems have been solved by De et al. (2009, 2010) by a relatively simple method based on Fourier analysis for the expansion of velocity potential function leading to solving Abel integral equations. The problem of water waves scattering by two vertical barriers extending infinitely downwards and each containing a narrow submerged aperture at the same level was studied by Evans (1975) by using an approximate method.

We consider two thin plane vertical barriers present in infinitely deep water, occupying the positions $x = \pm a,\ y \in L$ where in the y-axis is chosen vertically downwards into the fluid region with the plane $y = 0$ representing the mean position of the free surface. $L = L_1 = (h, \infty)$ when the barriers are submerged from a depth h below the free surface and extend infinitely downwards while $L = L_2 = (0, h)$ when the barriers are partially immersed up to a depth h below the free surface and $L = L_3 = (0, b) \cup (0, c)$ when the barriers are of the form of two thin walls with submerged gaps at the same level. Assuming linear theory and two-dimensional irrotational motion, a time harmonic progressive wave train described by the potential function $\mathrm{Re}\{\phi_0(x, y)e^{-i\sigma t}\}$ with $\phi_0(x, y) = e^{-Ky+iKx}$ from the direction of $x = -\infty$, is incident on the two barriers. If $\phi(x, y)$ denotes the complex-valued potential function describing the motion in the fluid region, then φ satisfies the boundary value problem described by

$$\nabla^2\phi = 0,\ y \geq 0, \tag{1.1}$$

$$K\phi + \phi_y = 0 \text{ on } y = 0, \tag{1.2}$$

$$\phi_x = 0 \text{ on } x = \pm a,\ y \in L, \tag{1.3}$$

$$r^{\frac{1}{2}}\nabla\phi = O(1) \text{ as } r = \{(x \mp a)^2 + y^2\}^{\frac{1}{2}} \to 0, \tag{1.4}$$

where r is the distance from a submerged and of the barriers,

$$\nabla\phi \to 0 \text{ as } y \to \infty, \tag{1.5}$$

$$\phi(x, y) \sim \begin{cases} T\,\phi_0(x, y) \text{ as } x \to \infty, \\ \phi_0(x, y) + R\,\phi_0(-x, y) \text{ as } x \to -\infty \end{cases} \tag{1.6}$$

where T and R respectively denote the unknown transmission and reflection coefficients, determination of which is the concern here.

The potential function $\phi(x, y)$ satisfying (1.1), (1.2), (1.5) and (1.6) has the following representation after using Havelock's expansion of water wave potential.

$$\phi(x, y) = \begin{cases} e^{-Ky}(e^{iKx} + Re^{-iKx}) + \dfrac{2}{\pi}\int_0^\infty A(\xi)M(\xi, y)e^{\xi x}d\xi,\ x < -a, \\ e^{-Ky}(\alpha e^{iKx} + \beta e^{-iKx}) + \dfrac{2}{\pi}\int_0^\infty \{B(\xi)e^{\xi x} + C(\xi)e^{-\xi x}\}M(\xi, y)d\xi,\ -a < x < a, \\ Te^{-Ky+iKx} + \dfrac{2}{\pi}\int_0^\infty D(\xi)M(\xi, y)e^{-\xi x}\,d\xi,\ x > a, \end{cases} \tag{1.7}$$

with

$$M(\xi, y) = \xi\cos\xi y - K\sin\xi y, \tag{1.8}$$

where α, β are unknown constants, $A(\xi)$, $B(\xi)$, $C(\xi)$ and $D(\xi)$ are unknown functions such that the integrals in (1.7) and in the subsequent mathematical analysis in which they appear, are convergent.

Using the condition of continuity of ϕ_x across the lines $x = \pm a, y > 0$, we obtain two relations involving, $A(\xi)$, $B(\xi)$, $C(\xi)$, $D(\xi)$ under integral sign and the four unknown constants α, β, R, T. Use of Havelock's inversion theorem to each of these two relations gives rise to four equations given by

$$e^{-iKa} - Re^{iKa} = \alpha e^{-iKa} - \beta Re^{iKa}, \tag{1.9a}$$

$$Te^{-iKa} = \alpha e^{iKa} - \beta Re^{-iKa}, \tag{1.9b}$$

and

$$A(\xi) = B(\xi) - C(\xi)e^{2\xi a}, \tag{1.10a}$$

$$D(\xi) = C(\xi) - B(\xi)e^{2\xi a}. \tag{1.10b}$$

Again, continuity of φ across the gaps $x = \mp a$, $y \in \bar{L} = (0, \infty) - L$ produces after using (1.9) and (1.10)

$$\frac{2}{\pi}\int_0^\infty C(\xi)e^{\xi a} M(\xi, y)d\xi = -(\beta - R)e^{-Ky+iKa}, y \in \bar{L}, \tag{1.11}$$

and

$$\frac{2}{\pi}\int_0^\infty B(\xi)e^{\xi a} M(\xi, y)d\xi = -\beta e^{-Ky+iKa}, y \in \bar{L}. \tag{1.12}$$

Using the condition that $\phi_x = 0$ on $x = \mp a$, $y \in L$, we obtain

$$\frac{2}{\pi}\int_0^\infty \xi\{B(\xi)e^{-\xi a} - C(\xi)e^{\xi a}\}M(\xi, y)d\xi = -iK(\alpha e^{-iKa} - \beta e^{iKa})e^{-Ky}, y \in L, \tag{1.13}$$

$$\frac{2}{\pi}\int_0^\infty \xi\{B(\xi)e^{\xi a} - C(\xi)e^{-\xi a}\}M(\xi, y)d\xi = -iK(\alpha e^{iKa} - \beta e^{-iKa})e^{-Ky}, y \in L, \tag{1.14}$$

Case 1: Two submerged barriers

For the case of two submerged barriers $L = L_1 = (h, \infty)$.

Multiplying both sides of (1.11) to (1.14) by e^{-Ky}, integrating with respect to y between 0 to $y(< h)$ in (1.11), (1.12) and between $y(> h)$ to ∞ in (1.13), (1.14), we obtain

$$\int_0^\infty C(\xi)e^{\xi a} \sin \xi y \, d\xi = \frac{\pi}{2K}(R - \beta)e^{iKa} \sinh Ky, 0 < y < h, \tag{1.15}$$

$$\int_0^\infty B(\xi)e^{\xi a} \sin \xi y \, d\xi = -\frac{\pi}{2K}\beta e^{-iKa} \sinh Ky, 0 < y < h, \tag{1.16}$$

$$\int_0^\infty \xi\{B(\xi)e^{-\xi a} - C(\xi)e^{\xi a}\}\sin \xi y \, d\xi = \frac{i\pi}{4}(\alpha e^{-iKa} - \beta e^{iKa})e^{-Ky}, y > h, \tag{1.17}$$

$$\int_0^\infty \xi \{B(\xi)e^{-\xi a} - C(\xi)e^{\xi a}\}\sin \xi y \, d\xi = -\frac{i\pi}{4}(\alpha e^{iKa} - \beta e^{-iKa})e^{-Ky}, \; y > h. \quad (1.18)$$

Let the left sides of (1.18) and (1.17) be equal to $g_1(y)$ and $g_2(y)$ respectively for $0 < y < h$ where $g_1(y)$ and $g_2(y)$ are unknown functions. Then by using the sine inversion formula, we find

$$\xi \{B(\xi)e^{\xi a} - C(\xi)e^{-\xi a}\} = \frac{2}{\pi}\int_0^h g_1(t) \sin \xi t \, dt + \frac{i}{2}(\alpha e^{iKa} - \beta e^{-iKa})\int_h^\infty e^{-Kt} \sin \xi t \, dt \quad (1.19)$$

and

$$\xi \{B(\xi)e^{-\xi a} - C(\xi)e^{\xi a}\} = \frac{2}{\pi}\int_0^h g_2(t) \sin \xi t \, dt + \frac{i}{2}(\alpha e^{-iKa} - \beta e^{iKa})\int_h^\infty e^{-Kt} \sin \xi t \, dt. \quad (1.20)$$

Solving for $B(\xi)$, $C(\xi)$ from (1.19) and (1.20), and substituting these in (1.16) and (1.15), we obtain two coupled integral equations for $g_1(t)$ and $g_2(t)$ given by

$$\frac{1}{\pi}\int_0^h g_1(t)\mathcal{K}_1(t, y)dt - \frac{1}{\pi}\int_0^h g_2(t)\mathcal{K}_2(t, y)dt = f_1(y), \; 0 < y < h, \quad (1.21)$$

$$\frac{1}{\pi}\int_0^h g_2(t)\mathcal{K}_2(t, y)dt - \frac{1}{\pi}\int_0^h g_2(t)\mathcal{K}_1(t, y)dt = f_2(y), \; 0 < y < h, \quad (1.22)$$

where

$$(\mathcal{K}_1, \mathcal{K}_2)(t, y) = \int_0^\infty \frac{\sin \xi t \sin \xi y}{\xi \sinh 2\xi a}(e^{2\xi a}, 1)d\xi,$$

so that

$$\mathcal{K}_1(t, y) = \frac{1}{2}\ln\left|\frac{(y+t)\sinh\frac{\pi(y+t)}{4a}}{(y-t)\sinh\frac{\pi(y-t)}{4a}}\right|, \quad (1.23a)$$

$$\mathcal{K}_2(t, y) = \frac{1}{2}\ln\left|\frac{\cosh\frac{\pi(y+t)}{4a}}{\cosh\frac{\pi(y-t)}{4a}}\right|, \quad (1.23b)$$

$$f_1(y) = -\frac{\pi}{2K}\beta e^{-iKa}\sinh Ky - \frac{i}{4}(\alpha e^{iKa} - \beta e^{-iKa})\int_h^\infty e^{-Kt}\mathcal{K}_1(t, y)dt$$
$$+ \frac{i}{4}(\alpha e^{-iKa} - \beta e^{iKa})\int_h^\infty e^{-Kt}\mathcal{K}_2(t, y)dt, \; 0 < y < h, \quad (1.24a)$$

$$f_2(y) = -\frac{\pi}{2K}(\beta - R)e^{iKa}\sinh Ky - \frac{i}{4}(\alpha e^{iKa} - \beta e^{-iKa})\int_h^\infty e^{-Kt}\mathcal{K}_2(t, y)\,dt$$

$$+\frac{i}{4}(\alpha e^{-iKa} - \beta e^{iKa})\int_h^\infty e^{-Kt}\mathcal{K}_1(t, y)dt,\ 0 < y < h. \tag{1.24b}$$

The coupled integral equations (1.21) and (1.22) are decoupled by addition and subtraction, and the decoupled equations are given by

$$\int_0^h g(t)\mathcal{K}(t, y)dt = \pi f(y),\ 0 < y < h, \tag{1.25}$$

$$\int_0^h G(t)\mathcal{L}(t, y)dt = \pi F(y),\ 0 < y < h \tag{1.26}$$

where

$$g(t) = g_1(t) - g_2(t),\ G(t) = g_1(t) + g_2(t), \tag{1.27}$$

$$\mathcal{K}(t, y) = \mathcal{K}_1(t, y) + \mathcal{K}_2(t, y),\ \mathcal{L}(t, y) = \mathcal{K}_1(t, y) - \mathcal{K}_2(t, y), \tag{1.28}$$

$$f(y) = f_1(y) - f_2(y),\ F(y) = f_1(y) + f_2(y). \tag{1.29}$$

It is important to note that $f(0) = 0$, $F(0) = 0$.

It follows from (1.23) that the kernels $\mathcal{K}(t, y)$ and $\mathcal{L}(t, y)$ are given by

$$\mathcal{K}(t, y) = \frac{1}{2}\ln\left|\frac{y+t}{y-t}\right| + \frac{1}{2}\ln\left|\frac{\tanh\frac{\pi y}{2a} + \tanh\frac{\pi t}{2a}}{\tanh\frac{\pi y}{2a} - \tanh\frac{\pi t}{2a}}\right|,$$

$$\mathcal{L}(t, y) = \frac{1}{2}\ln\left|\frac{y+t}{y-t}\right| + \frac{1}{2}\ln\left|\frac{\sinh\frac{\pi y}{2a} + \sinh\frac{\pi t}{2a}}{\sinh\frac{\pi y}{2a} - \sinh\frac{\pi t}{2a}}\right|. \tag{1.30}$$

Using the result (proved by elementary integration)

$$\ln\left|\frac{Y+T}{Y-T}\right| = \int_0^{\min(Y,T)} \frac{X}{[(Y^2 - X^2)(T^2 - X^2)]^{1/2}}\,dX$$

it is easy to show that if $\psi(y)$ is an increasing function, then

$$\frac{1}{2}\ln\left|\frac{\psi(y) + \psi(t)}{\psi(y) - \psi(t)}\right| = \int_0^{\min(y,t)} \frac{\psi'(\eta)\psi(\eta)}{[\{\psi^2(y) - \psi^2(\eta)\}\{\psi^2(t) - \psi^2(\eta)\}]^{1/2}}\,d\eta. \tag{1.31}$$

Using the result (1.31) in the logarithmic expressions for the kernel $\mathcal{K}(t, y)$, we find that the integral equation (1.25) is equivalent to the Abel type integral equation

$$\int_0^y \frac{\eta p(\eta)}{(y^2-\eta^2)^{1/2}}\,d\eta + \frac{\pi}{2a}\int_0^y \frac{q(\eta)\,\mathrm{sech}^2\frac{\pi\eta}{2a}\,\tanh\frac{\pi\eta}{2a}}{\left(\tanh^2\frac{\pi y}{2a}-\tanh^2\frac{\pi\eta}{2a}\right)^{1/2}}\,d\eta = f(y),\ 0<y<h, \quad (1.32)$$

where

$$p(\eta) = \int_\eta^h \frac{g(t)}{(t^2-\eta^2)^{1/2}}\,dt,\ q(\eta) = \int_\eta^h \frac{g(t)}{\left(\tanh^2\frac{\pi t}{2a}-\tanh^2\frac{\pi\eta}{2a}\right)^{1/2}}\,dt,\ 0<\eta<h, \quad (1.33)$$

so that $p(h)=0$, $q(h)=0$ for consistency. The functions $p(\eta)$, $q(\eta)$ are not independent and they satisfy the relation

$$\int_y^h \frac{\eta p(\eta)}{(\eta^2-y^2)^{\frac{1}{2}}}\,d\eta - \frac{\pi}{2a}\int_y^h \frac{q(\eta)\,\mathrm{sech}^2\frac{\pi\eta}{2a}\,\tanh\frac{\pi\eta}{2a}}{\left(\tanh^2\frac{\pi\eta}{2a}-\tanh^2\frac{\pi y}{2a}\right)^{\frac{1}{2}}}\,d\eta = 0,\ 0<y<h, \quad (1.34)$$

obtained by equating the two expressions of $g(y)$ found by inverting the two relations in (1.33).

Similarly, the integral equation (1.26) is equivalent to the Abel integral equation

$$\int_0^y \frac{\eta P(\eta)}{(y^2-\eta^2)^{\frac{1}{2}}}\,d\eta - \frac{\pi}{2a}\int_0^y \frac{Q(\eta)\cosh\frac{\pi\eta}{2a}\,\sinh\frac{\pi\eta}{2a}}{\left(\sinh^2\frac{\pi y}{2a}-\sinh^2\frac{\pi\eta}{2a}\right)^{\frac{1}{2}}}\,d\eta = F(y),\ 0<y<h, \quad (1.35)$$

where

$$P(\eta) = \int_\eta^h \frac{G(t)}{(t^2-\eta^2)^{\frac{1}{2}}}\,dt,\ Q(\eta) = \int_\eta^h \frac{G(t)}{\left(\sinh^2\frac{\pi t}{2a}-\sinh^2\frac{\pi y}{2a}\right)^{\frac{1}{2}}}\,dt,\ 0<\eta<h, \quad (1.36)$$

so that $P(h)=0$, $Q(h)=0$ for consistency. The functions $P(\eta)$, $Q(\eta)$ are not independent and they satisfy the relation

$$\int_y^h \frac{\eta P(\eta)}{(\eta^2-y^2)^{\frac{1}{2}}}\,d\eta - \frac{\pi}{2a}\int_y^h \frac{Q(\eta)\cosh\frac{\pi\eta}{2a}\,\sinh\frac{\pi\eta}{2a}}{\left(\sinh^2\frac{\pi\eta}{2a}-\sinh^2\frac{\pi y}{2a}\right)^{\frac{1}{2}}}\,d\eta = 0,\ 0<y<h. \quad (1.37)$$

Solution of $p(\eta)$, $P(\eta)$ (and hence $q(\eta)$, $Q(\eta)$) can be obtained utilizing a procedure given by De et al. (2009). In fact $p(\eta)$ is given by

$$p(\eta) = \frac{1}{2\pi}\int_0^\eta \frac{f'(x)}{(\eta^2-x^2)^{1/2}}\,dx,\ 0<\eta<h. \quad (1.38)$$

Since $p(h)=0$, we find that

$$\int_0^h \frac{f'(x)}{(h^2-x^2)^{1/2}}\,dx = 0 \quad (1.39)$$

Adopting a similar procedure $P(\eta)$ is obtained as

$$P(\eta) = \frac{1}{2\pi}\int_0^{\eta} \frac{F'(x)}{(\eta^2 - x^2)^{1/2}}\,dx,\ 0 < \eta < h. \tag{1.40}$$

Since $P(h) = 0$, we obtain the result

$$\int_0^{h} \frac{F'(x)}{(h^2 - x^2)^{1/2}}\,dx = 0. \tag{1.41}$$

The equations (1.39) and (1.41) together with (1.9a) and (1.9b) will produce the unknown constants α, β, R, T.

The functions $f(x)$ and $F(x)$ are defined by (1.29) where $f_1(x)$ and $f_2(x)$ are given by (1.24). Substituting the approximate expressions for $f(x)$ and $F(x)$ in (1.39) and (1.41), carrying out the necessary integrations, and substituting for α, β in terms of R, T from (1.9), we finally obtain

$$R, T = \frac{1}{2}\left[\frac{i\left\{\frac{\pi^2}{2}I_0(Kh) - \frac{U}{2}\sin 2Ka\right\} - U\sin^2 Ka}{i\left\{\frac{\pi^2}{2}I_0(Kh) - \frac{U}{2}\sin 2Ka\right\} + U\sin^2 Ka} \mp \frac{\frac{\pi^2}{2}I_0(Kh) + \frac{V}{2}\sin 2Ka + iV\cos^2 Ka}{\frac{\pi^2}{2}I_0(Kh) + \frac{V}{2}\sin 2Ka - iV\cos^2 Ka}\right] \tag{1.42}$$

where

$$(U, V) = e^{-Kh}\int_0^{\infty}\left[\frac{(\xi\cos\xi h + K\sin\xi h)}{(\xi^2 + K^2)\sinh 2\xi a}(1 \mp e^{2\xi a})J_0(\xi h)\right]d\xi. \tag{1.43}$$

It is straightforward to observe that R, T satisfy $|R|^2 + |T|^2 = 1$ which is the energy identity.

As observed by Jarvis (1971), there is an infinite sequence of values of Kh for which $|R|$ vanishes. These are determined by (using the same notation of Jarvis (1971).

$$\lambda\tan^2 Ka + 2\lambda\mu\tan Ka - \mu = 0 \tag{1.44}$$

where

$$\lambda = \frac{U}{\pi^2 I_0(Kh)},\ \mu = \frac{U}{\pi^2 I_0(Kh)}. \tag{1.45}$$

The roots of (1.44) have the same forms as those given by Jarvis (1971) since the form of (1.44) is exactly same but with apparently different forms for λ and μ.

It is possible to derive the results for a single submerged barrier from (1.42) by making $a \to 0$ but keeping Kh fixed. As $a \to 0$

$$U\sin Ka \to U_0,\ V \to \frac{\pi}{2}K_0(Kh)$$

where

$$U_0 = Ke^{-Kh}\int_0^{\infty}\left[\frac{(\xi\cos\xi h + K\sin\xi h)}{\xi(\xi^2 + K^2)}J_0(\xi h)\right]d\xi.$$

Using these results in (1.42), we find that as $a \to 0$

$$R \to R_0,\ T \to T_0$$

where

$$R_0 = \frac{K_0(Kh)}{K_0(Kh) + i\pi I_0(Kh)},\ T_0 = \frac{i\pi I_0(Kh)}{K_0(Kh) + i\pi I_0(Kh)}$$

which are well known results obtained long back by Dean (1945) and Ursell (1947).

The reflection and transmission coefficients $|R|$ and $|T|$ obtained in (1.42) are depicted graphically against Kh/π for different values of a/h in a number of figures to compare our results with those given in Jarvis (1971). In Fig. 2.1, $|R|$, $|T|$ are shown for small separation length ($a/h = 0.1$). The curves in this figure correspond to the reflection and transmission coefficients $|R_0|$ and $|T_0|$ respectively for a single barrier. As in Jarvis (1971), the difference in the result for $a/h = 0$ and $a/h = 0.1$ is not appreciable. In Figs. 2.2 to 2.4, $|R|$ and $|T|$ are shown against Kh/π for $a/h = 1, 5, 10$. It is observed that all

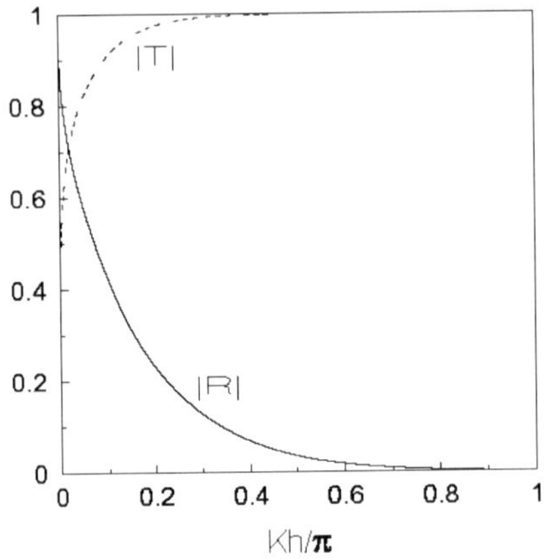

Fig. 2.1 $|R|$, $|T|$ for $a/h = 0.01$

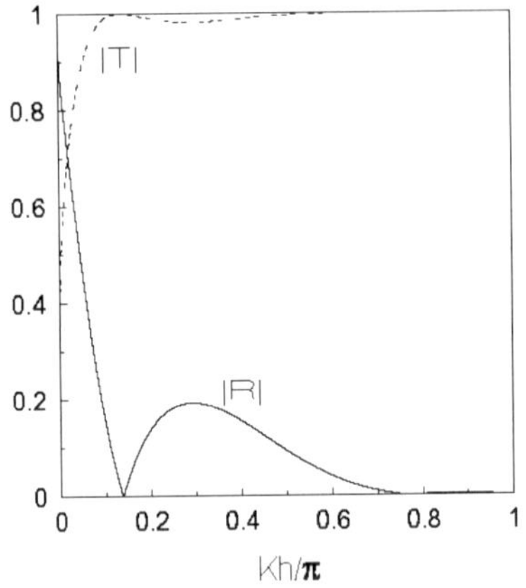

Fig. 2.2 $|R|$, $|T|$ for $a/h = 1.0$

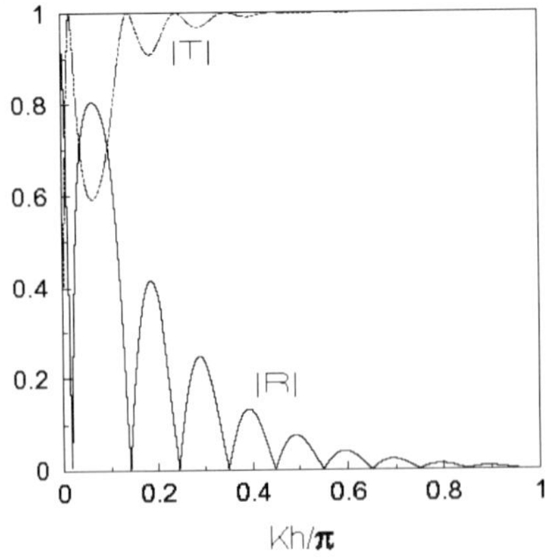

Fig. 2.3 $|R|$, $|T|$ for $a/h = 5.0$

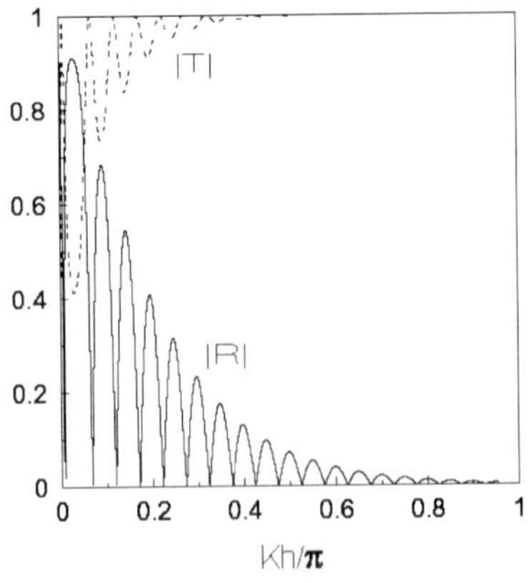

Fig. 2.4 $|R|$, $|T|$ for $a/h = 10.0$

the curves in the Figs. 2.1 to 2.4 coincide exactly with corresponding curves in Figs. 1 to 4 of Jarvis (1971). This shows the correctness of the results for the reflection and transmission coefficients obtained here.

Case 2: Two partially immersed barriers

For the case of two partially immersed barriers, $L = L_2 = (0, h)$. De et al. (2010) employed the method based on solution of Abel integral equations used for the case of two submerged barriers to solve this problem. Two sets of expressions for reflection and transmission coefficients were obtained in terms of computable integrals. These integrals are simpler than the integrals obtained by Levine and Rodemich (1958). When the two barriers are closely spaced, one set of expressions for the reflection and transmission coefficients reduces to those for a single barrier given by Ursell (1947) while the other set produces almost the same numerical results for a single barrier. Also when two barriers are widely separated, both sets of expressions for reflection and transmission coefficients reduce to the analytical results obtained originally by Levine and Rodemich (1958) and later by Evans and Morris (1972). Also the numerical results obtained by the present method almost coincide with those obtained by Evans and Morris (1972) by employing an approximate procedure.

The method of solution is similar to that described above for the case of two submerged barriers with appropriate modifications. Details are given in De et al. (2010). However, the two sets of expressions for the reflection coefficients mentioned above are given below.

One set is

$$R, T = \frac{1}{2}\left[\frac{\pi K_1(Kh) + 4e^{-iKa}U_1 \sin Ka}{\pi K_1(Kh) + 4e^{iKa}U_1 \sin Ka} \pm \frac{4ie^{-iKa}V_1 \cos Ka - \pi K_1(Kh)}{4ie^{iKa}V_1 \cos Ka + \pi K_1(Kh)}\right] \tag{1.46}$$

where

$$(U_1, V_1) = \frac{\pi}{2}\int_0^\infty \left[\frac{(K \cosh Kh \sin \xi h - \xi \sinh Kh \cos \xi h)}{(\xi^2 + K^2) \sinh 2\xi a}(e^{2\xi a} \pm 1)J_1(\xi h)\right] d\xi \tag{1.47}$$

while another set is

$$R, T = \frac{1}{2}\left[\frac{\pi W + 4e^{-iKa}U_2 \sin Ka}{\pi W + 4e^{iKa}U_2 \sin Ka} \pm \frac{4ie^{-iKa}V_2 \cos Ka - \pi W}{4ie^{iKa}V_2 \cos Ka + \pi W}\right] \tag{1.48}$$

where

$$(U_2, V_2) = \frac{\pi}{2}\int_0^\infty \frac{(K \cosh Kh \sin \xi h - \xi \sinh Kh \cos \xi h)}{(\xi^2 + K^2) \sinh 2\xi a}(e^{2\xi a} \pm 1) \times$$

$$\left\{\int_0^\infty \frac{\cos \xi x}{\left(\coth^2 \frac{\pi h}{2a} - \coth^2 \frac{\pi x}{2a}\right)} dx\right\} d\xi \tag{1.49a}$$

and

$$W = \int_h^\infty \frac{e^{-Kx}}{\left(\coth^2 \frac{\pi h}{2a} - \coth^2 \frac{\pi x}{2a}\right)} dx. \tag{1.49b}$$

For both the sets (1.46) and (1.48) it is straightforward to observe that R, T satisfy the energy identity

$$|R|^2 + |T|^2 = 1.$$

It may be noted that R and T given by (1.46) and (1.48) involve one modified Bessel function and two computable integrals U_1, V_1 given by (1.47) while R and T given by (1.48) involve three computable integrals U_2, V_2 and W given in (1.49). Expression for R, T given by (1.46) must coincide with the expressions for R, T given by (1.48). However, it is not easy to show this analytically. The ratios $\frac{U_1}{K1}, \frac{U_2}{W}$ and $\frac{V_1}{K1}, \frac{V_2}{W}$ are functions of Kh and a/h. De et al. (2010) computed these ratios for different values of Kh and fixed a/h, and found numerically that $\frac{U_1}{K1} \approx \frac{U_2}{W}$ and $\frac{V_1}{K1} \approx \frac{V_2}{W}$. Thus the two sets of expressions for R, T produce almost the same numerical results.

Approximation of R, T for small separation length

It is possible to derive the results for a single submerged barrier from (1.46) by making $a \to 0$ but keeping Kh fixed. As $a \to 0$, $U_1 \sin Ka \to U_{10}$, $V_1 \to -\frac{\pi^2}{4} I_1(Kh)$ where

$$U_{10} = -\frac{\pi K}{2} \int_0^h \sinh Kt \left\{ \int_0^\infty \frac{\sin \xi t \, J_1(\xi h)}{\xi} d\xi \right\} dt.$$

Using these results in (1.46) we find that as $a \to 0$,

$$R \to \frac{\pi I_1(Kh)}{\pi I_1(Kh) + iK_1(Kh)}, \quad T \to \frac{iK_1(Kh)}{\pi I_1(Kh) + iK_1(Kh)}$$

which coincide with the results obtained by Ursell (1947) for single partially immersed barrier.

A similar analysis to find the limit as $a \to 0$ from the results in (1.48) has been tried by De et al. (2010). It could not be possible to establish analytically that $U_2 \sin Ka/W \to U_{10}/K_1$ and $V_1/W \to -\frac{\pi^2}{4} I_1/K_1$ as $a \to 0$. However for small values of a/h, it has been shown numerically by De et al. (2010) that $U_2 \sin Ka/W \approx U_{10}/K_1$ and $V_1/W \approx -\frac{\pi^2}{4} I_1/K_1$.

Approximation of R, T for large separation length

We introduce two dimensionless parameters $\alpha = a/h$.

For large α

$$U_1, V_1 \approx -\frac{\pi^2}{2} I_1(Kh) \text{ and } \frac{U_2}{W}, \frac{V_2}{W} \approx -\frac{\pi^2}{2} \frac{I_1(Kh)}{K_1(Kh)}.$$

Using these approximations of U_1, V_1 and U_2/W, V_2/W for large α, in (1.46) and (1.48) we find

$$R \approx e^{-2i\alpha Kh} + \frac{2\pi \sin 2\alpha Kh\, K_1(Kh) I_1(Kh) - K_1^2(Kh) e^{-2i\alpha Kh}}{\Delta}$$

and

$$T \approx \frac{K_1^2(Kh)}{\Delta}$$

where

$$\Delta = \{K_1(Kh) - 2\pi \sin(\alpha Kh) e^{i\alpha Kh} J_1(Kh)\}\{K_1(Kh) - 2\pi i \cos(\alpha Kh) e^{i\alpha Kh} I_1(Kh)\}.$$

These results agree with those given in Levine and Rodemich (1958) (correcting a typographical error) and Evans and Morris (1971) for large separation length between the two barriers.

$|R|$ obtained in (1.46) (and also in (1.48)) is depicted graphically against Kh for different values of a/h in a number of figures to compare our results with those given in Evans and Morris (1972). In Figs. 2.5 to 2.7, $|R|$ is shown against Kh for $\frac{a}{h} = 1, 3, 5$. It is

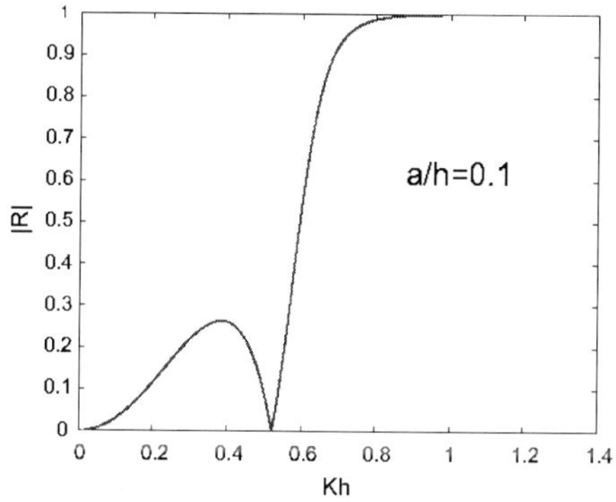

Fig. 2.5 $|R|$ against Kh

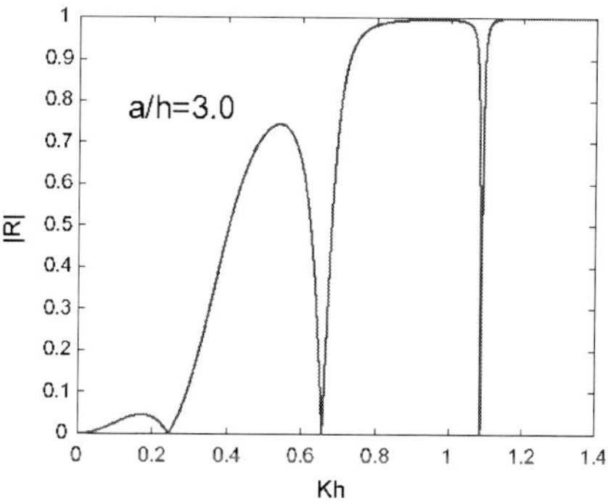

Fig. 2.6 $|R|$ against Kh

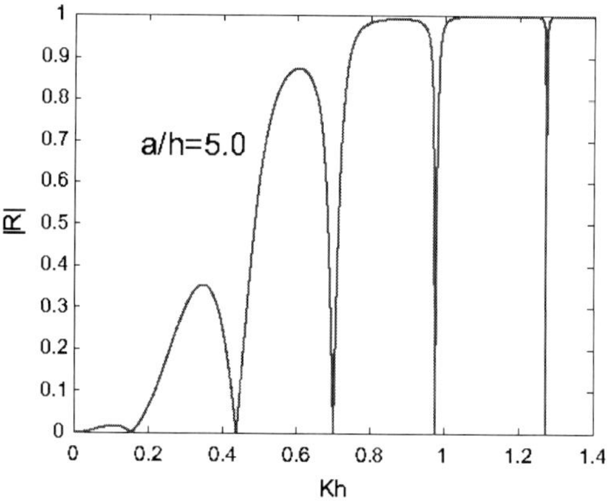

Fig. 2.7 $|R|$ against Kh

observed that the curves in all Figs. 2.5 to 2.7 coincide almost with the corresponding curves given in Evans and Morris (1971).

Case 3: Two thin vertical barriers with apertures

Let the two thin vertical barriers with submerged gaps at the same level occupy the position $x = \pm a,\ 0 < y < b,\ c < y < \infty$. In this case $L = (0, b) \cup (c, \infty)$ and $\bar{L} = (b, c)$. De et al. (2013) used the same method based on solution of Abel integral equations to obtain two sets of expressions for the reflection and transmission coefficients involving computable integrals. The two different expressions for each coefficient however produce the same numerical results although it appears difficult to prove analytically that the two expressions are equal. In this case the two sets of expressions for R and T are given by (cf. De et al. (2013))

$$(R, T) = \frac{1}{2}\left[\frac{\pi I + 4e^{-iKa}\,U \sin Ka}{\pi I + 4e^{iKa}\,U \sin Ka} \pm \frac{4ie^{-iKa}\,V \cos Ka - \pi I}{4ie^{iKa}\,V \cos Ka + \pi I}\right] \tag{1.50}$$

where

$$U = e^{Kb}\,U_1 - J_1 U_2,\ I = e^{-Kb}\,J_1 + e^{Kb}\,J_2,\ V = e^{Kb}\,V_1 - J_1 V_2,$$

with

$$(U_1, V_1) = \int_0^\infty \left[\frac{L(b,\xi) - \frac{1}{2}e^{-Kc}(\xi \cos \xi c + K \sin \xi c)}{(\xi^2 + K^2)\sinh 2\xi a}(e^{2\xi a} \pm 1)\int_b^c \frac{\cos \xi x}{(c^2 - x^2)^{\frac{1}{2}}}dx\right] d\xi,$$

$$(U_2, V_2) = \int_0^\infty \left[\frac{L(b,\xi) - \frac{1}{2}e^{-Kc}(\xi \cos \xi c + K \sin \xi c)}{(\xi^2 + K^2)\sinh 2\xi a}(e^{2\xi a} \pm 1)\frac{K \sin \xi c}{\xi}\right] d\xi,$$

$$L(b, \xi) = K \cosh Kb \sin \xi b - \xi \sinh Kb \cos \xi b,$$

$$(J_1, J_2) = \int_b^c \frac{(e^{Kx}, e^{-Kx})}{(c^2 - x^2)^{\frac{1}{2}}}\,dx. \tag{1.51}$$

and

$$(R, T) = \frac{1}{2}\left[\frac{\pi W_1 + 4e^{-iKa}\,X \sin Ka}{\pi I + 4e^{iKa}\,X \sin Ka} \pm \frac{4ie^{-iKa}\,Y \cos Ka - \pi W_2}{4ie^{iKa}\,Y \cos Ka + \pi W_2}\right] \tag{1.52}$$

with

$$X = e^{Kb}\,X_1 - U_2 W_{11},\ Y = e^{Kb}\,Y_1 - V_2 W_{22},\ W_1 = e^{Kb} W_{10} + e^{-Kb} W_{11},\ W_1 = e^{Kb} W_{20} + e^{-Kb} W_{21}$$

where

$$X_1 = \int_0^\infty \left[\frac{L(b,\xi) - \frac{1}{2}e^{-Kc}(\xi \cos \xi c + K \sin \xi c)}{(\xi^2 + K^2)\sinh 2\xi a}(e^{2\xi a} + 1)\int_b^c \frac{\cos \xi x}{\left(\tanh^2 \frac{\pi c}{2a} - \tanh^2 \frac{\pi x}{2a}\right)^{1/2}}dx\right] d\xi,$$

$$Y_1 = \int_0^{\infty} \left[\frac{L(b,\xi) - \frac{1}{2}e^{-Kc}(\xi\cos\xi c + K\sin\xi c)}{(\xi^2 + K^2)\sinh 2\xi a}(e^{2\xi a} + 1)\int_b^c \frac{\cos\xi x}{\left(\sinh^2\frac{\pi c}{2a} - \sinh^2\frac{\pi y}{2a}\right)^{1/2}}dx \right] d\xi,$$

$$(W_{10}, W_{11}) = \int_b^c \frac{(e^{-Kx}, e^{Kx})}{\left(\tanh^2\frac{\pi c}{2a} - \tanh^2\frac{\pi x}{2a}\right)^{1/2}}dx\,,$$

$$(W_{20}, W_{21}) = \int_b^c \frac{(e^{-Kx}, e^{Kx})}{\left(\sinh^2\frac{\pi c}{2a} - \sinh^2\frac{\pi x}{2a}\right)^{1/2}}dx. \tag{1.53}$$

For both the sets of expressions of R, T given by (1.50) and (1.52), it is straightforward to observe that R, T satisfy the energy identity

$$|R|^2 + |T|^2 = 1.$$

It is not easy to prove analytically that the two sets of expressions for each of R, T given by (1.50) and (1.52) coincide. However, numerical computations of $|R|$, $|T|$ as given by (1.50) and (1.52) produce almost same numerical results for various values of the different parameters. The ratios $\frac{U}{I}, \frac{X}{W_1}$ and $\frac{V}{I}, \frac{Y}{W_2}$ are functions of the wave number Kh, the separation parameter $\mu = \frac{c-b}{h}$ where $h\left(=\frac{b+c}{2}\right)$ is the depth of the middle of the aperture below the mean free surface. De et al. (2013) computed these ratios for different values of Kh and fixed $\frac{a}{h}$ and μ, and found numerically that $\frac{U}{I} \approx \frac{X}{W_1}$ and $\frac{V}{I} \approx \frac{Y}{W_2}$.

Approximation of R, T for small separation length

As in the case of two submerged or surface-piercing barriers, it is possible to derive the results for a single thin semi-infinite vertical plane barrier extending downwards from the surface and having a submerged aperture from (1.50) by making $a \to 0$. We find that

$$U_j \sin Ka \to U_{j0},\ V_j \to V_{j0}\ (j = 1,2) \text{ as } a \to 0$$

where

$$(U_{10}, U_{20}) = K\int_0^{\infty} \left[\frac{L(b,\xi) - \frac{1}{2}e^{-Kc}(\xi\cos\xi c + K\sin\xi c)}{\xi(\xi^2 + K^2)} \left(\int_b^c \frac{\cos\xi x}{(c^2 - x^2)^{\frac{1}{2}}}dx\,, \frac{K\sin\xi b}{\xi} \right) \right] d\xi,$$

$$(V_{10}, V_{20}) = \int_0^{\infty} \frac{L(b,\xi) - \frac{1}{2}e^{-Kc}(\xi\cos\xi c + K\sin\xi c)}{(\xi^2 + K^2)} \left(\int_b^c \frac{\cos\xi x}{(c^2 - x^2)^{\frac{1}{2}}}dx\,, \frac{K\sin\xi b}{\xi} \right) d\xi. \tag{1.54}$$

Using these results in (1.50), we find that as $a \to 0$

$$R \to R_1 \equiv \frac{4iW}{\pi I + 4iW} \text{ and } T \to T_1 \equiv \frac{\pi I}{\pi I + 4iW} \tag{1.55}$$

where

$$W = e^{Kc}V_{10} - J_1 V_{20}, \tag{1.56}$$

J_1 being given in (1.51). $|R_1|$, $|T_1|$ given by (1.55) are depicted graphically against $Kh(h = \frac{b+c}{2})$ for various values of the parameter $\mu(\mu = \frac{c-b}{h})$ in Fig. 2.8. Since $b = h(1 - \frac{\mu}{2}), c = h(1 + \frac{\mu}{2})$, we must have $0 < \mu < 2$. In Fig. 2.8, $|R_1|$ and $|T_1|$ are depicted against Kh for $\mu = 0.1, 0.5$, 1.0 and 1.5. The curves for $|R_1|$ and $|T_1|$ in Fig. 2.8 coincide with the corresponding curves given in Porter (1972) and Chakrabarti et al. (2003) who employed different methods to study the problem of a single barrier with a submerged gap. This establishes the correctness of the mathematical analysis employed here.

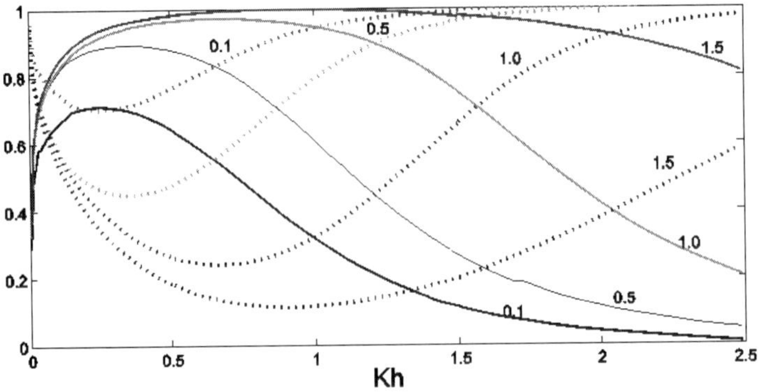

Fig. 2.8 $|R_1|(\cdots)$ and $|T_1|(-)$ against Kh for different values of μ

We can further approximate the results in (1.55) to obtain for a single barrier submerged in deep water by making $b \to 0$. We note that

$$I \to \pi I_0(Kc),\ W \to -\frac{\pi}{4} K_0(Kc) \text{ as } b \to 0$$

so that

$$R \to R_0 \equiv \frac{K_0(Kc)}{K_0(Kc) + i\pi I_0(Kc)},\ T \to T_0 \equiv \frac{i\pi I_0(Kc)}{K_0(Kc) + i\pi I_0(Kc)}.$$

These coincide with the classical results obtained by Ursell (1947) for a fixed vertical barrier submerged from a depth c below the mean free surface in deep water.

To compare with the results of Jarvis (1971) for two submerged barriers, $|R|$ and $|T|$ given by (1.50) are depicted in Fig. 2.9 graphically against $\frac{Kc}{\pi}$ for $\frac{b}{c} = 0.0001$ and $\frac{a}{c} = 1.0$. It is observed that the curves in Fig. 2.9 for $|R|$ and $|T|$ coincide exactly with the corresponding curves in Fig. 1 of Jarvis (1971).

The numerical results obtained by Evans (1975) for two barriers each having a *narrow* aperture at same level, can be recovered from our results. For this purpose $|T|^2$, the proportion of wave energy transmitted through the aperture is plotted against

Kh in Figs. 2.10 and 2.11 for $\mu = 0.05, 0.15$. These two figures almost coincide with the Figs. 2(a), 2(b) of Evans (1975).

The reflection coefficient $|R|$ is depicted against *Kh* for different values of the separation ratio $\frac{a}{h}$ and μ, the ratio of the width of the aperture to its mean depth ($0 < \mu < 2$). For $\mu = 1.0$ and $\frac{a}{h} = 5.0$ (moderate separation ratio), $|R|$ is depicted in Fig. 2.12. This figure shows that $|R|$ is oscillatory, it vanishes for some discrete values of the wave number *Kh*. It becomes almost unity for some other wave numbers. This means that a pair of barriers with apertures, completely reflects an incident wave train

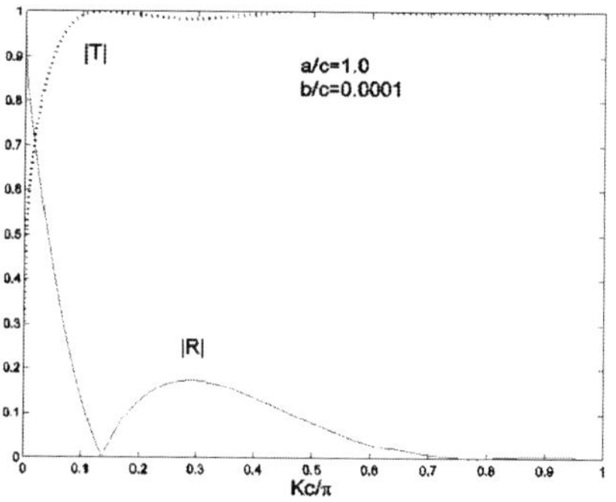

Fig. 2.9 $|R|$ and $|T|$ against $\frac{Kc}{\pi}$

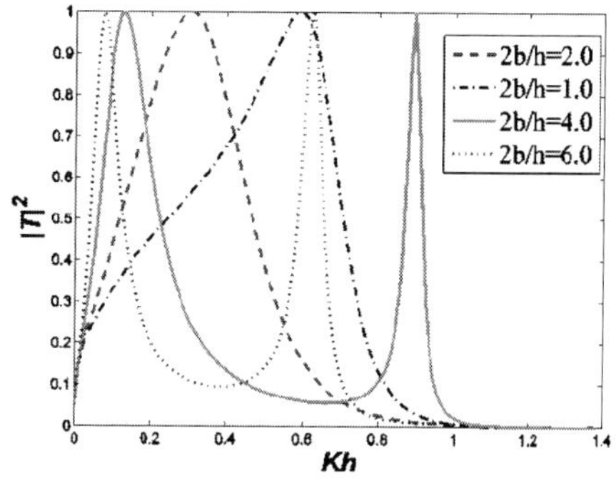

Fig. 2.10 $|T|^2$ against *Kh* for $\mu = 0.05$

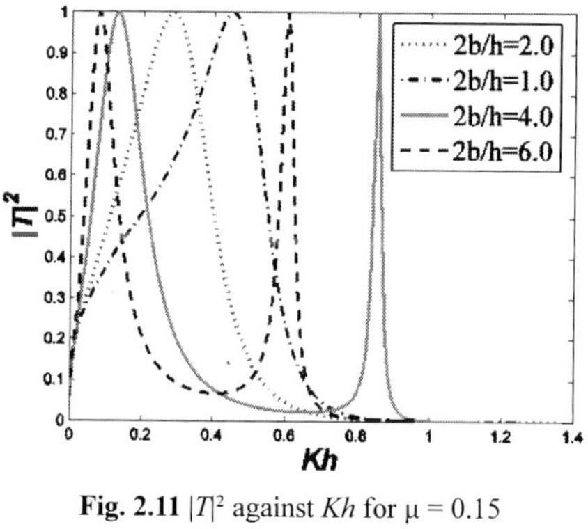

Fig. 2.11 $|T|^2$ against *Kh* for $\mu = 0.15$

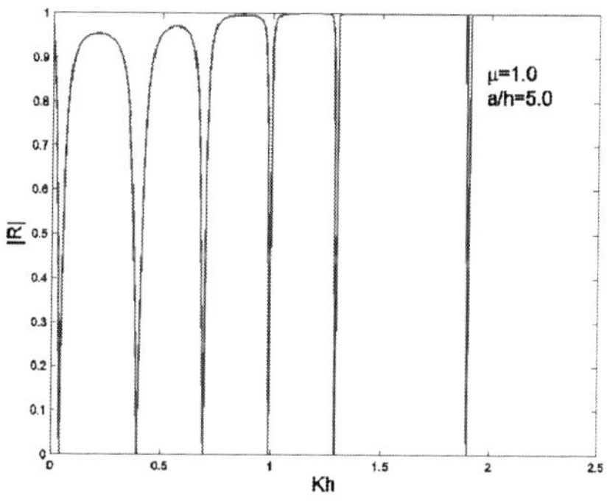

Fig. 2.12 $|R|$ against *Kh*

at certain discrete frequencies. The number of these frequencies (for which |R| vanishes) increases with the increase in the separation length of the two barriers.

Figure 1.12 depicts $|R|$ for $\mu = 1.0$ and $\frac{a}{h} = 10$ (large separation ratio). This figure for $|R|$ is seen to be more oscillatory. The oscillatory behavior is attributed due to multiple reflections by the two barriers. Also the number of zeros of the reflection coefficient $|R|$ in Fig. 2.13 is more than in Fig. 2.12. This means that a pair of barriers with apertures completely reflects an incident wave train at certain discrete frequencies. The number of these frequencies increases with the increase in the separation length of the two barriers.

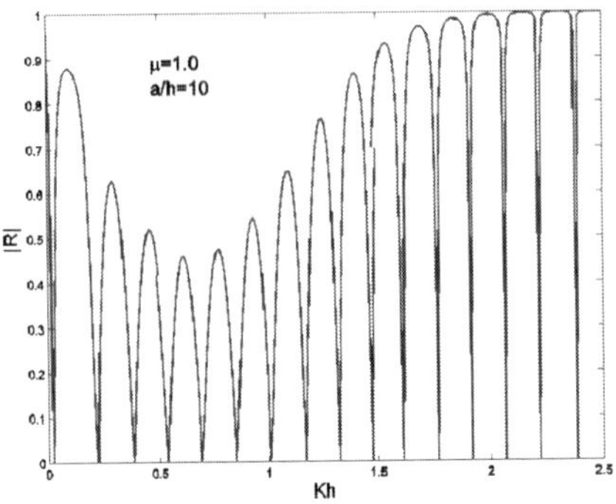

Fig. 2.13 $|R|$ against Kh

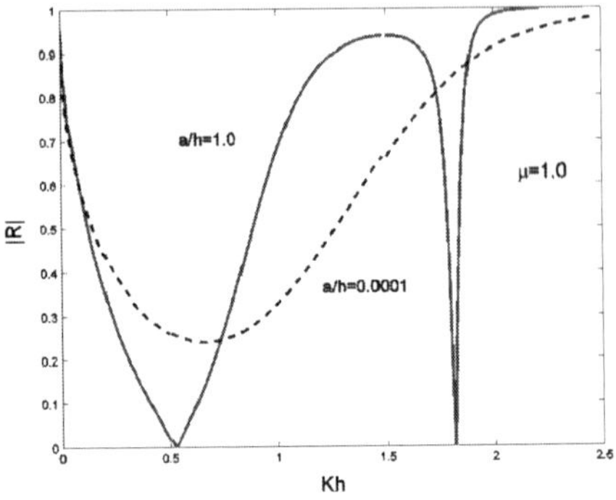

Fig. 2.14 $|R|$ against Kh

In Fig. 2.14, $|R|$ is plotted for $\mu = 1.0$ and $\frac{a}{h} = 1.0$ (moderate separation length), $\frac{a}{h} = 0.0001$ (very small separation length). This figure shows that as the separation length decreases the number of discrete frequencies for which $|R|$ vanishes, also decreases and the curve for $|R|$ becomes less oscillatory. This means that for moderate separation length, less multiple reflections by the barriers occur. The curve for $|R|$ for $a/h = 0.0001$ becomes identical with those given by Porter (1972) and Chakabarti et al. (2003), as was also stated earlier.

The case of two identical plates completely submerged in deep water can also be considered by this method and numerical results for the reflection and transmission coefficients can be obtained. The case of two non-identical thin vertical barriers occupying different geometrical configurations in infinitely deep water can be investigated by Galerkin approximations. This is however not discussed here.

2.2 Scattering by two nearly vertical barriers

This section involves two-dimensional problem of water wave scattering by two fixed identical nearly vertical barriers either submerged in deep water from the same depth below the mean free surface and extending infinitely downwards or partially immersed in deep water up to the same depth below the mean free surface. The problem of two nearly vertical barriers is investigated here by employing a simplified perturbation analysis used by Mandal and Chakrabarti (1989) for the case of a single nearly vertical barrier. The position of the mean free surface is given by $y = 0$, the y-axis being chosen vertically downwards into the fluid region, and x-axis along the direction of an incoming train of surface waves. The co-ordinates are non-dimensionalized with respect to the depth of the edges of the barriers. Let the configurations of the barriers be described by

$$x = \pm a + \epsilon c(y),\ y \in \mathcal{L}$$

where ϵ is a small non-dimensional number signifying the nearness of the thin vertical barriers and $c(y)$ is the shape function defined for $y \in \mathcal{L}$. For submerged barriers $\mathcal{L} = \mathcal{L}_1 = (1, \infty)$ and $c(y)$ satisfies the conditions that $c(1) = 0$ and is a bounded and continuous function of $y \in (1, \infty)$. For partially immersed barriers $\mathcal{L} = \mathcal{L}_2 = (0, 1)$ and $c(y)$ satisfies $c(0) = c(1) = 0$ and is bounded and continuous in $(0,1)$. The case $\mathcal{L} = \mathcal{L}_1$

$= (1, \infty)$ (i.e., two submerged nearly vertical barriers) was considered by Mandal and De (2006) and the case $\mathcal{L} = \mathcal{L}_2 = (0, 1)$ (i.e., two partially immersed nearly vertical barriers) by De et al. (2005).

Assuming the motion in the fluid to be irrotational and simple harmonic in time t with angular frequency σ, it can be described by a velocity potential $\mathrm{Re}\{\phi(x,y)e^{-i\sigma t}\}$. Then $\phi(x,y)$ satisfies

$$\nabla^2\phi = 0 \text{ in the fluid region,} \tag{2.1}$$

the linearised free surface condition

$$K\phi + \phi_y = 0 \text{ on } y = 0 \tag{2.2}$$

where $K = \frac{\sigma^2}{g}$, the barrier conditions

$$\frac{\partial \phi}{\partial n} = 0 \text{ on } x = \pm a + \epsilon c(y),\ y \in \mathcal{L} \tag{2.3}$$

where n denotes the normal to the surface of the curved barrier, the edge conditions

$$r^{1/2}\nabla\phi \to 0 \text{ is bounded as } r \to 0 \tag{2.4}$$

where r is the distance from the point $(\pm a, 1)$, the deep water conditions

$$\phi, \nabla\phi \to 0 \text{ as } y \to \infty \tag{2.5}$$

and finally, the infinity requirements given by

$$\phi(x, y) \sim \begin{cases} T\phi^{inc}(x, y) \text{ as } x \to \infty, \\ \phi^{inc}(x, y) + R\phi^{inc}(-x, y) \text{ as } x \to -\infty. \end{cases} \tag{2.6}$$

In the condition (2.6), R and T respectively denote the (complex) reflection and transmission coefficients respectively to be determined and

$$\phi^{inc}(x, y) = e^{-Ky+iKx} \tag{2.7}$$

denotes the incident wave potential propagating from $x = -\infty$.

The boundary conditions (2.3) on nearly vertical barriers can be expressed as

$$\phi_x(\pm a \pm 0, y) - \epsilon\frac{d}{dy}\{c(y)\phi_y(\pm a \pm 0, y)\} + O(\epsilon^2) = 0,\ y \in \mathcal{L}, \tag{2.8}$$

where ± 0 denote values on two sides of each barrier. The forms of approximate boundary conditions (2.8) suggest that ϕ, R and T have the following perturbational expansion, in terms of the small parameter ϵ:

$$\begin{aligned} \phi(x, y, \epsilon) &= \phi_0(x, y) + \epsilon\phi_1(x, y) + O(\epsilon^2), \\ R(\epsilon) &= R_0 + \epsilon R_1 + O(\epsilon^2), \\ T(\epsilon) &= T_0 + \epsilon T_1 + O(\epsilon^2). \end{aligned} \tag{2.9}$$

Substituting the expansion (2.9) in (2.1), (2.2), (2.3), (2.4), (2.5) and (2.6), we find, equating the coefficients of identical powers of ϵ upto first order from both sides of the results that ϕ_0 and ϕ_1 satisfy the two problems $\mathcal{P}_1$ and $\mathcal{P}_2$ respectively.

$\mathcal{P}_1$: The function ϕ_0 satisfies

i. $\nabla^2\phi_0 = 0$ in $y > 0, -\infty < x < \infty$,

ii. $K\phi_0 + \phi_{0y} = 0$ on $y = 0$,

iii. $\phi_{0x} = 0$ on $x = \pm a, y \in \mathcal{L}$,

iv. $r^{1/2}\nabla\phi_0$ is bounded as $r = \{(x \pm a)^2 + (y-1)^2\}^{1/2} \to 0$,

v. $\phi_0, \nabla\phi_0 \to 0$ as $y \to \infty$,

vi. $\phi_0(x, y) \sim \begin{cases} T_0 e^{-Ky+iKx} \text{ as } x \to \infty, \\ e^{-Ky+iKx} + R_0 e^{-Ky-iKx} \text{ as } x \to -\infty. \end{cases}$

$\mathcal{P}_2$: The function ϕ_1 satisfies

i. $\nabla^2\phi_1 = 0$ in $y > 0, -\infty < x < \infty$,

ii. $K\phi_1 + \phi_{1y} = 0$ on $y = 0$,

iii. $\phi_{1x}(\pm a \pm 0, y) = \frac{d}{dy}\{c(y)\phi_{0y}(\pm a \pm 0, y)\}, y \in \mathcal{L}$,

iv. $r^{1/2}\nabla\phi_1$ is bounded as $r = \{(x \pm a)^2 + (y-1)^2\}^{1/2} \to 0$,

v. $\phi_1, \nabla\phi_1 \to 0$ as $y \to \infty$,

vi. $\phi_1(x, y) \sim \begin{cases} T_1 e^{-Ky+iKx} \text{ as } x \to \infty, \\ R_1 e^{-Ky-iKx} \text{ as } x \to -\infty. \end{cases}$

The problem $\mathcal{P}_1$ corresponds to water wave scattering by two thin vertical barriers either completely submerged in deep water and extending infinitely downwards (for $\mathcal{L} = \mathcal{L}_1 = (1, \infty)$) or partially immersed (for $\mathcal{L} = \mathcal{L}_2 = (0, 1)$). Explicit solutions to the problem $\mathcal{P}_1$ for $\mathcal{L} = \mathcal{L}_1, \mathcal{L}_2$ were obtained by complex variable theory as mentioned earlier by Jarvis (1971) and Levine and Rodemich (1958) and by using Abel integral equation formulations by De et al. (2009, 2010).

Without solving the problem $\mathcal{P}_2$ fully, R_1 and T_1 can be obtained by employing Evans's (1976) idea.

We now treat the two case $\mathcal{L} = \mathcal{L}_1$ and $\mathcal{L} = \mathcal{L}_2$ separately.

Case 1: Two submerged nearly vertical barriers $\mathcal{L} = \mathcal{L}_1 = (1, \infty)$

To write the result of Jarvis (1971) for two submerged thin vertical barriers, we introduce a second complex unit j which does not interact with the complex unit i, and is used to denote the complex variable $z = x + jy$. Then the solution $\phi_0(x, y)$ of the problem $\mathcal{P}_1$ is given by

$$\phi_0(x, -y) = Re_j\, w(z) \tag{2.10}$$

where

$$w(z) = e^{-jKz}\int_{j\infty}^{z} \{A\zeta(u) + B\}f(\zeta(u))du + (C + jD)e^{-jKz}, \tag{2.11}$$

ζ and z being related by

$$z = \frac{2a}{\pi}\left[\frac{(a^2-1)\zeta}{(1-\zeta^2)^{1/2}} - j\log\{(1-\zeta^2)^{1/2} + j\zeta\}\right] \tag{2.12}$$

where α is a real number satisfying

$$\alpha(\alpha^2 - 1)^{1/2} + \log\{\alpha + (\alpha^2 - 1)^{1/2}\} = \frac{\pi}{2a} \tag{2.13}$$

and A, B, C, D are real constants with respect to j.

It may be noted that the relation (2.12) denotes a mapping of the whole complex z-plane cut from $\pm a \pm j$ to $\pm a \pm j\infty$ into the complex ζ plane cut from ± 1 to $\pm\infty$, and that branch of $(1 - \zeta^2)^{1/2}$ is chosen which is real and positive when $\zeta = 0$. The cuts in the z-plane are mapped into those in the ζ-plane and the x-axis is mapped onto the section $-1 \le Re\,\zeta \le 1$ of the real axis in the ζ-plane. Also the points $z = \pm a \pm j$ in the z-plane correspond to the point $\zeta = \pm a + j0$ in the ζ-plane.

The constants A, B, C, D (real with respect to j) are given by

$$A = (L\, e^{iKa} - I \operatorname{cosec} Ka)^{-1},\; B = (M\, e^{iKa} + iJ \sec Ka)^{-1},$$
$$C = -AI \operatorname{cosec} Ka,\; D = BJ \sec Ka \tag{2.14}$$

where

$$(I, J) = Re_j \int_{-1}^{1} (\zeta, 1) f(\zeta) e^{-Kv}\, dv \tag{2.15}$$

with $\zeta = \zeta(a + jv)$, $-1 \le v \le 1$,

$$(L, M) = e^{-jKa} \int_{\Gamma} (\zeta, 1) f(\zeta) e^{jKz}\, dz, \tag{2.16}$$

Γ being a loop around the cut $z = a + jv$ $(v > 1)$ in z-plane. It may be noted that L, M are real constants.

The reflection and transmission coefficients R_0, T_0 for the problem of submerged vertical barriers are given by

$$R_0 = -\, iLA + MB \cos Ka,\; T_0 = (L \cos Ka - I \sin Ka)A - MB \cos Ka \tag{2.17}$$

for which the energy identity $|R_0|^2 + |T_0|^2 = 1$ is satisfied.

To find R_1, we apply Green's integral theorem to the functions $\phi_0(x, y)$ and $\phi_1(x, y)$ in the region bounded by the lines $y = 0$, $-X \le x \le X$; $x = X$, $0 \le y \le Y$; $y = Y$, $a \le x \le X$; $x = a + 0$, $1 \le y \le Y$; $x = a - 0$, $1 \le y \le Y$; $y = Y$, $-a \le x \le a$; $x = -a + 0$, $1 \le y \le Y$; $x = -a - 0$, $1 \le y \le Y$; $y = Y$, $-X \le x \le -a$; $x = -X$, $0 \le y \le Y$ and circles of small radius δ with centers at $(\pm a, 1)$ and ultimately make X, Y to tend to infinity and δ to tend to zero. Using arguments similar to Evans (1976), we obtain

$$iR_1 = \int_1^{\infty} \{\phi_0(a + 0, y)\, \phi_{1x}(a + 0, y) - \phi_0(a - 0, y)\, \phi_{1x}(a - 0, y)\}dy$$
$$+ \int_1^{\infty} \{\phi_0(-a + 0, y)\, \phi_{1x}(a - 0, y) - \phi_0(-a - 0, y)\, \phi_{1x}(-a - 0, y)\}dy \tag{2.18}$$

Using the condition (iii) of $\mathcal{P}_2$ in the relation (2.18), integrating by parts and using $c(1) = 0$, we find that

$$iR_1 = \int_1^{\infty} \Big[\frac{d}{dy}\{\phi_0(a - 0, y) + \phi_0(-a + 0, y)\}\frac{d}{dy}\{\phi_0(a - 0, y) - \phi_0(-a + 0, y)\}$$
$$- \frac{d}{dy}\{\phi_0(a + 0, y) + \phi_0(-a - 0, y)\}\frac{d}{dy}\{\phi_0(a + 0, y) - \phi_0(-a - 0, y)\}\Big]dy. \tag{2.19}$$

Explicit expressions for $\phi_0(\pm a \pm 0), y > 1$ can be obtained from (2.10) and (2.11), and are given by

$$\phi_0(a+0,y) = -e^{-Ky}\{AP_1(y) + BQ_1(y) - C\cos Ka - D\sin Ka - AL_1 - BM_1\},$$

$$\phi_0(-a+0,y) = -e^{-Ky}\{AP_2(y) - BQ_2(y) - C\cos Ka + D\sin Ka - AL_1 + BM_1\}.$$

$\phi_0(a-0,y)$ is obtained from $\phi_0(a+0,y)$ with $P_1(y)$ replaced by $P_2(y)$ and $Q_1(y)$ replaced by $Q_2(y)$, while $\phi_0(-a-0,y)$ is obtained from $\phi_0(-a+0,y)$ with $P_2(y)$ replaced by $P_1(y)$ and $Q_2(y)$ replaced by $Q_1(y)$. The function $P_n(y), Q_n(y) (n=1,2)$ and the constants L_1, M_1 appearing (2.20) are given by

$$(P_n(y), Q_n(y)) = \int_{-1}^{-y} (\xi_n(v), 1) W(\xi_n(v)) e^{-Kv}\, dv \ (n=1,2), \tag{2.20}$$

$$(L_1, M_1) = \int_1^{\infty} (\xi_1(v), 1) W(\xi_1(v)) e^{-Kv}\, dv \tag{2.21}$$

where

$$W(\xi) = \frac{(\xi^2-1)^2}{\xi^2 - a^2}, \tag{2.22}$$

$\xi_1(v), \xi_2(v)\ (1 < \xi_2(v) < \alpha)$ being the two real roots of

$$(\alpha^2 - 1)\,\xi(\xi^2 - 1)^{-1/2} + \log\{\xi + (\xi^2-1)^{1/2}\} = \frac{\pi v}{2a}, \ v > 1. \tag{2.23}$$

Using the expressions for $\phi_0(\pm a \pm 0, y)$ given in (2.20), the relation (2.19) produces

$$\frac{iR_1}{4} = \int_1^{\infty} \begin{aligned}[t] & c(y)[A\{KP_2(y)e^{-Ky} - \xi_2(-y)W(\xi_2(-y))\} - HKe^{-Ky}] \\ & \times [B\{KQ_2(y)e^{-Ky} - W(\xi_2(-y))\} - GKe^{-Ky}]dy \end{aligned}$$

$$- \int_1^{\infty} \begin{aligned}[t] & c(y)[A\{KP_1(y)e^{-Ky} + \xi_1(-y)W(\xi_1(-y))\} + HKe^{-Ky}] \\ & \times [B\{KQ_1(y)e^{-Ky} + W(\xi_2(-y))\} + GKe^{-Ky}]dy \end{aligned} \tag{2.24}$$

where the constants G, H are given by

$$G = BM_1 + D\sin Ka,\ H = AL_1 + C\cos Ka. \tag{2.25}$$

The integrals appearing in (2.24) can be evaluated numerically once the form of $c(y)$ is known.

Again, to find T_1, we use the Green's integral theorem to the functions $\psi_0(x,y) = \phi_0(-x,y)$ and $\phi_1(x,y)$ in the region mentioned above to obtain finally,

$$iR_1 = \int_1^{\infty} \{\psi_0(a+0,y)\psi_{1x}(a+0,y) - \psi_0(a-0,y)\psi_{1x}(a-0,y)\}dy$$

$$+ \int_1^{\infty} \{\psi_0(-a+0,y)\psi_{1x}(a-0,y) - \psi_0(-a-0,y)\psi_{1x}(-a-0,y)\}dy.$$

Using the condition (iii) of $\mathcal{P}_2$, then integrating by parts and using $c(1) = 0$, we find that the above integrals vanish identically so that

$$T_1 \equiv 0. \tag{2.26}$$

Thus the first order correction to the transmission coefficient vanishes identically for two nearly vertical barriers also, as was the case for a single nearly vertical submerged barrier cf. Mandal and Chakrabarti (1989).

To produce the result for the first-order correction to the reflection coefficient for a single nearly vertical submerged barrier, we approximate R_1 given by (2.24) as $a \to 0$.

We first note from (2.13) that

$$\alpha \approx \left(\frac{\pi}{2a}\right)^{1/2} \text{ for small } a \tag{2.27}$$

so that $\alpha \to \infty$ as $a \to 0$.

For small value of a, i.e., large value of α, the real root $\xi_1(v)$ $(v > 1)$ of (2.23) can be approximated as

$$\xi_1(v) \approx \frac{1}{2}\, e^{1+\alpha^2(v-1)},\ v > 1 \tag{2.28}$$

and the other root $\xi_2(v)$ $(v > 1)$ of (2.23) can be approximated as

$$\xi_2(v) \approx \frac{v}{(v^2-1)^{\frac{1}{2}}}. \tag{2.29}$$

Using (2.28) and (2.29) it can be shown that for small values of a, $(v > 1)$

$$W(\xi_1) = \frac{2}{e^{1+\alpha^2(v-1)}};\ \xi_1 W(\xi_1) \approx 1;\ W(\xi_2) = \frac{2}{\alpha^2(v^2-1)^{\frac{1}{2}}};$$

$$\xi_2 W(\xi_2) = \frac{-v}{\alpha^2(v^2-1)^{1/2}}; \tag{2.30}$$

also

$$M \approx \frac{K_0(K)}{\alpha^2};\ J \approx \frac{\pi I_0(K)}{\alpha^2};\ L \approx \frac{e^{-K}}{K} + \frac{U}{\alpha^2};\ I \approx \frac{V}{\alpha^2},$$

where $I_0(K)$, $K_0(K)$ are modified Bessel functions and

$$U = \int_1^{\infty} \frac{ye^{-Ky}}{y^2-1}\,dy;\ V = \int_{-1}^{1} \frac{ye^{-Ky}}{y^2-1}\,dy. \tag{2.31}$$

Again for small a

$$A \approx \frac{\pi K}{\pi e^{-K} - 2V};\ B \approx \frac{\alpha^2}{K_0(K) + i\pi I_0(K)};\ C \approx \frac{-2V}{\pi e^{-K} - 2V};\ D \approx \frac{\pi I_0(K)}{K_0(K) + i\pi I_0(K)}, \tag{2.32}$$

also for small a,

$$P_1(y) \approx \frac{1}{K}(e^{K} - e^{Ky});\ P_2(y) \approx \frac{1}{\alpha^2}\int_{-1}^{y} \frac{ve^{-Kv}}{v^2-1}\,dv;\ Q_1(y) \approx \frac{2}{e(\alpha^2-K)}\left[\frac{1}{e^{\alpha_2(y-1)-Ky}} - \frac{1}{e^{2\alpha^2-K}}\right];$$

$$Q_2(y) \approx -\frac{1}{\alpha^2}\int_{-1}^{-y} \frac{e^{-Kv}}{v^2-1}\,dv;\ G \approx \frac{1}{K_0(K) + i\pi I_0(K)}\,\frac{\pi^2 K I_0(K)}{2\alpha^2} \text{ and } H \approx 1. \tag{2.33}$$

Using these expressions in (2.24), and noting that the second integral contributes nothing as $a \to 0$ and using the contributions of the first integral as $a \to 0$, we find

$$\frac{iR_1}{4K}(K_0(K) + i\pi I_0(K)) \approx K\int_1^{\infty} c(y)e^{-2Ky}\left(\int_1^{y} \frac{e^{Kv}}{(v^2-1)^{\frac{1}{2}}}dv\right)dy - \int_1^{\infty} c(y)\frac{e^{-Ky}}{(y^2-1)^{\frac{1}{2}}}dy. \tag{2.34}$$

This result coincides with the result obtained by Mandal and Chakarbarti (1989) for a single completely submerged nearly vertical barrier.

Case 2: Two partially immersed nearly vertical barriers $\mathcal{L} = \mathcal{L}_2 = (0, 1)$

To find R_1, we apply Green's integral theorem to the functions $\phi_0(x, y)$ and $\phi_1(x, y)$ in the region bounded by the lines $y = 0$, $a \le x \le X$; $x = X$, $0 \le y \le Y$; $y = Y$, $-X \le x \le X$; $x = -X$, $0 \le y \le Y$; $y = 0$, $-X \le x \le -a$; $x = -a - 0$, $0 \le y \le 1$; $x = -a + 0$, $0 \le y \le 1$; $y = 0$, $-a \le x \le a$; $x = a - 0$, $0 \le y \le 1$; $x = a + 0$, $0 \le y \le 1$ and circles of small radius δ with centers at $(\pm a,1)$ and ultimately make X, Y tend to infinity and δ tend to zero. Using arguments similar to Evans (1976), we obtain

$$iR_1 = \int_0^1 \{\phi_0(a+0,y)\phi_{1x}(a+0,y) - \phi_0(a-0,y)\,\phi_{1x}(a-0,y)\}dy$$
$$+ \int_0^1 \{\phi_0(-a+0,y)\phi_{1x}(-a+0,y) - \phi_0(-a-0,y)\,\phi_{1x}(-a-0,y)\}dy. \tag{2.35}$$

Using the condition (iii) of $\mathcal{P}_2$ for $\mathcal{L} = \mathcal{L}_2 = (0,1)$ in the relation (10), integrating by parts and using $c(0) = c(1) = 0$, we find that

$$iR_1 = \int_0^1 c(y)[\{\phi_{0y}(a-0,y) + \phi_{0y}(-a+0,y)\}\{\phi_{0y}(a-0,y) - \phi_{0y}(-a+0,y)\}$$
$$- \{\phi_{0y}(a+0,y) + \phi_{0y}(-a-0,y)\}\{\phi_{0y}(a+0,y) - \phi_{0y}(-a-0,y)\}]dy. \tag{2.36}$$

Following Levine and Rodemich (1958), $\phi_{0y}(x, y)$ is given by

$$\phi_{0y}(x,y) = e^{-Ky}\int_0^y e^{Kv}\,\mathrm{Re}f(x+iv)dv - Ke^{-Ky}\,\lambda_{\pm}(x) \text{ for } x > a \text{ and } x < -a,$$
$$\phi_{0y}(x,y) = e^{-Ky}\int_0^y e^{Kv}\,\mathrm{Re}f(x+iv)dv - Ke^{-Ky}\,\mu(x) \text{ for } -a < x < a. \tag{2.37}$$

where the functions $f(z)$, $\lambda_{\pm}(x)$ and $\mu(x)$ are given below.

$$f(z) = (C-D)f_1(z) + (C+D)f_2(z)$$

with

$$f_1(z) = \frac{3\beta\zeta^2 + \beta^2}{(\zeta^2-\beta^2)^3} + \frac{2p\beta^2}{(\zeta^2-\beta^2)^2},$$

$$f_2(z) = \frac{\zeta^3 + 3\beta^2\zeta}{(\zeta^2-\beta^2)^3} + \frac{2p\beta\zeta}{(\zeta^2-\beta^2)^2} \tag{2.38}$$

where C, D, p are constants cf. Levine and Rodemic (1958), z and ζ are related by

$$z = \frac{\alpha\gamma}{\beta^2}\int_0^{\zeta} \frac{\beta^2 - t^2}{(\alpha^2 - t^2)^{\frac{1}{2}}(\gamma^2 - t^2)^{\frac{1}{2}}}\, dt, \tag{2.39}$$

where α, β, γ are determined by the following relations

$$\frac{\beta^2}{\gamma^2} = \frac{E(k)}{K(k)},$$

$$\alpha[K(k)E(k,\theta) - E(k)F(k,\theta)] = E(k),$$

$$\frac{\pi}{2}\alpha = aE(k)$$

with

$$k = \left(1 - \frac{\alpha^2}{\gamma^2}\right)^{\frac{1}{2}} \text{ and } \theta = \sin^{-1}\frac{1}{k}\left(1 - \frac{\beta^2}{\gamma^2}\right)^{\frac{1}{2}}, \tag{2.40}$$

$F(k, \theta)$, $E(k, \theta)$, $K(k)$ and $E(k)$ being elliptic integrals. Also $\lambda_{\pm}(x)$ and $\mu(x)$ in (2.37) are given by

$$\lambda_{\pm}(x) = -\frac{1}{K}\int_{\pm\infty}^{x} \sin K(x-u) f(u)du + \begin{cases} T_0 e^{iKx}, & \text{for } x > a, \\ e^{iKx} + R_0 e^{-iKx}, & \text{for } x < -a, \end{cases} \tag{2.41}$$

$$\mu(x) = -\frac{1}{K}\int_0^x \sin K(x-u) f(u)du - \frac{\cos K(a+x)}{K \sin 2Ka}\int_0^a \cos K(a-u) f(u)du$$

$$- \frac{\cos K(a-x)}{K \sin 2Ka}\int_{-a}^{0} \cos K(a-u) f(u)du, \text{ for } -a < x < a, \tag{2.42}$$

where

$$T_0 = \frac{1}{iK} e^{-iKa}[(C-D)\{I_1(1) - \sin Ka I_2(1)\} + (C+D)\{I_1(2) - i\cos Ka I_2(2)\}], \tag{2.43}$$

$$R_0 = e^{-2iKa} + \frac{1}{iK} e^{-iKa}[(C-D)\{I_1(1) - \sin Ka I_2(1)\} - (C+D)\{I_1(2) - i\cos Ka I_2(2)\}], \tag{2.44}$$

with

$$I_1(j) = \int_0^a \cos K(a-x) f_j(x)dx, \quad I_2(j) = \int_0^{\infty} e^{-Ky} f_j(iy)dy$$

$$I_3(j) = \int_{C_1} [\cos K(a-z) - \cosh K] f_j(z)dz \ (j = 1,2) \tag{2.45}$$

where C_1 is a curve from $a + 0$ to $a - 0$ in the cut half plane occupied by the fluid.

Using (2.37) and (2.36), we find

$$iR_1 = \int_0^1 c(y)\left[e^{-Ky}\int_0^y e^{Kv}\{\mathrm{Re}f(a-0+iv)+\mathrm{Re}f(-a+0+iv)\}dv - Ke^{-Ky}\{\mu(a)+\mu(-a)\}\right]$$

$$\times\left[e^{-Ky}\int_0^y e^{Kv}\{\mathrm{Re}f(a-0+iv)-\mathrm{Re}f(-a+0+iv)\}dv - Ke^{-Ky}\{\mu(a)+\mu(-a)\}\right]dy$$

$$-\int_0^1 c(y)\left[e^{-Ky}\int_0^y e^{Kv}\{\mathrm{Re}f(a+0+iv)+\mathrm{Re}f(-a-0+iv)\}dv - Ke^{-Ky}\{\lambda_+(a)+\lambda_-(-a)\}\right]$$

$$\times\left[e^{-Ky}\int_0^y e^{Kv}\{\mathrm{Re}f(a+0+iv)-\mathrm{Re}f(-a-0+iv)\}dv - Ke^{-Ky}\{\lambda_+(a)-\lambda_-(-a)\}\right]dy. \tag{2.46}$$

The integrals appearing in (2.46) can be evaluated numerically, once the form of $c(y)$ is known.

Again, to find T_1, we use Green's integral theorem to the functions $\psi_0(x, y) = \phi_0(-x, y)$ and $\phi_1(x, y)$ in the region mentioned above to obtain finally,

$$\begin{aligned} iT_1 &= \int_0^1 \{\psi_0(a+0,y)\phi_{1x}(a+0,y) - \psi_0(a-0,y)\phi_{1x}(a-0,y)\}dy \\ &+ \int_0^1 \{\psi_0(-a+0,y)\phi_{1x}(-a+0,y) - \psi_0(-a-0,y)\phi_{1x}(-a-0,y)\}dy \\ &= \int_0^1 \{\phi_0(-a-0,y)\phi_{1x}(a+0,y) - \phi_0(-a+0,y)\phi_{1x}(a-0,y)\}dy \\ &+ \int_0^1 \{\phi_0(a-0,y)\phi_{1x}(-a+0,y) - \phi_0(a+0,y)\phi_{1x}(-a-0,y)\}dy. \end{aligned}$$

Using the condition (iii) of $\mathcal{P}_2$, then integrating by parts and using $c(0) = c(1) = 0$, we find that the above integral vanishes identically so that

$$T_1 \equiv 0. \tag{2.47}$$

Thus the first order correction to the transmission coefficient vanishes identically for two nearly vertical barriers also, as was the case for a single nearly vertical barrier partially immersed in deep water cf. Shaw (1985), Mandal and Chakrabarti (1989).

As $a \to 0$, the two partially immersed barriers merge to a single barrier. So if we make $a \to 0$ in (2.46), R_1 for a single barrier should be recovered. To show this, approximations of various quantities as $a \to 0$ appearing in the expression for R_1 is now obtained.

Since the elliptic integral $E(k)$ is given by

$$E(k) = \int_0^1 \frac{(1-k^2x^2)^{\frac{1}{2}}}{(1-x^2)^{\frac{1}{2}}}dx,$$

use of the first mean value theorem of integral calculus produces

$$E(k) = \frac{\pi}{2}(1-k^2\xi^2),\ 0 < \xi < 1.$$

Thus from the relation $\frac{\pi}{2}\alpha = aE(k)$ we find

$$\alpha = a(1 - k^2\xi^2)^{\frac{1}{2}}$$

so that

$$1 - \frac{\alpha^2}{\gamma^2} = \frac{\gamma^2 - \alpha^2}{\gamma^2 - \alpha^2\xi^2} \approx 1 \text{ as } a \to 0$$

and hence

$$k \approx 1 \text{ as } a \to 0.$$

Again since

$$K(k) \approx \ln \frac{4}{(1 - k^2)^{\frac{1}{2}}} \approx \infty \text{ as } a \to 0, \tag{2.48}$$

we find from the relation $\frac{\beta^2}{\gamma^2} = \frac{E(k)}{K(k)}$ that

$$\frac{\beta^2}{\gamma^2} \approx 0 \text{ as } a \to 0. \tag{2.49}$$

Also from the relation (2.40) we find that

$$\theta \approx \frac{\pi}{2} \text{ as } a \to 0. \tag{2.50}$$

Thus we find from the relation $\frac{\pi}{2}\alpha = aE(k)$ that

$$\alpha \approx \frac{2a}{\pi}, \beta^2 \approx 0 \text{ as } a \to 0. \tag{2.51}$$

Now using $\frac{\beta^2}{\gamma^2} = \frac{E(k)}{K(k)}$ in the relation

$$\alpha[K(k)E(k, \theta) - E(k)F(k, \theta)] = E(k)$$

we find that

$$\frac{\beta^2}{\alpha} = \frac{\gamma^2 E(k, \theta)}{1 + \alpha F(k, \theta)}. \tag{2.52}$$

Since as $a \to 0$, $k \to 1$ and $\theta \to \frac{\pi}{2}$, we find that

$$E(k, \theta) \approx 1. \tag{2.53}$$

Using (2.48) we find that as $a \to 0$

$$\alpha F(k, \theta) \approx \alpha K(k) \approx \alpha \ln\left(\frac{4\gamma}{\alpha}\right) \approx \alpha \ln\left(\frac{4\gamma_0}{\alpha}\right) \approx 0 \tag{2.54}$$

since, as $\gamma > 0$ always, we can write

$$\gamma \approx \gamma_0 \text{ as } a \to 0 \tag{2.55}$$

where $\gamma_0 > 0$. Using (2.53), (2.54) and (2.55) in (2.52), we find that

$$\frac{\beta^2}{\alpha} \approx \gamma_0^2. \tag{2.56}$$

Now from (2.39)

$$\frac{dz}{d\zeta} = \frac{\alpha\gamma}{\beta^2} \frac{\beta^2 - t^2}{(\alpha^2 - t^2)^{\frac{1}{2}}(\gamma^2 - t^2)^{\frac{1}{2}}}.$$

As $a \to 0$, using (2.51), (2.55) and (2.56), we find that

$$\frac{dz}{d\zeta} \approx \frac{1}{\gamma_0} \frac{\zeta}{(\zeta^2 - \gamma_0^2)^{\frac{1}{2}}}. \tag{2.57}$$

But we know that for a single barrier ($a = 0$), (cf. Levine and Rodemich (1958))

$$\frac{dz}{d\zeta} = \frac{\zeta}{(\zeta^2 - 1)^{\frac{1}{2}}}. \tag{2.58}$$

Comparing (2.57) and (2.58) we find

$$\gamma_0 = 1.$$

Using this and (2.51), (2.55) and (2.56), we find that as $a \to 0$

$$\alpha \approx \frac{2a}{\pi}, \beta \approx \left(\frac{2a}{\pi}\right)^{\frac{1}{2}}, \gamma \approx 1. \tag{2.59}$$

The approximations in (2.59) are now used to find the approximations for $f_1(z)$, $f_2(z), f(z)$ and the various integrals for $a \approx 0$. From (2.38) we get

$$f_1(z) \approx \beta\left(\frac{3}{\zeta^4} + \zeta\right), f_2(z) \approx \frac{2}{\zeta^3} \text{ as } a \to 0 \tag{2.60}$$

where ζ is given by the transformation (2.58). From (2.45) we have cf. Levine and Rodemich (1958)

$$I_1(1) - \sin Ka\, I_2(1) = -\frac{1}{2} KK_1\,(K, a),$$
$$I_1(2) - i \cos Ka\, I_2(2) = -\frac{1}{2} KK_2\,(K, a),$$
$$I_j(K, a) = \frac{2i}{\pi K} I_3(j), j = 1,2$$

where

$$K_j(K, a) = \frac{2}{K} \int_a^\infty \cos K(x - a) f_j(x) dx, j = 1,2.$$

As $a \to 0$, we find

$$K_1(K, a) \approx \frac{2\beta}{K} \int_0^\infty \cos Kx \left[\frac{3}{(1 + x^2)^2} + (1 + x^2)^{\frac{1}{2}}\right] dx,$$

$$I_1(K, a) \approx -\frac{2\beta}{\pi K} \int_0^1 (\cosh Ky - \cosh K) \left[\frac{3}{(1 + y^2)^2} + (1 - y^2)^{\frac{1}{2}}\right] dy,$$

$$K_2(K, a) \approx 4K_1(K), I_2\,(K, a) \approx 2I_1\,(K) \tag{2.61}$$

where $I_1(K)$ and $K_1(K)$ are modified Bessel functions. From (2.43) and (2.44) we find after using the approximations (2.61)

$$T_0 \approx \frac{iK_1(K)}{\Delta_1}, \; R_0 \approx \frac{iI_1(K)}{\Delta_1} \tag{2.62}$$

where $\Delta_1 = \pi I_1(K) + iK_1(K)$.

Again from (2.38) we find that as $a \rightarrow 0$

$$f(z) \approx +\frac{1}{\Delta_1}\frac{1}{(1+z^2)^{\frac{3}{2}}} \text{ for } a > 0,$$

$$f(z) \approx -\frac{1}{\Delta_1}\frac{1}{(1+z^2)^{\frac{3}{2}}} \text{ for } a < 0 \tag{2.63}$$

so that from (2.41) and (2.42), we obtain as $a \rightarrow 0$,

$$\lambda_+(a) + \lambda_-(a) \approx 2,$$

$$\lambda_+(a) - \lambda_-(a) \approx \frac{2}{K\Delta_1}\int_0^\infty \frac{\sinh Kv}{(1-v^2)^{\frac{3}{2}}}\,dv - \frac{2\pi I_1(K)}{\Delta_1} \tag{2.64}$$

and

$$\mu(a) + \mu(-a) \approx 0,$$

$$\mu(a) - \mu(-a) \approx -\frac{2}{\Delta_1 K^2}. \tag{2.65}$$

Using the results given in (2.63), (2.64) and (2.65) in the expression for R_1 given in (2.46), and noting that the second integral in (2.46) contributes nothing, and using the contributions for the first integral in (2.46), we obtain as $a \rightarrow 0$,

$$\frac{iR_1}{4K}(\pi I_1(K) + iK_1(K)) \approx -K\int_0^1 c(y)e^{-2Ky}\left(\int_1^y \frac{ve^{Kv}}{(v^2-1)^{1/2}}\,dv\right)dy + \int_0^1 c(y)\frac{e^{-Ky}}{(y^2-1)^{1/2}}\,dy. \tag{2.66}$$

This result coincides with the result, obtained by Shaw (1985) and Mandal and Chakrabarti (1989) for a single partially immersed nearly vertical barrier.

2.3 Scattering by two thin inclined plates

In this section the problem of water wave scattering by two symmetric thin inclined plates submerged in deep water has been investigated. The plates are symmetrically situated with respect to the vertical passing through the mid point of the line segment joining the mid points of the two plates. Exploiting the geometrical symmetry, the problem is split into two separate problems involving the symmetric and the antisymmetric potential functions describing the motion in the fluid region. Appropriate uses of Green's integral theorem followed by utilization of the boundary conditions on the plates produce two integro-differential equations, for the discontinuities for the symmetric and the antisymmetric potential functions across one of the plates. The two hypersingular integral equations are solved numerically by approximating the discontinuities of the potential functions across one of the plates in terms of two finite series involving Chebyshev polynomials of second kind followed by a collocation method. The reflection and transformation coefficients are then computed numerically using these solutions. The numerical results for the reflection coefficient, showing variations of the depth of the mid points of the plates, their angle of inclination, separation length between their mid points, are depicted graphically against the wave number in a number of figures. The main feature of the numerical results is the occurrence of zeros of the reflection coefficients as a function of the wave number for the case of inclined and horizontal submerged plates. The results for a single plate are also recovered as limiting cases.

A rectangular Cartesian co-ordinate system is chosen, in which the y-axis is taken vertically downwards into the fluid region and the plane $y = 0$ is the position of the undisturbed free surface. Let two thin inclined plates of length $2b$ be submerged in deep water and they be placed symmetrically with respect to the y-axis with their mid points situated at $(\pm a, d)$ as illustrated in the Fig. 2.15. The positions of the plates are described by $\Gamma_i (i = 1,2)$ where

$$\Gamma_1, \Gamma_2 : (x = \pm(a + bt \sin a), y = d - bt \cos a, -1 \leq t \leq 1) \tag{3.1}$$

with $a > b \sin \alpha$, $d > b \cos \alpha$, α being the angle made by the plates with the vertical. Then d is the depth of submergence of the mid points of the plates below the mean free surface and $2a$ is the separation length between the mid points. It may be noted that the plates are infinitely long in the z-direction and as their submergence is independent of z, the problem is assumed to be two-dimensional in x, y.

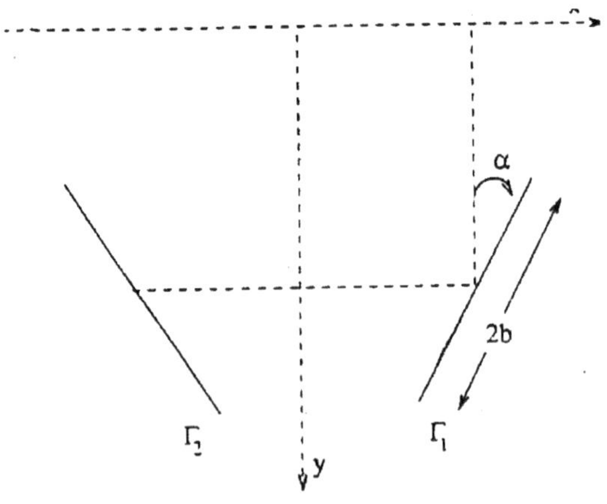

Fig. 2.15 Geometry of the plates

Let a train of surface water waves represented by the velocity potential $Re\{\phi_0(x, y)\, e^{-i\sigma t}\}$ be incident on the plates from the direction of $x = \infty$, with $\phi_0(x, y)$ given by

$$\phi_0(x, y) = 2\, e^{-Ky - iK(x-a)} \tag{3.2}$$

where $K = \sigma^2/g$, g being the acceleration due to gravity and σ being the circular frequency of the incoming wave train. Let the resulting motion in the fluid be described by the potential function $\text{Re}\{\phi(x, y)\, e^{-i\sigma t}\}$, then $\phi(x, y)$ satisfies

$$\nabla^2\phi = 0 \text{ in the fluid region,} \tag{3.3}$$

$$K\phi + \phi_y = 0 \text{ on } y = 0, \tag{3.4}$$

$$\phi_n = 0 \text{ for } (x, y) \in \Gamma_i \ (i = 1,2) \tag{3.5}$$

where ϕ_n denotes the normal derivative of ϕ at a point on Γ_i $(i = 1,2)$,

$$r^{1/2}\, \nabla\phi \text{ is bounded as } r \to 0 \tag{3.6}$$

where r denotes the distance from a submerged edge of the plates,

$$\phi, \nabla\phi \to 0 \text{ as } y \to \infty \tag{3.7}$$

and

$$\phi(x, y) \to \begin{cases} \phi_0(x, y) + R\phi_0(x, y) \text{ as } x \to \infty, \\ T\phi_0(x, y) \text{ as } x \to -\infty \end{cases} \tag{3.8}$$

where R and T denote the reflection and transmission coefficients respectively, and their determination is the principal aim here.

Because of geometrical symmetry of the positions of the two plates about the y-axis, the potential function $\phi(x, y)$ can be split into its symmetric and antisymmetric parts $\phi^s(x, y)$ and $\phi^a(x, y)$ respectively so that

$$\phi(x, y) = \phi^s(x, y) + \phi^a(x, y) \tag{3.9}$$

where

$$\phi^s(-x, y) = \phi^s(x, y),\ \phi^a(-x, y) = -\phi^a(x, y). \tag{3.10}$$

The relations (3.10) suggest that the analysis can be confined in the region $x > 0$ only. Now $\phi^{s,a}(x, y)$ satisfy (3.3) to (3.7) together with

$$\phi^s_x(0, y) = 0,\ \phi^a(0, y) = 0,\ y > 0. \tag{3.11}$$

Let the behavior of $\phi^{s,a}(x, y)$ for large x be represented by

$$\phi^{s,a}(x, y) \to e^{-Ky}\,\{e^{-iK(x-a)} + R^{s,a}\, e^{iK(x-a)}\} \tag{3.12}$$

where R^s and R^a are unknown constants. Then, by using (3.8) it is found that

$$R, T = \frac{1}{2}(R^s \pm R^a)e^{-2iKa}. \tag{3.13}$$

To obtain representations of $\phi^{s,a}(\xi, \eta)$ at a point (ξ, η) in the fluid region, the fundamental potential function $G(x, y; \xi, \eta)$ due to a line source situated at the point (ξ, η) is required. This is given by

$$G(x, y; \xi, \eta) = \ln\frac{r}{r'} - 2\int_C \frac{e^{-k(y+\eta)}\cos k(x-\xi)}{(k-K)}\, dk \tag{3.14}$$

where

$$r, r' = \{(x-\xi)^2 + (y \mp \eta)^2\}^{1/2}$$

and C denotes the path of integration along the positive real axis in the complex k-plane, indented below the pole at $k = K$, to ensure the outgoing nature of G as $|x - \xi| \to \infty$. The Green's integral theorem is applied to $\psi^{s,a}(x, y) = \phi^{s,a}(x, y) - e^{-Ky-iK(x-a)}$ and

$$\mathcal{G}^{s,a}(x, y; \xi, \eta) = G(x, y; \xi, \eta) \pm G(-x, y; \xi, \eta) \tag{3.15}$$

in the region bounded by the lines $y = 0, 0 \le x \le X; x = X, 0 \le y \le Y; y = Y, 0 \le x \le X; x = 0, 0 \le y \le Y$; a circle of small radius ϵ with the centre at (ξ, η) and a contour enclosing Γ_1. Ultimately making $X, Y \to \infty, \epsilon \to 0$ and shrinking the contour enclosing Γ_1 to the two sides of Γ_1, the following representations are obtained :

$$\phi^{s,a}(q) = 2e^{-K\eta+iKa}(\cos K\xi, -i \sin K\xi) - \frac{1}{2\pi}\int_{\Gamma_1} F^{s,a}(p)\frac{\partial \mathcal{G}^{s,a}}{\partial n_p}(p; q)ds_p \tag{3.16}$$

where $q \equiv (\xi, \eta)$ is a point in the fluid region and $p \equiv (x, y)$ is a point in Γ_1 and $F^{s,a}(p)$ denote the discontinuities of the potential functions $\phi^{s,a}(x, y)$ across Γ_1 at p and $\frac{\partial}{\partial np}$ denotes the normal derivative at the point p on Γ_1. The functions $F^{s,a}(p)$ are unknown, and they vanish at the end points of Γ_1. Use of boundary conditions $\frac{\partial \phi^{s,a}}{\partial np} = 0$ on Γ_1 where $q \equiv (\xi, \eta)$ now another point of Γ_1, produces the integro-differential equations

$$\frac{\partial}{\partial n_q}\int_{\Gamma_1} F^{s,a}(p)\frac{\partial \mathcal{G}^{s,a}}{\partial n_p}(p; q)ds_p = 2\pi\frac{\partial}{\partial n_q}[2e^{-K\eta+iKa}(\cos K\xi, -i \sin K\xi)] \tag{3.17}$$

for the unknown functions $F^{s,a}(p)$. The order of differentiation and integration in the equations (3.17) can be interchanged, provided the integrals are interpreted as Hadamard finite part integrals. This leads to the hypersingular integral equations

$$\times\!\!\!\!\!\!\int_{\Gamma_1} F^{s,a}(p)\mathcal{G}^{s,a}_{n_p n_q}(p; q)ds_p = 2\pi\frac{\partial}{\partial n_q}[2\, e^{-K\eta+iKa}(\cos K\xi, -i \sin K\xi)],\ q \in \Gamma_1 \tag{3.18}$$

where the cross on the integral sign indicates that the integral is to be interpreted as Hadamard finite part integral. The equations (3.18) are to be solved subject to the conditions that $F^{s,a}(p)$ vanish at the end points of Γ_1.

If $\boldsymbol{n}_p$ and $\boldsymbol{n}_q$ denote the unit outward at the points p and q on Γ_1 respectively, then $\boldsymbol{n}_p = \boldsymbol{n}_q = (\cos\alpha, \sin\alpha)$. Also, the points $p \equiv (x, y)$ and $q \equiv (\xi, \eta)$ on Γ_1 can be represented parametrically as

$$x = (a + bt\sin\alpha),\ y = d - bt\cos a,\ -1 \le t \le 1,$$
$$\xi = (a + b\tau\sin\alpha),\ y = d - b\tau\cos a,\ -1 \le \tau \le 1.$$

The hypersingular integral equations (3.18) then can be rewritten as

$$\times\!\!\!\!\!\!\int_{-1}^{1} f^{s,a}(t)\left[-\frac{1}{(\tau-t)^2} + \mathcal{K}^{s,a}(\tau, t)\right]dt = h^{s,a}(\tau),\ -1 \le \tau \le 1, \tag{3.19}$$

where the notations $f^{s,a}(t)$ are used for $F^{s,a}(p)$, $p \in \Gamma_1$, and

$$K^{s,a}(\tau, t) = b^2\left[-\left\{\frac{Y^2 - X^2}{(X^2 + Y^2)^2} + 2K\frac{Y}{X^2 + Y^2} + 2K^2\Phi_0(X, Y)\right\}\left\{\frac{Z^2 - X_1^2}{(X_1^2 + Z^2)^2} - 2\sin 2\alpha\left(\frac{KX_1}{X_1^2 + Y^2} + K^2\Psi_0(X_1, Y) + \frac{X_1 Y}{(X_1^2 + Y^2)^2}\right) + \cos 2\alpha\left(\frac{Y^2 - X_1^2}{(X_1^2 + Y^2)^2} + 2K\frac{Y}{X_1^2 + Y^2} + 2K^2\Phi_0(X_1, Y)\right)\right\}\right] \quad (3.20)$$

with

$$X = b(t - \tau)\sin\alpha,\ X_1 = 2a + b(\tau + t)\sin\alpha,\ Y = 2d - b(t + \tau)\cos\alpha,\ Y = -b(t - \tau)\cos\alpha, \quad (3.21)$$

$$\Phi_0(X, Y), \Psi_0(X_1, Y) = \int_0^\infty \frac{e^{-kY}}{k - K}(\cos kX, \sin kX)dk, \quad (3.22)$$

and

$$h^{s,a}(\tau) = -4\pi K e^{-K\eta + iK\alpha}(\sin(K\xi + \alpha), i\cos(K\xi + \alpha)). \quad (3.23)$$

Following Yu and Ursell (1961), the function Φ_0 and Ψ_0 in (3.22) can be expanded as

$$\Phi_0(X, Y), \Psi_0(X_1, Y) = e^{-kY}[\{\ln K(X^2 + Y^2)^{1/2} - i\pi + 0.5772\}(\cos KX, \sin KX) + (\sin KX, -\cos KX)\tan^{-1}\frac{X}{Y}]$$

$$+\sum_{m=1}^{\infty}\frac{\{K(X^2 + Y^2)^{1/2}\}^m}{m!}\left(1 + \frac{1}{2} + \cdots + \frac{1}{m}\right)(\cos, -\sin)\left(m\tan^{-1}\frac{X}{Y}\right). \quad (3.24)$$

Thus the kernels $K^{s,a}(\tau, t)$ in (3.19) can be expanded, for the purpose of numerical computations, by inserting the expansions (3.24) in (3.20).

Since $f^{s,a}(t)$ must satisfy

$$f^{s,a}(\pm 1) = 0, \quad (3.25)$$

for solving the equations (3.19), $f^{s,a}(t)$ are approximated as

$$f^{s,a}(t) = (1 - t^2)^{1/2}\sum_{n=0}^{N} a_n^{s,a} U_n(t) \quad (3.26)$$

where $U_n(t)$ is a Chebyshev polynomial of the second kind and $a_n^{s,a}$ ($n = 0,1, \ldots, N$) are unknown coefficients to be found. Substituting the expressions (3.26) in the equations (3.19), it is found that

$$\sum_{n=0}^{N} a_n^{s,a} A_n^{s,a}(\tau) = h^{s,a}(\tau),\ -1 < \tau < 1 \quad (3.27)$$

where

$$A_n^{s,a} = \pi(n + 1)U_n(\tau) + \int_{-1}^{1}(1 - t^2)^{1/2} K^{s,a}(\tau, t)U_n(t)dt,\ -1 < \tau < 1. \quad (3.28)$$

To find the unknown constants $a_n^{s,a}$ $(n = 0,1, \dots, N)$, $\tau = \tau_j$'s $(j = 0,1, \dots, N)$ are put in the relations (3.27) to obtain the following systems of linear equations:

$$\sum_{n=0}^{N} a_n^{s,a} A_n^{s,a}(\tau_j) = h^{s,a}(\tau_j), j = 0,1, \dots, N. \tag{3.29}$$

The collocation points τ_j $(j = 0,1, \dots, N)$ are chosen as the zeros of $T_{N+1}(\tau)$, the Chebyshev polynomial of the first kind, and are given by

$$\tau_j = \cos\left(\frac{2j+1}{2N+2}\pi\right), j = 0,1, \dots, N. \tag{3.30}$$

The two quantities $R^{s,a}$ are now obtained by making $\xi \to \infty$ in the representations (3.16) and comparing with the infinity conditions (3.12), (with (x, y) replaced by (ξ, η)). For this the following results are required:

$$\mathcal{G}^{s,a}(x, y; \xi, \eta) \to -4\pi e^{-K(y+\eta)+iK\xi}(i \cos Kx, \sin Kx) \text{ as } \xi \to \infty. \tag{3.31}$$

Thus $R^{s,a}$ are obtained as

$$R^{s,a} = \pm e^{2iKa} + 2e^{iKa}\int_{\Gamma_1} F^{s,a}(p)\frac{\partial}{\partial n_q}[e^{-Ky}(i \cos Kx, \sin Kx)]ds_p$$

$$= \pm e^{2iKa} + 2Kbe^{iKa}$$

$$\sum_{n=0}^{N} a_n^{s,a}\int_{-1}^{1}(1-t^2)^{1/2}U_n(t)\left[e^{-Kb\left(\frac{d}{b}-t\cos\alpha\right)}\begin{pmatrix}-i\sin\\ \cos\end{pmatrix}\left\{Kb\left(\frac{a}{b}+t\sin\alpha+\frac{\alpha}{Kb}\right)\right\}\right]dt. \tag{3.32}$$

Once the constant $a_n^{s,a}$ $(n = 0,1, \dots, N)$ are obtained by solving the linear systems, $R^{s,a}$ can be evaluated numerically using (3.32) for different values of the angle of inclination α, the parameters d/b, a/b and the wave number Kb. Having found $R^{s,a}$, R and T can be obtained numerically by using the relations (3.13). Since R and T must satisfy the energy identity

$$|R|^2 + |T|^2 = 1, \tag{3.33}$$

the relations (3.33) can be utilized to check partially the correctness of the numerical results obtained for $|R|$ and $|T|$.

The reflection and the transmission coefficients $|R|$ and $|T|$ are computed numerically for various values of the angle of inclination α, the wave number Kb and the depth and separation parameters d/b and a/b. In the Table 2.1 a representative set of values of $|R|$ for $Kb = 1$, $d/b = 1$, $a/b = 1$, $\alpha = \pi/3$, $\pi/6$ are presented to show its convergence with respect to N, the truncation size. It is observed from this table that the truncation size N somewhat depends on the inclination of the plates in this case. This is somewhat expected since, by looking at the hypersingular integral equations (3.19), it is obvious that the convergence of the series (3.26) depends on the kernel $K(\tau, t)$, and when everything is kept fixed except the angle of inclination, the convergence will obviously depend on this parameter.

Table 2.1 Convergence of $|R|$ on truncation size

$(d/b = 1, a/b = 1, Kb = 1.5)$

$|R|$

N	$\alpha = \frac{\pi}{3}$	$\alpha = \frac{\pi}{6}$
0	0.35827	0.73623
1	0.30340	0.60316
2	0.32765	0.55945
3	0.32659	0.55640
4	0.32658	0.55637
5	0.32658	0.55652
6		0.55658
7		0.55660
8		0.55660

In the Table 2.2, a representative set of values of $|R|$, $|T|$, and $|R|^2 + |T|^2$ is given. It is observed from this table that $|R|^2 + |T|^2$ coincides with unity, as it should be, for different values of the wave number Kb and a fixed angle of inclination $\alpha = \frac{\pi}{3}$, $d/b = .51$, $a/b = 1.5$. This gives a partial check on the correctness of the numerical results obtained here. In fact, in all the numerical results for $|R|$ and $|T|$, this check is always made. Because of the energy identity (3.3) henceforth the discussion will be confined to numerical results for $|R|$ only.

Table 2.2 Reflection and transmission coefficients

$(d/b = .51, a/b = 1.5, \alpha = \pi/3)$

Kb	$\|R\|$	$\|T\|$	$\|R\|^2 + \|T\|^2$
0.391	0.943810	0.270759	1.00000
1.651	0.723799	0.4691496	1.00000
1.741	0.6958449	0.9676188	1.00000

The Fig. 2.16 depicts $|R|$ against the wave number Kb for different separation length parameter a/b but fixed depth parameter d/b and the angle of inclination ($d/b = 1$, $\alpha = \pi/6$). It is observed that the main feature of the curves $|R|$ for is the occurrence of zeros of $|R|$ as a function of wave number. As a/b increases, the zeros are shifted toward the left. This is due to the fact that as a/b increases, multiple reflection of waves by the two plates increases and this results in quicker occurrence of zeros of $|R|$ as a function of the wave number.

In the Fig. 2.17, $|R|$ is depicted against Kb for the fixed inclination $\alpha = \pi/6$ and separation parameter $a/b = 1.5$ but different depth parameters $d/b = 1, 1.2, 1.5$. As the depth increase, the overall reflection coefficients decrease and the zeros of $|R|$ are slightly shifted towards the right. The decrease of overall values of $|R|$ for the increase

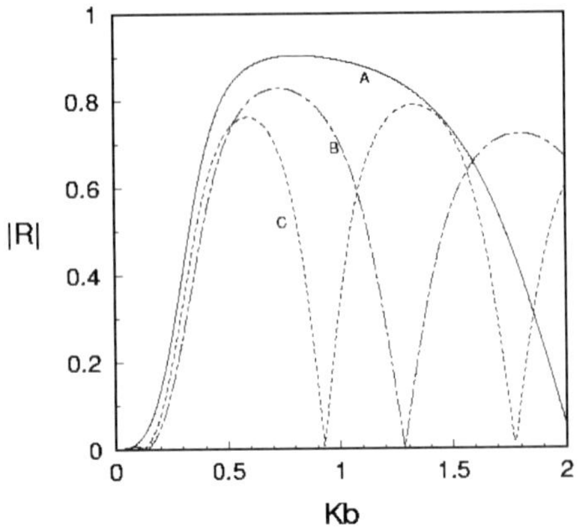

Fig. 2.16 Reflection coefficient for different separation lengths $\alpha = \pi/6$, $d/b = 1$, $a/b = .5(A)$, $.1(B)$, $1.5(C)$

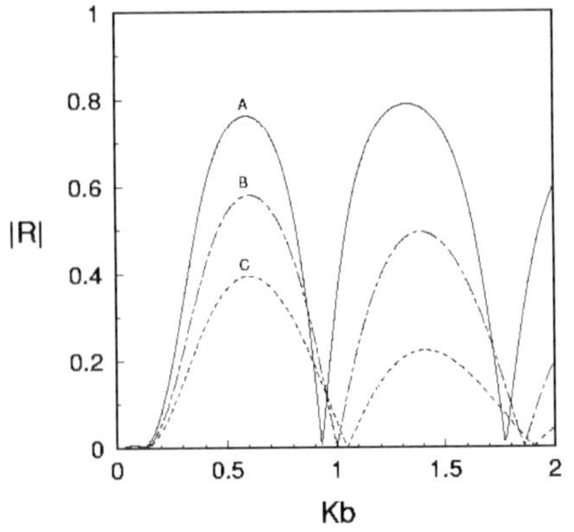

Fig. 2.17 Reflection coefficient for different depths $\alpha = \pi/6$, $a/b = 1.5$, $d/b = 1(A)$, $1.2(B)$, $1.5(C)$

of depth is plausible since the plates encounter less amount of the incident wave energy as their depth below the free surface increases, resulting in less reflection.

In the Fig. 2.18, the effect of the angle of inclination α on $|R|$ is illustrated. In this figure $|R|$ is drawn against Kb keeping d/b and a/b as fixed ($a/b = 1$, $d/b = 1$) and taking $= \pi/6, \pi/4, \pi/3$. It is observed from the figure that overall $|R|$ decreases with the increases of α and also the zeros of $|R|$ are shifted towards the right. The shifting is more as the wave number increases. As α increases, the upper ends of the plates move further down from the surface resulting in less encounter of the plates with incident wave field which in turn produces less reflection.

The Fig. 2.19 depicts $|R|$ against a new wave number $Kl = K(d + b)$ for the vertical plates kept very close to each other ($a/b = 0.1$) for the different values of the parameter $\mu = \frac{d-b}{d+b} < \mu = .01, .05, .25$) for the purpose of comparing with known results for a single vertical plate submerged in deep water available in the literature. It is observed

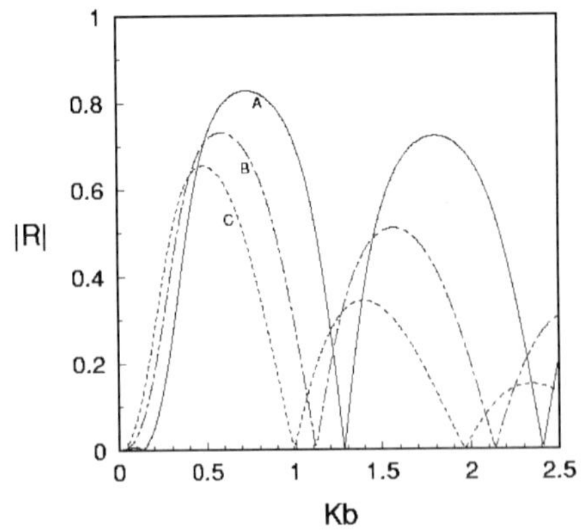

Fig. 2.18 Reflection coefficient for different inclinations $a/b = 1.5$, $b/b = 1$, $\alpha = \pi/6(A)$, $\pi/4(B)$, $\pi/3(C)$

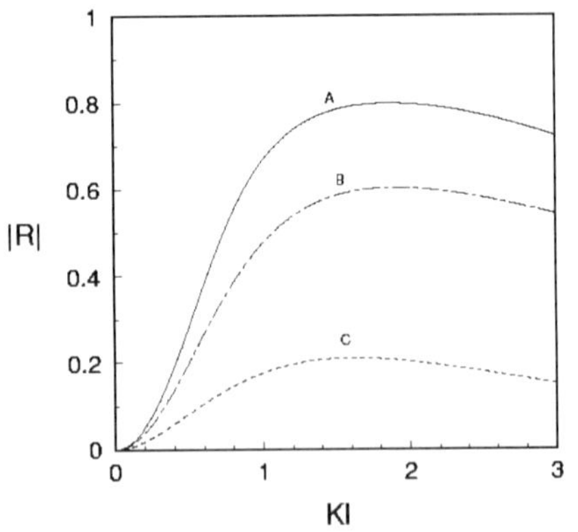

Fig. 2.19 Reflection coefficient for closed vertical plates $\alpha = 0$, $a/b = .1$, $l = d + b$, $\mu = \frac{d-b}{d+b} = .01(A), .05(B), .25(C)$

that the curves of $|R|$ in this figure have almost similar features as those for a single vertical plate obtained earlier in Evans (1970) and Parsons and Martin (1992).

The choices $\alpha = 90°$, $a/b = 1$ and $\alpha = 89.1°$, $a/b = \sin 89.1°$ reduce the two-plate configuration to a single horizontal plate and an almost horizontal plate respectively submerged in deep water. The Fig. 2.20 depicts $|R|$ for the horizontal plate and has almost same features as the curve for $|R|$ given in Parsons and Martin (1992). The curve for an almost horizontal plate depicted also in the Fig. 2.20 possesses similar features as given in Parsons and Martin (1992) obtained there, however for $a/b = .2(1 + \cos 89.1°)$.

Again, the choice $a/b = \cos \alpha$ produces an open wedge shaped obstacle. If α is positive, then the vertex of the wedge is downwards, and if α is negative, then the vertex of the wedge is upwards. The Fig. 2.21 shows $|R|$ for such an open wedge when $\alpha = \pi/3, -\pi/3$ but fixed depth parameter d/b. This figure shows that the overall reflection coefficient for a wedge with downward vertex is much higher than that for the corresponding wedge with upward vertex. Almost similar results are obtained for wedges of other angles, although these are not depicted here. This is quite expected since more waves are reflected by the two sides of an open wedge with downward vertex than the case when the wedge has upward vertex provided d/b for both configurations remains fixed. Wave scattering by such wedge-like structure appears to have not been studied in the literature.

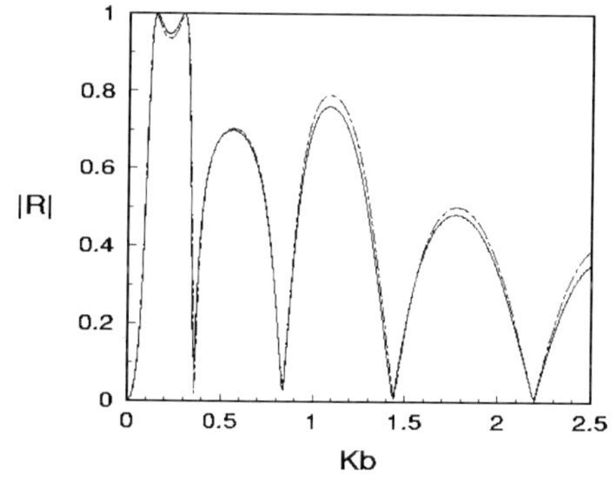

Fig. 2.20 Reflection coefficient for horizontal and nearly horizontal plates (– – –): $\alpha = \pi/2$, $a/b = 1$, $d/b = .2$ (– – –): $\alpha = 89.1°$, $a/b = \sin 89.1°$, $d = .2$

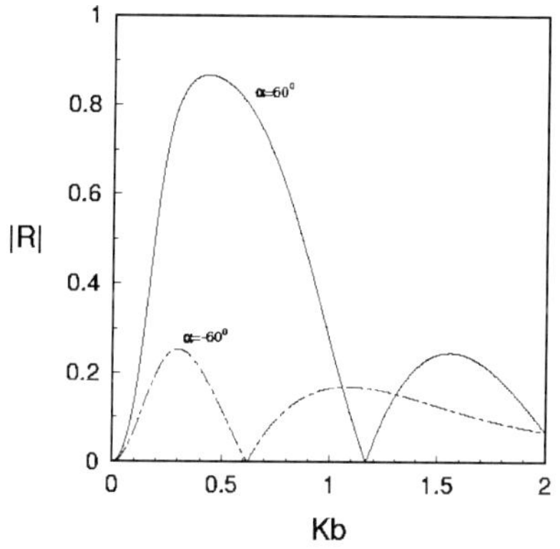

Fig. 2.21 Reflection coefficient for wedge shaped obstacle, $a/b = \sin 60°\ 1$, $d/b = 1$

The choice $\alpha = \pi/2$, $a/b = 1$ reduces the two-plate configuration to a submerged single plate and the Fig. 2.22 displays $|R|$ against Kb for such a configuration with different depth parameters $d/b = .5, .3, .1$. Wave scattering by a submerged horizontal plate has been investigated by Burke (1964). However, no numerical results were given. The occurrence of zeros of $|R|$ as a function of the wave number is the main feature in the Fig. 4.8 as is the case with inclined plates. As the depth parameter decreases the number of zeros increases. This is due to more multiple reflections of the incident wave field between the free surface and the submerged horizontal plate as its depth decreases. The most notable feature is perhaps the occurrence of total reflection ($|R|$

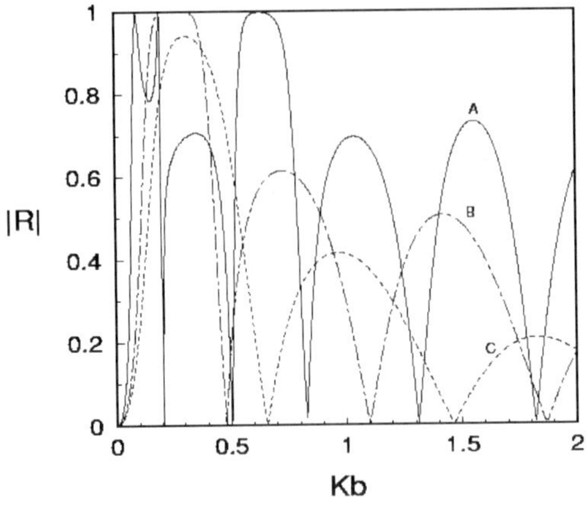

Fig. 2.22 Reflection coefficient for horizontal plate at different depths, $\alpha = \pi/b$, $a/b = 1$, $d/b = .1(A)$, $.3(B)$, $.5(C)$

= 1) when the plate is submerged somewhat near the free surface. Another feature is that the number of crests and troughs in the reflection curve increases with the decrease of depth.

Finally, the Fig. 3.9 depicts $|R|$ for a horizontal plate with a slit obtained here by choosing $\alpha = \pi/2$. Here also the occurrence of zeros of $|R|$ as a function of the wave number is the main feature. The zeros are shifted towards the right as the slit length parameter $\frac{a}{b} - 1$ decreases.

2.4 Scattering by a submerged circular arc-shaped plate

A water wave lens can be constructed from a system of submerged bodies, called elements, to focus waves for the purpose of extracting wave energy from them (Mclver and Urka, 1995). The wave focussing mechanism is similar to what governs the focussing of light waves. The system is designed in such a way that incident wave trains experience phase shifts without much reflection as they pass over it and are transformed by the lens into converging wave trains. Thus, each element of the wave lens should be such that an incoming surface wave train experiences little reflection by it. It is well known that a circular cylinder with horizontal axis is transparent to normally incident waves of all frequencies when submerged in water of infinite depth (Dean, 1948; Ursell, 1950) and it can be used as an ideal element of a wave lens. However, there are some practical difficulties in using it as a lens element and other bodies such as thin horizontal plates, chevron shaped plates, etc. have been employed (Mclver and Urka, 1995). Mclver (1985) considered a submerged horizontal flat plate moored to the seabed to study the feasibility of its use as a lens element. There are instances of total transmission of normally incident waves past submerged obstacles at isolated frequencies. Examples of such obstacles include submerged long two-dimensional bodies, bottom mounted submerged rectangular barriers, submerged rectangular blocks Kanoria et al. (1999), and submerged thin inclined plates Parsons and Martin (1992). Thus, study of water wave scattering problems involving submerged obstacles of different geometrical shapes has significant relevance in the construction of wave lens to use them as its elements.

The case of a surface piercing thin plate inclined at an angle $\frac{(n-1)}{2n}\pi$ (n being a nonnegative integer) with the vertical was considered by John (1948) by using complex variable theory for explicit solutions. However, the method is unwieldy for $n \geq 2$. For arbitrary inclination of the thin plate Parsons and Martin (1992) devised a hypersingular integral equation formulation of the problem for the case when the plate is submerged in infinitely deep water to obtain very accurate numerical estimates for the reflection and transmission coefficients. The surface piercing inclined plate problem was also investigated by Parsons and Martin (1994) by the same technique with some appropriate modifications. This technique of hypersingular integral equation can also be applied for a thin curved plate and Parsons and Martin (1994) used this technique to study the case of a submerged circular-arc-shaped thin plate submerged in deep water, the plate being convex upwards and symmetric about the vertical passing through the centre of the arc. For studying the feasibility of using a circular-arc-shaped plate as a wave lens element Mclver and Urka (1995) investigated the symmetric circular-arc-shaped plate problem of Parsons and Martin (1994) by two methods, one based on the method of matched series expansions and the other based on Schwinger variational approximation. They found that there is very little reflection for plates which occupy half a circle or more, and such plates are good candidates for use as lens elements.

In this section, we consider a submerged circular-arc-shaped plate placed arbitrarily, i.e., the plate is, in general, not symmetric with respect to the vertical through its centre and not necessarily convex upward, and study its reflective properties fro the purpose of its feasibility to use as a lens elements as given in Kanoria and Mandal (2002). As in Parsons and Martin (1994), a hypersingular integral equation formulation of the problem is derived. The integral equation contains a discontinuity of the velocity potential across the plate. It is solved numerically by approximating the discontinuity in terms of a finite series involving Chebyshev polynomials of the second kind. The unknown constants appearing as coefficients in the finite series are determined numerically by using two methods. The first one is based on collocation as has been used by Parsons and Martin (1994). The second is a Galerkin method based on utilization of the orthogonal property of the Chebyshev polynomials. Both the methods produce very accurate numerical estimates for the reflection coefficient, which are depicted graphically against the wave number for various configurations of the circular-arc plate. Some results are compared with those obtained by Mclver and Urka (1995). A good agreement is seen to be achieved. When the circular-arc approaches a full circle, the reflection coefficient is almost zero, which agrees with the classical result concerning a circular cylinder mentioned above. Again, when the position of the circular-arc plate is reversed with respect to the vertical through the centre of the circle, the reflection coefficient remains unchanged. This is in accordance with the principle of complementarity theorem.

Numerical results for the reflection coefficient show that when the arc length of the plate is more than half (or less than quarter) of a circle, then its upward (or downward) convex configuration is a good candidate for use as a lens element. A semi-circular-arc-shaped plate with horizontal diameter can also be used as lens element for the low-(or high) frequency range when it is convex downward (or upward).

Let a circular-arc-shaped thin plate Γ be submerged in infinitely deep water and its configuration be described by using Cartesian coordinates with x- and z-axis lying on the mean free surface and y-axis directed vertically downwards and passing through the centre of the circular-arc.

The vertical section of the plate is in the form of an arc of a circle of radius b with centre at depth $d + b$ below the mean free surface and let α and β (> α) be the angles made with the upward vertical by the radii at two end points of the arc (see Fig. 2.23).

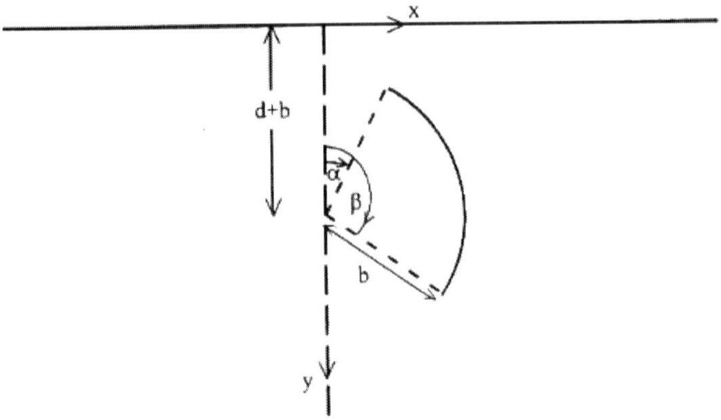

Fig. 2.23 Configuration of a circular-arc-shaped plate

Thus, a point (x, y) on the plate Γ has the parametric representation

$$x = b \sin \Theta t, \; y = d + b - b \cos \Theta t \left(\frac{\alpha}{\Theta} \le t \le \frac{\beta}{\Theta}\right) \tag{4.1}$$

where $\Theta = \beta - \alpha$. With the usual assumptions that water waves are linear phenomena, the fluid is incompressible and inviscid, and its motion is irrotational, there exists a velocity potential which can be written as $\mathrm{Re}\{\frac{-igb_0}{\sigma}\phi(x, y)e^{-i\sigma\tau}\}$, where b_0 is the amplitude of the incoming surface wave train of angular frequency σ and described by $\mathrm{Re}\{\frac{-igb_0}{\sigma}$ $\phi_0(x, y)e^{-i\sigma\tau}\}$ with

$$\phi_0(x, y) = e^{-Ky+ikx},$$

g being the acceleration due to gravity, τ the time and $K = \sigma^2/g$. It is assumed that the incoming wave train propagates along a direction normal to the axis of the cylinder whose cross section is the circle, a part of which is Γ, so that the problem is two-dimensional and independent of z. Then, the function $\phi(x, y)$ satisfies, in the fluid region, that

$$\nabla^2\phi = 0 \tag{4.2}$$

where ∇^2 is the two-dimensional Laplace operator with the free surface condition

$$K\phi + \frac{\partial \phi}{\partial y} = 0 \text{ on } y = 0 \tag{4.3}$$

and the condition on the plate

$$\frac{\partial \phi}{\partial y} = 0 \text{ on } \Gamma \tag{4.4}$$

where $\partial/\partial n$ denotes the normal derivative at a point on Γ. The edge conditions are given by

$$r^{1/2}\nabla\phi \text{ is bounded as } r \to 0, \tag{4.5}$$

where r is the distance from the submerged edges of Γ, the infinite-depth condition by

$$\nabla\phi \to 0 \text{ as } y \to \infty, \tag{4.6}$$

and the condition at infinity by

$$\phi(x, y) \to \begin{cases} T\phi_0(x, y) \text{ as } x \to \infty, \\ \phi_0(x, y) + R\phi_0(-x, y) \text{ as } x \to \infty \end{cases} \tag{4.7}$$

where R and T denote, respectively, the reflection and transmission coefficients, and are to be determined.

In order to obtain a representation of the function $\phi(x, y)$ we need the source potential $G(x, y; \xi, \eta)$, $(\eta > 0)$ which satisfies.

$$\nabla^2 G = 0 \text{ in the fluid region except at. } (\xi, \eta),$$

$$G \to \ln r \text{ as } r = \{(x-\xi)^2 + (y-\eta)^2\}^{1/2} \to 0$$

$$KG + \frac{\partial G}{\partial y} = 0 \text{ on } y = 0$$

$$\nabla G \to 0 \text{ as } y \to \infty,$$

$$G(x, y; \xi, \eta) \to \text{const } e^{-K(y+\eta)+iK|x-\xi|} \text{ as } |x-\xi| \to \infty.$$

Then G is given by cf. Mandal and Chakrabarti (2000), p. 28

$$G(x, y; \xi, \eta) = \ln\frac{r}{r'} - 2\int_C \frac{e^{-K(y+\eta)}}{k-K}\cos(x-\zeta)dk, \tag{4.8}$$

where $r, r' = \{(x-\xi)^2 + (y \mp \eta)^2\}^{1/2}$ and C denotes a path along the positive real axis in the complex k-plane indented below the pole at $k = K$.

We now apply the Green's integral theorem to the functions $\phi(x, y) - \phi_0(x, y)$ and $G(x, y; \xi, \eta)$ in the region bounded by the lines $y = 0$, $-X \le x \le X$; $x = \pm X$, $0 \le y \le Y$; $y = Y$, $-X \le x \le X$, a closed curve enclosing Γ and a circle of small radius ε with centre at (ξ, η), and ultimately make $X \to \infty$, $Y \to \infty$, and the two sides of the closed curve enclosing Γ shrinking almost to Γ and $\varepsilon \to 0$. Then, we obtain

$$\phi(\xi, \eta) = \phi_0(\xi, \eta) - \frac{1}{2\pi}\int_\Gamma F(q)\frac{\partial G}{\partial n_q}(x, y; \xi, \eta)ds_q, \tag{4.9}$$

where $q \equiv (x, y)$ is a point on Γ, $F(q)$ denotes the discontinuity of $\phi(x, y)$ across Γ and $\partial/\partial n_q$ the normal derivative at the point q. It should be noted that the unknown function $F(q)$ vanishes at the end points of Γ while its derivative has square root singularity there.

Use of boundary conditions (4.4) rewritten as

$$\frac{\partial \phi}{\partial n_q} = 0 \text{ on } \Gamma$$

where $p \equiv (\xi, \eta)$ is another point of Γ leads to the integro-differential equation

$$\frac{\partial}{\partial n_q} \int_\Gamma F(q) \frac{\partial G}{\partial n_q}(q; p) ds_q = 2\pi \frac{\partial \phi_0}{\partial n_p}(\xi, \eta)\ (p \in \Gamma), \tag{4.10}$$

where $F(q)$ vanishes at the end points of Γ. The order of differentiation and integration in (4.10) can be interchanged provided the integral is interpreted as a Hadamard finite part integral, and this leads to the hypersingular integral equation.

$$\int_\Gamma F(q) \frac{\partial^2 G}{\partial n_p \partial n_q}(q; p) ds_q = 2\pi \frac{\partial \phi_0}{\partial n_p}\ (p \in \Gamma) \tag{4.11}$$

where the integral is in the sense of Hadamard finite part integral of order two.

To obtain the actual expression of the kernel in (4.11), we note that the unit normals n_p and n_q at the points p and q on Γ are given by

$$n_p = (\sin \Theta u, -\cos \Theta u),\ n_q = (\sin \Theta t, -\cos \Theta t), \tag{4.12}$$

where u and t denote the parametric co-ordinates of p and q, respectively. Using (4.12), we find that

$$\frac{\partial^2 G}{\partial n_p \partial n_q}(q; p) = -\frac{1}{b^2\Theta^2(u-t)^2} - K(u, t) \left(\frac{\alpha}{\Theta} < u,t < \frac{\beta}{\Theta}\right) \tag{4.13}$$

where

$$K(u, t) = \frac{1}{4b^2} \frac{1}{\sin^2\frac{\Theta}{2}(u-t)^2} - \frac{1}{b^2\Theta^2(u-t)^2}$$

$$+ \cos \Theta(u-t) \left[\frac{Y^2 - X^2}{(X_1^2 + Y^2)^2} + \frac{2KY}{(X^2+Y^2)} + 2K^2 \int_C \frac{e^{-kY}}{k-K} \cos kX\, dk\right]$$

$$+ \sin \Theta(u-t) \left[\frac{2XY}{(X^2+Y^2)^2} + \frac{2KX}{(X^2+Y^2)} + 2K^2 \int_C \frac{e^{-kY}}{k-K} \sin kX\, dk\right] \tag{4.14}$$

with

$$X \equiv X(u, t) = b(\sin \Theta t - \sin \Theta u),\ Y \equiv Y(u, t) = 2(d+b) - b(\cos b\Theta t + \cos\Theta u). \tag{4.15}$$

Following Yu and Ursell (1961), the integrals in (4.12) can be expanded as

$$\int_c \frac{e^{-kY}}{k-K} \begin{pmatrix}\cos \\ \sin\end{pmatrix} kX\, dk = -e^{KY} \left\{(\ln Kr_1 - i\pi + \gamma \begin{pmatrix}\cos \\ \sin\end{pmatrix} KX) \pm \theta_1 \begin{pmatrix}\sin \\ \cos\end{pmatrix} KX\right\} \tag{4.16}$$

$$+ \sum_{m=1}^{\infty} \frac{(-Kr_1)^m}{m!} \left(1 + \frac{1}{2} + \frac{1}{3} + \cdots + \frac{1}{m}\right) \begin{pmatrix}\cos \\ -\sin\end{pmatrix} m\theta_1,$$

where $\gamma = 0.5772...$ is Euler's constant $r_1 = (X^2 + Y^2)^{1/2}$ and $\theta_1 = \tan^{-1}(X/Y)$.

Also, the right-hand side of (4.11) can be expressed as

$$2\pi \frac{\partial \phi_0}{\partial n_p}(p) = \frac{2\pi i g b_0}{\sigma} h(u)\left(\frac{\alpha}{\Theta} < u < \frac{\beta}{\Theta}\right), \tag{4.17}$$

where

$$h(u) = -Ke^{-k\eta(u)+iK(\xi(u)+\Theta u)}.$$

Thus (4.11) becomes

$$\int_{\alpha/\Theta}^{\beta/\Theta} g_0(t)\left[\frac{1}{(u-t)^2} + b^2\Theta^2 \mathrm{K}(u,t)\right]dt = -2\pi b^2\Theta^2 h(u)\left(\frac{\alpha}{\Theta} < u < \frac{\beta}{\Theta}\right) \tag{4.18}$$

where

$$g_0(t) = -\frac{\sigma}{igb_0}F(t) \tag{4.19}$$

so that $g_0(t)$ may vanish at the end points $t = \alpha/\Theta, \beta/\Theta$ of the interval Replacing t and u by

$$\frac{1}{2}\left(\frac{\alpha+\beta}{\Theta}+t\right) \text{ and } \frac{1}{2}\left(\frac{\alpha+\beta}{\Theta}+u\right)$$

respectively, we can reduce (4.18) to the hypersingular integral equation of standard form in the interval $(-1, 1)$ as

$$\int_{-1}^{1} g_1(t)\left[\frac{1}{(u-t)^2} + \mathrm{K}_1(u,t)\right]dt = h_1(u)\,(-1<u<1), \tag{4.20}$$

where

$$\mathrm{K}_1(u,t) = \frac{1}{4}b^2\Theta^2\mathrm{K}\left(\frac{\alpha+\beta}{2\Theta}+\frac{u}{2}, \frac{\alpha+\beta}{2\Theta}+\frac{t}{2}\right),$$

$$g_1(t) = g_0\left(\frac{\alpha+\beta}{2\Theta}+\frac{t}{2}\right),$$

$$h_1(u) = -\pi b^2\Theta^2 h\left(\frac{\alpha+\beta}{2\Theta}+\frac{u}{2}\right), \tag{4.21}$$

and $g_1(t)$ satisfies the end conditions

$$g_1(\pm 1) = 0. \tag{4.22}$$

To solve (4.20) we approximate $g_1(t)$ as

$$g_1(t) = (1-t^2)^{1/2}\sum_{n=0}^{N} a_n U_n(t), \tag{4.23}$$

where N is an integer, $U_n(t)$ is the nth order Chebyshev polynomial of the second kind, and $a_n (n = 0,1, ..., N)$ are unknown complex constants. The square root factor in (4.12) ensures that $g_1(t)$, or rather $F(q)$, has the correct behaviour at the ends of the plate. The unknown constants $a_n(n = 0,1, ..., N)$ will be determined by using the collocation and the Galerkin methods.

Substitution of (4.23) into (4.20) leads to

$$\sum_{n=0}^{N} a_n A_n(u) = h_1(u) \quad (-1 < u < 1), \tag{4.24}$$

where

$$A_n(u) = -\pi(n+1)U_n(u) + \int_{-1}^{1} (1-t^2)^{1/2} \mathrm{K}_1(u,t) U_n(t)\, dt. \tag{4.25}$$

The collocation points are chosen as

$$u_j = \cos\frac{j+1}{N+2}\pi \quad (j = 0, 1,.. \ , N). \tag{4.26}$$

Putting $u = u_j(j = 0,1, ..., N)$ we obtain a system of linear equations,

$$\sum_{a=0}^{N} a_n A_n(u_j) = h_1(u_j) \quad (j = 0, 1, \ldots, N) \tag{4.27}$$

for the determination of the constants $a_n(n = 1, ..., N)$. We use the Gauss Jordan method to solve (4.27) numerically.

The orthogonal properties of the Chebyshev polynomials are used in the Galerkin method. We multiply both sides of (4.24) by $(1 - u^2)^{1/2} U_m(u)$ $(m = 0,1, ..., N)$ and integrate with respect to u over -1 to 1 to obtain another system of linear equations.

$$\sum_{n=0}^{N} a_n P_{nm} = d_m \quad (m = 0, 1, \ldots, N) \tag{4.28}$$

where

$$P_{nm} = -\frac{\pi^2}{2}(n+1)\delta_{nm} + \int_{-1}^{1} (1-u^2) U_m(u) \left\{ \int_{-1}^{1} \mathrm{K}_1(u,t)(1-t^2)^{1/2} U_m(t)\, dt \right\} du$$

and

$$d_m = \int_{-1}^{1} (1-u^2)^{1/2} U_m(u) h_1(u)\, du.$$

Linear equations (4.28) are also solved by the Gauss Jordan method.

Now the reflection and transmission coefficients R and T are obtained by taking the limits $\xi \to \mp\infty$ in (4.9) for $\phi(\xi, \eta)$. For this purpose, we require the asymptotic result, Mandal and Chakrabarti (2000)

$$G(x, y; \xi, \eta) \to 2\pi i e^{-K(y+\eta) \pm iK(\xi - x)} \text{ as } \xi \to \pm\infty$$

Making $\xi \to -\infty$ in (4.9) after using (4.30) and noting (4.7) with (x, y) replaced by (ξ, η), we find that

$$R = iK \int_{r} g_0(q) e^{-Ky+iKx+i\Theta t} ds_q$$

$$= \frac{iKb\Theta}{2} \sum_{n=0}^{M} a_n \int_{-1}^{1} (1-t^2)^{1/2} U_n(t) e^{-K(b+d-b\cos\Theta t')+i(Kb\sin\Theta+\Theta t')} dt,$$

with, $t' = (\alpha + \beta)/2\Theta + t/2$. Similarly, the limit $\xi \to \infty$ leads to

$$T = 1 + \frac{iKb\Theta}{2} \sum_{n=0}^{N} a_n \int (1-t^2)^{1/2} U_n(t) e^{-K(b+d-b\cos\Theta t')-i(Kb\sin\Theta+\Theta t')} dt.$$

Thus, once $a_n (n = 0,1, \ldots, N)$ are found numerically by solving linear system (4.27) or (4.28) $|R|$ and $|T|$ can be computed from (4.31) and (4.32) respectively. Here, the integrals in (4.31) and (4.32) are evaluated numerically by Gauss quadrature. The relation

$$|R|^2 + |T|^2 = 1$$

is used as a partial check on the correctness of the numerical results.

For a particular set of values of the depth parameter compared for different values of the truncation size N between the Galerkin and collocation methods in Table 2.3. It is seen that slightly smaller N is sufficient for the Galerkin method compared to the $d/b = 0.1$, wave number $Kb = 0.2$ and angles $\alpha = \pi/4$, $\beta = 3\pi/2$, the numerical results of $|R|$ are collocation methods to obtain numerical estimates for $|R|$ correct upto 4–5 decimal places. However, more computational time is needed for the Galerkin method compared to the collocation method because we have to compute a double integral at every stage in the former while a single integral in the latter. Thus the collocation method is advantageous. The Galerkin method may be used to check the correctness of the results computed by using the collocation method. It should be mentioned here that the truncation size N which gives the same accuracy is different depending upon the arc length, depth parameter and the wave number. For all the data presented here, N is chosen in such a way that the result is correct upto 4–5 decimal places.

Table 2.3 Reflection coefficient
$|R|$ ($d/b = 0.1$, $\alpha = \frac{\pi}{4}, \beta = \frac{3\pi}{2}$; $Kb = 0.2$)

N	**Collocation method**	**Galerkin method**
0	0.24096	0.18433
1	0.13232	0.12150
2	0.12353	0.12612
3	0.12644	0.12612
4	0.12636	0.12634
5	0.12633	
6	0.12634	
7	0.12634	

Table 2.4 shows a representative set of values of $|R|$, $|T|$ and $|R|^2 + |T|^2$ for $\alpha = \pi/6$, $\beta = \pi/2$, $d/b = 0.1$ and for three different values of Kb. It is observed that relation $|R|^2 + |T|^2 =$ is satisfied up to 5 decimal places.

Table 2.4 Reflection and transmission coefficient $|R|$ and $|T|$ ($d/b = 0.1$, $\alpha = \frac{\pi}{6}$, $\beta = \pi/2$)

Kb	$\|R\|$	$\|T\|$	$\|R\|^2 + \|T\|^2$
1.0	0.25213	0.96769	1.00000
1.6	0.26491	0.96536	1.00000
2.2	0.22521	0.97431	1.00000

Since the main concern here is the reflective properties of the submerged circular-arc-shaped plate for an incoming wave train, we depict $|R|$ against the wave number for various configurations of the plate. For each data point used in the plots of various graphs for $|R|$, the transmission coefficient $|T|$ is computed so as to ensure energy equality (4.33). To visualize the dependence of the depth of submergence of the plate, the reflection coefficient $|R|$ is depicted in Fig. 2.24 against the wave number Kb for $\alpha = 45°$, $\beta = 180°$ and the depth parameter $d/b = 0.1, 0.3, 05$. It is observed that for a fixed wave number, $|R|$ decreases as d/b increases, i.e., the more is the depth of the submergence of the plate below the free surface the less is the reflection. This is plausible since less energy is reflected by the plate if it is submerged more below the free surface.

Again, to visualize the effect of the arc length of the plate, |R| is plotted in Fig. 2.25 against Kb for a number of plates whose one end is kept fixed at $\alpha = 30°$ and its centre is kept at a fixed depth ($d/b = 0.1$). The arc lengths of the plates are taken as $\beta = 90°, 180°, 270°, 330°, 345°, 360°, 375°$ and $389°$. As the arc length increases, the overall reflection coefficient initially increases (from $\beta = 90°$ to $270°$ here), takes a maximum around $\beta = 330°$ and then decreases. For the case of an almost full circle ($\alpha = 30°$, $\beta = 389°$), the reflection is seen to be quite insignificant for all wave numbers. This is in conformity with the classical result that a long horizontal circular cylinder submerged in deep water experiences no reflection when an incoming wave train is normally incident on it. This may also be regarded as another partial check on the correctness of the numerical method employed here.

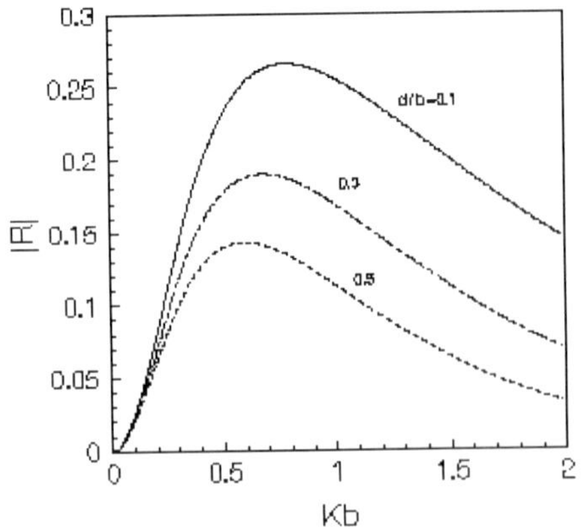

Fig. 2.24 Reflection coefficient vs wave number, $\alpha = 45°$, $\beta = 180°$

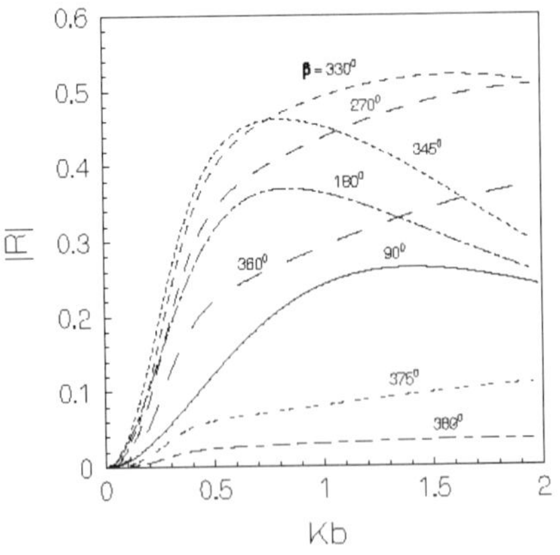

Fig. 2.25 Reflection coefficient vs. wave number, $d/b = 0.1$, $\alpha = 30°$

For the pupose of comparison with the results obtained by the present method and those obtained by McIver and Urka (1995) by employing the method of matched series solutions, $|R|$ is depicted in Fig. 2.26 for the symmetric configuration of the plate by choosing $d/b = 0.25$ and $\alpha = -54°$, $\beta = 54°$. The solid curve displays the present results and the circles the ones estimated from the Fig. 3 in McIver and Urka (1995). The agreement between the two is excellent. However, some differences are observed if our results are compared with theirs obtained by using the method of variational approximation and given also in Fig. 3 there (not shown here). This difference may be attributed to the fact that the method of variational approximation produces good results only when the arc length of the plate is shorter than the wavelength and the plate occupies a small fraction of a circle. It should also be noted that the method of matched series expansion employed by them produces accurate results but quite a large number of multipole potential functions are needed, e.g., 256 multipole potentials are required to ensure numerical accuracy to two decimal places. In contrast, the present method usually requires only 8–12 terms in the truncated series expansion (4.23) to get numerical accuracy upto 4–5 decimal places.

Figure 2.27 depicts $|R|$ against Kb for a number of configurations of the circular-arc plate. Some configurations are symmetric and some are not, but the depth parameter $d/b = 0.1$ is common to all of them. The curve denoted by I shows $|R|$ for an upward convex symmetric circular-arc in the form of a quarter of a circle ($\alpha = -45°$, $\beta = 45°$) while the curve denoted by II its reflection about the horizontal diameter of the circle ($\alpha = 135°$, $\beta = 225°$). It is observed by comparing these two curves that a symmetric upward convex circular-arc produces more reflection compared to a symmetric downward convex circular-arc of the same arc length and having the same centre. This is expected since the downward convex arc lying below the upward convex arc encounters less disturbances due to the incident surface wave train. The curve I′ and II′ represent $|R|$ for a symmetric upward convex circular-arc plate of arc length greater than the quarter circle but less than the half circle ($\alpha = -80°$, $\beta = 80°$) and its reflection about the horizontal diameter ($\alpha = 100°$, $\beta = 260°$).

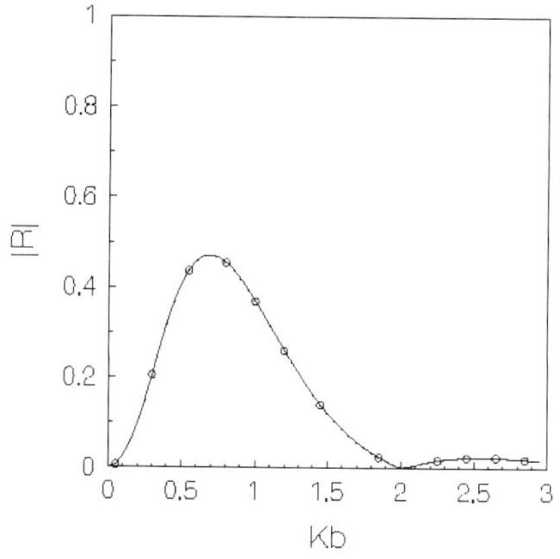

Fig. 2.26 Reflection coefficient vs wave number, $d/b = 0.25$, $\alpha = -54°$, $\beta = 54°$ – – –, present: o, McIver and Urka (1995)

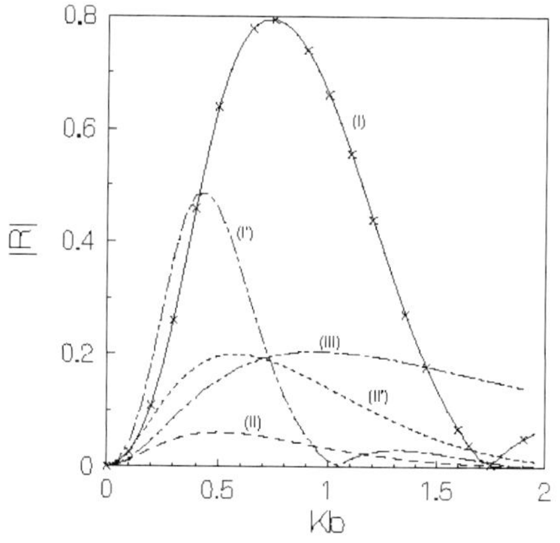

Fig. 2.27 Reflection coefficient vs wave number, $d/b = 0.01(I)\alpha = -45°$, $\beta = 45°(I')\alpha = -80°$, $\beta = 80°$,$(II)\alpha = 135°$, $\beta = 260°$, $(III)\alpha = 45°$, $\beta = 135°$; $\alpha = 225°$, $\beta = 315°$, × estimated Parson and Martin for I

For these plates also, the upward convex arc produces more reflection compared to the downward convex arc. Curve III depicts $|R|$ for a circular-arc which is convex towards positive x direction as well as symmetric about a horizontal line passing through its centre ($\alpha = 45^\circ$, $\beta = 135^\circ$). The reversal of the position of this arc with respect to the vertical through its centre ($\alpha = 225^\circ, \beta = 315^\circ$) produces the same curve III. The same is true for an arc which is almost a full circle with a small opening facing an incoming wave train ($\alpha = -89^\circ, \beta = 269^\circ$) and for the reversed arc ($\alpha = 90^\circ, = -269^\circ$). In this case, the curves for $|R|$ are very small for all wave numbers. This phenomenon follows from the so-called complementarity theorem which states that if the scattering body is reversed but the incident field is left unchanged, then the magnitudes of the reflection and transmission coefficients are unaltered.

Figure 2.28a shows six different shapes of the circular-arc-shaped plate corresponding to (*I*) $\alpha = -45^\circ, \beta = 45^\circ$; (*II*) $\alpha = 45^\circ, \beta = 135^\circ$; (*III*) $\alpha = 15^\circ, \beta = 105^\circ$; (*IV*) $\alpha = -135^\circ, \beta = 135^\circ$; (*V*) $\alpha = 105^\circ, \beta = 195^\circ$ and (*VI*) $\alpha = 135^\circ, \beta = 225^\circ$. Figure 2.28b displays $|R|$ against Kb for the shapes (I) to (VI), shown in Fig. 2.28a, with the same depth parameter $d/b = 0.2$. It is observed that the reflection is much reduced when the arc is convex upward compared to the case when it is convex downward and the plate length is more than half a circle. For plates occupying less than half a circle, on the other hand, the downward convex plate produces less reflection compared to the upward convex plate with the same arc length. Thus circular-arc plates occupying more than half a circle are good candidates for use as elements for a water wave lens when they are convex upward. This has also been confirmed by McIver and Urka (1995) for symmetric arcs. However, circular-arc plates occupying less than half a circle and convex downward, are also good candidates for use as lens elements. Figure 2.29 depicts $|R|$ for two semi-circular-arcs with horizontal diameter, one is convex upward ($\alpha = -90^\circ, \beta = 90^\circ$) and the other is convex downward ($\alpha = 90^\circ, \beta = 270^\circ$) for $d/b = 0.2$. It is observed that the upward convex semi-circular plate produces more

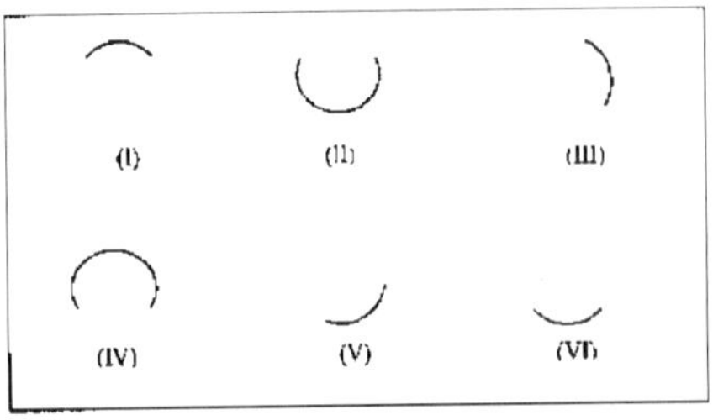

(a)

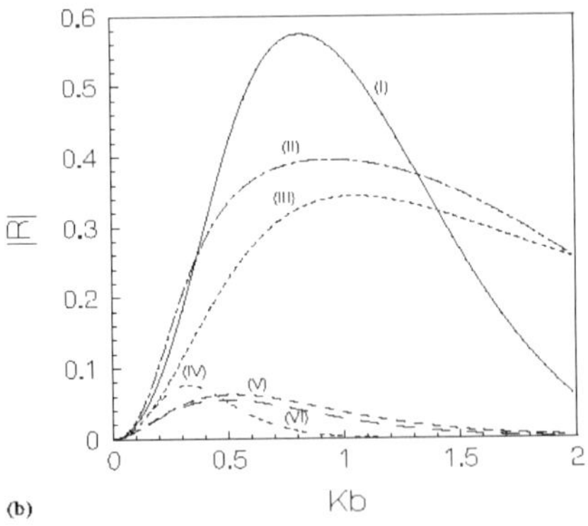

(b)

Fig. 2.28 (a) Different shapes of circular-arc-shaped plate
(α, β) (*I*)(–45°, 45°), (*II*)(45°, 315°), (*III*)(15°, 105°), (*IV*)(–135°, 135°), (*V*)(105°, 195°), (*VI*)(135°, 225°)

Fig. 2.28 (b) Reflection coefficient vs. Wave number, $d/b = 0.2$, shapes (I)–(VI)

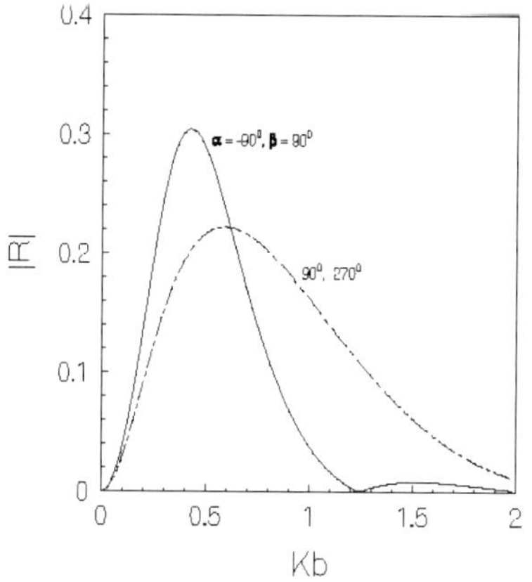

Fig. 2.29 Reflection coefficient vs. wave number, $d/b = 0.2$

reflection in the low-frequency range and less reflection in the complementary range compared to the downward convex semi-circular plate. Thus their use as lens elements is dependent on the frequency range of the incoming wave train.

2.5 Scattering by two symmetric circular arc-shaped thin plates

This section is concerned with a generalization of the water wave scattering problem involving a single arc-shaped thin plate considered in section 2.4 to two symmetric circular arc shaped thin plates submerged in deep water. This problem was investigated by Mandal and Gayen (2002). The line joining the centers of the circles, whose arcs assume the positions of the two plates, is horizontal, and the two arcs are symmetrically placed with respect to the vertical through the midpoint of this line. Exploiting the geometrical symmetry, the scattering problem is split into two separate problems involving the symmetric and anti-symmetric potential functions describing the resulting motion in the fluid due to an incoming surface water wave train incident on the plates. Appropriate use of Green's integral theorem followed by utilization of the boundary condition on the plates produces two integro-differential equations, which are interpreted as equivalent to two hypersingular integral equations in the discontinuities of the symmetric and anti-symmetric potential functions across one of the two plates. These hypersingular integral equations are solved numerically by approximating discontinuities of the potential functions across the plate in terms of two finite series involving Chebyshev polynomials of the second kind followed by a collocation method. The reflection and transmission coefficients are then computed numerically by using these solutions.

Numerical results for the reflection coefficients showing variation of the depth of submergence of the plates, arc lengths of the plates, are depicted graphically against the wave number in a number of figures. For the case of semicircular plates with vertical diameters, the reflection coefficient is seen to become very small when the separation between the centers, i.e., the distance between the vertical diameters of the semi circular plates, is made very small. However, in this case, the two semicircles almost assume the form of a full circle, and the phenomenon of very small reflection is consistent with the classical result obtained long back by Dean (1948) and Ursell (1950), namely that a horizontal circular cylinder submerged in deep water, experiences no reflection by a normally incident incoming train of surface water waves. Again, numerical results are obtained for a submerged obstacle in the form of a convex lens whose two sides consist of intersecting circular arcs of the same radius and are symmetric about the vertical mid section. For some frequencies the phenomena of total reflection and total transmission are observed to occur for such an obstacle. Somewhat similar phenomena are also seen to occur when the effect of transition from a circular arc shaped plate which is symmetric about the vertical, to a horizontal plate of the same arc length on the reflection coefficient is considered.

A Cartesian co-ordinate system is chosen in which the y-axis is taken vertically downwards into the fluid region and the plane $y = 0$ is the rest position of the free surface. Let two symmetric circular arc shaped thin plates $\Gamma_i (i = 1,2)$ be submerged in deep water and occupy the positions described by arcs of two circles of the same radius b with centers at $(\pm a, d + b)$, as illustrated in the Fig. 2.30.

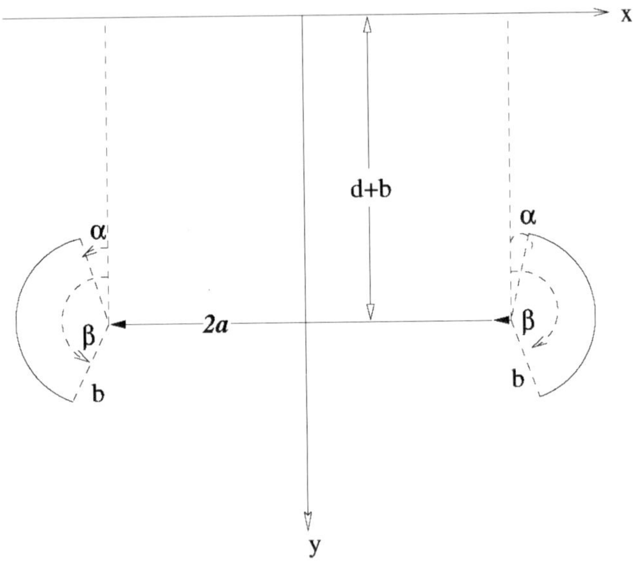

Fig. 2.30 Geometrical sketch of the problem

A point (x, y) on the plate $\Gamma_i (i = 1,2)$ can be expressed as

$$x = \pm(a + b \sin \theta), y = d + b(1 - \cos \theta), \alpha \leq \theta \leq \beta, \tag{5.1}$$

where α, β are the angles made by the radii through the upper and lower ends of the plates with the vertical, and further, a, b and the angles α, β must satisfy the inequality

$$a \geq \max(-b \sin \alpha, -b \sin \beta) \tag{5.2}$$

Assuming linear theory, incompressible and inviscid fluid, and irrotational motion, an incoming surface wave train can be described by the potential function $\mathrm{Re}\{\phi^{inc}(x, y)e^{-i\sigma t'}\}$ where t' is the time, σ is the frequency and

$$\phi^{inc}(x, y) = 2\, e^{-Ky-iK(x-a)} \tag{5.3}$$

with $= \frac{\sigma^2}{g}$, g being the acceleration due to gravity. Let this incoming wave train be incident on the two arc shaped plates from the direction of $x = \infty$. Let the resulting motion in the fluid be described by the potential function $\mathrm{Re}\{\phi(x, y)e^{-i\sigma t'}\}$; then $\phi(x, y)$ satisfies

$$\nabla^2\phi = 0 \quad \text{in the fluid region} \tag{5.4}$$

where ∇^2 is the two-dimensional Laplacian operator,

$$K\phi + \phi_y = 0 \text{ on } y = 0 \tag{5.5}$$

$$\phi_n = 0 \text{ on } \Gamma_i (i = 1,2); \tag{5.6}$$

where ϕ_n denotes the normal derivative at a point on $\Gamma_i (i = 1,2)$.

$$r^{\frac{1}{2}}\nabla\phi \text{ is bounded as } r \to 0,$$

where r is the distance from any submerged edge of Γ_i,

$$\nabla\phi \to 0 \text{ as } y \to \infty, \tag{5.8}$$

and

$$\phi(x, y) = \begin{cases} \phi^{inc}(x, y) + R\phi^{inc}(-x, y) \text{ as } x \to \infty, \\ T\phi^{inc}(x, y) \text{ as } x \to \infty \end{cases} \tag{5.9}$$

where R and T denote respectively the reflection and transmission coefficients (complex) and their numerical estimation is the principal aim here.

Because of geometrical symmetry of the positions of the two plates about the y-axis, the potential function $\phi(x, y)$ can be split into its symmetric and anti-symmetric parts $\phi^s(x, y)$ and $\phi^a(x, y)$ respectively so that

$$\phi(x, y) = \phi^s(x, y) + \phi^a(x, y) \tag{5.10}$$

where

$$\phi^s(-x, y) = \phi^s(x, y),\ \phi^a(-x, y) = -\phi^a(x, y). \tag{5.11}$$

Thus, we can restrict our analysis to the region $x \geq 0$ only. Now the functions $\phi^{s,a}(x, y)$ satisfy (5.4) to (5.8) together with

$$\phi^s_x(0, y) = 0,\ \phi^a(0, y) = 0,\ y > 0. \tag{5.12}$$

Let the behaviour of $\phi^{s,a}(x, y)$ for large x be represented by

$$\phi^{s,a}(x, y) \to e^{-Ky}\{e^{-iK(x-a)} + R^{s,a}e^{iK(s-a)}\} \text{ as } x \to \infty, \tag{5.13}$$

where $R^{s,a}$ are unknown constants, and because of (5.9), are related to R, T by

$$R, T = \frac{1}{2}(R^s \pm R^a)e^{-2iKa}. \tag{5.14}$$

For obtaining representation of $\phi^{s,a}(x, y)$ we now apply Green's integral theorem to the functions $\phi^{s,a} - e^{-Ky-iK(x-a)}$ and $\mathcal{G}^{s,a}(x, y; \xi, \eta)$ where

$$\mathcal{G}^{s,a}(x, y; \xi, \eta) = G(x, y; \xi, \eta) \pm G(-x, y; \xi, \eta) \tag{5.15}$$

(G being given by (4.8)) in the region bounded by the lines $y = 0, 0 \leq x \leq X; x = X, 0 \leq y \leq Y; y = Y, 0 \leq x \leq X; x = 0, 0 \leq y \leq Y$; a small circle of radius ϵ with centre at (ξ, η) and a contour enclosing arc Γ_1 and ultimately making $X, Y \to \infty, \epsilon \to 0$ and shrinking the contour enclosing Γ_1 into the two sides of Γ_1 to obtain

$$\phi^{s,a}(\xi, \eta) = 2e^{-K\eta+iKa}(\cos K\xi, -i\sin K\xi) - \frac{1}{2\pi}\int_{\Gamma_1} F^{s,a}(p)\mathcal{G}^{s,a}_{np}(x, y; \xi, \eta)ds_p, \tag{5.16}$$

where $p \equiv (x, y)$ is a point on Γ_1, $F^{s,a}(p)$ are the discontinuities of $\phi^{s,a}(x, y)$ across Γ_1 at p, and $\mathcal{G}^{s,a}_{np}$ denote the normal derivative of $\mathcal{G}^{s,a}$ at the point $p \in \Gamma_1$. It should be noted that the unknown functions $F^{s,a}(p)$ vanish at the end points of while their derivatives have square-root singularities at the end points.

Use of the boundary condition (5.6) rewritten as

$$\phi^{s,a}_{nq} = 0 \text{ on } \Gamma_1,$$

where $q \equiv (\xi, \eta)$ is a point on Γ_1, produces the integro-differential equations

$$\frac{\partial}{\partial n_q}\int_{\Gamma_1} F^{s,a}(p)\, \mathcal{G}^{s,a}_{n_p}(p; q)ds_p = 2\pi\frac{\partial}{\partial n_q}\left[2\, e^{-K\eta+iKa}(\cos K\xi\,, -i\sin K\xi)\right],\ q \in \Gamma_1. \tag{5.17}$$

As in the section 2.4 these leads to the hypersingular integral equations

$$\times\!\!\!\!\!\!\int_{\Gamma_1} F^{s,a}(p)\mathcal{G}^{s,a}_{n_p,n_q}(p; q)ds_p = 2\pi\frac{\partial}{\partial n_q}\left[2\, e^{-K\eta+iKa}(\cos K\xi\,, -i\sin K\xi)\right], q \in \Gamma_1 \tag{5.18}$$

where the cross on the integral sign indicates that the integrals are to be interpreted as Hadamard finite-part integrals.

Let $\boldsymbol{n}_p$ and $\boldsymbol{n}_q$ denote the unit normals at the points p and q, respectively, on Γ_1, then

$$\boldsymbol{n}_p = (\sin\theta_t, -\cos\theta_t), \boldsymbol{n}_q = (\sin\theta_\tau, -\cos\theta_\tau),$$

where

$$\theta_{t,\tau} = \frac{\alpha+\beta}{2} + \frac{\beta-\alpha}{2}(t,\tau), -1 < t,\tau < 1, \tag{5.19}$$

the co-ordinates of the points $p \equiv (x, y)$ and $q \equiv (\xi, \eta)$ on Γ_1 being parametrically expressed as

$$x = a + b\sin\theta_t, y = d + b(1 - \cos\theta_t), -1 \leq t \leq 1,$$

$$\xi = a + b\sin\theta_\tau, \eta = d + b(1 - \cos\theta_\tau), -1 \leq \tau \leq 1. \tag{5.20}$$

The hypersingular equations (5.17) are now rewritten as

$$\times\!\!\!\!\!\!\int_{-1}^{1} f^{s,a}(t)\left[-\frac{1}{(\tau-t)^2} + \mathcal{K}^{s,a}(\tau,t)\right]dt = h^{s,a}(\tau), \quad -1 < \tau < 1, \tag{5.21}$$

where we have used the notation $f^{s,a}(t)$ for $F^{s,a}(p)$, and

$$\mathcal{K}^{s,a} = -\frac{\Theta^2}{4}\left[\frac{1}{\sin^2\frac{\Theta}{2}(\tau-t)} - \frac{4}{\Theta^2(\tau-t)^2}\right]$$

$$+b^2\Theta^2\Bigg[-\cos(\theta_\tau-\theta_t)\left\{\frac{Y^2-X^2}{(X^2+Y^2)^2}+\frac{2KX}{X^2+Y^2}+2K^2\Phi_0(X,Y)\right\}$$

$$-2\sin(\theta_\tau-\theta_t)\left\{\frac{XY}{(X^2+Y^2)^2}+\frac{KX}{X^2+Y^2}+K^2\Psi_0(X,Y)\right\}$$

$$\pm\cos(\theta_\tau-\theta_t)\frac{Z^2-X_1^2}{(X_1^2+Z^2)^2}\pm\sin(\theta_\tau-\theta_t)\frac{2X_1Z}{(X_1^2+Z^2)^2}$$

$$\pm 2\sin(\theta_\tau+\theta_t)\left\{\frac{X_1Y}{(X_1^2+Y^2)^2}+\frac{KX_1}{X_1^2+Y^2}+K^2\Psi_0(X,Y)\right\}\Bigg] \qquad (5.22)$$

with

$$\Theta = \frac{1}{2}(\beta-\alpha), X = b(\sin\theta_t-\sin\theta_\tau), X_1 = 2a+b(\sin\theta_t+\sin\theta_\tau),$$

$$Y = 2(d+b)-b(\cos\theta_t-\cos\theta_\tau), Z = b(\cos\theta_t-\cos\theta_\tau) \qquad (5.23)$$

$$\Phi_0(X,Y), \Psi_0(X,Y) = \int_C \frac{e^{-kY}}{k-K}(\cos kX, \sin kX)dk,$$

and

$$h^{s,a}(\tau) = 4\pi Kb\Theta e^{-K\eta+iKa}(\cos(K\xi+\theta_\tau), -i\sin(K\xi+\theta_\tau). \qquad (5.24)$$

We have to solve the hypersingular integral equations (5.21) keeping in mind that $f^{s,a}(\pm1) = 0$.

As in Parsons and Martin (1994) the integrals in (5.23) can be expanded as

$$\int_C \frac{e^{-kY}}{k-K}\binom{\cos}{\sin}kX\Big)dk = -e^{-KY}\left\{(\log Kr_1 - i\pi+\gamma)\binom{\cos}{\sin}KX\Big)+\theta_1\binom{\sin}{-\cos}KX\Big)\right\}$$

$$\sum_{m=1}^{\infty}\frac{(-Kr_1)^m}{m!}\left(\frac{1}{1}+\frac{1}{2}+\cdots+\frac{1}{m}\right)\binom{\cos}{-\sin}m\theta_1\Big) \qquad (5.25)$$

where $r_1^2 = X^2+Y^2$, $\theta_1 = \tan^{-1} X/Y$ and $\gamma = 0.5772$ ... is Euler's constant.

We now approximate $f^{s,a}(t)$ as

$$f^{s,a}(t) = (1-t^2)^{1/2}\sum_{n=0}^{N} a_n^{s,a}U_n(t), \qquad (5.26)$$

where $U_n(t)$ is the Chebyshev polynomial of the second kind and $a_n^{s,a}(n = 0,1, ..., N)$ are unknown constants to be found. The square-root factor in (5.26) ensures that $f^{s,a}(t)$, i.e., $F^{s,a}(p)$, have the correct behavior at the end points. Using the expressions (5.24) in (5.21) we obtain

$$\sum_{n=0}^{N} a_n^{s,a} A_n^{s,a}(\tau) = h^{s,a}(\tau), -1 < \tau < 1 \tag{5.27}$$

where

$$A_n^{s,a}(\tau) = \pi(n+1)U_n(\tau) + \int_{-1}^{1} (1-t^2)^{\frac{1}{2}} \mathcal{K}^{s,a}(\tau,t)U_n(t)dt. \tag{5.28}$$

To find the unknown constants $a_n^{s,a}(n = 0,1, ..., N)$, we put $\tau = \tau_j(j = 0,1, ..., N)$ in (5.22) to obtain the linear systems

$$\sum_{n=0}^{N} a_n^{s,a} A_n^{s,a}(\tau_j) = h^{s,a}(\tau_j), j = 0,1, \dots, N \tag{5.29}$$

where τ_j's are collocation points and are chosen as cf. Parsons and Martin (1994)

$$\tau_j = \cos\frac{2j+1}{2N+2}\pi, j = 0,1, \dots, N. \tag{5.30}$$

The two linear systems (5.29) can be solved numerically by any standered method to determine $a_n^{s,a}(n = 0,1, ..., N)$ numerically. Here we have used Gauss Jordan method.

To find the reflection and transmission coefficients $|R|$ and $|T|$ we first obtained the quantities $R^{s,a}$ by making $\xi \to \infty$ in the representations (5.16) for $\phi^{s,a}(\xi, \eta)$ and comparing with (5.13), with (x, y) replaced by (ξ, η). For this we require the asymptotic results

$$\mathcal{G}^{s,a}(x, y; \xi, \eta) \to -4\pi e^{-K(y+\eta)+iK\xi}(i \cos kx, \sin kx) \text{ as } \xi \to \infty. \tag{5.31}$$

Thus we find that

$$R^{s,a} = \pm e^{2iKa} + 2e^{iKa} \int_{\Gamma_1} F^{s,a}(p)\frac{\partial}{\partial n_p}[e^{-k\eta y}(i\cos kx, \sin kx)]ds_p = \pm e^{2iKa}$$

$$+2Kb\Theta e^{iKa} \sum_{n=0}^{N} a_n^{s,a} \int_{-1}^{1} (1-t^2)^{\frac{1}{2}} U_n(t) e^{-K(d+b(1-\cos\theta_t))} \begin{pmatrix} i\cos \\ \sin \end{pmatrix} (Ka + Kb\sin\theta_t) \Bigg) dt. \tag{5.32}$$

The integrals in (5.32) can be evaluated numerically by standard methods. Thus once $a_n^{s,a}(n = 0,1, ..., N)$ are found numerically by solving the linear systems (5.29), $R^{s,a}$ can be computed numerically from (5.32) for different values of the parameters Kb, d/b, a/b and Θ. Having found $R^{s,a}$, we can obtain the reflection and transmission coefficients R and T numerically by using (3.5). Also, as $|R|$ and $|T|$ must satisfy the identity

$$|R|^2 + |T|^2 = 1, \tag{5.33}$$

we can use this as a partial check on the correctness of the numerical results obtained for $R^{s,a}$ by using (5.32).

The reflection coefficient $|R|$ is computed numerically for various values of the different parameters. In Table 2.5 we display the numerical results for $|R|$ showing its convergence with the truncation size N of the finite series (5.26) for different arc lengths of the two plates by choosing $d/b = 0.5$, $a/b = 1.0$, $Kb = 1.5$ and $\alpha = 0, \beta = \frac{\pi}{2}$

Table 2.5 Reflection coefficient $|R|$
($d/b = 0.5$, $a/b = 1.0$, $Kb = 1.5$, $\alpha = 0$)

N	$\beta = \pi/2$	$\beta = \pi$
0	0.17198	0.72171
2	0.30619	0.15170
4	0.29101	0.28328
6	0.29089	0.27598
8	0.29089	0.25734
10		0.27541
12		0.27541

and π. It is observed from the table that the truncation size depends on $\beta - \alpha$, i.e., the arc lengths of the plates when d/b and a/b are kept fixed. However, if the arc length is kept fixed and the separation parameter a/b is varied keeping d/b fixed or the depth parameter is varied keeping a/b fixed, then for a fixed wave number, the corresponding tabular values of $|R|$ (not shown here) would produce different truncation sizes for the convergence. Thus, the truncation size depends on all the different parameters associated with the geometrical position of the plates as well as the wave number Kb.

In our numerical computations for every data appropriate safeguard has been taken on the truncation size, so as to produce numerical results that are correct up to almost five decimal places.

In the Table 2.6, a representative set of values $|R|$, $|T|$ and $|R|^2 + |T|^2$ against Kb for $d/b = 0.1$, $a/b = 1.5$, $\alpha = 0$, $\beta = \pi$ are given. It is observed that $|R|^2 + |T|^2$ almost coincides with unity for different Kb. Thus the reflection and transmission coefficients $|R|$ and $|T|$ computed by using the formulae (5.14) where $R^{s,a}$ are computed by using the relations in (5.32), satisfy the energy equality $|R|^2 + |T|^2 = 1$. This may provide a partial check on the correctness of the numerical results obtained here, although this cannot be regarded as a sufficient requirement for the correctness of the numerical results. However, some other checks are also provided later in which the results obtained by following the present numerical procedures for the limiting cases of a submerged full circle and a submerged horizontal plate, produce known numerical results existing in the literature obtained by following some other methods.

Table 2.6 $|R|$ and $|T|$
($d/b = 0.1$, $a/b = 1.5$, $\alpha = 0$, $\beta = \pi$)

Kb	$\|R\|$	$\|T\|$	$\|R\|^2 + \|T\|^2$
0.5	0.567566	0.823327	1.0000
1.0	0.783176	0.621799	1.0000
1.5	0.462642	0.886545	1.0000

Figure 2.31 depicts $|R|$ against Kb for two semi-circular arc shaped plates with vertical diameters ($\alpha = 0$, $\beta = \pi$) and constant depth parameter $d/b = 0.5$ for different separation lengths. The main feature of the curves for $|R|$ is the occurrence of zeros of $|R|$ as a function of the wave number. As the separation length a/b decreases, the zeros of $|R|$ are shifted towards the right. Another important feature is the overall decrease of $|R|$ with the decrease of a/b. When the separation length becomes negligibly small

($a/b = 0.0001$), then $|R|$ also becomes very small for all wave numbers. This result is consistent with the classical result that |R| is identically zero for a full circular arc, i.e., for a circular cylinder submerged in deep water irrespective of the depth and the frequency of the incident wave field, obtained first by Dean (1948) and established rigorously by Ursell (1950) soon afterwards, since the two semi-circular arc shaped plates with vertical diameters assume almost a full circle when a/b becomes very small. This result also provides another partial check on the correctness of the numerical results obtained here.

Figure 2.32 depicts $|R|$ against Kb for fixed depth ($d/b = 0.5$) and separation length ($a/b = 1.5$) but different arc lengths of the plates ($\alpha = 70°, \beta = 340°, 280°, 220°$). It shows the effect of different arc lengths on $|R|$ for fixed depth and separation length. Here also the occurrence of zeros of $|R|$ as a function of the wave number is the principal feature. As the arc length decreases the zeros of $|R|$ are shifted towards the right. Also, the overall reflection coefficient decreases with the decrease of the arc length. This is quite plausible since in general, less energy is reflected by the plates when the arc length decreases, due to the incident wave train facing less resistance.

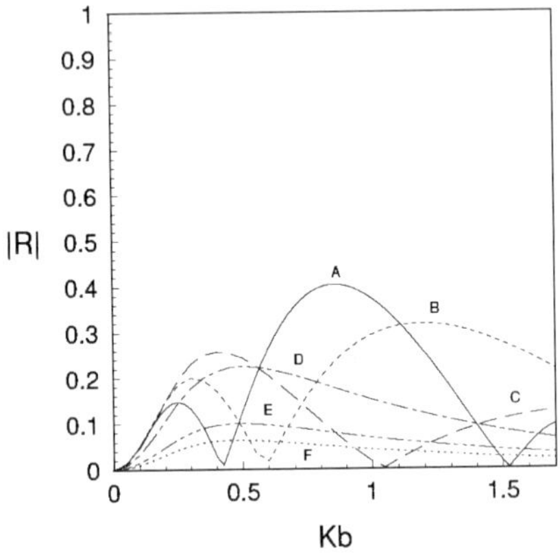

Fig. 2.31 Reflection coefficient for two half-circles ($\alpha = 0, \beta = \pi$), $d/b = 0.5$, $a/b = 1.5(A)$, $1.0(B)$ $0.5(C)$, $0.1(D).001(E)$, $0.00001(F)$

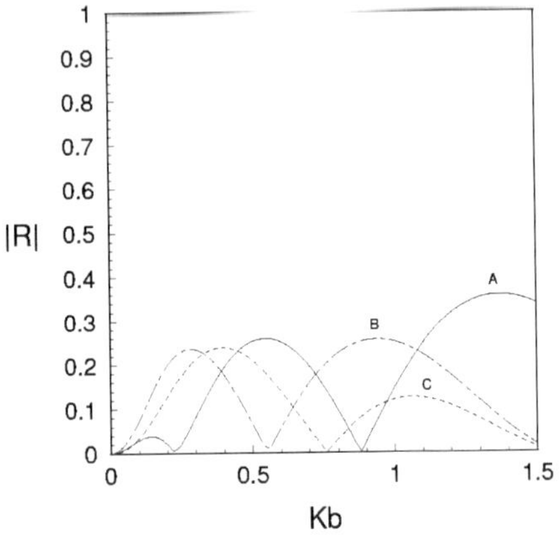

Fig. 2.32 Reflection coefficient for different arc lengths $d/b = 0.5$, $a/b = 1.5$, $\alpha = 70$, $\beta = 340(A)$, $280(B)$, $220(C)$

Figure 2.33 displays the dependence of $|R|$ for a fixed pair of circular arc shaped plates ($\alpha = 5°, \beta = 150°$) on the depth parameter ($d/b = 0.1, 0.5, 1.0$). It is observed from this figure that the overall reflection coefficient decreases with the increase of the depth, and the shifting of zeros of $|R|$ with the increase of d/b is of not much significance. The overall decrease of $|R|$ with the increase of d/b is expected since the disturbance created due to the incident wave train does not penetrate much below the free surface and as such less energy is reflected by an obstacle whose depth below the free surface is considerable.

The reflection coefficient for two submerged almost full circles (two circular cylinders) is shown in the Fig. 2.34. For $a/b = 0.5$, $d/b = 1.5$, $|R|$ for almost two full circles is obtained by choosing $\alpha = 0°$, $\beta = 359°$ and also by choosing $\alpha = -90°$, $\beta = 269°$, and the two curves for $|R|$ for these two sets of values of α and β almost coincide.

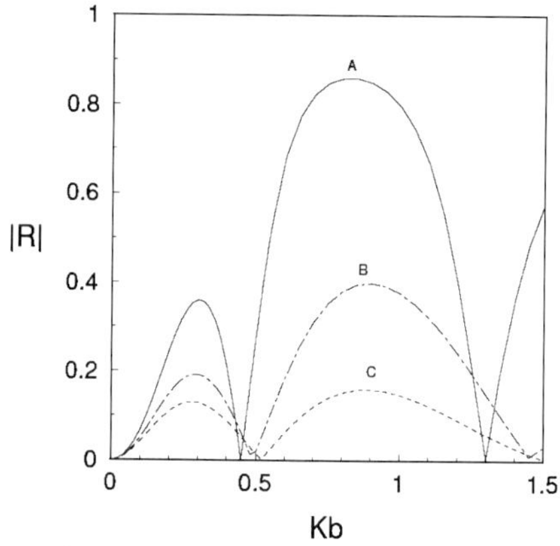

Fig. 2.33 Reflection coefficient for different depths a/b = 1.5, $\alpha = 5$, β = 150,0.1(A), 0.5(B), 1.0(C)

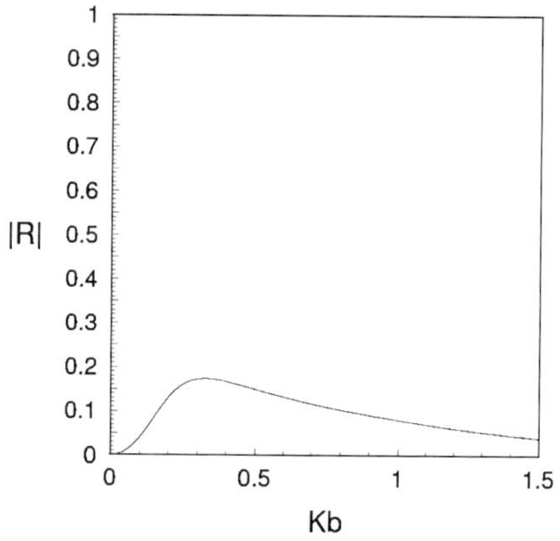

Fig. 2.34 Reflection coefficient for two almost full circles $d/b = 0.5$, $a/b = 1.5$, $\alpha = 0$, $\beta = 359$; $\alpha = -90$, $\beta = 269$

This can be regarded as another partial check on the correctness of the numerical method utilized here. Again, this figure demonstrates that, although a single submerged cylinder experiences no reflection, submerged twin cylinders do experience reflection for an incoming surface wave train.

The result for a submerged lens shaped body whose two sides consist of intersecting circular arcs of the same radius and symmetric about the vertical mid section of the lens, can be obtained by suitable choices of α, β and a/b. One such choice is $\alpha = -\pi/6$, $\beta = \pi + \pi/6$ and $a/b = \sin(\pi/6) = 0.5$. Figure 2.35 displays $|R|$ for such a lens shaped obstacle for different values of the depth parameter d/b (d/b = 0.5, 0.1, 0.05, 0.01, 0.0). For moderate values of the depth parameter (d/b = 0.5, 0.1) $|R|$, regarded as a function of the wave number Kb, first increases as Kb increases from zero, attains a maximum value and then decreases as Kb further increases. However, for a small value of d/b ($d/b = 0.05$), $|R|$ sharply increases from zero value to almost unity and then decreases as Kb further increases. As d/b is further decreased (d/b = 0.01), $|R|$ sharply increases almost to unity as before but then again decreases sharply to zero and again almost becomes unity as the wave number further increases. This is perhaps due to interaction of waves between the free surface and the sharp upper edge of the convex lens shaped body because of its proximity to the free surface. For $d/b = 0$ the upper edge is still below the free surface, but it is nearer to the free surface compared to the previous situations. In this case $|R|$ initially increases sharply from zero to unity, then oscillates near the unit value for moderate values of the wave number and becomes almost unity as the wave number further increases. This type of a lens shaped body submerged not much below the free surface appears to possess the property of almost total reflection for most values of the wave number, and thus may act as an efficient breakwater. Also, the phenomenon of occurrence of total reflection for some frequency may have some bearing on the search for trapped modes in the presence of submerged obstacles.

The effect of transition from a circular arc shaped plate which is symmetric about the vertical, to a horizontal plate of same arc length, on $|R|$, can be visualized by putting $a/l = 0$ where $2l$ is the fixed arc length of the plate, and by decreasing the difference $\beta - \alpha$ but increasing b such that $b(\beta - \alpha)$ has the constant value l. The Fig. 2.36 displays $|R|$ against the new wave number Kl (when $d/l = 0.1$) for $\alpha = 0°$ and the choices $\beta = 36°$, $18°$ and $1°$ when $l = b\beta$ is kept fixed. The configuration $\alpha = 0°$, $\beta = 1°$ can be regarded as an almost straight horizontal plate. The qualitative features of the curves for $|R|$ due to the transition from a circular arc shaped plate to a horizontal plate are observed to be the same as given by Parsons and Martin (1994), who however, investigated water wave scattering by a single circular arc shaped plate. This observation may also be regarded as another partial check on the correctness of the numerical results obtained here.

For various configurations of the two circular arc shaped plates (including the situations when they assume the form of an almost full circle or a horizontal plate) submerged in deep water, it is observed that the long wave limit of the reflection coefficient is zero. This is in conformity with the observation of Martin and Dalrymple (1988) and McIver (1994) who confirmed by using the method of matched asymptotic expansions, that the reflection coefficient becomes zero in the long-wave limit.

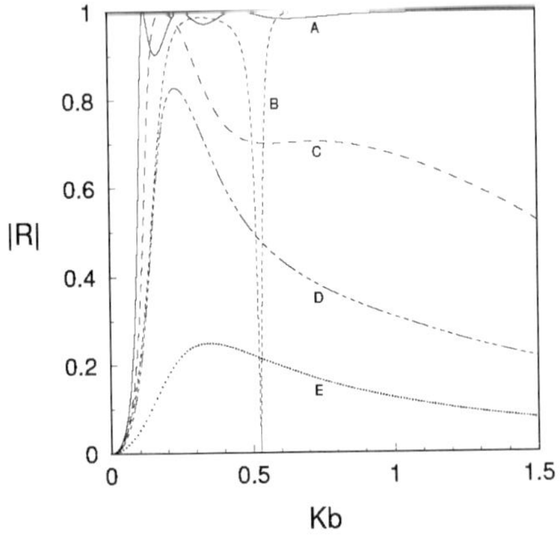

Fig. 2.35 Reflection coefficient for lens shaped obstacle $\alpha = -\pi/6$, $\beta = 7\pi/6$, $a/b = 0.5$, $d/b = 0(A)$, $0.01(B)$, $0.05(C)$, $0.1(D)$, $0.5(E)$

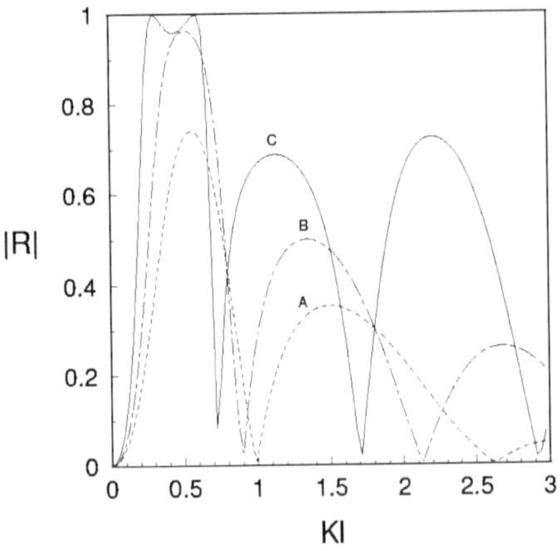

Fig. 2.36 Reflection coefficient for small $\beta - \alpha$, $d/l = 0.1$, $a/l = 0.0$, $\alpha = 0$, $\beta = \pi/5(A)$, $\pi/10(B)$, $\pi/180(C)$

2.6 Scattering by a thin vertical barrier submerged beneath an ice-cover

During the last two decades, there has been a considerable interest amongst the ocean engineers and applied mathematicians to investigate various types of surface water wave problems in the presence of a thin ice-sheet floating on water, the ice-sheet being modelled as a thin elastic plate. Due to considerable increase in different types of scientific and industrial activities in polar oceans towards the latter half of the last century, study of wave propagation problems in polar oceans covered by floating ice has become very attractive to some researchers. Quite a number of models for the ice-cover have been proposed for the purpose of mathematical investigation of wave propagation problems. In the case when the ice-sheet is thin and continuous, it is modeled as a floating thin elastic plate with uniform elastic properties. This model has become very attractive for the purpose of mathematical analysis of water wave problems in water with an ice-cover, since under the assumption of linear theory, the ice-cover condition involves fifth order derivative of the potential function while the free surface condition involves only the first order derivative.

As already mentioned, various water wave scattering problems involving bodies in the form of thin vertical or curved barriers submerged or partially immersed in water with a free surface, have been studied in the literature quite extensively. However, there is no such study of similar problems in water with an ice-cover. The reason behind this may be attributed to the linearized ice-cover condition which involves fifth-order derivative in contrast with the free surface condition which involves only the first order derivative. It is felt that such wave scattering problems for a free surface which have been solved earlier can be tackled in the presence of an ice-cover if the hypersingular integral equation formulation developed by Parsons and Martin (1992, 1994) with Green's function appropriate for the ice-covered region, is employed. In this section, this is demonstrated by considering the problem of wave scattering by an inclined thin rigid semi-infinite barrier submerged in deep water with an ice-cover as given in Maiti and Mandal (2010). When the angle of inclination with the vertical is made zero, it becomes an extension of Ursell's (1947) classical problem of water wave scattering by a thin vertical barrier submerged in deep water with a free surface. The integral equation formulation of the problem is achieved by employing Green's integral theorem in the fluid region to the scattered potential function and an appropriate Green's function. This provides a representation of the potential function describing the motion in the fluid in terms of an integral involving the unknown difference of potential across the inclined barrier. The use of vanishing of velocity normal to the barrier produces the desired hypersingular integral equation. By an appropriate change of variable, the range of this equation is made equal to $(-1, 1)$. Approximating the unknown function of the integral equation by a finite series involving Chebyshev polynomials of second kind, the unknown coefficients of the finite series are determined from a linear system obtained by collocation method, and thus the integral equation is solved numerically. The reflection and transmission coefficients are determined directly from the representation of the velocity potential in terms of an integral involving the difference of potential across the barrier by noting its behaviours at the two infinities on two sides of the barrier. Their numerical estimates are then computed from the numerical solution of the hypersingular integral equation. Numerical results are presented here for a submerged vertical barrier although these can be obtained for any submerged inclined barrier. That the reflection and transmission coefficients satisfy the energy identity is seen to be true from their numerical estimates for various values of the wave number. For very small values of the ice-cover parameters D, ε (non-dimensionalised appropriately, D, ε being defined later), the numerical estimates of the reflection coefficient computed by the present method for different values of the wave number, almost coincide with the corresponding free surface results obtained from the exact expression given by Ursell (1947). The reflection and transmission coefficients are depicted graphically against the wave number for different values of the non-dimensional parameters associated with the ice-cover in a number of figures so as to visualize the effect of the presence of ice-cover on them. These figures show that the reflection coefficient increases and the transmission coefficient decreases due to the presence of the ice-cover.

A rectangular Cartesian coordinate system is used in which the y-axis is chosen vertically downwards into the region of an infinitely deep water and the (x, z)-plane denotes the rest position of a thin sheet of ice floating on the water. The thin ice sheet is modelled as a thin elastic plate of infinite extent. A thin rigid plane semi-infinite barrier Γ inclined at angle θ $(0 \le \theta \le \pi/2)$ with the downward vertical is submerged beneath the ice-cover from a depth a below it. The barrier Γ is assumed to be infinitely long along the z-direction and thus we consider the two-dimensional problem of water wave scattering by it. The geometry of Γ is given in Fig. 2.37

The point $(x, y) \in \Gamma$ is given by

$$x = (t - a) \sin \theta,\ y = a + (t - a) \cos \theta,\ a < t < \infty \tag{6.1}$$

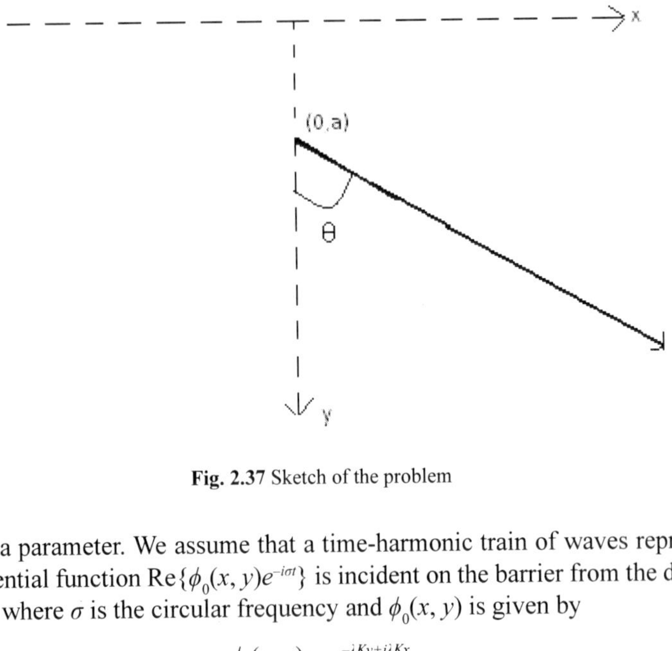

Fig. 2.37 Sketch of the problem

where t is a parameter. We assume that a time-harmonic train of waves represented by the potential function $\text{Re}\{\phi_0(x, y)e^{-i\sigma t}\}$ is incident on the barrier from the direction of $x = -\infty$, where σ is the circular frequency and $\phi_0(x, y)$ is given by

$$\phi_0(x, y) = e^{-\lambda Ky + i\lambda Kx}. \tag{6.2}$$

In (6.2), λK is the unique positive real root of the dispersion equation

$$(Dk^4 + 1 - \varepsilon K)k - K = 0 \tag{6.3}$$

where $K = \sigma^2/g$, g being the acceleration due to gravity, $D = \frac{Eh_0^3}{12(1 - \nu^2)\rho g}$, E being Young's modulus and ν being Poisson's ratio of the elastic material of the ice-cover, ρ is the density of water and h_0 is the very small thickness of the ice-cover, $\varepsilon = \frac{\rho_0}{\rho} h_0$ where ρ_0 is the density of floating ice. The other roots of (6.3) are $\lambda_1 K$, $\bar{\lambda}_1 K$ and $\lambda_2 K$, $\bar{\lambda}_2 K$ where $\text{Re}\,\lambda_1 > 0$, and $\text{Re}\,\lambda_2 < 0$.

The ensuing motion in the fluid region is described by the velocity potential $\mathrm{Re}\{\phi(x, y)e^{-i\sigma t}\}$ where $\phi(x, y)$ satisfies Laplace's equation

$$\nabla^2\phi = 0 \text{ in the fluid region,} \tag{6.4}$$

the linearized ice-cover condition

$$\left\{D\frac{\partial^4}{\partial x^4} + (1-\varepsilon K)\right\}\phi_y + K\phi = 0 \quad \text{on} \quad y = 0, \tag{6.5}$$

the condition on the barrier Γ given by

$$\frac{\partial\phi}{\partial n} = 0 \text{ on } \Gamma \tag{6.6}$$

where $\frac{\partial}{\partial n}$ denotes the normal derivative on the barrier, the edge condition

$$r^{1/2}\,\nabla\phi = 0(1) \text{ as } r = \{x^2 + (y-a)^2\}^{1/2} \to 0, \tag{6.7}$$

the condition at infinite depth

$$\nabla\phi \to 0 \text{ as } y \to \infty, \tag{6.8}$$

and the condition at infinity described by

$$\phi(x, y) \to \begin{cases} \phi_0(x, y) + R\phi_0(-x, y) \text{ as } x \to -\infty, \\ \qquad T\phi_0(x, y) \text{ as } x \to \infty \end{cases} \tag{6.9}$$

where R and T denote respectively the complex reflection and transmission coefficients. Our problem here is to determine approximate numerical estimates for $|R|$ and $|T|$.

Let

$$\psi(x, y) = \phi(x, y) - \phi_0(x, y) \tag{6.10}$$

so that $\psi(x, y)$ denotes the velocity potential due to the presence of the barrier. Then $\psi(x, y)$ satisfies the differential equation

$$\nabla^2\psi = 0 \qquad \text{in the fluid region} \tag{6.11}$$

with boundary conditions

$$\left(D\frac{\partial^4}{\partial x^4} + 1 - \varepsilon K\right)\psi_y + K\psi = 0 \quad \text{on} \quad y = 0 \tag{6.12}$$

and

$$\frac{\partial\psi}{\partial n}(x, y) = -\frac{\partial\phi_0}{\partial n}(x, y) \quad \text{for} \quad (x, y) \in \Gamma, \tag{6.13}$$

the edge condition

$$r^{1/2}\,\nabla\psi = O(1) \text{ as } r = \{x^2 + (y-a)^2\}^{1/2} \to 0, \tag{6.14}$$

condition at infinite depth

$$\nabla\psi \to 0 \text{ as } y \to \infty \tag{6.15}$$

and the infinity requirements

$$\psi(x, y) \to \begin{cases} R\phi_0(-x, y) \text{ as } x \to -\infty, \\ (T-1)\phi_0(x, y) \text{ as } x \to \infty. \end{cases} \tag{6.16}$$

so that $\psi(x, y)$ satisfies the radiation condition at infinity.

Let $G(x, y; \xi, \eta)$ denote the Green's function which satisfies

$$\nabla^2 G = 0 \text{ in the fluid region except at } (\xi, \eta)$$

$$G : \ln \rho \text{ as } \rho = \{(x-\xi)^2 + (y-\eta)^2\}^{1/2} \to 0$$

$$\left(D\frac{\partial^4}{\partial x^4} + 1 - \varepsilon K\right)G_y + KG = 0 \text{ on } y = 0$$

$$\nabla G \to 0 \text{ as } y \to \infty.$$

and
G satisfies the radiation condition as $|x - \xi| \to \infty$.

G can be constructed as Thorne (1953) and as given by

$$G(x,y;\xi,\eta) = \frac{1}{2}ln\frac{(x-\xi)^2+(y-\eta)^2}{(x-\xi)^2+(y+\eta)^2} - 2\int_0^\infty \frac{(Dk^4+1-\varepsilon K)e^{-k(y+\eta)}}{(Dk^4+1-\varepsilon K)k-K}\cos k(x-\xi)dk \tag{6.17}$$

where the contour in the integral is indented below the pole at $k = \lambda K$ on the real k-axis to take care of the radiation condition as $|x - \xi| \to \infty$. An alternative representation of $G(x, y; \xi, \eta)$ in which its behaviour as $|x - \xi| \to \infty$ is evident, is given by

$$G(x,y;\xi,\eta) = -2\int_0^\infty \frac{L(k,y)L(k,\eta)}{k\{(Dk^4+1-\varepsilon K)^2k^2+K^2\}}e^{-k|x-\xi|}dk$$

$$-2\pi i\frac{e^{-\lambda K(y+\eta)+i\lambda K|x-\xi|}}{\lambda(5D\lambda^4K^4+1-\varepsilon K)} - 2\pi i\frac{e^{-\lambda_1 K(y+\eta)+i\lambda_1 K|x-\xi|}}{\lambda_1(5D\lambda_1^4K^4+1-\varepsilon K)} - 2\pi i\frac{e^{-\overline{\lambda_1}K(y+\eta)-i\overline{\lambda_1}K|x-\xi|}}{\overline{\lambda_1}(5D\overline{\lambda_1}^4K^4+1-\varepsilon K)}. \tag{6.18}$$

By an appropriate use of Green's integral theorem to $\psi(x, y)$ and $G(x, y; \xi, \eta)$ in the fluid region, a representation of $\psi(\xi, \eta)$ is obtained as

$$\psi(\xi,\eta) = -\frac{1}{2\pi}\int_\Gamma F(q)\frac{\partial G}{\partial n_q}(x,y;\xi,\eta)ds_q. \tag{6.19}$$

where $q \equiv (x, y)$ is a point on Γ, $F(q)$ denotes the discontinuity of $\psi(x, y)$ across Γ (and hence the discontinuity of $\phi(x, y)$ across Γ) and $\frac{\partial}{\partial n_q}$ denotes the normal derivative at the point q on Γ. Thus by using (6.10) we find

$$\phi(\xi,\eta) = \phi_0(\xi,\eta) - \frac{1}{2\pi}\int_\Gamma F(q)\frac{\partial G}{\partial n_q}(x,y;\xi,\eta)ds_q. \tag{6.20}$$

Condition (6.6) on the barrier Γ, rewritten as

$$\frac{\partial \phi}{\partial n_q} = 0 \text{ on } \Gamma$$

where $p \equiv (\xi, \eta)$ is another point on Γ, produces the integro-differential equation

$$\frac{\partial}{\partial n_p}\int_\Gamma F(q)\frac{\partial G}{\partial n_q}(q;p)ds_q = 2\pi\frac{\partial\phi_0}{\partial n_p}(p),\ \ p\in\Gamma. \tag{6.21}$$

The order of differentiation and integration in (6.21) can be interchanged provided the integral is interpreted as a Hadamard finite part integral of order two. This leads to the hypersingular integral equation

$$\int_\Gamma F(q)\frac{\partial^2 G}{\partial n_p \partial n_q}(p;q)ds_q = 2\pi\frac{\partial\phi_0}{\partial n_p}(p),\ \ p\in\Gamma \tag{6.22}$$

where the integral is in the sense of Hadamard finite part integral of order 2. The end conditions to be satisfied by $F(q)$ are that it vanishes at the edge of Γ and as the point q goes to infinity on Γ.

To find the expression of the kernel of the integral equation (6.22), we note that the normals n_q and n_q at the points p and q on Γ are given by $n_p = n_q = (\cos\theta, -\sin\theta)$. Also the coordinates of the points $p, q \in \Gamma$ can be parametrically represented by

$$\xi = (u-a)\sin\theta,\ \eta = a + (u-a)\cos\theta,\ a < u < \infty,$$

$$x = (t-a)\sin\theta,\ y = a + (t-a)\cos\theta,\ a < t < \infty,$$

Thus we find

$$\frac{\partial^2 G}{\partial n_p \partial n_q}(p;q) = -\frac{1}{(t-u)^2} - L(u,t) \tag{6.23}$$

where

$$L(u,t) = -\frac{\cos 2\theta}{X^2+Y^2} + \frac{2\sin^2\theta(Z^2-Y^2)}{(X^2+Y^2)^2} + 2\frac{Y^2-X^2}{(X^2+Y^2)^2} + 2K\chi(u,t) \tag{6.24}$$

with

$$X = (t-u)\sin\theta,\ Y = (t+u)\cos\theta + 2a(1-\cos\theta),\ Z = (t-u)\cos\theta \tag{6.25}$$

$$\chi(u,t) = \int_0^\infty \frac{ke^{-kY}\cos kX}{k(Dk^4+1-\varepsilon K)-K}dk. \tag{6.26}$$

Thus the integral equation (6.22) becomes

$$\frac{1}{2\pi}\int_a^\infty\left[\frac{1}{(t-u)^2} + L(u,t)\right]f(t)dt = h(u),\ a \le u \le \infty \tag{6.27}$$

where

$$f(t) \equiv F(q(t)), \tag{6.28a}$$

$$h(u) = -\lambda K i e^{-\lambda K\{a+(u-a)e^{-i\theta}\}-i\theta} \tag{6.28b}$$

and $f(t)$ satisfies

$$f(t) \to 0 \quad \text{as } t \to a \text{ and } f(t) \to 0 \quad \text{as } t \to \infty. \tag{6.29}$$

Transferring u, t to r, s where

$$u = \frac{2a}{1+r}, \qquad t = \frac{2a}{1+s}, \qquad -1 \le r, s \le 1 \tag{6.30}$$

Eq. (6.27) becomes

$$\int_{-1}^{1} f_1(s)\left[\frac{1}{(s-r)^2} + L_1(r,s)\right] ds = h_1(r), -1 \le r \le 1 \tag{6.31}$$

where

$$f_1(s) = f\left(\frac{2a}{1+s}\right), \tag{6.32}$$

$$h_1(r) = \frac{4a\pi}{(1+r)^2} h\left(\frac{2a}{1+r}\right) \tag{6.33}$$

and

$$L_1(r,s) = \frac{4a^2}{(1+r)^2(1+s)^2} L\left(\frac{2a}{1+r}, \frac{2a}{1+s}\right). \tag{6.34}$$

The end conditions (6.29) become

$$f_1(\pm 1) = 0. \tag{6.35}$$

Thus we have to solve the equation (6.31) with the end conditions (6.35). For this, we write

$$f_1(s) = (1 - s^2)^{1/2} f_2(s) \tag{6.36}$$

where $f_2(s)$ is a bounded function in [−1, 1]. This ensures that the end conditions (4.13) are satisfied. We now approximate $f_2(s)$ as

$$f_2(s) \approx \sum_{n=0}^{N} a_n U_n(s), -1 \le s \le 1 \tag{6.37}$$

where $U_n(s)$ is the Chebyshev polynomial of the second kind and $a_n (n = 0,1,2 \ldots N)$ are unknown constants. Using the expansion (6.37) in (6.36) and substituting in (6.31), we obtain

$$\sum_{n=0}^{N} a_n A_n(r) = h_1(r), \qquad -1 \le r \le 1 \tag{6.38}$$

where

$$A_n(r) = -\pi(n+1)U_n(r) + \int_{-1}^{1} (1 - s^2)^{1/2} U_n(s) L_1(r, s) ds. \tag{6.39}$$

To find the unknown constants $a_n (n = 0,1,2,\ldots N)$, we put $r = r_j (j = 0,1,\ldots N)$, $-1 < r_j < 1$ in the relation (6.38) to obtain the linear system

$$\sum_{n=0}^{N} a_n A_{nj} = h_{1j}, \qquad j = 0,1,2 \cdots N \tag{6.40}$$

where

$$A_{nj} = A_n(r_j), \qquad h_{1j} = h_1(r_j). \tag{6.41}$$

The collocation points r_j can be chosen suitably. Here we have chosen

$$r_j = \cos\frac{(j+1)\pi}{N+2}, \qquad j = 0,1,2,\ldots N. \tag{6.42}$$

The linear system (6.40) is solved by any standard method to obtain the constants $a_n (n = 0,1, \ldots N)$ and thus the solution of the hypersingular integral equation (6.31) and hence (6.27) is obtained.

The reflection and transmission coefficients can be found approximately in terms of series involving the constants $a_n (n = 0,1, \ldots N)$ defined in the approximation above. This is achieved by making $\xi \to \mp\infty$ in $\phi(\xi, \eta)$ given in (6.20) and using the conditions at infinity given by (6.9) with (x, y) replaced by (ξ, η)). Thus we find the formulae

$$R = -\frac{Ke^{-i\theta}}{5D\lambda^4K^4 + 1 - \varepsilon K}\int_a^{\infty} f(t)e^{-\lambda K\{a+(t-a)\cos\theta\}+i\lambda K(t-a)\sin\theta}dt, \tag{6.43}$$

and

$$T = 1 - R. \tag{6.44}$$

Thus R, T are expressed in terms of integrals involving the discontinuity of the potential function across the barrier. Putting $t = \frac{2a}{1+s}$ and using (6.32) and (6.36) and approximation (6.37) we find approximately

$$R \simeq R_a = -\frac{2Kae^{-i\theta}}{5D\lambda^4K^4 + 1 - \varepsilon K}\sum_{n=0}^{N} a_n f_n \tag{6.45}$$

where

$$f_n = \int_{-1}^{1}\frac{(1-s)^{1/2}}{(1+s)^{3/2}}U_n(s)e^{-\lambda K\{a+(\frac{2a}{1+s}-a)\cos\theta\}+i\lambda K(\frac{2a}{1+s}-a)\sin\theta}ds \tag{6.46}$$

and

$$T \simeq T_a = 1 - R_a. \tag{6.47}$$

While the approximate analytical expressions for the reflection and transmission coefficients R and T given by (6.45) and (6.47), respectively are for any submerged barrier inclined with the downward vertical at an angle $\theta (0 \le \theta \le \pi/2)$, for simplicity, we present numerical results only for a vertical barrier, i.e., $\theta = 0$. Numerical values of $|R_a|$ and $|T_a|$ (with $\theta = 0$) are computed for different values of non-dimensionalised parameters $D' = D/a^4$, $\varepsilon' = \varepsilon/a$ and the wave number Ka. For numerical computation, we have used $N = 10$ in (6.38). The different integrals are evaluated using Mathematica. Almost the same numerical results are obtained if N is taken as 15. The energy identity $|R|^2 + |T|^2 = 1$ is used as a particular check on the correctness of the numerical results obtained here.

In Table 2.7, numerical estimates for $|R|$ are presented for different values of D' and ε' (e.g., D', ε' = 0.0001, 0.0001; 0.5, 0.01; 1.0, 0.01; 1.5, 0.01) against the wave

Table 2.7 Values of $|R|$

Ka	$D' = 0$ $\varepsilon' = 0$	$D' = 0.0001$ $\varepsilon' = 0.0001$	$D' = 0.5$ $\varepsilon' = 0.01$	$D' = 1$ $\varepsilon' = 0.01$	$D' = 1.5$ $\varepsilon' = 0.01$
0.1	0.610408	0.612669	0.611764	0.611179	0.610662
0.5	0.266668	0.266681	0.24406	0.231421	0.222682
1.	0.105264	0.105223	0.0866619	0.086297	0.0867835
1.5	0.0412931	0.0412542	0.0483265	0.0511354	0.0540978
2.	0.0159016	0.0158881	0.0317988	0.0371543	0.04059

number *Ka*. Also are presented, the exact numerical values of $|R|$ ($\equiv |R_0|$) obtained from the exact free surface result obtained by Ursell (1947). It is observed from this table that $|R|$ computed from (6.38) for very small values of D', $\varepsilon'(D' = 0.0001$, $\varepsilon' = 0.0001)$ and Ursell's exact free surface result coincide within 3 to 4 decimal places. This demonstrates the effectiveness of the numerical scheme based on hypersingular integral equation formulation of the problem.

Again, in Table 2.8, numerical estimates for $|T|$ are presented for the same set of values of D' and ε' used in Table 2.7. From this table it is observed that numerical estimates for $|T|$ computed from (6.40) for very small values of D', $\varepsilon'(D' = 0.0001$, $\varepsilon' = 0.0001)$ and from the Ursell's exact free surface result coincide within 3–4 decimal places in most cases. $|R|$ and $|T|$ are plotted in Fig. 2.38 and Fig. 2.40 respectively against the wave number *Ka* for the range of values of *Ka* between 0.1 and 1.0 taking $D', \varepsilon' = 0.0001, 0.0001; 0.5, 0.01; 1.0, 0.01; 1.5, 0.01$. From these figures it is observed that $|R|$ decreases with *Ka* while $|T|$ increases with *Ka*. In Figs. 2.39 and 2.41, $|R|$ and $|T|$ are plotted for same set of values of D' and ε' used in Fig. 2.38 and Fig. 2.40 for the range of values of *Ka* between 1.0 *and* 5.0. It is observed from these figures that $|R|$ increases with D' while $|T|$ decreases with D'. Also it is observed that for small values of D', $\varepsilon' = 0.0001, 0.0001$, the result almost coincides with the Ursell's (1947) free surface results.

Table 2.8 Values of $|T|$

Ka	$D' = 0$ $\varepsilon' = 0$	$D' = 0.0001$ $\varepsilon' = 0.0001$	$D' = 0.5$ $\varepsilon' = 0.01$	$D' = 1$ $\varepsilon' = 0.01$	$D' = 1.5$ $\varepsilon' = 0.01$
0.1	0.792087	0.79034	0.791041	0.791493	0.791892
0.5	0.963789	0.963785	0.96976	0.972854	0.974891
1.	0.994444	0.994449	0.996238	0.996269	0.996227
1.5	0.999147	0.999149	0.99891	0.998692	0.998536
2.	0.999874	0.999874	0.999494	0.99931	0.999176

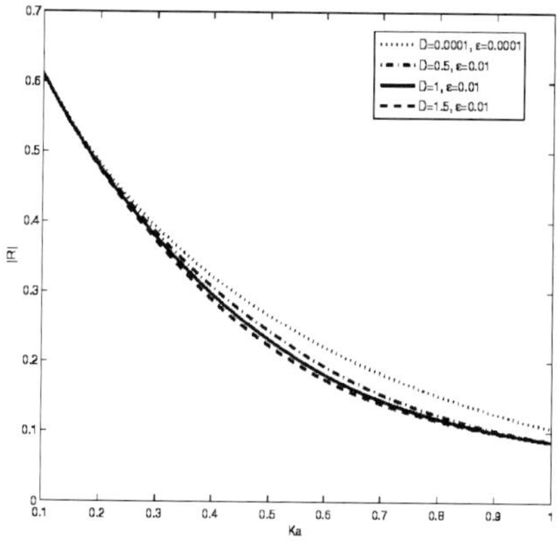

Fig. 2.38 Reflection coefficient

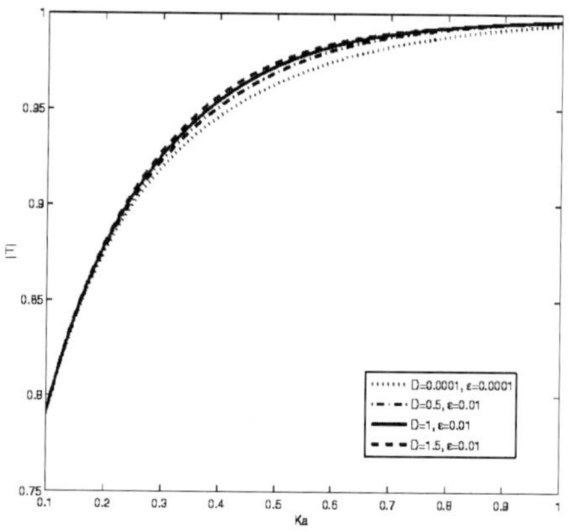

Fig. 2.39 Transmission coefficient

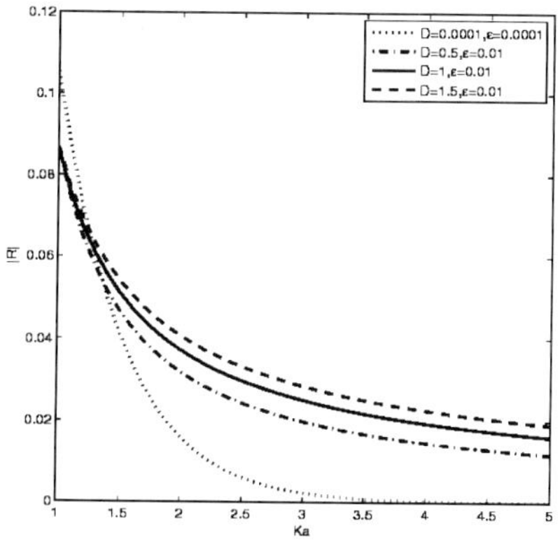

Fig. 2.40 Reflection coefficient

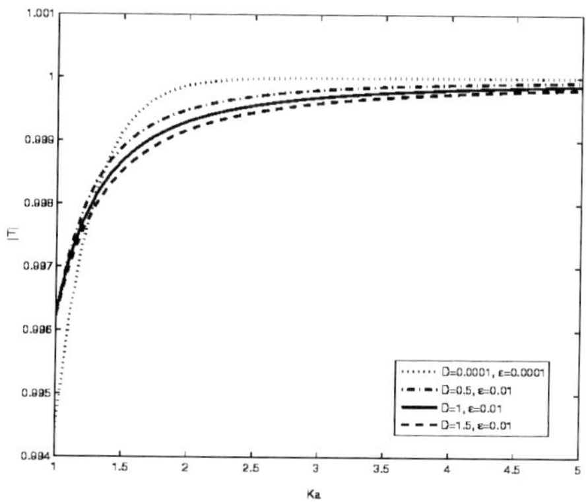

Fig. 2.41 Transmission coefficient

2.7 Scattering by a submerged thin vertical elastic plate

Within the framework of linearized theory of water waves, problems of wave scattering by thin vertical fixed rigid plates of some four basic configurations present in infinitely deep water admit of explicit solutions when a train of surface water waves is normally incident on the plates. Ursell (1947) employed Havelock's expansion of water wave potential to solve explicitly the problem of water wave scattering by a thin vertical rigid plate partially immersed in deep water or by a completely submerged thin vertical rigid barrier extending infinitely downwards. Evans (1970) used a method based on the theory of complex variable to solve the problem of wave scattering by a thin vertical rigid plate of finite vertical height submerged in deep water. Porter (1972) employed two methods, one based on a reduction procedure and the other based on an integral equation formulation, to solve the problem of wave scattering by a semi-infinite thin vertical rigid barrier, extending downwards from the free surface and containing a gap of arbitrary width at some depth. Both the methods reduce to solving the same Reimann-Hilbert problem of complex variable theory. As thin vertical rigid barriers are structurally simple in the engineering design of breakwaters constructed to protect a sheltered area by reflecting back the incident waves into the rough sea, the corresponding wave scattering problems were studied in the literature quite extensively for different situations employing various mathematical techniques. A good account of these techniques can be found in the monograph by Mandal and Chakrabarti (2000).

In the aforesaid mentioned problems, the plates or barriers are assumed to be rigid so that flexural response is not included. In many cases, a floating or submerged body is capable of considerable flexure. Thus, it is important to consider the interaction of water waves with a floating or submerged elastic body. Meylan (1995) generalized the surface-piercing thin vertical rigid plate problem of Ursell (1947) to a thin vertical elastic plate. He formulated the problem in terms of non-dimensional quantities that are non-dimensionalized using the wavelength of the incoming incident wave train. From the Bernoulli–Euler equation of the elastic plate, he derived the boundary condition on the plate as its normal velocity to be prescribed in terms of an integral involving the difference in velocity potentials across the plate multiplied by an appropriate Green's function. The reflection and transmission coefficients were then obtained in terms of integrals involving combinations of the unknown velocity potentials on the two sides of the plate, and the normal derivative of the potential on the plate satisfies three simultaneous linear integral equations that were solved numerically. The reflection and transmission coefficients were computed numerically for various values of different non-dimensional parameters and presented graphically in a number of figures. However, Ursell's (1947) results for a thin vertical rigid plate (particularly the curve for the reflection coefficient in Fig. 1 of Ursell (1947)) could not be retrieved from the figures for a free plate given by Meylan (1995) as the mass and stiffness parameters and the depth of the elastic plate were non-dimensionalized using the wavelength.

It is checked that, if the Meylan's (1995) problem is formulated without non-dimensionalizing different quantities as in Meylan (1995), but the same mathematical technique is followed, the simultaneous integral equations being solved numerically, and curves for reflection and transmission coefficients are drawn against the wave number for various values of different parameters, then Ursell's curves are retrieved for very small values of mass and stiffness parameters (non-dimensionalized using the depth of the plate) so that the plate becomes almost a non-elastic rigid plate. These curves for a surface-piercing thin vertical elastic plate however, are not presented here. Instead, the problem of wave scattering by a thin vertical elastic plate submerged in infinitely deep water is considered here and the method used by Meylan (1995) for the partially immersed elastic plate problem is followed as has been done by Chakraborty and Mandal (2014a). The energy balance relation (or the energy identity) connecting the unknown reflection and transmission coefficients for this problem is derived here by a straightforward application of Green's identity involving the complex valued velocity potential and its complex conjugate (cf. Chakrabarti and Martha (2009); Das et al. (2008)). All the data for the reflection and transmission coefficients computed here for various values of different parameters are seen to satisfy this energy identity.

The reflection and transmission coefficients are depicted graphically against the wave number for various values of the different parameters. A very good agreement is seen to have been achieved when the curves drawn for very small values of mass and stiffness parameters are compared with the figures given by Evans (1970) for a submerged rigid plate. Also, Ursell's (1947) curves are almost recovered when the mass and stiffness parameters and the depth of the upper submerged end of the plate are chosen to be very small (so that it almost becomes a partially immersed rigid plate).

We consider two-dimensional motion under the action of gravity only. Water is assumed to be inviscid, incompressible, and homogeneous liquid of density ρ. A rectangular Cartesian coordinate system (x,y,z) is chosen wherein y-axis is taken vertically downwards into the fluid region and the plane $y = 0$ represents the undisturbed free surface. The motion is independent of the coordinate z. A thin vertical elastic plate occupying the position $x = 0$, $a < y < b$, $-\infty < z < \infty$ is submerged in water. Assuming linear theory and the motion to be irrotational, it can be described by a velocity potential function $\Phi(x,y,t)$ where t denotes the time. Φ satisfies

$$\frac{\partial^2\Phi}{\partial x^2} + \frac{\partial^2\Phi}{\partial y^2} = 0 \text{ in the fluid region,} \tag{7.1}$$

the linearized free surface condition

$$\frac{\partial^2\Phi}{\partial t^2} = g\frac{\partial\Phi}{\partial y} \text{ on } y = 0 \tag{7.2}$$

where g is the acceleration due to gravity, and the bottom condition

$$\nabla\Phi \to 0 \text{ as } y \to \infty. \tag{7.3}$$

If $w(t)$ denotes the displacement of the elastic plate from its mean vertical position $x = 0$, then the Bernoulli–Euler equation of motion for the submerged thin vertical elastic plate is

$$\frac{\partial^2 w}{\partial t^2} + \mu^2\frac{\partial^4 w}{\partial y^4} = \frac{p_1 - p_2}{\rho' h'} \quad a < y < b \tag{7.4}$$

where ρ is the density of the material of the elastic plate and h is its small thickness,

$$p_{1,2} = p(\mp 0, y, t),\ a < y < b, \tag{7.5}$$

$p(x, y, t)$ being the pressure at a point (x, y) of the fluid region at time t,

$$\mu^2 = \frac{Eh'^2}{12\rho'(1 - \nu^2)} \tag{7.6}$$

where E is the Young's modulus and ν is the Poission's ratio of the material of the elastic plate. The pressure $p(x, y, t)$ is given by the Bernoulli's equation

$$\frac{\partial\Phi}{\partial t} = gy - \frac{p}{\rho}$$

so that

$$p_1 - p_2 = -\rho\left(\frac{\partial\Phi_1}{\partial t} - \frac{\partial\Phi_2}{\partial t}\right)$$

where

$$\Phi_{1,2} = \Phi(\mp 0, y, t), a < y < b. \tag{7.7}$$

Since

$$\frac{\partial w}{\partial t}(y,t) = \frac{\partial \Phi}{\partial x}(0, y, t), a < y < b, \tag{7.8}$$

we find from (7.4) the condition on the plate, after using (7.7) and (7.8), as

$$\frac{\partial^2 \Phi_x}{\partial t^2} + \mu^2 \frac{\partial^4 \Phi_x}{\partial y^4} = -\frac{\rho}{\rho' h'} \frac{\partial^2}{\partial t^2}(\Phi_1 - \Phi_2), a < y < b, \tag{7.9}$$

where

$$\Phi_x = \frac{\partial \Phi}{\partial x}(0, y, t), a < y < b. \tag{7.10}$$

If the two ends of the submerged elastic plate are assumed to be free, then the conditions at the two ends of the submerged plate are

$$\frac{\partial^2 \Phi_x}{\partial y^2} = 0 = \frac{\partial^3 \Phi_x}{\partial y^3} \text{ at } y = a, b. \tag{7.11}$$

If the motion in water is assumed to be time-harmonic with frequency σ, then $\Phi(x, y, t)$ can be expressed as $\text{Re}\varphi(x, y)e^{-i}$, and $\varphi(x, y)$ then satisfies the Laplace's equation

$$\frac{\partial^2 \varphi}{\partial x^2} + \frac{\partial^2 \varphi}{\partial y^2} = 0 \text{ in the fluid region,} \tag{7.12}$$

the free surface condition

$$K\varphi + \frac{\partial \varphi}{\partial y} = 0 \text{ on } y = 0 \tag{7.13}$$

and the bottom condition

$$\nabla \varphi \to 0 \text{ as } y \to \infty. \tag{7.14}$$

The condition (7.9) on the plate reduces to

$$D\frac{d^4 \varphi_x}{dy4} - \varepsilon K \varphi_x = K(\varphi_1 - \varphi_2) \text{ on } x = 0, a < y < b \tag{7.15}$$

where

$$D = \frac{Eh'^3}{12\rho(1-\nu^2)g}, \varepsilon = \frac{\rho'}{\rho}h', \; K = \frac{\sigma^2}{g}. \tag{7.16}$$

It may be noted that K and ε have the dimension of length while D has the dimension (length) 4. For the problem of a surface piercing elastic plate ($a = 0$ in (7.15)), Meylan non-dimensionalized x, y, b, ε, and D using the wave length λ $(= \frac{2\pi}{K})$.

The conditions (2.11) at the two ends of the submerged plate become

$$\frac{d^2\varphi_x}{dy^2} = 0 = \frac{d^3\varphi_x}{dy^3} \text{ at } y = a,\, b. \tag{7.17}$$

Now, if $\varphi_0(x, y)$ represents the velocity potential for a train of surface water waves incident on the plate from the direction of negative x-axis, then

$$\varphi_0(x, y) = e^{-Ky+iKx}. \tag{7.18}$$

For the scattering problem the velocity potential $\varphi(x, y)$ is required to satisfy the requirement that

$$\varphi(x, y) \to \begin{cases} \varphi_0(x, y) + R\varphi_0(-x, y) \text{ as } x \to -\infty, \\ T\varphi_0(x, y) \text{ as } x \to \infty \end{cases} \tag{7.19}$$

where R and T are the reflection and transmission coefficients respectively and our primary task is to determine these physical quantities.

The problem is to solve for $\varphi(x, y)$ satisfying (7.12), (7.13), (7.14), (7.15) and (7.19). The condition (7.15) on the plate $x = 0,\ a < y < b$, can be rewritten as

$$\frac{d^4\varphi_x}{dy^4} - \alpha^4\varphi_x = \frac{K}{D}(\varphi_1 - \varphi_2) \text{ on, } a < y < b \tag{7.20}$$

where

$$\alpha^4 = \frac{\varepsilon K}{D}. \tag{7.21}$$

We now construct a Green's function $g(\eta, y)$ satisfying the ordinary differential equation

$$\frac{d^4 g}{d\eta^4} - \alpha^4 g = \delta(\eta - y),\ a < y,\ \eta < b \tag{7.22}$$

with boundary conditions

$$\frac{d^2 g}{d\eta^2} = 0 = \frac{d^3 g}{d\eta^3} \text{ at } \eta = a,\, b, \tag{7.23}$$

continuity conditions

$$g,\ g_\eta,\ g_{\eta\eta} \text{ to be continuous at } \eta = y,$$

while

$$g_{\eta\eta\eta}(y + 0, y) - g_{\eta\eta\eta}(y - 0, y) = -1, \tag{7.24}$$

and g is to be symmetric in y and η. Then $g(\eta, y)$ is found to be

$$g(\eta, y) = \begin{cases} A_1 e^{i\alpha\eta} + B_1 e^{-i\alpha\eta} + C_1 e^{\alpha\eta} + D_1 e^{-\alpha\eta},\ a < \eta < y < b, \\ A_2 e^{i\alpha\eta} + B_2 e^{-i\alpha\eta} + C_2 e^{\alpha\eta} + D_2 e^{-\alpha\eta},\ a < y < \eta < b, \end{cases} \tag{7.25}$$

where $A_1, A_2, B_1, B_2, \ldots$ are functions of y only. Using the end conditions (7.23) and the matching and jump conditions (7.24), we find that A_1, B_1, C_1 and D_1 are determined from the linear system

$$\begin{bmatrix} -e^{i\alpha a} & -e^{-i\alpha a} & e^{\alpha a} & e^{-\alpha a} \\ -ie^{i\alpha a} & ie^{-i\alpha a} & e^{\alpha a} & -e^{-\alpha a} \\ -e^{i\alpha b} & -e^{-i\alpha b} & e^{\alpha b} & e^{-\alpha b} \\ -ie^{i\alpha b} & ie^{-i\alpha b} & e^{\alpha b} & -e^{-\alpha b} \end{bmatrix} \begin{bmatrix} A_1 \\ B_1 \\ C_1 \\ D_1 \end{bmatrix} = \frac{1}{4\alpha^3} \begin{bmatrix} 0 \\ 0 \\ -ie^{i\alpha(b-y)} + ie^{-i\alpha(b-y)} + e^{\alpha(b-y)} - e^{-\alpha(b-y)} \\ e^{i\alpha(b-y)} + e^{-i\alpha(b-y)} + e^{\alpha(b-y)} + e^{-\alpha(b-y)} \end{bmatrix} \tag{7.26}$$

while A_2, B_2, C_2 and D_2 are given by

$$A_2 = A_1 - \frac{i}{4\alpha^3} e^{-i\alpha y},\ B_2 = B_1 + \frac{i}{4\alpha^3} e^{i\alpha y},\ C_2 = C_1 - \frac{1}{4\alpha^3} e^{-\alpha y},\ D_2 = D_1 + \frac{1}{4\alpha^3} e^{\alpha y}. \tag{7.27}$$

Thus the condition (7.20) on the plate is reduced to

$$\varphi_x(0,\, y) = \frac{K}{D} \int_a^b g(\eta,\, y)(\varphi_1(\eta) - \varphi_2(\eta))d\eta,\ a < y < b. \tag{7.28a}$$

Also since $\varphi(x, y)$ is continuous across the gaps above and below the plate, we have

$$\varphi_1(y) = \varphi_2(y) \text{ for } 0 < y < a \text{ and } b < y < \infty \tag{7.28b}$$

and hence we can write

$$\varphi_x(0,\, y) = \frac{K}{D} \int_0^\infty g(\eta,\, y)(\varphi_1(\eta) - \varphi_2(\eta))d\eta, 0 < y < \infty. \tag{7.28c}$$

Now employing Green's integral theorem to the functions $\varphi(\xi, \eta) - e^{-K\eta + iK\xi}$ and $G(\xi, \eta; x, y)(x < 0)$ in the region in the (ξ, η)-plane bounded externally by the lines $\eta = 0, -X \le \xi \le 0$; $\xi = 0,\ 0 \le \eta \le Y$; $\eta = Y,\ -X \le \xi \le 0$; $\xi = -X,\ 0 \le \eta \le Y$ and internally by a circle of very small radius δ with center at (x, y) and ultimately making $X, Y \to \infty$ and $\delta \to 0$, we obtain

$$\varphi(x,\, y) = e^{-Ky + iKx} + \int_0^\infty \left[\frac{\partial G}{\partial \xi}(0,\ \eta; x,\, y)\varphi_1(\eta) - G(0,\ \eta; x,\, y)\varphi_x(\eta) \right] d\eta \text{ for } x < 0, \tag{7.29}$$

where

$$G(\xi,\ \eta; x,\, y) = \frac{1}{4\pi} \ln \frac{(\xi - x)^2 + (\eta - y)^2}{(\xi - x)^2 + (\eta + y)^2}$$
$$- \frac{1}{\pi} \int_0^\infty \frac{e^{-k|\xi - x|}}{k^2 + K^2} \{k \cos k(\eta + y) - K \sin k(\eta + y)\} dk - ie^{-K(\eta + y) + iK|\xi - x|}. \tag{7.30}$$

It may be noted that G is symmetric in (ξ, η) and (x, y) and behaves as an outgoing wave as $|\xi - x| \to \infty$. (7.29) gives a representation of $\varphi(x, y)$ for $x < 0$, and hence the reflection coefficient R is obtained by making $x \to -\infty$ in (7.29) and comparing with (7.19). This produces

$$R = \int_0^\infty \{K\varphi_1(\eta) - i\varphi_x(\eta)\}e^{-K\eta}d\eta. \tag{7.31}$$

If we choose $x = -0$, then in the region of use of the Green's integral theorem mentioned above, we have to take a half circle of small radius δ with center at $(-0, y)$, so that instead of (7.29) we obtain

$$\frac{1}{2}\varphi_1(y) = \frac{1}{2}e^{-Ky} + \int_0^\infty [\frac{\partial G}{\partial \xi}(0, \eta; 0, y)\varphi_1(\eta) - G(0, \eta; 0, y)\varphi_x(\eta)]d\eta, 0 < y < \infty. \tag{7.32}$$

This is in fact an integral equation for $\varphi_1(y)$ in $(0, \infty)$.

Similarly, applying Green's integral theorem to the functions $\varphi(\xi, \eta)$ and $G(\xi, \eta; x, y)$ $(x > 0)$ in the region in the (ξ, η)- plane bounded externally by the lines $\eta = 0, 0 \leq \xi \leq X$; $\xi = 0, 0 \leq \eta \leq Y$; $\eta = Y, 0 \leq \xi \leq X$; $\xi = X, 0 \leq \eta \leq Y$ and internally by a circle of very small radius δ with center at (x, y) and ultimately making $X, Y \to \infty$ and $\delta \to 0$, we obtain

$$\varphi(x, y) = \int_0^\infty [\frac{\partial G}{\partial \xi}(0, \eta; x, y)\varphi_2(\eta) - G(0, \eta; x, y)\varphi_x(\eta)]d\eta \text{ for } x > 0. \tag{7.33}$$

(7.33) gives a representation of $\varphi(x, y)$ for $x > 0$, and hence the reflection coefficient T is obtained by making $x \to \infty$ in (7.33) and comparing with (7.19). This gives

$$T = \int_0^\infty \{K\varphi_2(\eta) + i\varphi_x(\eta)\}e^{-K\eta}d\eta. \tag{7.34}$$

Again, if we choose $x = +0$, then in the region of the use of the Green's integral theorem we have to take a half circle of small radius δ with center at $(+0, y)$, so that instead of (7.33) we obtain

$$\frac{1}{2}\varphi_2(y) = -\int_0^\infty [\frac{\partial G}{\partial \xi}(0, \eta; 0, y)\varphi_2(\eta) - G(0, \eta; 0, y)\varphi_x(\eta)]d\eta, 0 < y < \infty. \tag{7.35}$$

This serves as an integral equation for $\varphi_2(y)$ in $(0, \infty)$. We can regard (7.32), (7.35), (7.28a) and (7.28b) together, i.e., (7.28c) as three simultaneous integral equations for the unknowns φ_1, φ_2 and φ_x in $(0, \infty)$. These integral equations are solved here numerically and the reflection and transmission coefficients $|R|$ and $|T|$ are then estimated numerically for various values of different parameters.

To derive the form of the energy identity for the present problem, the Green's integral theorem is employed to $\varphi(x, y)$ and its complex conjugate cf. Das et al. (2008) and Chakrabarti and Martha (2009) in the region bounded by the lines $y = 0$ $(-X \leq x \leq X)$; $x = X$ $(0 \leq y \leq Y)$, $y = Y$ $(-X \leq x \leq X)$, $x = -X$ $(0 \leq y \leq Y)$, and $x = \pm 0$ $a \leq y \leq b$ and making $X, Y \to \infty$ ultimately. The contribution from the line $x = X$ $(0 \leq y \leq Y)$ is

$$i\,|T|^2. \tag{7.36}$$

There is no contribution from the line $y = Y$ $(-X \leq x \leq X)$ because of the bottom condition (when $y \to \infty$) and also the line $y = 0$ $(-X \leq x \leq X)$ has no contribution because of the free surface condition.

The contribution from the line $x = -X(0 \le y \le Y)$ is

$$i(|R|^2-1). \tag{7.37}$$

Lastly the contribution from $x = \pm 0,\ a \le y \le b$ is

$$\frac{K}{D}\int_a^b\left[\int_a^b\left\{L^*(\eta, y)(\varphi_1(y)-\varphi_2(y)) - L(\eta, y)(\varphi^*_1(y)-\varphi^*_2(y))\right\}d\eta\right]dy \tag{7.38}$$

where $L(\eta, y) = g(\eta, y)(\varphi_1(\eta) - \varphi_2(\eta))$ and L^*, $\varphi_1{}^*, \varphi_2{}^*$ are the complex conjugates of the functions L, φ_1, φ_2 respectively.

Thus, combining all these results, the energy balance relation is obtained as

$$|T|^2 + |R|^2 + \kappa = 1 \tag{7.39}$$

where

$$\kappa = -\frac{Ki}{D}\int_a^b\left[\int_a^b\{L^*(\eta, y)(\varphi_1(y)-\varphi_2(y)) - L(\eta, y)(\varphi_1^*(y)-\varphi_2^*\ (y))\}d\eta\right]dy. \tag{7.40}$$

It may be noted that κ is real and positive and it occurs due to elastic property of the plate. It can be evaluated numerically using the Gauss quadrature formula repeatedly. For a rigid plate, the energy identity is $|R|^2 + |T|^2 = 1$, which readily follows from the above Green's integral theorem by noting

$$\varphi_x(\pm 0) = 0,\ a < x < b,$$

for a rigid plate. It is not apparent that κ in (7.39) vanishes if D and ε are made zero. However, this term becomes negligible numerically when the non-dimensional parameters D/b^4 and ε/b are made very small so as to make the elastic plate to be almost rigid. The presence of κ in the energy identity (7.39) is due to the elasticity of the plate. In fact when a surface wave is incident on the elastic plate, some energy is dissipated to deform the elastic plate. However for a rigid plate, no deformation occurs and the energy identity $|R|^2 + |T|^2 = 1$ is satisfied. Some representative numerical values of $|R|$, $|T|$, κ and the left side of (7.39) are given in Table 2.9 for different values of Kb with fixed values of a/b, ε/b, D/b^4. It becomes obvious form this table that the energy identity is satisfied. In fact, each data for $|R|$ and $|T|$ obtained later, is checked to satisfy this identity.

Table 2.9 Energy identity for $a/b = 0.001$, $D/b^4 = 2$, $\varepsilon/b = 0.0001$

Kb	$\|R\|$	$\|T\|$	κ	$\|R\|^2 + \|T\|^2 + \kappa$
0.1	0.003465	0.996535	0.00690616	0.999346
0.5	0.164863	0.835137	0.275367	0.992988
1.0	0.697152	0.302848	0.422262	0.992782
1.5	0.830982	0.169018	0.280902	0.992956
2.0	0.854596	0.145404	0.248524	0.993175
2.5	0.870957	0.129043	0.224781	0.993389
3.0	0.883684	0.116316	0.205573	0.993598

The integrals in (7.32), (7.35) and (7.28c) have the range $(0, \infty)$ and the integrands involve water wave potential and its derivative, which can be assumed to decay exponentially cf. Meylan (1995). Thus we can use Gauss Laguerre quadrature rule in the form cf. Ralston (1965)

$$\int_0^\infty f(\eta)d\eta \approx \sum_{j=1}^{n} w_j f(y_j) \tag{7.40}$$

in each of the integrals in (7.32), (7.35) and (7.28c), after non-dimensionalizing the variables y, η with respect to b. In (7.40), $y_j (j = 1,2, \ldots, n)$ are the zeros of Laguerre polynomial $L_n(y)$ of degree n and $w_j (j = 1,2, \ldots, n)$ are the corresponding weight functions given by

$$w_j = \frac{(j!)^2}{L_n'(y_j)L_{n+1}(y_j)},$$

$L_n'(y)$ being the derivative of $L_n(y)$. There will be $3n$ unknowns $\varphi_1(y_j)$, $\varphi_2(y_j)$ and $\varphi_x(y_j)$ $(j = 1, 2, \ldots, n)$ in the approximations of integrals in (7.32), (7.35) and (7.28c). Substituting $y = y_k (k = 1, 2, \ldots, n)$ in each of (7.32), (7.35) and (7.28c) we obtain $3n$ linear equations for the determination of $\varphi_1(y_j)$, $\varphi_2(y_j)$ and $\varphi_x(y_j)$ $(j = 1, 2, \ldots, n)$. These are used to evaluate R, T and κ from (7.31), (7.34) and (7.40) respectively by using the integration rule (7.41). It is observed that the energy identity (7.39) is always satisfied numerically. Evans (1970) investigated the scattering problem involving a rigid vertical plate submerged in infinitely deep water and he plotted the curves of $|R|$ against the wave number Kb in Fig. 2 of his paper, for different values of a/b. To check the correctness of our present numerical results, the curves of $|R|$ from Evans's results and the curves of $|R|$ from our results for very small values of stiffness and mass parameters (viz. $D/b^4 = 0.0001$, $\varepsilon/b = 0.0001$) (so that the plate becomes almost a rigid plate), are drawn side by side in Fig. 2.42. It is seen from this figure that Evans's curves

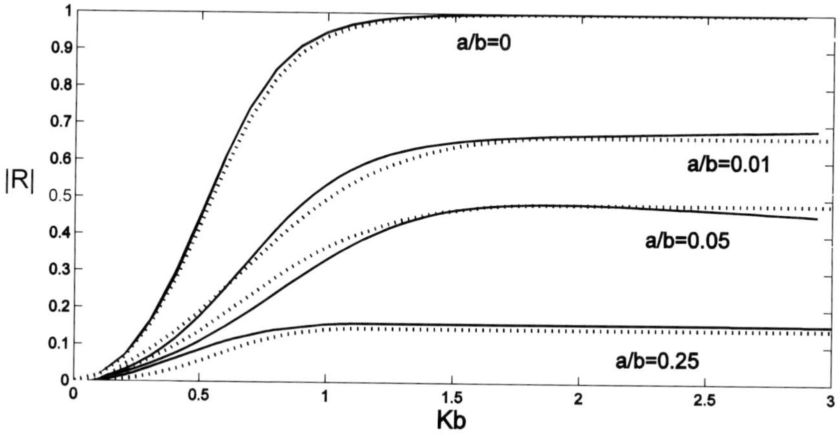

Fig. 2.42 Plot of $|R|$ against Kb for $\frac{\varepsilon}{b} = 0.0001$, $\frac{D}{b^4} = 0.0001$ (. . . . Evans method, ----- present method)

for $|R|$ almost coincide with our curves. Also the curve in this figure corresponding to $a/b = 0$, appears to have almost similar feature with the curve given by Ursell (1947) in Fig. 1 of his paper. In Figs. 2.43(a)–2.43(d), $|R|$ is depicted against the wave number Kb for different values of a/b (viz. (a) $a/b = 0$, (b) $a/b = 0.01$, (c) $a/b = 0.05$, (d) $a/b = 0.25$) and D/b^4 (viz. $D/b^4 = 1, 2, 3, 4$) and fixed ε/b (= 0.01). From these figures it is observed that $|R|$ decreases as a/b increases for fixed values of D/b^4 and ε/b. This is plausible since less energy is reflected as the depth of the upper edge of the plate increases. Again, for fixed a/b and ε/b, $|R|$ decreases as D/b^4 increases. Thus the effect of elasticity of the plate is to reduce the reflection of the incident wave energy. Curves for the transmission coefficient $|T|$ are depicted in Figs. 2.44, 2.45a to 2.45d for the same values of parameters a/b, D/b^4 and ε/b corresponding to the Figs. 2.42, 2.43a to 2.43d. Similar to the Fig. 1, the curves of $|T|$ from Evans's (1970) result cf. Fig. 2 of Evans (1970) and the curves of $|T|$ from our results for very small values of D/b^4 and ε/b, are drawn side by side in Fig. 2.44. It is seen that both the curves match with each other. From the Figs. 2.42 and 2.44, we conclude that for small values of stiffness and mass parameters, the curves of reflection and transmission coefficients of our problem almost coincide with the curves of the coefficients for rigid plate problem. Also the curve in Fig. 2.44 of $|T|$ for $a/b = 0$ almost similar with that given in Fig. 1 of Ursell

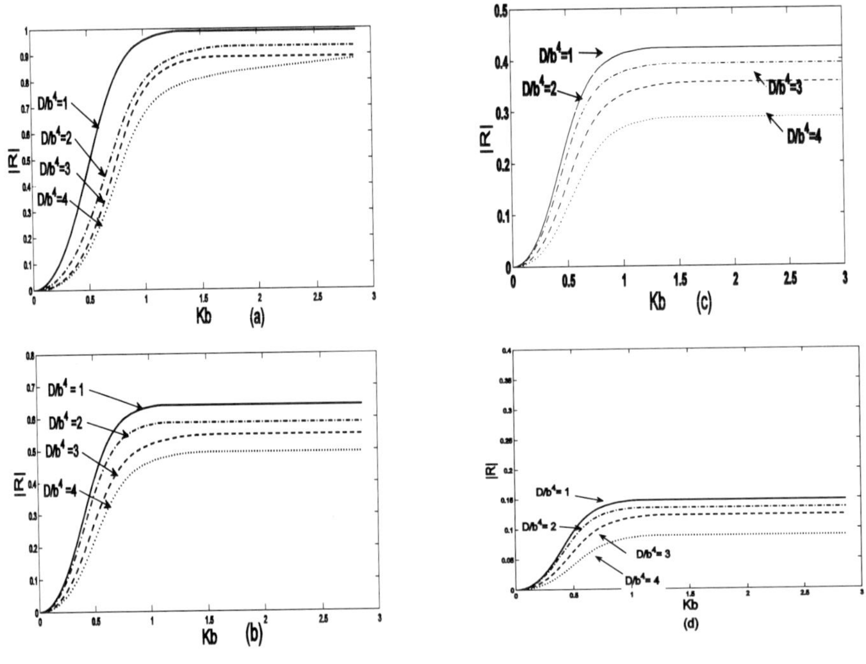

Fig. 2.43 Plot of $|R|$ against Kb for different values of $\frac{a}{b}$ (a) $\frac{a}{b} = 0.0001$, (b) $\frac{a}{b} = 0.01$, (c) $\frac{a}{b} = 0.05$, (d) $\frac{a}{b} = 0.25$ and $\frac{\varepsilon}{b} = 0.01$

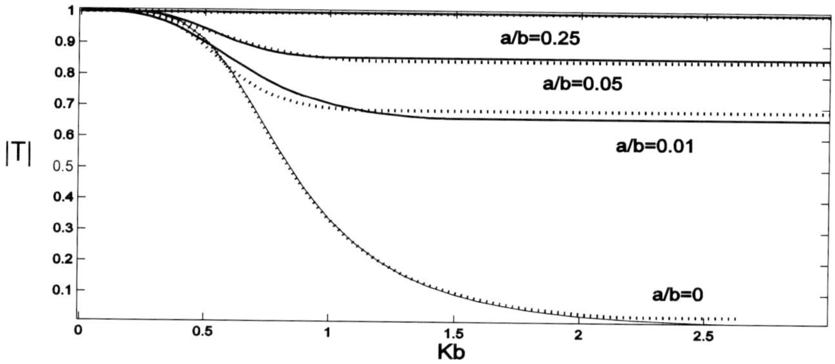

Fig. 2.44 Plot of $|T|$ against Kb for $\frac{\varepsilon}{b} = 0.0001$, $\frac{D}{b^4} = 0.005$ (· · · · Evans method ----- present method)

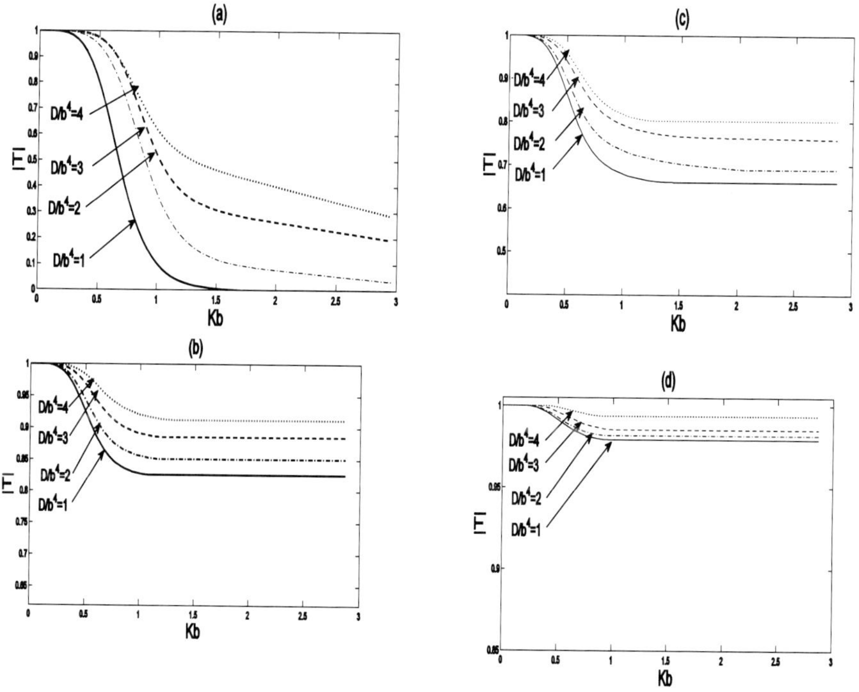

Fig. 2.45 Plot of $|T|$ against Kb for different values of $\frac{a}{b}$ (a) $\frac{a}{b} = 0.0001$, (b) $\frac{a}{b} = 0.01$, (c) $\frac{a}{b} = 0.05$ (d) $\frac{a}{b} = 0.25$ and $\frac{\varepsilon}{b} = 0.01$

(1947) for partially immersed rigid barrier. The curves 7.4(a)–7.4(d) depict $|T|$ against the non-dimensional wave number Kb for different values of D/b^4 (= 1, 2, 3, 4), (viz. (a) $a/b = 0$, (b) $a/b = 0.01$, (c) $a/b = 0.05$, (d) $a/b = 0.25$) and for fixed ε/b (= 0.01). For fixed a/b and ε/b, $|T|$ decreases as D/b^4 decreases.

CHAPTER III

Scattering by Rectangular Trench

The problem of water wave scattering by obstacles of various geometrical shapes present in infinitely deep water or in uniform finite depth water are being studied in the literature of linearized theory of water waves for more than last six decades by employing a variety of interesting mathematical techniques. The method of Galerkin approximations has been widely used to investigate water wave scattering problems involving thin vertical barriers cf. Porter and Evans (1995), Evans and Fernyhough (1995), Banerjea et al. (1996), Das et al. (1997) or thick vertical barriers with rectangular cross sections Kanoria et al. (1999), Mandal and Kanoria (2000). There is another important class of wave scattering problems involving water of variable depth in which the depth is constant except for variations over a finite interval. Kreisel (1949) first investigated wave propagation over variable depth geometries by reducing the fluid domain into a rectangular strip using an appropriate conformal mapping, and then converting the boundary value problem of the potential function to an integral equation which was solved by iteration. Mei and Black (1969) employed a variational formulation to obtain numerical estimates for the reflection co-efficient for water wave scattering by a bottom-standing thick rectangular barrier. Lassiter (1972) studied the complementary problem of scattering of water waves normally incident on a rectangular trench where the water depths before and after the trench are constants but not necessarily equal. He obtained the reflection and transmission coefficients after formulating the problem in terms of complementary variational integrals of Schwinger's type using the condition that the velocity potential and horizontal velocity must be continuous along the vertical lines before and after the trench. Lee and Ayer (1981) employed a matching procedure to solve the rectangular trench problem by writing solutions in two subregions comprising of an infinite rectangular region of constant depth and the finite rectangular region of the trench and obtained numerically the reflection and transmission coefficients. They have also conducted a series of laboratory experiments in a wave tank and compared the experimental results with the theoretical results. Shortly afterwards, Miles (1982) employed a conformal mapping algorithm to solve the trench problem for normal incidence of the wave train after formulating it for long waves. He also solved the problem for oblique incidence through variational formulation and obtained long wave limits for the reflection and transmission coefficients. Kirby and Dalrymple (1983) also investigated the trench problem for obliquely incident waves over an asymmetric trench for which the water depths in its two sides are unequal but constant. They solved the problem numerically by solving a set of integral equations derived by matching the truncated eigenfunction expansions for each subregion of constant depth along two vertical boundaries and also compared the numerical results with the data obtained from small-scale wave-tank experiments.

Different types of trench problems were considered by Bender and Dean (2003), Xie et al. (2011) and Liu et al. (2013). Bender and Dean (2003) investigated the problem of wave propagation over two-dimensional trenches and shoals. They employed three solution methods, namely, the step method valid in arbitrary depth of water, the slope method and the numerical method which are valid in the shallow water region. Xie et al. (2011) studied the problem of long-wave reflection by a rectangular obstacle with two scour trenches of power function profile and they obtained the reflection coefficient in closed form involving first and second kind Bessel functions. Also Liu et al. (2013) investigated wave reflection by a rectangular breakwater with two scour trenches by formulating the problem in terms of modified mild-slop equation (MMSE) and the reflection coefficient was obtained analytically.

In this chapter, we reinvestigate the problem of water wave scattering by a rectangular submarine trench by employing the multi-term Galerkin approximation method. Exploiting the geometrical symmetry of the rectangular trench about its center line taken as y-axis, the problem is split into two separate problems involving the symmetric and antisymmetric potential functions describing the resultant motion in the fluid region as was done by Kanoria et al. (1999) for the problem of water wave scattering by a thick vertical barrier of rectangular cross section having four different geometrical shapes. Use of eigenfunction expansions of the potential functions along with Havelock inversion formula followed by a matching process, produces integral equations for the corresponding unknown horizontal velocity components across the vertical line through the corner point of the trench. The integral equations are approximated by using multi-term Galerkin approximations involving ultraspherical Gegenbauer polynomials. Numerical estimates for the reflection and transmission coefficients are then obtained for various values of different parameters involved in the problem. These are seen to satisfy the energy identity. These coefficients obtained by the present method are also depicted graphically against the wave number. Some of the curves for these coefficients are compared with those found in the literature obtained by using other methods and also by laboratory experiments cf. Lee and Ayer (1981). Also, numerical results for the reflection and transmission coefficients are tabulated to compare with corresponding results given by Kirby and Dalrymple (1983). A very good agreement in both the cases is seen to have been achieved. These establish the correctness of the numerical results obtained here. It is also found that the width of the trench affects the reflection and transmission coefficients significantly and there exists an infinite number of discrete wave frequencies at which waves are completely transmitted as was also observed by Lee and Ayer (1981). We consider first the case of normally incident waves and then obliquely incident waves. These have been considered by Chakraborty and Mandal (2014b, 2015).

3.1 Normally incident waves

A rectangular submarine trench of width $2b$ is present at the bottom of an ocean of uniform finite depth h and the depth of the trench from the mean free surface is c (cf. Fig. 3.1). Water is assumed to be a homogeneous, inviscid and incompressible fluid and the motion in the fluid is two-dimensional and under the action of gravity only. Let (x, y) denote a Cartesian co-ordinate system where the y-axis is taken vertically downwards into the fluid region and $y = 0$ is the undisturbed free surface. Under the assumption of the linearized theory of water waves, a normally incident wave train is described by the velocity potential function $\text{Re}\{\varphi^{\text{inc}}(x, y)e^{-i\sigma t}\}$ where

$$\varphi^{\text{inc}} = \frac{2\cosh k_0(h-y)e^{-ik_0(x-b)}}{\cosh k_0 h}, \tag{1.1}$$

k_0 being the real positive root of the transcendental equation

$$k \tanh kh = K, \; K = \frac{\sigma^2}{g}, \tag{1.2}$$

σ being the circular frequency of the incoming wave train, g being the acceleration due to gravity. Let the resulting motion in the fluid be described by the velocity potential $\text{Re}\{\varphi(x, y)e^{-i\sigma t}\}$, then $\varphi(x, y)$ satisfies the boundary value problem,

$$\nabla^2\varphi = 0 \text{ in the fluid region} \tag{1.3}$$

$$K\varphi + \varphi_y = 0 \text{ on } y = 0, |x| < \infty \tag{1.4}$$

$$\varphi_x = 0 \text{ on } x = \pm b, y \in (h, c), (c > h) \tag{1.5}$$

$$r^{\frac{1}{3}}\nabla\varphi \text{ is bounded as } r \to 0, \tag{1.6}$$

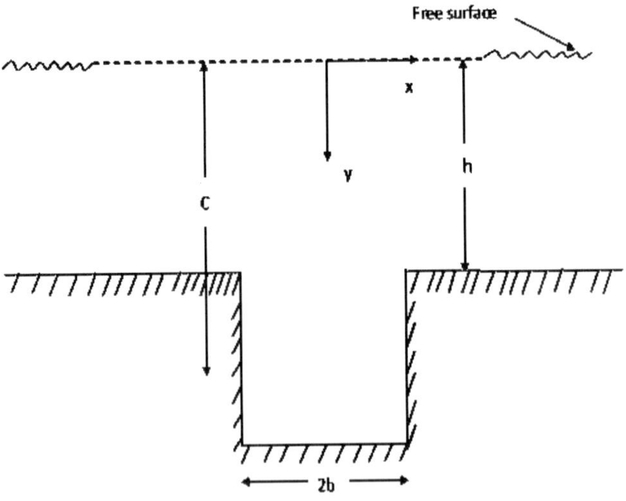

Fig. 3.1 Schematic diagram of the submarine trench

r being the distance from a submerged edge of the trench,

$$\varphi_y = 0 \text{ on } y = h,\ |x| > b, \tag{1.7}$$

$$\varphi_y = 0 \text{ on } y = c,\ |x| < b, \tag{1.8}$$

and finally

$$\varphi(x, y) \sim \begin{cases} \varphi^{inc}(x, y) + R\varphi^{inc}(-x, y) \text{ as } x \to \infty, \\ T\varphi^{inc}(x, y) \text{ as } x \to -\infty \end{cases} \tag{1.9}$$

where R and T are the unknown reflection and transmission coefficients respectively and are to be determined.

Due to geometrical symmetry of the rectangular trench about $x = 0$, we spilt the potential function $\varphi(x, y)$ into symmetric and antisymmetric parts $\varphi^s(x, y)$ and $\varphi^a(x, y)$ respectively, so that

$$\varphi(x, y) = \varphi^s(x, y) + \varphi^a(x, y) \tag{1.10}$$

where

$$\varphi^s(-x, y) = \varphi^s(x, y) \text{ and } \varphi^a(-x, y) = -\varphi^a(x, y). \tag{1.11}$$

Thus, we may restrict our analysis to the region $x \geq 0$ only. Now $\varphi^s(x, y)$ and $\varphi^a(x, y)$ satisfy the equations (1.3) to (1.8) together with

$$\varphi^s_x(0, y) = 0 \text{ and } \varphi^a(0, y) = 0,\ 0 < y < c. \tag{1.12}$$

Let the behavior of $\varphi^{s,a}(x, y)$ for large x be represented by

$$\varphi^{s,a}(x, y) \sim \frac{\cosh k_0(h-y)}{\cosh k_0 h}\{e^{-ik_0(x-b)} + R^{s,a}e^{ik_0(x-b)}\} \text{ as } x \to \infty \tag{1.13}$$

where R^s and R^a are unknown constants. Using (1.9) we find that R^s and R^a are related to R and T by

$$R,\ T = \frac{1}{2}(R^s \pm R^a)e^{-2ik_0 b}. \tag{1.14}$$

Now the eigenfunction expansions of $\varphi^{s,a}(x, y)$ satisfying the equations (1.3) to (1.5), (1.7), (1.8), (1.12) and (1.13) (for $x > b$) are given below.

Region I ($x > b$, $0 < y < h$):

$$\varphi^{s,a}(x, y) = \frac{\cosh k_0(h-y)}{\cosh k_0 h}\{e^{-ik_0(x-b)} + R^{s,a}e^{ik_0(x-b)}\} + \sum_{n=1}^{\infty} A_n^{s,a} \cos k_n(h-y)e^{-k_n(x-b)} \tag{1.15}$$

where k_n ($n = 1, 2, \ldots$) are the real positive roots of the equation

$$k \tan kh + K = 0 \tag{1.16}$$

Region II $(0 < x < b, 0 < y < c)$:

$$\begin{pmatrix} \varphi^s(x,y) \\ \varphi^a(x,y) \end{pmatrix} = \begin{pmatrix} B_0^s \cos\alpha_0 x \\ B_0^a \sin\alpha_0 x \end{pmatrix} \frac{\cosh\alpha_0(c-y)}{\cosh\alpha_0 c} + \sum_{n=1}^{\infty} \begin{pmatrix} B_n^s \cos\alpha_n x \\ B_n^a \sin\alpha_n x \end{pmatrix} \cos\alpha_n(c-y) \quad (1.17)$$

where $\pm\alpha_0$, $\pm i\alpha_n (n = 1, 2, ...)$ are the roots of the equation

$$\alpha \tanh \alpha c = K. \quad (1.18)$$

Now we have the matching conditions

$$\varphi_x^{s,a}(b+0,y) = \varphi_x^{s,a}(b-0,y),\ 0 < y < h. \quad (1.19)$$

We define

$$\varphi^{s,a}(b+0,y) = f^{s,a}(y),\ 0 < y < h, \quad (1.20)$$

$$\varphi_x^s(b-0,y) = g^{s,a}(y),\ 0 < y < h, \quad (1.21)$$

then

$$g^{s,a}(y) = \begin{cases} f^{s,a}(y), & 0 < y < h, \\ 0 & h < y < c. \end{cases} \quad (1.22)$$

Due to the edge condition (1.6), we find that

$$f^{s,a}(y) = O(|y-h|^{-\frac{1}{3}}) \text{ as } y \to h. \quad (1.23)$$

Using the expansions of $\varphi^{s,a}(x, y)$ in (1.15) into (1.20), we obtain

$$\frac{ik_0 \cosh k_0(h-y)}{\cosh k_0 h}\{R^{s,a} - 1\} + \sum_{n=1}^{\infty} A_n^{s,a}(-k_n)\cos k_n(h-y) = f^{s,a}(y),\ 0 < y < h. \quad (1.24)$$

Using Havelock's inversion formula in (1.24), we find

$$1 - R^{s,a} = \frac{4i}{\delta_0}\cosh k_0 h \int_0^h f^{s,a}(y) \cosh k_0(h-y)dy, \quad (1.25)$$

with $\delta_0 = 2k_0h + \sinh 2k_0h$ and

$$A_n^{s,a} = -\frac{4}{\delta_n}\int_0^h f^{s,a}(y) \cos k_n(h-y)dy \quad (1.26)$$

with $\delta_n = 2k_nh + \sin 2k_nh$ $(n = 1,2, ...)$.

Now using (1.17) into (1.21) we obtain

$$B_0^{s,a}\alpha_0(-\sin\alpha_0 b,\ \cos\alpha_0 b)\cosh\alpha_0(c-y)\cosh\alpha_0 c + \sum_{n=1}^{\infty}(\sinh\alpha_n b,\ \cosh\alpha_n b)B_n^{s,a}\alpha_n\cos\alpha_n(c-y)$$
$$= g^{s,a}(y), 0 < y < c. \quad (1.27)$$

Applying Havelock's inversion formula and noting (1.22), we obtain

$$B_0^{s,a} = \frac{4\cosh\alpha_0 c}{\gamma_0(-\sin\alpha_0 b, \cos\alpha_0 b)} \int_0^h f^{s,a}(y)\cosh\alpha_0(c-y)dy$$

with

$$\gamma_0 = 2\alpha_0 c + \sinh 2\alpha_0 c. \tag{1.28}$$

and

$$B_n^{s,a} = \frac{4}{\gamma_n(\sinh\alpha_n b, \cosh\alpha_n b)} \int_0^h f^{s,a}(y)\cos\alpha_n(c-y)dy$$

with

$$\gamma_n = 2\alpha_n c + \sinh 2\alpha_n c. \tag{1.29}$$

Now matching of $\varphi^{s,a}(x, y)$ across the line $x = b$ gives

$$\frac{\cosh k_0(h-y)}{\cosh k_0 h}\{1 + R^{s,a}\} + \sum_{n=0}^{\infty} A_n^{s,a}\cos k_n(h-y) = B_0^{s,a}(\cos\alpha_0 b, \sin\alpha_0 b)\frac{\cosh\alpha_0(c-y)}{\cosh\alpha_0 c}$$
$$+\sum_{n=1}^{\infty}(\cosh\alpha_n b, \sinh\alpha_n b)B_n^{s,a}\cos\alpha_n(c-y), 0 < y < h \tag{1.30}$$

which ultimately produces the first kind integral equations

$$\int_0^{\infty} F^{s,a}(u)M^{s,a}(y, u)du = \frac{\cosh k_0(h-y)}{\cosh k_0 h}, 0 < y < h, \tag{1.31}$$

where

$$F^{s,a}(y) = \frac{4\cosh^2 k_0 h}{\delta_0(1 + R^{s,a})} f^{s,a}(y), 0 < y < h, \tag{1.32}$$

and

$$M^{s,a}(y, u) = \frac{\delta_0}{\cosh^2 k_0 h}\sum_{n=1}^{\infty}\Big[\frac{\cos k_n(h-y)\cos k_n(h-u)}{\delta_n} + (\coth\alpha_n b, \tanh\alpha_n b)\frac{\cos\alpha_n(c-y)\cos\alpha_n(c-u)}{\gamma_n}$$
$$+(-\cot\alpha_0 b, \tan\alpha_0 b)\frac{\cosh\alpha_0(c-y)\cosh\alpha_0(c-u)}{\gamma_0}\Big], 0 < y, u < h. \tag{1.33}$$

It is obvious that $M^{s,a}(y, u)(0 < y, u < h)$ are real and symmetric in y and u. Now if we define

$$C^{s,a} = -i\frac{1 - R^{s,a}}{1 + R^{s,a}}, \tag{1.34}$$

then by using (1.25) into (1.31), we obtain

$$C^{s,a} = \int_0^h F^{s,a}(y)\frac{\cosh k_0(h-y)}{\cosh k_0 h}dy, \tag{1.35}$$

and $F^{s,a}(y)$ and $C^{s,a}$ are real quantities. Thus if the integral equations (1.31) are solved, then these solutions can be used to evaluate $C^{s,a}$ from the relations (1.35), and these produce the actual reflection and transmission coefficients $|R|$ and $|T|$ by using

$$|R| = \frac{|1+C^s C^a|}{\Delta} \text{ and } |T| = \frac{|C^s - C^a|}{\Delta} \tag{1.36}$$

with

$$\Delta = \{1 + (C^s)^2 + (C^s C^a)^2\}^{\frac{1}{2}} \tag{1.37}$$

which are obtained from equations (1.34) and (1.14).

Now we consider the Galerkin approximation method to solve the integral equations (1.31). The unknown functions $F^{s,a}(y)$ are approximated as

$$F^{s,a}(y) \approx \mathcal{F}^{s,a}(y),\ 0 < y < h \tag{1.38}$$

where $\mathcal{F}^{s,a}(y)$ are the multi-term Galerkin expansions in terms of suitable basis functions given by

$$\mathcal{F}^{s,a}(y) = \sum_{n=0}^{N} a_n^{s,a} f_n^{s,a}(y) \tag{1.39}$$

$a_n^{s,a}$ being unknown constants. Following similar arguments as given in Kanoria et al. (1999), the basis functions $f_m^{s,a}(y)$ are found to be

$$f_m^{s,a}(y) = -\frac{d}{dy}[e^{-ky}\int_y^h \hat{f}_m(t)dt],\ 0 < y < h, \tag{1.40}$$

where $\hat{f}_m(y)$ is chosen in terms of ultraspherical Gegenbauer polynomials of order 1/6 as

$$\hat{f}_m(y) = \frac{2^{\frac{7}{6}}\Gamma(\frac{1}{6})(2m)!}{\pi\Gamma(2m+13)h^{\frac{1}{3}}(h^2-y^2)^{\frac{1}{3}}} C_{2m}^{\frac{1}{6}}(\frac{y}{h}),\ 0 < y < h. \tag{1.41}$$

Now using (1.41) into (1.40) and substituting these in to (1.39), we get the approximate forms of $F^{s,a}(y)$. Using these approximate forms in (1.31), multiplying both sides by $f_m^{s,a}(y)$ and integrating over $(0, h)$, we obtain the linear systems

$$\sum_{n=0}^{N} a_n^{s,a} K_{mn}^{s,a} = d_m^{s,a}\ m = 0,1, \ldots, N \tag{1.42}$$

where

$$K_{mn}^{s,a} = \int_0^h\int_0^h M^{s,a}(y, u) f_n^{s,a}(u) f_m^{s,a}(y)du\, dy,\ m, n = 0,1,2, \ldots, N, \tag{1.43}$$

and

$$d_m^{s,a} = \int_0^h \frac{\cosh k_0(h-y)}{\cosh k_0 h} f_m^{s,a}(y)dy,\ m = 0,1, \ldots, N. \tag{1.44}$$

The integrals (1.43) and (1.44) can be evaluated explicitly, as in Kanoria et al. (1999) by using the different properties and standard results on Gegenbauer polynomials. Thus we find

$$K_{mn}^{s,a} = \frac{\delta_0}{\cosh^2 k_0 h}[4(-1)^{m+n}\sum_{r=1}^{\infty}\frac{\cos^2 k_r h}{\delta_r (k_r h)^{\frac{1}{3}}}(J_{2m+\frac{1}{6}}(k_r h)J_{2n+\frac{1}{6}}(k_r h)$$

$$+\frac{(\coth\alpha_r b, \tanh\alpha_r b)}{\gamma_r(\alpha_r h)^{\frac{1}{3}}}\cos 2\alpha_r c J_{2m+\frac{1}{6}}(\alpha_r h)J_{2n+\frac{1}{6}}(\alpha_r h)$$

$$+\frac{(-\cot\alpha_0 b, \tan\alpha_0 b)}{\gamma_0(\alpha_0 h)^{\frac{1}{3}}}\cosh^2\alpha_0 c\, I_{2m+\frac{1}{6}}(\alpha_0 h)I_{2n+\frac{1}{6}}(\alpha_0 h)\}],\ m,\ n = 0,1,\ ...,N, \tag{1.45}$$

and

$$d_m^{s,a} = \frac{I_{2m+\frac{1}{6}}(k_0 h)}{(k_0 h)^{\frac{1}{6}}},\ m = 0,1,\ ...,N. \tag{1.46}$$

The constants $a_n^{s,a}$ $(n = 0,1, ..., N)$ are now obtained by solving the linear systems (1.42), and then the relations (1.35) produce

$$C^{s,a} = \sum_{n=0}^{N} a_n^{s,a} d_n^{s,a} \tag{1.47}$$

so that $C^{s,a}$ are now found and $|R|$ and $|T|$ are evaluated from the relations (1.36).

To solve the linear system (1.42) numerically we need to truncate the infinite series involving $K_{mn}^{s,a}$ (equation (1.36)). As in Kanoria et al. (1999), here also a six-figure accuracy is achieved by taking 200 terms in each series of $K_{m,n}^{s,a}$. Also the accuracy can be further increased by taking more terms in the series (1.36). A representative set of numerical estimates of $|R|$ is given in Table 3.1, taking $N = 1, 2, 3, 4$ and 5 in the $(N+1)$-term Galerkin approximations for different values of the non-dimensional wave number Kh and for some particular values of non dimensional parameter b/h and c/h. From this table it is seen that the results for $|R|$ converges very rapidly with N. For $N \geq 2$, an accuracy of almost five decimal places is achieved. Thus the present procedure for the numerical computation of $|R|$ (and also $|T|$) is quite efficient.

Table 3.1 Reflection coefficient against Kh for $b/h = 1.2$; $c/h = 1.5$

Kh	*N*=1	*N*=2	*N*=3	*N*=4	*N*=5
0.2	0.530422	0.530441	0.530591	0.530461	0.530463
0.6	0.238629	0.238631	0.238694	0.238705	0.238707
1.0	0.159541	0.159554	0.159564	0.159569	0.159572
1.4	0.106001	0.106050	0.106107	0.106119	0.106127
1.8	0.050003	0.050011	0.050012	0.050051	0.050005

The reflection and transmission coefficients $|R|$ and $|T|$ are evaluated numerically for different values of the parameters b/h, c/h and the wave number Kh. It is observed that the energy identity $|R|^2 + |T|^2 = 1$ is always satisfied numerically. A representative set of values of $|R|$ and $|T|$ and $|R|^2 + |T|^2$ for $b/h = 1.2$, $c/h = 1.2$ and $Kh = 0.1, 0.3, 0.5, 0.7$ and 0.9 are given in Table 3.2. This table provides a partial check on the correctness of the results obtained by the present method.

The numerical values of $|R|$ and $|T|$ estimated by the present method with those given by Kirby and Dalrymple (1983)] for $c/h = 3$, $b/h = 5$ and $k_0h = 0.341, 0.723, 1.296$ ($k_0h = k_1h_1$) in Kirby and Dalrymple (1983)) obtained by using the eigenfunction matching method are displayed in Table 3.3 for the purpose of direct comparison. From this table, it is observed the results coincide up to 2 to 3 decimal places. This also provides another check on the correctness of the results.

Table 3.2 Energy identity for $b/h = 1.2$; $c/h = 1.2$

Kh	$\|T\|$	$\|R\|$	$\|R\|^2+\|T\|^2$
0.1	0.872057	0.489404	1.0000
0.3	0.997393	0.0721599	1.000
0.5	0.989156	0.146869	1.000
0.7	0.996946	0.0780884	1.000
0.9	0.999952	0.00984322	1.000

Table 3.3 Transmission and Reflection co-efficient for $c/h = 3$, $b/h = 5$

	Kirby and Dalrymple		**present method**	
k_0h	$\|T\|$	$\|R\|$	$\|T\|$	$\|R\|$
0.341	0.8881	0.4596	0.888926	0.469061
0.723	0.9552	0.2960	0.9560	0.2956
1.296	0.9995	0.0306	0.9985	0.0312

We have also compared our results with those given by Lee and Ayer (1981) obtained by a different method. Lee and Ayer (1981) plotted |R| and |T| against the ratio of the depth of water and incident wave length h/λ ($\equiv k_0h/2\pi$ here) in their Figs. 2 to 7 for different values of d (the depth of the trench below the uniform finite depth h, here d $(+(c-h))$, and l (length of the trench, here $l = 2b$). In their paper they have considered dimensional quantities (dimension of length in inches). In their Fig. 2, Lee and Ayer (1981) presented $|R|$ against h/λ ($\equiv k_0h/2\pi$) for $h = 4$", $d(\equiv c - h) = 4$", $l(\equiv 2b) = 20$" so that $c/h = 2$, $b/h = 2.5$. In Fig. 3.2 here, $|R|$ is depicted against $k_0h/2\pi$ for different values of the ratios c/h and b/h obtained from Lee and Ayer's (1981) data. For better comparison the curve for $|R|$ given in Fig. 2 of Lee and Ayer (1981), is displayed in Fig. 3.2 here together with the curve for $|R|$ drawn by the present method. These two curves of $|R|$ match quite well with each other.

Again in Fig. 3 of Lee and Ayer (1981), $|T|$ is depicted against h/λ for $h = 4''$, $c - h = 26.5''$, $2b = 21.125''$. This is also displayed in Fig. 3.3 here wherein $|T|$ (calculated by the present method) is also depicted against $k_0h/2\pi$ for $c/h = 7.625$, $b/h = 2.640625$, and the two curves match quite well. Also the curve in Fig. 4 of Lee and Ayer (1981) depicting $|R|$ against h/λ for the same values of h, $c - h$ and $2b$ is displayed in Fig. 3.4 along with $|R|$ (calculated by the present method) against $k_0h/2\pi$ for the same values of c/h and b/h as in Fig. 3.3. The two curves also match very well.

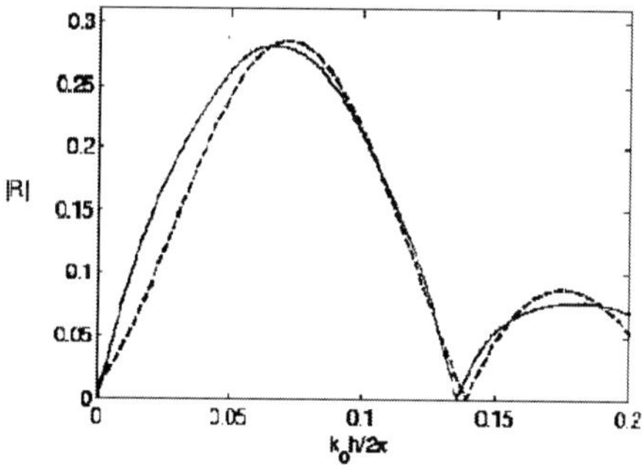

Fig. 3.2 $|R|$ against $k_0h/2\pi$ with $c/h = 2$, $b/h = 2.5$ (solid line Lee and Ayer, dotted line present method)

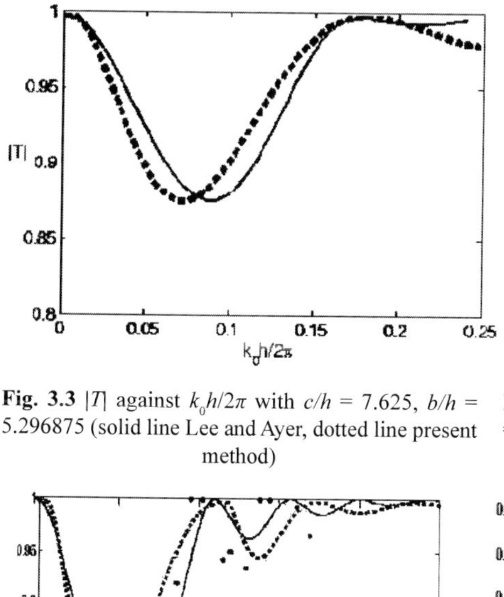

Fig. 3.3 $|T|$ against $k_0h/2\pi$ with $c/h = 7.625$, $b/h = 5.296875$ (solid line Lee and Ayer, dotted line present method)

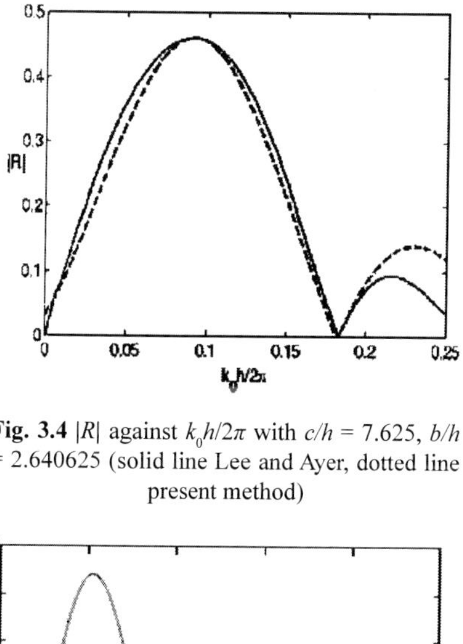

Fig. 3.4 $|R|$ against $k_0h/2\pi$ with $c/h = 7.625$, $b/h = 2.640625$ (solid line Lee and Ayer, dotted line present method)

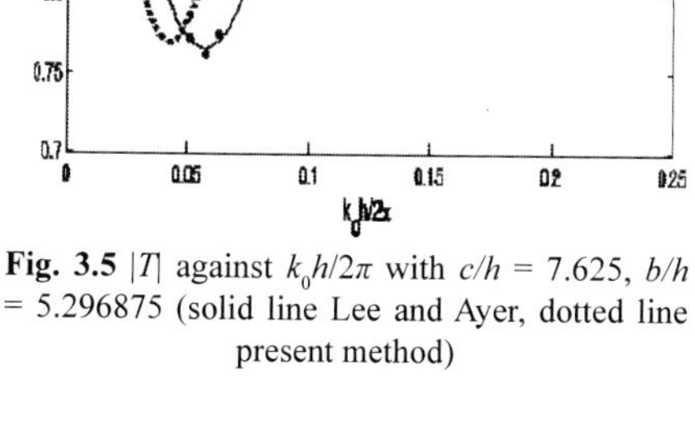

Fig. 3.5 $|T|$ against $k_0h/2\pi$ with $c/h = 7.625$, $b/h = 5.296875$ (solid line Lee and Ayer, dotted line present method)

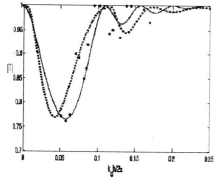

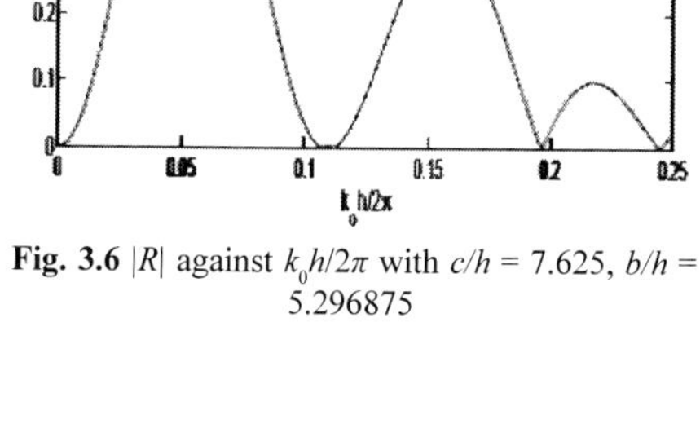

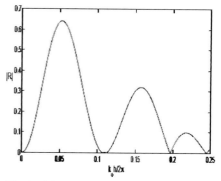

Fig. 3.6 $|R|$ against $k_0h/2\pi$ with $c/h = 7.625$, $b/h = 5.296875$

Finally, the curve and the experimental results in Fig. 5 of Lee and Ayer (1981) depicting $|T|$ against h/λ for $h = 4''$, $c-h = 26.5''$, $2b = 42.375''$ are displayed in Fig. 3.5 in which $|T|$, calculated by the present method, is also depicted against $k_0h/2\pi$ for values of c/h and b/h obtained from Lee and Ayer's data for their Fig. 5, i.e., $c/h = 7.625$, $b/h = 5.296875$. The two curves and the experimental data of Lee and Ayer (1981) match more or less satisfactorily. The idea to draw different figures keeping h, $c - h$ fixed and increasing the trench length is to find the effect of the trench length on $|R|$ and $|T|$. As the trench length increases, both $|R|$ and $|T|$ become more oscillatory and the number of zeros of $|R|$ increases. The Fig. 3.6 here depicts $|R|$ against $k_0h/2\pi$ for the same values of c/h and b/h as in the Fig. 3.5 (the corresponding figure was however not given in Lee and Ayer (1981)) and this figure clearly demonstrates the phenomenon of occurrence of more zeros of $|R|$ and more oscillatory behavior of $|R|$ against $k_0h/2\pi$ as the trench length increases. The Fig. 3.7 is the same as Fig. 3.6 with enlarged range of $k_0h/2\pi$, i.e. (0, 1.0) and it demonstrates that the maxima of the amplitude of $|R|$ become small as $k_0h/2\pi$ increases. This means that for waves of small wavelengths, most of the incident wave energy is transmitted over the trench since the waves are mostly confined near the free surface.

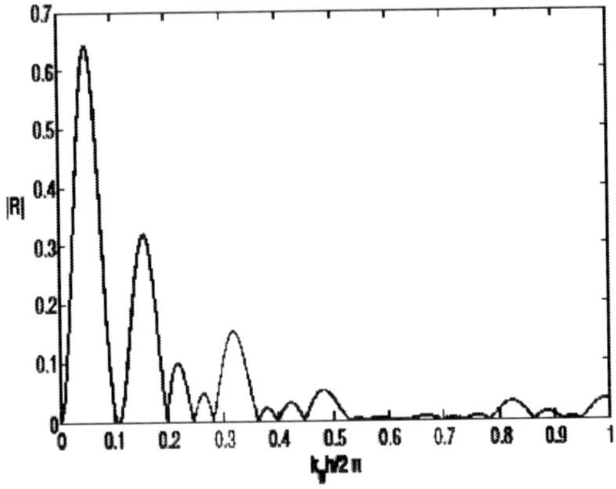

Fig. 3.7 $|R|$ against $k_0h/2\pi$ with $c/h = 7.625$, $b/h = 5.296875$

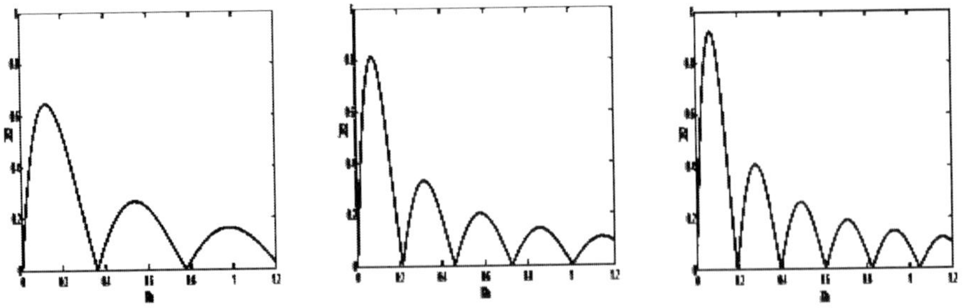

(a) (b) (c)

Fig. 3.8 $|R|$ against Kh with (a) $c/h = 1.5$, $b/h = 1.2$, (b) $c/h = 1.5$, $b/h = 1.5$, (c) $c/h = 1.5$, $b/h = 2$

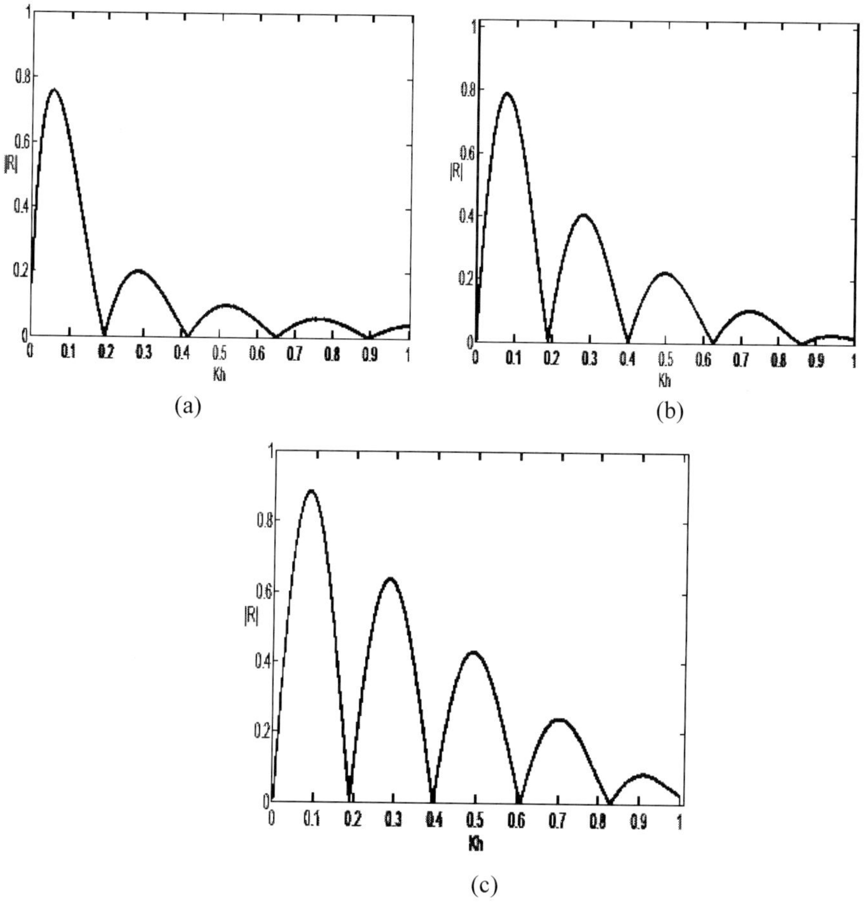

Fig. 3.9 $|R|$ against Kh with (a) $c/h = 1.5$, $b/h = 2$, (b) $c/h = 5$, $b/h = 2$, (c) $c/h = 10$, $b/h = 2$

In Fig. 3.8, $|R|$ is depicted against the wave number Kh for $c/h = 1.5$ and three different values of b/h, viz. $b/h = 1.2$, 1.5, 2.0. From these figures it is obvious that the number of zeros of $|R|$ increases as the trench length increases. This was also demonstrated by Lee and Ayer (1981) theoretically as well as experimentally. It may be noted that the zero reflection phenomena holds for symmetric trenches only, i.e., when the two sides of the trench are of same depth, then the reflection coefficient becomes zero for some discrete values of Kh, and the number of zeros increases with the depth and length of the trench.

The phenomenon of zero reflection was studied in some detail in the papers by Xie et al. (2011) and Liu et al. (2013), Xie et al. (2011) showed that zero reflection at long-wave range (small values of Kh or $k_0h/2\pi$ here) occurs only if the trench is symmetrical. This phenomena has also been observed here. However, the results obtained here are valid for all wave ranges in case of symmetric trench only. The periodic oscillatory behavior of the reflection co-efficient observed by Liu et al. (2013)

in the long wave range has also been observed here. In Fig. 3.9, $|R|$ is drawn against Kh keeping $b/h = 2$ as fixed and taking $c/h = 1.5, 5, 10$ respectively to visualize the effect of the trench depth. It is observed that amplitude of $|R|$ increases to some extent as the trench depth increases.

3.2 Obliquely incident waves

In this section the oblique wave scattering problem involving a rectangular submarine symmetric trench is investigated by employing multi-term Galerkin approximation method. Due to geometrical symmetry of the rectangular trench about the central line $x = 0$, the problem is spilt into two separate problems involving the symmetric and antisymmetric potential functions describing the resultant motion in the fluid region as was done in section 3.1 for the case of normally incident waves. Using eigenfunction expansions of the potential functions along with Havelock inversion formula followed by a matching procedure, integral equations are obtained for the corresponding unknown horizontal velocity components across the vertical line through the corner point of the trench. By using the multi-term Galerkin approximation method involving ultraspherical Gegenbauer polynomials as basis functions, the integral equations are solved approximately. Numerical estimates for the reflection and transmission coefficients are obtained for different angles of incidence and different values of various parameters associated with the problem. These coefficients are seen to satisfy the energy identity. These are also depicted graphically against the wave number in a number of figures. Some of the curves for these coefficients are compared with those available in the literature obtained by using other methods and also by laboratory experiments and a very good agreement is seen to have been achieved. From the numerical results it can be seen that the depth and the width of the trench affects the reflection and transmission co-efficients significantly. It is also observed that for large angles of incidence of the waves, reflection is more while transmission is less, which is plausible.

We consider the irrotational motion in water regarded as an incompressible, inviscid and homogeneous fluid, with free surface over a rectangular submarine trench of width $2b$. The trench is at the bottom of an ocean of uniform finite depth h and the depth of the trench from the mean free surface is c (cf. Fig. 3.1). Here we consider the case of a train of surface waves obliquely incident on the right hand side of the trench with angle of incidence θ. The obliquely incident wave is represented by the velocity potential

$$\mathrm{Re}\{\varphi^{inc}(x, y)e^{i(vz-\sigma t)}\} \tag{2.1}$$

where

$$\varphi^{inc} = \frac{2\cosh k_0(h-y)e^{-i\mu(x-b)}}{\cosh k_0 h} \tag{2.2}$$

k_0 being the real positive root of the transcendental equation

$$k\tanh kh = K, K = \frac{\sigma^2}{g}, \tag{2.3}$$

σ being the circular frequency of the incoming wave train, g being the acceleration due to gravity, and

$$\mu = k_0 \cos\theta, \nu = k_0 \sin\theta \ (0 < \theta < \frac{\pi}{2}). \tag{2.4}$$

Due to geometrical symmetry of the rectangular trench, the z− dependence term can be eliminated by assuming the velocity potential describing the motion in the fluid to be of the form $\mathrm{Re}\{\varphi(x, y)e^{i(\nu z-\sigma t)}\}$. Henceforth the factor $e^{i(\nu z-\sigma t)}$ will always be suppressed. Then $\varphi(x, y)$ satisfies the boundary value problem,

$$(\nabla^2 - \nu^2)\varphi = 0 \text{ in the fluid region} \tag{2.5}$$

$$K\varphi + \varphi_y = 0 \text{ on } y = 0, |x| < \infty, \tag{2.6}$$

$$\varphi_x = 0 \text{ on } x = \pm\, b, y \in (h, c), (c > h), \tag{2.7}$$

$$r^{\frac{1}{3}}\nabla\varphi \text{ is bounded as } r \to 0, \tag{2.8}$$

r being the distance from a submerged edge of the trench,

$$\varphi_y = 0 \text{ on } y = h, |x| > b, \tag{2.9}$$

$$\varphi_y = 0 \text{ on } y = c, |x| < b, \tag{2.10}$$

and finally

$$\varphi(x, y) \sim \begin{cases} \varphi^{inc}(x, y) + R\varphi^{inc}(-x, y) \text{ as } x \to \infty, \\ T\varphi^{inc}(x, y) \text{ as } x \to -\infty \end{cases} \tag{2.11}$$

where R and T are the unknown reflection and transmission coefficients respectively and are to be determined.

Due to geometrical symmetry of the rectangular trench about $x = 0$, we spilt the potential function $\varphi(x, y)$ into symmetric and antisymmetric parts $\varphi^s(x, y)$ and $\varphi^a(x, y)$ respectively, so that

$$\varphi(x, y) = \varphi^s(x, y) + \varphi^a(x, y) \tag{2.12}$$

where

$$\varphi^s(-x, y) = \varphi^s(x, y) \text{ and } \varphi^a(-x, y) = -\varphi^a(x, y). \tag{2.13}$$

Thus, we may restrict our analysis to the region $x \geq 0$ only. Now $\varphi^s(x, y)$ and $\varphi^a(x, y)$ satisfy the equations (2.5) to (2.10) together with

$$\varphi^s_x(0, y) = 0 \text{ and } \varphi^a(0, y) = 0, 0 < y < c. \tag{2.14}$$

Let the behavior of $\varphi^{s,a}(x, y)$for large x be represented by

$$\varphi^{s,a}(x, y) \sim \frac{\cosh k_0(h-y)}{\cosh k_0 h}\{e^{-i\mu(x-b)} + R^{s,a}e^{i\mu(x-b)}\} \text{ as } x \to \infty \tag{2.15}$$

where R^s and R^a are unknown constants. Using (1.9) we find that R^s and R^a are related to R and T by

$$R,\ T = \frac{1}{2}(R^s \pm R^a)e^{-2i\mu b}. \tag{2.16}$$

Now the eigenfunction expansions of $\varphi^{s,a}(x, y)$ satisfying the equations (2.6) to (2.7), (2.9), (2.10) (for $x > 0$) (in different regions) are given below.

Region I ($x > b,\ 0 < y < h$):

$$\varphi^{s,a}(x, y) = \frac{\cosh k_0(h-y)}{\cosh k_0 h}\{e^{-i\mu(x-b)} + R^{s,a}e^{i\mu(x-b)}\} + \sum_{n=1}^{\infty} A_n^{s,a}\cos k_n(h-y)e^{-s_n(x-b)} \tag{2.17}$$

where $k_n (n = 1, 2, \ldots)$ are the real positive roots of the equation

$$k \tan kh + K = 0. \tag{2.18}$$

and

$$s_n = (k_n^2 + \nu^2)^{\frac{1}{2}} \tag{2.19}$$

Region II ($0 < x < b,\ 0 < y < c$):

$$\begin{pmatrix}\varphi^s(x, y)\\ \varphi^a(x, y)\end{pmatrix} = \begin{pmatrix} C_0^s \cos(\alpha_0^2 - \nu^2)^{\frac{1}{2}}x \\ C_0^a \sin(\alpha_0^2 - \nu^2)^{\frac{1}{2}}x \end{pmatrix}\frac{\cosh\alpha_0(c-y)}{\cosh\alpha_0 c} + \sum_{n=1}^{\infty}\begin{pmatrix} C_n^s \cos t_n x \\ C_n^a \sin t_n x\end{pmatrix}\cos\alpha_n(c-y) \tag{2.20}$$

where $\pm\alpha_0$, $\pm i\alpha_n (n = 1, 2, \ldots)$ are the roots of the equation

$$\alpha \tanh \alpha c = K.$$

and

$$t_n = (\alpha_n^2 + \nu^2)^{\frac{1}{2}},\ n = 1,2,\ldots \tag{2.21}$$

Now we have the matching conditions

$$\varphi_x^{s,a}(b + 0, y) = \varphi_x^{s,a}(b - 0, y),\ 0 < y < h. \tag{2.22}$$

We define

$$\varphi^{s,a}(b + 0, y) = f^{s,a}(y),\ 0 < y < h, \tag{2.23}$$

$$\varphi_x^{s,a}(b - 0, y) = g^{s,a}(y),\ 0 < y < c, \tag{2.24}$$

then

$$g^{s,a}(y) = \begin{cases} f^{s,a}(y), & 0 < y < h, \\ 0, & h < y < c. \end{cases} \tag{2.25}$$

Due to the edge condition (2.8), we find that

$$f^{s,a}(y) = O(|y - h|^{-\frac{1}{3}}) \text{ as } y \to h. \tag{2.26}$$

Using the expansions of $\varphi^{s,a}(y)$ in (2.17) into (2.23), we obtain

$$\frac{ik_0 \cosh k_0(h-y)}{\cosh k_0 h}\{R^{s,a}-1\}+\sum_{n=1}^{\infty} A_n^{s,a}(-k_n)\cos k_n(h-y) = f^{s,a}(y),\ 0<y<h. \tag{2.27}$$

Using Havelock's inversion formula in (2.27), we find

$$1-R^{s,a}=\frac{4i}{\delta_0\mu}\cosh k_0 h\int_0^h f^{s,a}(y)\cosh k_0(h-y)dy \tag{2.28}$$

with $\delta_0 = 2k_0h + \sinh 2k_0h$ and

$$A_n^{s,a}=-\frac{4}{\delta_n s_n}\int_0^h f^{s,a}(y)\cos k_n(h-y)dy \tag{2.29}$$

with

$$\delta_n = 2k_nh + \sin 2k_nh\ (n = 1, 2, ...). \tag{2.30}$$

Now using (2.20) into (2.24) we obtain

$$C_0^{s,a}(\alpha_0^2-\nu^2)^{\frac{1}{2}}(-\sin(\alpha_0^2-\nu^2)^{\frac{1}{2}}b,\ \cos(\alpha_0^2-\nu^2)^{\frac{1}{2}}b)\cosh\alpha_0(c-y)\cosh\alpha_0 c+$$
$$\sum_{n=1}^{\infty}(\sinh t_n b,\ \cosh t_n b)C_n^{s,a}t_n\cos\alpha_n(c-y)=g^{s,a}(y), 0<y<c. \tag{2.31}$$

Applying Havelock's inversion formula and noting (2.25), we obtain

$$C_0^{s,a}=\frac{4\alpha_0\cosh\alpha_0 c}{\gamma_0(\alpha_0^2-\nu^2)^{\frac{1}{2}}(-\sin(\alpha_0^2-\nu^2)^{\frac{1}{2}}b,\cos(\alpha_0^2-\nu^2)^{\frac{1}{2}}b)}\int_0^h f^{s,a}(y)\cosh\alpha_0(c-y)dy$$

with

$$\gamma_0 = 2\alpha_0c + \sinh 2\alpha_0c. \tag{2.32}$$

and

$$C_n^{s,a}=\frac{4}{\gamma_n t_n(\sinh t_n b,\cosh t_n b)}\int_0^h f^{s,a}(y)\cos\alpha_n(c-y)dy$$

with

$$\gamma_n = 2\alpha_nc + \sinh 2\alpha_nc. \tag{2.33}$$

Now matching of $\varphi^{s,a}(x, y)$ across the line x = b gives

$$\frac{\cosh k_0(h-y)}{\cosh k_0 h}\{1+R^{s,a}\}+\sum_{n=0}^{\infty}A_n^{s,a}\cos k_n(h-y)=C_0^{s,a}(\cos(\alpha_0^2-\nu^2)^{\frac{1}{2}}b,\sin(\alpha_0^2-\nu^2)^{\frac{1}{2}}b)$$
$$\frac{\cosh\alpha_0(c-y)}{\cosh\alpha_0 c}+\sum_{n=1}^{\infty}(\cosh t_n b,\sinh t_n b)C_n^{s,a}cos\alpha_n(c-y), 0<y<h \tag{2.34}$$

which ultimately produces the first kind integral equations

$$\int_0^\infty F^{s,a}(u)M^{s,a}(y,\ u)du = \frac{\cosh k_0(h-y)}{\cosh k_0 h}, 0 < y < h, \tag{2.35}$$

$$F^{s,a}(y) = \frac{4\cosh^2 k_0 h}{\delta_0(1+R^{s,a})} f^{s,a}(y), 0 < y < h, \tag{2.36}$$

and

$$M^{s,a}(y,\ u) = \frac{\delta_0}{\cosh^2 k_0 h}\sum_{n=1}^{\infty}\Big[\frac{\cos k_n(h-y)\ \cos k_n(h-u)}{\delta_n} + (\coth\alpha_n b, \tanh\alpha_n b)$$
$$\frac{\cos\alpha_n(c-y)\ \cos\alpha_n(c-u)}{\gamma_n} + (-\cot\alpha_0 b, \tan\alpha_0 b)$$
$$\frac{\cosh\alpha_0(c-y)\ \cosh\alpha_0(c-u)}{\gamma_0}\Big], 0 < y,\ u < h. \tag{2.37}$$

It is obvious that $M^{s,a}$ $(y, u)(0 < y, u < h)$ are real and symmetric in y and u. Now if we define

$$C^{s,a} = -i\frac{1-R^{s,a}}{1+R^{s,a}}, \tag{2.38}$$

then by using (1.25) into (1.31), we obtain

$$C^{s,a} = \int_0^h F^{s,a}(y)\frac{\cosh k_0(h-y)}{\cosh k_0 h}dy, \tag{2.39}$$

and $F^{s,a}(y)$ and $C^{s,a}$ are real quantities. Thus if the integral equations (1.31) are solved, then these solutions can be used to evaluate $C^{s,a}$ from the relations (1.35), and these produce the actual reflection and transmission coefficients |R| and |T| by using

$$|R| = \frac{|1 + C^s C^a|}{\Delta} \text{ and } |T| = \frac{|C^s - C^a|}{\Delta} \tag{2.40}$$

$$\text{with } \Delta = \{1 + (C^s)^2 + (C^a)^2 + (C^s C^a)^2\}^{\frac{1}{2}}, \tag{2.41}$$

which are obtained from equations (1.34) and (1.14).

Now we consider the Galerkin approximation method to solve the integral equations (1.31). The unknown functions $F^{s,a}(y)$ are approximated as

$$F^{s,a}(y) \approx \mathcal{F}^{s,a}(y),\ 0 < y < h, \tag{2.42}$$

where $F^{s,a}(y)$ have the multi-term Galerkin expansions in terms of suitable basis functions given by

$$\mathcal{F}^{s,a}(y) = \sum_{n=0}^{N} a_n^{s,a} f_n^{s,a}(y), \qquad 0 < y < h, \tag{2.43}$$

$a_n^{s,a}$ being unknown constants. Following similar arguments as given in Kanoria et al. (1999), the basis functions $f_m^{s,a}(y)$ are found to be

$$f_m^{s,a}(y) = -\frac{d}{dy}\ [e^{-Ky}\int_y^h \hat{f}_m(t)dt],\ 0 < y < h, \tag{2.44}$$

where $\hat{f}_m(y)$ is chosen in terms of ultraspherical Gegenbauer polynomials of order $\frac{1}{6}$ as

$$\hat{f}_m(y) = \frac{2^{\frac{7}{6}}\Gamma(\frac{1}{6})(2m)!}{\pi\Gamma(2m+13)h^{\frac{1}{3}}(h^2-y^2)^{\frac{1}{3}}} C_{2m}^{\frac{1}{6}}(\frac{y}{h}), 0 < y < h. \quad (2.45)$$

Now using (1.41) into (1.40) and substituting these into (1.39), we get the approximate forms of $F^{s,a}(y)$. Using these approximate forms in (1.31), multiplying both sides by $f_m^{s,a}(y)$ and integrating over (0, h), we obtain the linear systems

$$\sum_{n=0}^{N} a_n^{s,a} K_{mn}^{s,a} = d_m^{s,a}, \; m = 0,1, ..., N \quad (2.46)$$

where

$$K_{mn}^{s,a} = \int_0^h \int_0^h M^{s,a}(y, u) f_n^{s,a}(u) f_m^{s,a}(y) du\, dy, \; m, n = 0,1,2, ..., N, \quad (2.47)$$

and

$$d_m^{s,a} = \int_0^h \frac{coshk_0(h-y)}{coshk_0 h} f_m^{s,a}(y)dy, \; m = 0,1, ..., N. \quad (2.48)$$

The integrals (1.43) and (1.44) can be evaluated explicitly, as in Kanoria et al. (1999) by using the different properties and standard results on Gegenbaur polynomials. Thus we find

$$K_{mn}^{s,a} = \frac{\delta_0}{\cosh^2 k_0 h}[4(-1)^{m+n} \sum_{r=1}^{\infty} \frac{\cos^2 k_r h}{\delta_r (k_r h)^{\frac{1}{3}}} (J_{2m+\frac{1}{6}}(k_r h) J_{2n+\frac{1}{6}}(k_r h)$$

$$+ \frac{(\coth \alpha_r b, \tanh \alpha_r b)}{\gamma_r (\alpha_r h)^{\frac{1}{3}}} \cos 2\alpha_r c J_{2m+\frac{1}{6}}(\alpha_r h) J_{2n+\frac{1}{6}}(\alpha_r h)$$

$$+ \frac{(-\cot \alpha_0 b, \tan \alpha_0 b)}{\gamma_0 (\alpha_0 h)^{\frac{1}{3}}} \cosh^2 \alpha_0 c\, I_{2m+\frac{1}{6}}(\alpha_0 h) I_{2n+\frac{1}{6}}(\alpha_0 h)\}], \; m, n = 0,1, ..., N, \quad (2.49)$$

and

$$d_m^{s,a} = \frac{I_{2m+\frac{1}{6}}(k_0 h)}{(k_0 h)^{\frac{1}{6}}}, \; m = 0,1, ..., N. \quad (2.50)$$

The constants $a_n^{s,a} (n = 0,1, ..., N)$ are now obtained by solving the linear systems (1.42), and then the relations (1.35) produce

$$C^{s,a} = \sum_{n=0}^{N} a_n^{s,a} d_n^{s,a} \quad (2.51)$$

so that $C^{s,a}$ are now found and $|R|$ and $|T|$ are evaluated from the relations (2.38).

To solve the linear system (2.46) a suitable choice of N is $N = 40$ to obtain numerical results correct up to almost six decimal places. The reflection and transmission co-efficients $|R|$ and $|T|$ are evaluated numerically for different values of Kh. These coefficients are also depicted graphically for various values of the parameters b/h, c/h and for different incident angles in a number of figures. As a check on the correctness of the results obtained here, we have compared our results with those given by Kirby and Dalrymple (1983) obtained by employing a different method. Kirby and Dalrymple (1983) plotted $|T|$ against the non-dimensional wave number $k_1h_1(= k_0h$ here) in their Figs. 4a and 4b for various values of h_2/h_1 ($\equiv c/h$ here) and a fixed value of the ratio of the trench length and water depth in the left side of the trench ($\equiv 2b/h$ here) and different values of incident angle θ. In the Fig. 4a of Kirby and Dalrymple (1983), $|T|$ was depicted against $k_1h_1 \equiv k_0h$ for $2b/h = 10$, $c/h = 2$ with $\theta = 0°$ and $45°$. In Fig. 3.10 here $|T|$ isdepicted against k_0h for the same values of $2b/h$, c/h and θ. Also, the curves in Fig. 4b of Kirby and Dalrymple (1983) depicting $|T|$ against k_1h_1 for $2b/h = 10$, $c/h = 3$, $\theta = 0°$ and $45°$, almost coincide with the corresponding curves in Fig. 3.10 here which are plots of $|T|$ against k_0h for $b/h = 5$, $c/h = 3$, $\theta = 0°$ and $45°$. Also $\theta = 0°$ corresponds to the case of normal incidence and Fig. 3.12 depicts $|R|$ against $k_0h/2\pi$ with $c/h = 2$, $b/h = 2.5$ and $\theta = 0°$. This curve is compared with Lee and Ayer's (1981) Fig. 2, in which $|R|$ was plotted against h/λ ($\equiv k_0h/2\pi$ here) with the same values of c/h and b/h for the case of normal incidence. Here it is observed that Lee and Ayer's (1981) result (Fig. 2 of Lee and Ayer) is recovered from our present obliquely incident result (Fig. 3.12 here) by taking incident angle θ to be $0°$.

Some more figures of $|R|$ and $|T|$ are drawn against the non-dimensional wave number Kh in Figs. 3.13a, 3.13b, 3.14, 3.15. In Figs. 3.13a and 3.13b, $|R|$ and $|T|$ are depicted respectively with fixed values of $c/h = 2$, $b/h = 3$ and different values of the incident angle θ as $\theta = 30°, 45°, 60°$. In these figures it is observed that the amplitude of $|R|$ gradually increases and the number of zeros of $|R|$ gradually decreases with increasing values of incident angle. The value of $|T|$ gradually decreases and also the number of zeros of $|T|$ becomes less with the increase of θ. Thus more energy is reflected and less transmitted with increase of the incident angle.

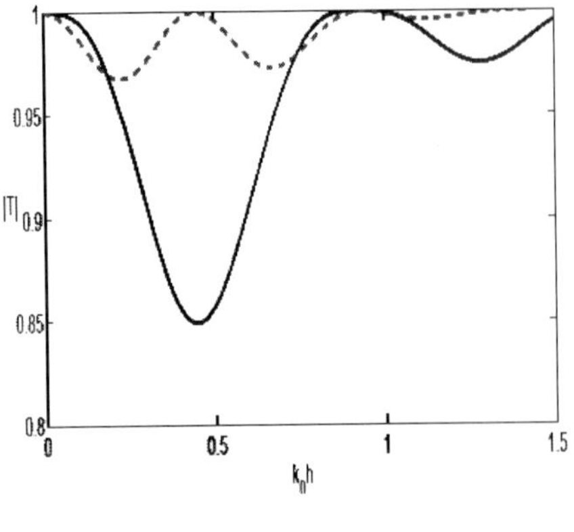

Fig. 3.10 $|T|$ against K_0h with fixed values of $c/h = 2$, $b/h = 5$ and different values of $\theta = 0°(- - -)$, $45°(-)$

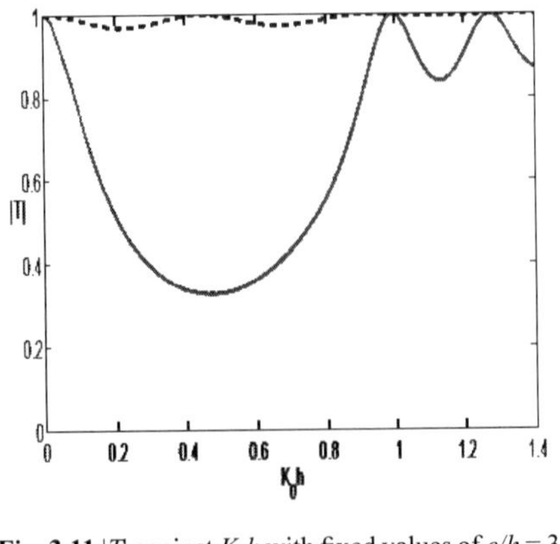

Fig. 3.11 $|T|$ against K_0h with fixed values of $c/h = 3$, $b/h = 5$ and different values of $\theta = 0°(- - -)$, $45°(-)$

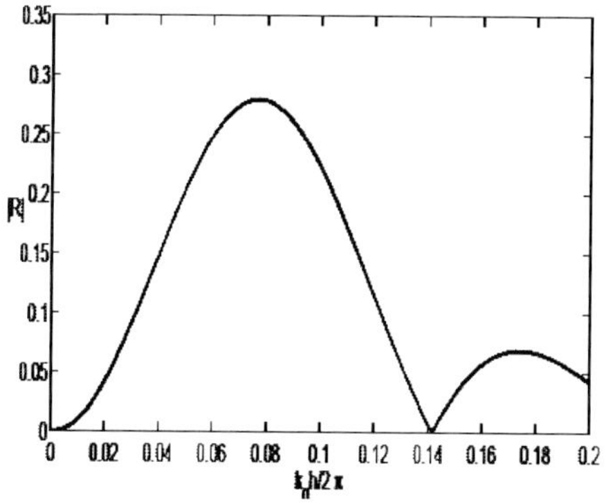

Fig. 3.12 $|R|$ against $K_0h/2\pi$, $c/h = 2$, $b/h = 2.5$, $\theta = 0^{\circ}$

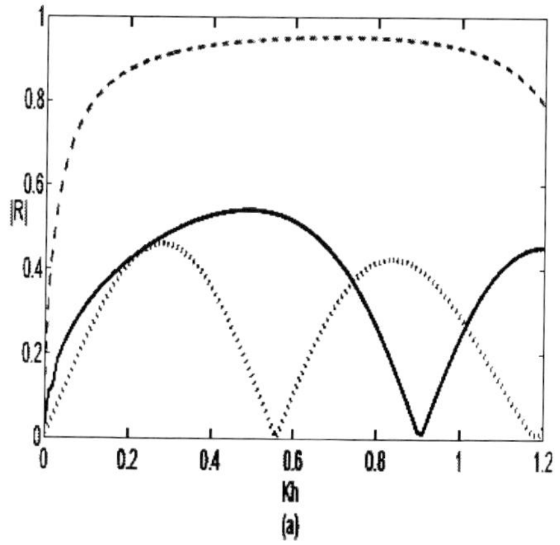

Fig. 3.13a $|R|$ against Kh with fixed $\frac{b}{h} = 3$, $\frac{c}{h} = 2$ different values of $\theta = 30^{\circ}$(– –), 45°(...), 60°(–)

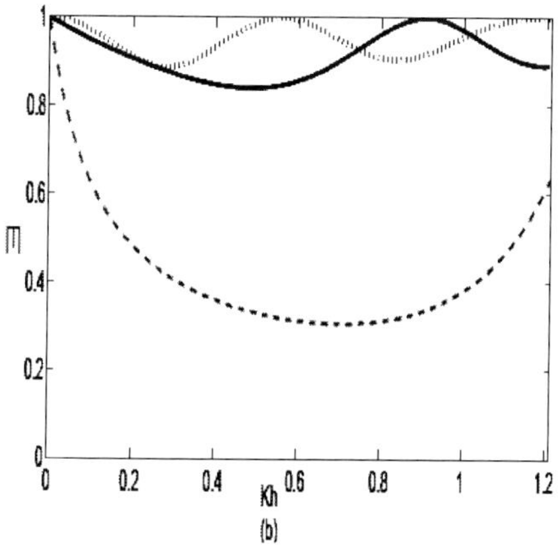

Fig. 3.13b $|T|$ against Kh with fixed $bh = 3$, $c = 2$, different values of $\theta = 30^{\circ}$(– –), 45°(...), 60°(–)

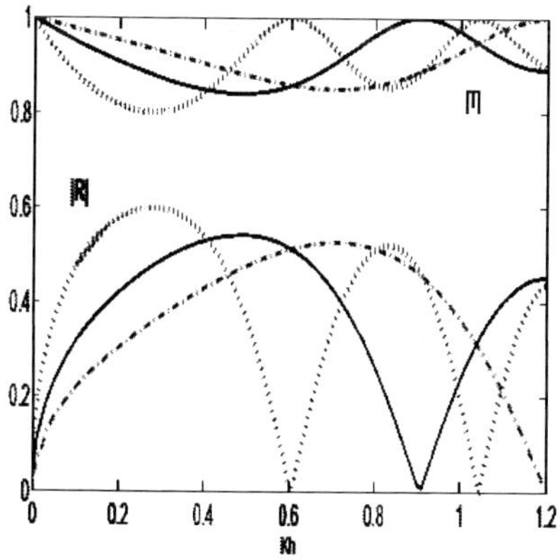

Fig. 3.14 $|R|$ and $|T|$ against Kh with fixed $c/h = 2$ and different values of $b/h = 2$(– –), 3(–)

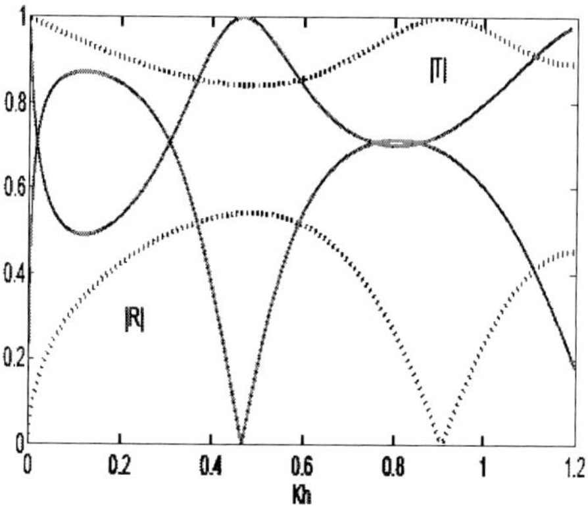

Fig. 3.15 $|R|$ and $|T|$ against Kh with fixed $\theta = 45^{\circ}$, $b/h = 4$ and different values of $c/h = 1$(...), 2(–)

In Fig. 3.14, $|R|$ and $|T|$ are plotted with fixed values of $c/h = 2$ and $\theta = 45°$ for different values of b/h as $b/h = 2, 3, 5$. From these figures it is clear that as the trench width becomes large, more energy is reflected and less energy is transmitted. The Fig. 3.15 depicts $|R|$ and $|T|$ against Kh. Here the values of b/h (= 4), θ(= 45°) remain fixed and c/h(= 1, 2) varies. From this figure, it is observed that, as c/h increases $|R|$ gradually decreases and $|R|$ gradually increases. Also the energy identity relation $|R|^2 + |T|^2 = 1$ is always seen to have been satisfied numerically. This is also used as a partial check in the correctness of the numerical results.

CHAPTER IV

Scattering by a Semi-infinite Dock

The dock problem (cf. Friedrichs and Lewy (1949); Holford I, II (1964)), which is that of understanding the scattering of surface water waves by a thin semi-infinite rigid plate floating on the free surface of water of infinite depth, gives rise to the following mixed boundary value problem, involving Laplace's equation in two dimensions, with (x, y) representing the rectangular Cartesian coordinates, assuming linearized theory of water waves:

$$\nabla^2\phi = 0, -\infty < x < \infty, y > 0, \tag{1}$$

$$K\phi + \phi_Y = 0 \text{ on } y = 0, x < 0, \tag{2}$$

$$\phi_Y = 0 \text{ on } y = 0, x > 0, \tag{3}$$

$$r\frac{\partial\phi}{\partial r} = 0 \text{ as } r = (x^2 + y^2)^{1/2} \to 0, \tag{4}$$

$$\phi \text{ and } \phi_x \text{ are continuous at } x = 0, y > 0, \tag{5}$$

$$\phi \to 0 \text{ as } y \to \infty, \tag{6}$$

$$\phi \to \begin{cases} e^{-Ky+iKx} + Re^{-Ky-iKx} & \text{as } x \to -\infty, \\ 0 & \text{as } x \to \infty. \end{cases} \tag{7}$$

Here $Re\{(g^2/\sigma^3)\phi(x, y)e^{-i\sigma t}\}$ denotes the velocity potential (actual) describing the fluid motion assumed irrotational, where σ is the circular frequency and g is the acceleration due to gravity, $K = \sigma^2/g$, R is the unknown reflection coefficient due to a progressive wave train described by the complex velocity potential function $e^{-Ky+iKx}$ incident on the semi-infinite rigid plate occupying the position $y = 0$, and $x \geq 0$.

A direct use of a Fourier type of analysis cf. Ursell (1947) has been shown by Chakrabarti et al. (2005) to reduce the above boundary value problem to either of two possible singular integral equations of the Carleman type over a semi-infinite range. It is then shown that the closed form solutions of both of these Carleman equations are possible, giving rise to closed form solution of the problem under consideration.

The associated reflection coefficient R is determined and is found to agree with the known result. The free surface profile and the pressure distribution on the dock are depicted graphically at initial time $t = 0$ against the distance. These figures coincide with those given in Friedrichs and Lewy (1948).

The present analysis is believed to be more straightforward and simple, as compared to the existing methods of Friedrichs and Lewy (1948) and Holford I, II (1964) to handle this class of problems in the theory of surface water waves.

Using Havelock's expansion of water potential, we look for the following representations of the function $\phi(x, y)$ in the regions $x < 0$ and > 0 $(y > 0)$, satisfying (1), (2), (3), (6) and (7):

$$\phi(x,y)= e^{-Ky+iKx} + Re^{-Ky+iKx} + \frac{2}{\pi}\int_0^\infty \frac{A(\xi)}{\xi^2+K^2} L(\xi,y)e^{\xi x}\,d\xi\,, \text{for } x<0. \quad (8)$$

$$\phi(x,y) = \frac{2}{\pi}\int_0^\infty \frac{B(\xi)}{\xi}\cos\xi y\; e^{-\xi x}\,d\xi\,, \quad \text{for } x>0. \quad (9)$$

where

$$L(\xi, y) = \xi \cos \xi y - K \sin \xi y, \quad (10)$$

and $A(\xi)$, $B(\xi)$ are two unknown functions to be determined along with the unknown reflection coefficient R.

We emphasize, at this stage itself that the representation (9) demands that we must have

$$B(0) = 0 \quad (11)$$

to ensure that the integral in (9) converges.

The condition (5) gives the following relations:

$$\begin{aligned}(1+R)e^{-Ky} + \frac{2}{\pi}\int_0^\infty \frac{A(\xi)}{\xi^2+K^2}L(\xi,y)d\xi &= \frac{2}{\pi}\int_0^\infty \frac{B(\xi)}{\xi}\cos\xi y\; d\xi\,, y>0\\ iK(1-R)e^{-Ky} + \frac{2}{\pi}\int_0^\infty \frac{\xi A(\xi)}{\xi^2+K^2}L(\xi,y)d\xi &= -\frac{2}{\pi}\int_0^\infty B(\xi)\cos\xi y\; d\xi\,, y>0. \end{aligned} \quad (12)$$

Now there are two possible ways of handling the two relations (12) as far as the application of Fourier analysis is concerned. The first possibility is to use a Fourier cosine inversion formula to both the relations (12) to determine $B(\xi)$ in terms of $A(\xi)$, and the second possibility is to use the Havelock's expansion theorem cf. Ursell (1947) to determine $A(\xi)$ in terms of $B(\xi)$.

Then by using the first of the above two approaches, we obtain that

$$\frac{B(\xi)}{\xi} = \frac{(1+R)K}{\xi^2+K^2} + \frac{\xi A(\xi)}{\xi^2+K^2} - \frac{2K}{\pi}\int_0^\infty \frac{uA(u)}{(u^2-\xi^2)(u^2+K^2)}du\,, \xi>0, \quad (13)$$

$$-B(\xi) = \frac{i(1-R)K}{\xi^2+K^2} + \frac{\xi^2 A(\xi)}{\xi^2+K^2} - \frac{2K}{\pi}\int_0^\infty \frac{u^2A(u)}{(u^2-\xi^2)(u^2+K^2)}du\,, \xi>0, \quad (14)$$

which on elimination of $B(\xi)$, give rise to the following singular integral equation of the Carleman type:

$$\xi C(\xi) - \frac{K}{\pi}\int_0^\infty \frac{C(u)}{u-\xi}du = -K\left(\frac{1}{\xi - iK} + \frac{R}{\xi + iK}\right), \xi > 0, \tag{15}$$

where

$$C(\xi) = \frac{2\xi A(\xi)}{\xi^2 + K^2}. \tag{16}$$

We observe that (15) contains the unknown constant R (the unknown reflection coefficient), and this can be determined by utilizing the convergence criterion (11).

Also, by using the second approach, we obtain that

$$A(\xi) = B(\xi) - \frac{2K\xi}{\pi}\int_0^\infty \frac{B(u)}{u(\xi^2 - u^2)}du, \xi > 0, \tag{17}$$

provided that

$$1 + R = \frac{4K^2}{\pi}\int_0^\infty \frac{B(u)}{(u^2 + K^2)}du, \tag{18}$$

$$\xi A(\xi) = -\xi B(\xi) + \frac{2K\xi}{\pi}\int_0^\infty \frac{B(u)}{\xi^2 - u^2}du, \xi > 0 \tag{19}$$

provided that

$$1 - R = \frac{4iK}{\pi}\int_0^\infty \frac{B(u)}{u^2 + K^2}du. \tag{20}$$

The following generalized identities cf. Chakrabarti (2000a) have been utilized in deriving the results (13), (14), (17) and (19):

$$\begin{aligned} \lim_{\epsilon\to 0}\textstyle\int_0^\infty e^{-\epsilon y}\cos uy \cos \xi y\, dy &= \tfrac{\pi}{2}\{\delta(\xi - u) + \delta(\xi + u)\}, \\ \lim_{\epsilon\to 0}\int_0^\infty e^{-\epsilon y}\sin uy \sin \xi y\, dy &= \frac{\pi}{2}\{\delta(\xi - u) - \delta(\xi + u)\}, \\ \lim_{\epsilon\to 0}\int_0^\infty e^{-\epsilon y}\sin uy \cos \xi y\, dy &= \frac{u}{u^2 - \xi^2}, \end{aligned} \tag{21}$$

where $u, \xi > 0$ and $\delta(x)$ is the Dirac delta function.

Also the singular integral occurring in (15) and elsewhere in this chapter is to be understood as its Cauchy principal value.

Eliminating $A(\xi)$ between the relations (17) and (19), we obtain a second singular integral equation of the Carleman type, as given by

$$\xi B(\xi) + \frac{K}{\pi}\int_0^\infty \frac{B(u)}{u - \xi}du = c, \text{ (say)}\xi > 0, \tag{22}$$

where c can be regarded as an unknown constant to be determined, along with the other unknown constant R (the unknown reflection coefficient), by using the two constraints (18) and (20).

Now both the integral equations (15) and (22) are of the same type and each of them can be cast into a Riemann-Hilbert problem involving the complex-plane, with a cut along the positive real axis, which can finally be solved by using standard techniques available in Muskhelishvilli (1953) or Gakhov (1966).

The two Riemann-Hilbert problems for the two integral equations (15) and (22) are given by

$$\Phi^+(\xi) - \frac{\xi + iK}{\xi - iK}\Phi^-(\xi) = -K\left\{\frac{R}{\xi^2 + K^2} + \frac{1}{(\xi - iK)^2}\right\}, \xi > 0,$$

$$\Lambda^+(\xi) - \frac{\xi - iK}{\xi + iK}\Lambda^-(\xi) = \frac{c}{\xi + iK}, \quad \xi > 0 \tag{23}$$

respectively, involving the two sectionally analytic functions $\Phi(\zeta)$ and $\Lambda(\zeta)$, ($\zeta = \xi + i\eta$), in the cut ζ plane, where

$$\begin{aligned}\Phi(\zeta) &= \frac{1}{2\pi i}\int_0^\infty \frac{C(u)}{u - \zeta}du, \\ \Lambda(\zeta) &= \frac{1}{2\pi i}\int_0^\infty \frac{B(u)}{u - \zeta}du,\end{aligned} \quad \eta \neq 0 \tag{24}$$

with $\Phi^+(\xi) = \Phi(\xi + i0)$, $\Phi^-(\xi) = \Phi(\xi - i0)$, $\Lambda^+(\xi) = \Lambda^+(\xi + i0)$, and $\Lambda^-(\xi) = \Lambda(\xi - i0)$.

The solutions of the above two Riemann-Hilbert problems are straightforward cf. Gakhov (1966) and we find that

$$\Phi(\zeta) = \frac{K}{2\pi i}\Phi_0(\zeta)\int_0^\infty \left\{\frac{R}{u^2 + K^2} + \frac{1}{(u - iK)^2}\right\}\frac{1}{\Phi_0^+(u)(u - \zeta)}du,$$

$$\Lambda(\zeta) = \frac{\Lambda_0(\zeta)}{2\pi i}\int_0^\infty \frac{c}{\Lambda_0^+(u)(u + iK)(u - \zeta)}du \tag{25}$$

with

$$\Phi_0(\zeta) = \exp\left[\frac{1}{2\pi i}\int_0^\infty \frac{\ln(u + iK)/(u - iK)}{u - \zeta}du\right], \tag{26a}$$

$$\Lambda_0(\zeta) = \exp\left[\frac{1}{2\pi i}\int_0^\infty \frac{\ln\frac{(u + iK)}{(u - iK)} - 2\pi i}{u - \zeta}du\right]. \tag{26b}$$

The solutions of the integral equations (15) and (22) can be finally determined by using the Plemelj's formulae as given by

$$C(\xi) = \Phi^+(\xi) - \Phi^-(\xi), \tag{27}$$

$$B(\xi) = \Lambda^+(\xi) - \Lambda^-(\xi). \tag{28}$$

The various integrals appearing in the relations (25) can be evaluated by using standard techniques involving contour integration.

Determination of $C(\xi)$

Here we describe the contour integration procedure to obtain the value of $C(\xi)$. The relation (27) gives

$$C(\xi) = \Phi^+(\xi) - \Phi^-(\xi) = -\frac{\xi K}{\xi + iK}\left\{\frac{R}{\xi^2 + K^2} + \frac{1}{(\xi - iK)^2}\right\}$$

$$-\frac{K^2\Phi_0^+(\xi)}{\pi(\xi+iK)}\int_0^\infty\left\{\frac{R}{u^2+K^2}+\frac{1}{(u-iK)^2}\right\}\frac{1}{\Phi_0^+(u)(u-\xi)}du, \xi>0. \tag{29}$$

The integrals appearing in (29) can be evaluated by considering integrals of the form

$$I(\xi)=\int_\Gamma\frac{P(\tau)}{Q(\tau)}\frac{1}{\Phi_0(\tau)(\tau-\zeta)}d\tau\,, \tag{30}$$

with Γ a positively oriented closed contour consisting of a loop around the positive real axis and a circle of large radius with centre at the origin, in the complex τ-plane, and $P(\tau)$ and $Q(\tau)$ are polynomials in τ. If these polynomials are such that the contribution to the integral in (30) over the circle of large radius vanishes, then

$$I(\zeta)=\int_\Gamma\frac{P(u)}{Q(u)}\left\{\frac{1}{\Phi_0^+(u)}-\frac{1}{\Phi_0^-(u)}\right\}\frac{1}{u-\zeta}du=-2iK\int_0^\infty\frac{P(u)}{Q(u)}\frac{1}{\Phi_0^+(u)(u+iK)(u-\zeta)}du \tag{31}$$

after using the relation

$$\Phi_0^+(\xi)=\frac{\xi+iK}{\xi-iK}\Phi_0^-(\xi)\,. \tag{32}$$

Then by using $P(\tau)=1$ and $Q(\tau)=-2iK(\tau+iK)$, it is observed that

$$I_1(\zeta)=\int_0^\infty\frac{1}{(u^2+K^2)\Phi_0^+(u)(u-\zeta)}du$$

$$=-\frac{1}{2iK}\int\frac{d\tau}{(\tau+iK)\Phi_0(\tau)(\tau-\zeta)}=-\frac{\pi}{K}\frac{1}{\zeta+iK}\left\{\frac{1}{\Phi_0(\zeta)}-\frac{1}{\Phi_0(-iK)}\right\}. \tag{33}$$

Similarly, by choosing $P(\tau)=1$ and $Q(\tau)=-2iK(\tau-iK)$ in (31)

$$I_2(\zeta)=\int_0^\infty\frac{1}{(u-ik)^2\Phi_0^+(u)(u-\zeta)}du$$

$$=-\frac{1}{2iK}\int\frac{d\tau}{(\tau-iK)\Phi_0(\tau)(\tau-\zeta)}=-\frac{\pi}{K}\frac{1}{\zeta-iK}\left\{\frac{1}{\Phi_0(\zeta)}-\frac{1}{\Phi_0(iK)}\right\}. \tag{34}$$

Using Plemelj's formulae,

$$\int_0^\infty\frac{1}{(u^2+K^2)\Phi_0^+(u)(u-\xi)}du=\frac{1}{2}\{I_1^+(\xi)+I_1^-(\xi)\}$$

$$=\frac{\pi}{K}\frac{1}{\xi+iK}\left\{\frac{1}{\Phi_0(-iK)}-\frac{\xi}{(\xi-iK)\Phi_0^+(\xi)}\right\}, \tag{35}$$

$$\int_0^\infty\frac{1}{(u-iK)^2\Phi_0^+(u)(u-\xi)}du=\frac{1}{2}\{I_2^+(\xi)+I_2^-(\xi)\}$$

$$=\frac{\pi}{K}\frac{1}{\xi-iK}\left\{\frac{1}{\Phi_0(iK)}-\frac{\xi}{(\xi-iK)\Phi_0^+(\xi)}\right\}. \tag{36}$$

Using (35) and (36) in (29), $C(\xi)$ is obtained as

$$C(\xi) = -\frac{K}{\xi^2 + K^2}\frac{\Phi_0^+(\xi)}{\Phi_0(iK)} - \frac{KR}{(\xi + iK)^2}\frac{\Phi_0^+(\xi)}{\Phi_0(-iK)}, \xi > 0. \tag{37}$$

We now find $B(\xi)$ by using (13), (16) along with (37) and obtain

$$B(\xi) = \frac{(1+R)K\xi}{\xi^2 + K^2} - \frac{K\xi}{2(\xi^2 + K^2)}\left\{\frac{\Phi_0^+(\xi)}{\Phi_0(iK)} + R\frac{\Phi_0^-(\xi)}{\Phi_0(-iK)}\right\}$$

$$+\frac{K^2\xi}{\pi}\int_0^\infty \frac{1}{(u^2 - \xi^2)(\xi^2 + K^2)}\left\{\frac{\Phi_0^+(\xi)}{\Phi_0(iK)} + R\frac{\Phi_0^-(\xi)}{\Phi_0(-iK)}\right\}du. \tag{38}$$

The integrals in (38) can be evaluated by considering the integral

$$J(\zeta) = \int_\Gamma \frac{P(\tau)}{Q(\tau)}\frac{\Phi_0(\tau)}{(\tau^2 - \zeta^2)}d\tau \tag{39}$$

where Γ is the same as in (30) and $P(\tau)$, $Q(\tau)$ are polynomials such that the contribution to the integral in (39) from the circle of large radius vanishes. We obtain

$$\int_0^\infty \frac{\Phi_0^+(u)}{(u^2 + K^2)(u^2 - \zeta^2)}du = \frac{\pi}{K}\left\{\frac{\Phi_0(\zeta)}{2\zeta(\zeta - iK)} + \frac{\Phi_0(-\zeta)}{2\zeta(\zeta + iK)} - \frac{\Phi_0(iK)}{(\zeta^2 + K^2)}\right\}, \tag{40}$$

$$\int_0^\infty \frac{\Phi_0^-(u)}{(u^2 + K^2)(u^2 - \zeta^2)}du = \frac{\pi}{K}\left\{\frac{\Phi_0(\zeta)}{2\zeta(\zeta + iK)} + \frac{\Phi_0(-\zeta)}{2\zeta(\zeta - iK)} - \frac{\Phi_0(-iK)}{(\zeta^2 + K^2)}\right\}, \tag{41}$$

where $\zeta = \xi + i\eta (\xi > 0)$. Hence the use of Plemelj's formulae produces

$$\int_0^\infty \frac{\Phi_0^+(u)}{(u^2 + K^2)(u^2 - \xi^2)}du = \frac{\pi}{K}\left\{\frac{\Phi_0^+(\xi)}{2(\xi^2 + K^2)} + \frac{\Phi_0(-\xi)}{2\xi(\xi + iK)} - \frac{\Phi_0(iK)}{(\xi^2 + K^2)}\right\}, \tag{42}$$

$$\int_0^\infty \frac{\Phi_0^-(u)}{(u^2 + K^2)(u^2 - \xi^2)}du = \frac{\pi}{K}\left\{\frac{\Phi_0^-(\xi)}{2(\xi^2 + K^2)} + \frac{\Phi_0(-\xi)}{2\xi(\xi - iK)} - \frac{\Phi_0(-iK)}{(\xi^2 + K^2)}\right\}. \tag{43}$$

Using the result of (43) in (38) ultimately $B(\xi)$ is obtained as

$$B(\xi) = \frac{K}{2}\Phi_0(-\xi)\left\{\frac{1}{(\xi + iK)\Phi_0(iK)} + \frac{R}{(\xi - iK)\Phi_0(-iK)}\right\}. \tag{44}$$

$B(\xi)$ can also be obtained by (28) and we find

$$B(\xi) = \frac{cD_1}{\pi}\frac{\Lambda_0^+(\xi)}{\xi - iK} \tag{45}$$

where D_1 is an unknown constant.

Evaluation of R

From (26a), we find that

$$\ln \Phi_0(z) = \frac{1}{\pi}\int_0^{\pi/2} \ln\left(\frac{z - K\tan\theta}{z}\right)d\theta \tag{46}$$

so that

$$\ln\left\{\frac{\Phi_0(iK)}{\Phi_0(-iK)}\right\} = \frac{1}{\pi}\int_0^{\pi/2} \ln\left(\frac{i - \tan\theta}{i + \tan\theta}\right)d\theta = \frac{i\pi}{4}. \tag{47}$$

Now using the condition (11) in (44) we find that

$$R = \frac{\Phi_0(iK)}{\Phi_0(-iK)} = \exp\left(-\frac{i\pi}{4}\right). \tag{48}$$

To determine the values of $\frac{cD_1}{\pi} = D$(say) in (45), the relation (45) is used in the relations (18) and (20). This gives rise to the relations

$$\begin{aligned} 1 + R &= \frac{4DK^2}{\pi} \int_0^\infty \frac{\Lambda_0^+(\xi)}{\xi(\xi - iK)(\xi^2 + K^2)} d\xi, \\ 1 - R &= \frac{4DK^2}{\pi} \int_0^\infty \frac{\Lambda_0^+(\xi)}{(\xi - iK)(\xi^2 + K^2)} d\xi . \end{aligned} \tag{49}$$

The integrals in (49) can be evaluated by considering integrals of the form (30). Thus,

$$\begin{aligned} 1 + R &= \frac{2D}{K}\{\Lambda_0(iK) + \Lambda_0(-iK)\}, \\ 1 - R &= \frac{2D}{K}\{\Lambda_0(-iK) - \Lambda_0(iK)\} \end{aligned} \tag{50}$$

so that

$$R = \frac{\Lambda_0(iK)}{\Lambda_0(-iK)}, \qquad D = \frac{cD_1}{\pi} = \frac{K}{2}\frac{1}{\Lambda_0(-iK)} \tag{51}$$

Equivalence of (44) and (45)

Here we prove the following results:

$$\frac{\Lambda_0(-iK)}{\Phi_0(iK)} = \frac{1}{2}, \tag{52}$$

$$\frac{\Lambda_0^+(\xi)}{\Phi_0(-\xi)} = \frac{\xi}{\xi + iK} \tag{53}$$

so that the expressions as given by the relations (44) along with (51) and (44) along with (48) represent the same function $B(\xi)$.

To show (52), it may be noted from (26a) and (26b) that

$$\begin{aligned} \Phi_0(iK) &= \exp\left[\frac{1}{2\pi i}\int_0^\infty \frac{\ln(u + iK)/(u - iK)}{u - iK} du\right], \\ \Lambda_0(-iK) &= \exp\left[\frac{1}{2\pi i}\int_0^\infty \frac{\ln\dfrac{(u + iK)}{(u - iK)} - 2\pi i}{u + iK} du\right]. \end{aligned} \tag{54}$$

Using the result

$$\ln\left(\frac{u - iK}{u + iK}\right) + \ln\left(\frac{u + iK}{u - iK}\right) = 2\pi i, \tag{55}$$

it is found that

$$\frac{\Lambda_0(-iK)}{\Phi_0(iK)} = \exp\left[-\frac{1}{\pi i}\int_0^\infty \ln\left(\frac{u - iK}{u + iK}\right)\frac{u}{u^2 + K^2} du\right]$$

$$= \exp\left[-\frac{2}{\pi}\int_0^{\pi/2}\left(\frac{\pi}{2}-\theta\right)\tan\theta d\theta\right] = \frac{1}{2}. \tag{56}$$

To show (53), the results in (26b) and (55) are used to obtain

$$\Lambda_0^+(\xi) = \exp\left[\frac{1}{2}\ln\left(\frac{\xi - iK}{\xi + iK}\right) - \frac{1}{2\pi i}\int_0^\infty \frac{\ln\dfrac{(u+iK)}{(u-iK)}}{u-\xi}du\right], \xi > 0. \tag{57}$$

Also, from (26a), it is seen that

$$\Phi_0(-\xi) = \exp\left[\frac{1}{2\pi i}\int_0^\infty \frac{\ln(u+iK)/(u-iK)}{u+\xi}du\right], \quad \xi > 0. \tag{58}$$

Hence

$$\frac{\Lambda_0^+(\xi)}{\Phi_0(-\xi)} = \frac{(\xi - iK)^{1/2}}{(\xi + iK)}\exp\left[-\frac{1}{\pi i}\int_0^\infty \ln\left(\frac{u+iK}{u-iK}\right)\frac{u}{u^2-\xi^2}du\right]. \tag{59}$$

The integral in (59) can be evaluated and its value is $2\pi i \ln\left\{\frac{(\xi^2+K^2)^{1/2}}{\xi}\right\}$. Substituting this value in (59), the result in (52) is obtained.

We thus find that the mixed boundary value problem under consideration gets solved completely either by the aid of the relations (44), (48), (37), and (16), or by the aid of the relations (45), (48), (51) and (19), which determine the unknown functions $A(\xi)$ and $B(\xi)$ and the unknown reflection coefficient R completely so that the complete knowledge of the potential $\phi(x, y)$ can be obtained by using the relations (8) and (9).

We find that the value of R is $e^{-i\pi/4}$, as obtained by Holford (1964), by using a completely different analysis.

It is rather interesting to verify that the two representations of $B(\xi)$, as given by the relations (45) along with (51) and (44) along with (48), are identical.

The exact form of $\phi(x, y)$ can be obtained from (8) for $x < 0$ and from (9) for $x > 0$ after substituting $A(\xi)$ and $B(\xi)$ in terms of $\Phi(\xi)$. This should coincide with the result for the potential function (except for a multiplying constant) given by Friedrichs and Lewy (1948) (Re $\chi^R(z)$ given there). However, this is not verified here directly. Instead we obtain here the free surface depression $\eta(x, t)$ and the pressure $p(x, t)$ on the dock by using Bernoulli's equation, and depict them graphically against the nondimensional distance Kx at time $t = 0$ (actually $K\eta(x, 0)$ and $Kp(x, 0)/\rho g$) for $x < 0$ and $x > 0$, respectively.

Using Bernoulli's equation, the free surface distribution $\eta(x, t)$ $(x < 0)$ is obtained as

$$\eta(x,t) = -\frac{1}{K}\mathrm{Re}\{i\phi(x,0)e^{-i\sigma t}\}$$
$$= \frac{1}{K}\left[\sin(Kx - \sigma t) - \sin\left(\frac{\pi}{4} + Kx + \sigma t\right) + \frac{\sqrt{2}}{\pi}\sin\left(\frac{\pi}{8} + \sigma t\right)I(Kx)\right] \tag{60}$$

where

$$I(s) = \int_0^{\pi/2}\cos\theta \exp\left\{s\cot\theta + \frac{\sin\theta}{\pi}\int_0^{\pi/2}\frac{\alpha}{\sin\alpha\sin(\theta-\alpha)}d\alpha\right\}d\theta. \tag{61}$$

Similarly, the pressure on the dock $p(x, t)$ $(x > 0)$ is obtained as

$$p(x,t) = \frac{\rho g}{K} \mathrm{Re}\left\{i\phi(x,0)e^{-i\sigma t}\right\} = \frac{\rho g}{K}\frac{\sqrt{2}}{\pi} \sin\left(\frac{\pi}{8} + \sigma t\right) J(Kx) \tag{62}$$

where

$$J(s) = \int_0^{\pi/2} \exp\left\{-s\cot\theta + \frac{\sin\theta}{\pi}\int_0^{\pi/2} \frac{\alpha}{\sin\alpha\sin(\theta+\alpha)} d\alpha\right\} d\theta. \tag{63}$$

Figures 4.1 and 4.2 depict, respectively, the free surface profile $\eta(x, 0)$ against x $(x < 0)$ and the pressure distribution $p(x, 0)$ on the dock also against x $(x > 0)$; η, p, x being nondimensionalized as $K\eta$, $\frac{Kp}{\rho g}$, Kx. These curves in Figs. 4.1 and 4.2 can

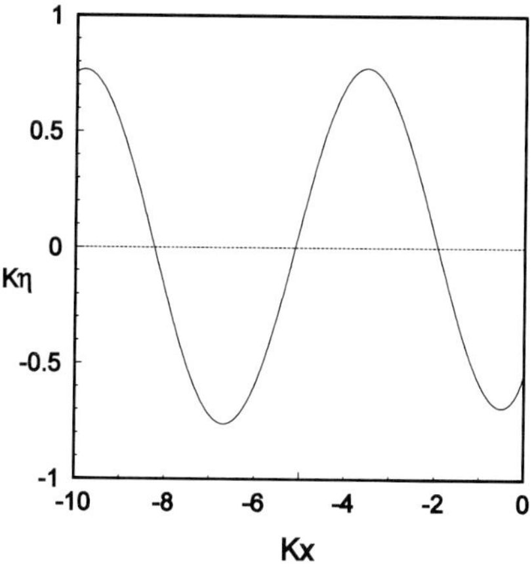

Fig. 4.1 Free surface profile at t = 0

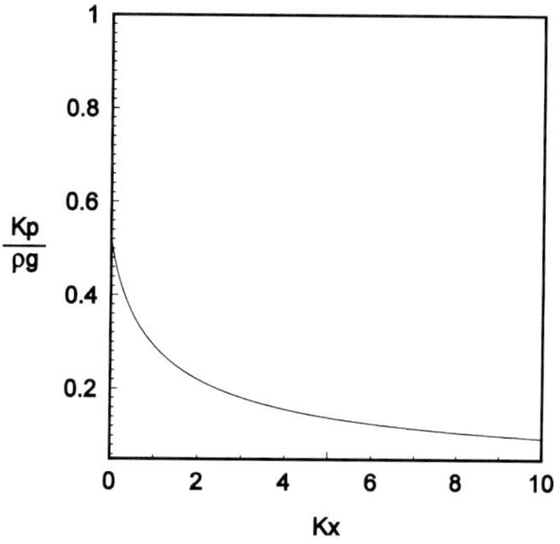

Fig. 4.2 Distribution of pressure on the dock at t = 0

be identified with the curves given by Friedrichs and Lewy (1948), obtained from the potential function Re $\chi^R(z)$ given there. This indirectly verifies that the solution for the potential function obtained here coincides with the solution Re $\chi^R(z)$ given in Friedrichs and Lewy (1948) (except for a multiplying constant).

CHAPTER V

Surface Discontinuities

5.1 Scattering by a semi-infinite inertial surface

If a part of a surface of deep water s covered by an inertial surface composed of a thin but uniform distribution of non-interacting particles (e.g., broken ice, unstretched mat) and the remaining part is free, then the surface boundary condition becomes discontinuous in the sense that there are one condition on the free surface and another condition on the inertial surface. The line separating the free surface and the inertial surface becomes a line of discontinuity. Peters (1950), Weitz and Keller (1950) first developed this model to study wave-ice interaction. When half the surface of water is covered by an inertial surface and the other half of the surface is free, Peters (1950) investigated the case when waves from the free surface region are normally incident on the straight line separating the free surface and the inertial surface. Weitz and Keller (1950) treated the same problem for water of uniform finite depth and oblique incidence of waves.

Gabov et al. (1989) considered two infinitely extended, immiscible superposed fluids for which half of the interface is covered by an inertial surface and the half of the interface is a free separating boundary of the two fluids, and investigated the scattering of the interface waves travelling from the free surface region and normally incident on the line separating the free interface and the inertial interface.

The two physical problems discussed above are mathematically equivalent. This can be shown by considering the case of the deep water and the case of two superposed fluids separately as has been shown by Kanoria et al. (1999).

Case (a). Deep water: We first consider deep water and choose the y-axis vertically downwards into the water so that its surface at the rest position coincides with the plane $y = 0$. Let the semi-infinite plane represented by $y = 0, x < 0$ be the free surface and the semi-infinite plane $y = 0, x > 0$ be covered by an inertial surface of area density σ, say. Let $Re\ \psi(x, y)e^{-i\omega t}$ represent the velocity potential describing the irrotational motion in the fluid where t denotes the time. Let the factor $e^{-i\omega t}$ be always suppressed henceforth. Then the complex-valued potential function $\psi(x, y)$ is harmonic in the fluid region. The linearized free surface condition is

$$\psi_y + a\psi = 0 \quad \text{on } y = 0, \qquad x < 0, \tag{1.1}$$

where

$$a = \omega^2/g, \tag{1.2}$$

g being the acceleration due to gravity. The condition of no motion at infinite depth gives

$$\psi, \nabla\psi \to 0 \quad \text{as } y \to \infty. \tag{1.3}$$

Thus an incident progressive surface wave field is represented by the potential function $e^{-ay+iax}$. Again the linearized inertial surface condition is

$$\psi_y + b\psi = 0 \quad \text{on } y = 0, \quad x > 0, \tag{1.4}$$

where

$$b = \frac{\rho\omega^2}{\rho g - \omega^2\sigma}, \tag{1.5}$$

ρ being the density of the water. It may be noted that for $\sigma > \rho g/\omega^2$ (i.e., $b > 0$), the form of (1.4) is merely a modification of the usual free surface condition (1.1) corresponding to $\sigma = 0$, and it allows progressive waves at the inertial surface. However, for $\sigma \leq \rho g/\omega^2$ (i.e., $b < 0$ or $|b| = \infty$), the form of (1.4) is different and does not allow progressive waves at the inertial surface Rhodes-Robinson (1984).

Let $\phi(x, y)$ denote the scattered potential function due to the incident wave field $e^{-ay+iax}$ propagating from infinity along the free surface and normally incident on the line separating the free surface and the inertial surface. Then $\phi(x, y)$ is harmonic in the fluid region and satisfies the boundary conditions

$$\phi_y + a\phi = 0 \quad \text{on } y = 0, \quad x < 0, \tag{1.6a}$$

$$\phi_y + b\phi = -(b - a)e^{iax} \quad \text{on } y = 0, \quad x > 0, \tag{1.6b}$$

so that $x = 0$ is a point of discontinuity in the boundary condition on the boundary $y = 0$. ϕ also satisfies the same condition as ψ satisfies as $y \to \infty$. Moreover, it also satisfies some edge conditions as $(x^2 + y^2)^{1/2} \to 0$ and infinity requirements involving the unknown complex-valued reflection and transmission coefficients as $|x| \to \infty$. These will be stated later. These unknown coefficients form a part of the problem.

Case (b). Two superposed fluid: Again, we consider two superposed fluids of densities ρ_1, ρ_2 where ρ_1 is the density of the lower fluid and $\rho_2 (< \rho_1)$ is the density of the upper fluid. Let the y-axis be chosen vertically downwards into the lower fluid so that the rest position of the common interface is $y = 0$. Let the half plane represented by $y = 0, x < 0$ be the free separating boundary of the two fluids and the half plane $y = 0, x > 0$ be covered by an inertial surface of the area density σ. Let $Re\, \psi_1(x, y)e^{-i\omega t}$, $Re\, \psi_2(x, y)e^{-i\omega t}$, respectively, denote velocity potential in the lower and upper fluids describing irrotational motion. Then ψ_1 is harmonic in $y > 0$ and ψ_2 is harmonic in $y < 0$. The linearized conditions at the separating boundary are

$$\rho_1 (g\psi_{1y} + \omega^2\psi_1) = \rho_2 (g\psi_{2y} + \omega^2\psi_2) \quad \text{on } y = 0, x < 0, \tag{1.7}$$

$$\psi_{1y} = \psi_{2y} \quad \text{on } y = 0, \quad x < 0, \tag{1.8}$$

and the conditions at the bottom and top are

$$\psi_1, \Delta\psi_1 \to 0 \quad \text{as } y \to \infty, \; \psi_2, \nabla\psi_2 \to 0 \quad \text{as } y \to -\infty. \tag{1.9}$$

Thus an incident progressive interface wave field propagating on the free separating boundary is represented by

$$\psi_1^{\text{inc}}(x, y) = e^{-ay+iax}, \, y > 0, \; \psi_2^{\text{inc}}(x, y) = -e^{-ay+iax}, \, y > 0, \tag{1.10}$$

where now

$$a = \frac{\rho_1 + \rho_2}{\rho_1 - \rho_2} \frac{\omega^2}{g}. \tag{1.11}$$

Again, the linearized conditions at the inertial interface are

$$\rho_1 (g\psi_1 y + \omega^2\psi_1) - \rho_2 (g\psi_2 y + \omega^2\psi_2) = \sigma\omega^2\psi_{1y} \tag{1.12}$$
$$= \sigma\omega^2\psi_{2y} \; \text{ on } y = 0, \, x > 0.$$

Let $\chi_j(x, y)(j - 1,2)$ denote the scattered potential due to the incident wave field represented by $\psi_j^{\text{inc}}(x, y)(j = 1,2)$ propagating from infinity along the free separating boundary and normally incident on the line separating the free interface and inertial interface. Then $\chi_j(x, y)(j = 1,2)$ satisfy

$$\nabla^2\chi_1 = 0 \quad \text{in } y > 0, \quad \nabla^2\chi_2 = 0 \quad \text{in } y < 0, \tag{1.13}$$

the interface conditions

$$\rho_1(g\chi_{1y} + \omega^2\psi_1) - \rho_2 (g\chi_{2y} + \omega^2\chi_2) = \begin{cases} 0 & \text{on } y = 0, x < 0, \quad (1.14) \\ \sigma\omega^2\chi_{1y} - ae^{iax} & \text{on } y = 0, x > 0, \quad (1.15) \end{cases}$$

$$\chi_{1y} = \chi_{2y} \quad \text{on } y = 0, \tag{1.16}$$

and the bottom and top conditions

$$\chi_1, \nabla\chi_1 \to 0 \quad as\; y \to \infty, \quad \chi_2, \nabla\chi_2 \to 0 \quad as\; y \to -\infty. \tag{1.17}$$

We now show that

$$\chi_2(x, y) = -\chi_1(x, -y), \qquad y < 0. \tag{1.18}$$

To show this, let

$$\chi(x, y) = \chi_1(x, y) + \chi_2(x, -y), \qquad y > 0. \tag{1.19}$$

then $\chi(x, y)$ satisfies the boundary value problem described by

$$\nabla^2\chi = 0 \quad in\; y > 0, \quad \chi_y = 0 \text{ on } y = 0, \chi, \nabla\chi \to 0 \text{ as } y \to \infty \tag{1.20}$$

where Eq. (1.13) and conditions (1.16) and (1.17) have been utilized. Now, by uniqueness theorem of harmonic functions, we find that

$$\chi(x, y) \equiv 0. \tag{1.21}$$

Thus Eq. (1.18) is proved. Hence it is sufficient to solve for the function $\chi_1(x, y)$ which is harmonic in the region $y > 0$. Then boundary conditions satisfied by $\chi_1(x, y)$ on $y = 0$ are obtained from (1.14) and (1.15) as

$$\chi_{1y} + a\chi_1 = 0 \quad \text{on } y = 0, \quad x < 0, \tag{1.22}$$

$$\chi_{1y} + b\chi_1 = -(b - a)e^{iax} \quad \text{on } y = 0, \quad x > 0, \tag{1.23}$$

where a is given by (1.11) and

$$b = \frac{(\rho_1 + \rho_2)\omega^2}{(\rho_1 - \rho_2)g - a\omega^2}. \tag{1.24}$$

We note that, in the absence of the upper fluid, a, b assume the values given by (1.2) and (1.5), respectively.

Thus the problem for deep water and the problem for two superposed fluids are mathematically similar, the only difference being that the constants a, b for deep water and for two superposed fluids are to be two different sets of constants.

Boundary value problem

The mathematical formulation of the wave scattering problems involving a discontinuity in the boundary condition is now stated in the form of a boundary value problem (BVP) described below.

To solve the Laplace equation

$$\phi_{xx} + \phi_{yy} = 0, \qquad y > 0, \qquad -\infty < x < \infty, \tag{1.25}$$

along with the surface boundary conditions, as given by

$$\phi_y + a\phi = 0 \quad \text{on } y = 0, \text{for } x < 0, \text{ with } a > 0, \tag{1.26}$$

$$\phi_y + b\phi = -(b - a)e^{iax} \quad \text{on } y = 0, \text{for } x > 0, \text{with } -\infty < b < \infty \tag{1.27}$$

producing a discontinuity at the origin in the surface boundary condition, the edge conditions

$$\phi = O(1) \text{ and } \nabla\phi = O(1) \text{ as } r = (x^2 + y^2)^{1/2} \to 0 \text{ for finite } |b|,$$

or

$$\phi = O(1) \text{ and } r^{1/2}\nabla\phi = O(1) \text{ as } r = (x^2 + y^2)^{1/2} \to 0 \text{ for } |b| = \infty, \tag{1.28}$$

which are given by the physics of the problem, the conditions

$$\phi, \nabla\phi \to 0 \quad as\ y \to \infty, \tag{1.29}$$

and the conditions as $|x| \to \infty$, as given by

$$\phi(x,y) + e^{-ay+iax} \sim A_1 e^{-ay-iax} + e^{-ay+iax} \quad as\ x \to -\infty,$$

$$\sim \begin{cases} B_1 e^{-by+ibx} & \text{as } x \to \infty \text{ for } b > 0 \\ 0 & \text{as } x \to \infty \text{ for } b < 0 \ or\ |b| = \infty. \end{cases} \tag{1.30}$$

In (1.30) A_1 and B_1 are two unknown complex constants to be determined, and $|A_1|$, $|B_1|$ represent the reflection and transmission coefficients, respectively, corresponding to the incident surface (or interface) wave field represented by $e^{-ay+iax}$ propagating in the region $x < 0$. Hence $\phi(x, y)$ denotes the scattered field so that the total field is

$$\phi^t(x,y) = \phi(x,y) + e^{-ay+iax}. \tag{1.31}$$

The BVP can be viewed as a special case of a more general problem of wave scattering by a surface discontinuity tackled for its solution by Gabov et al. (1989).

We observe that both the BVP can be handled for its solution by the aid of the Wiener-Hopf technique after generalizing the Laplace equation (2.1) to the Helmholtz equation

$$\phi_{xx} + \phi_{yy} + \varepsilon^2\phi = 0, \qquad y > 0, \quad -\infty < x < \infty, \tag{1.32}$$

where ε is a complex number with small positive real and imaginary parts as well as by generalizing some conditions (as described shortly) and ultimately passing on to the limit as $\varepsilon \to 0$ in a manner similar to Gabov et al. (1989). Thus the generalized form of the BVP is to solve the Helmholtz equation (1.32) along with the boundary conditions

$$\phi_y + a\phi = 0 \quad \text{on } y = 0, \qquad \text{for } x < 0, \tag{1.33}$$

$$\phi_y + b\phi = -(b-a)e^{ikx} \quad \text{on } y = 0, \ \text{for } x > 0, \tag{1.34}$$

where

$$k = (a^2 + \varepsilon^2)^{1/2} \ \text{ with } k = a \ \text{ for } \varepsilon = 0, \tag{1.35}$$

the edge conditions

$$\phi = O(1) \text{ and } \nabla\phi = O(1) \text{ as } r = (x^2 + y^2)^{1/2} \to 0 \text{ for finite } |b| \tag{1.36a}$$

or

$$\phi = O(1) \text{ and } r^{1/2}\nabla\phi = O(1) \text{ as } r = (x^2 + y^2)^{1/2} \to 0 \text{ for finite } |b| = \infty, \tag{1.36b}$$

and the conditions at infinity, as given by

$$|\phi| + |\nabla\phi| \le \text{const.}\, e^{-\delta(\varepsilon)r} \quad \text{as } r = (x^2 + y^2)^{1/2} \to \infty, \tag{1.37}$$

where $\delta(\varepsilon)$ is such that

$$\delta(\varepsilon) \to 0+ \qquad \text{as } \varepsilon \to 0. \tag{1.38}$$

With this as background, we proceed to present the Wiener-Hopf technique applied to generalized BVP satisfying the Helmholtz equation (1.32) involving the complex parameter ε, the surface conditions (1.33) and (1.34), the edge conditions (1.36) and the infinity requirement (1.37) as has been worked out by Kanoria et al. (1999). The reflection and the transmission coefficients have been obtained explicitly. The reflection coefficient is depicted graphically against the wave number aL for $b > 0$ to visualize the effect of the inertial surface on the incident wave train progressing along the free surface, where L is a characteristic length used to non-dimensionalise a, $b(> 0)$ and σ/ρ for deep water.

The Wiener-Hopf technique

Let $\phi(x, y) = \phi(x, y; \varepsilon)$ denote the function satisfying the generalized BVP described by (1.32)–(1.34), (1.36) and (1.37). We introduce the Fourier transform $\Phi(\alpha, y)$ is defined by

$$\Phi(\alpha, y) = \int_{-\infty}^{\infty} \phi(x, y) e^{i\alpha x}\, dx,$$

where $\alpha = \sigma + i\tau$, σ and τ being real. Then

$$\Phi(\alpha, y) = \Phi_{-}(\alpha, y) + \Phi_{+}(\alpha, y),$$

where

$$\Phi_{-}(\alpha, y) = \int_{-\infty}^{0} \phi(x, y) e^{i\alpha x}\, dx, \quad \Phi_{+}(\alpha, y) = \int_{0}^{\infty} \phi(x, y) e^{i\alpha x}\, dx. \tag{1.39}$$

By using condition (1.37) it is observed that the functions $\Phi_{+}(\alpha, y)$ and $\Phi_{-}(\alpha, y)$ are analytic functions of α in the overlapping half-planes $\tau > -\delta(\varepsilon)$ and $\tau < \delta(\varepsilon)$, respectively, and by using the edge conditions (1.36) along with the Abelian theorem, see Noble (1958), it can be shown that

$$\begin{aligned} |\phi_{+}(\alpha, y)| &= O(|\alpha|^{-1}) \text{ as } |\alpha| \to \infty \quad \text{in } \tau > -\delta(\varepsilon), \\ |\phi_{-}(\alpha, y)| &= O(|\alpha|^{-1}) \text{ as } |\alpha| \to -\infty \quad \text{in } \tau < \delta(\varepsilon). \end{aligned} \tag{1.40}$$

To use Wiener-Hopf procedure, conditions (1.33) and (1.34) are written as

$$\phi_y + a\phi = \begin{cases} 0 & \text{on } y = 0 \text{ for } x < 0, \\ f(x) & \text{on } y = 0 \text{ for } x > 0, \end{cases} \tag{1.41}$$

and

$$\phi_y + b\phi = \begin{cases} g(x) & \text{on } y = 0 \text{ for } x < 0, \\ -(b - a)e^{ikx} & \text{on } y = 0 \text{ for } x > 0, \end{cases} \tag{1.42}$$

where $f(x)$ (for $x > 0$) and $g(x)$ (for $x < 0$) are unknown functions having the behaviors

$$f(x) = O(1) \quad \text{as } x \to +0,$$

and

$$g(x) = O(1) \quad \text{as } x \to -0 \text{ for finite } |b|,$$

while

$$g(x) = O\left(|x|^{-\frac{1}{2}}\right) \quad \text{as } x \to -0 \text{ for } |b| = \infty, \tag{1.43}$$

obtained from the edge condition (1.36).

Now applying the Fourier transform to the PDE (1.32), we obtained that

$$\frac{d^2\phi(\alpha, y)}{dy^2} - \gamma^2 \phi(\alpha, y) = 0, \qquad y \geq 0,$$

with $\gamma^2(\alpha) = \alpha^2 - \varepsilon^2$, whose appropriate solution is given by

$$\phi(\alpha, y) = D_1(\alpha)e^{-\gamma y}, \quad y \geq 0, \tag{1.44}$$

where $D_1(\alpha)$ is an arbitrary function of the transform parameter α, and we denote by $\gamma(\alpha)$ that branch of the function $(\alpha^2 - \varepsilon^2)^{\frac{1}{2}}$ that takes the value for $-i\varepsilon$ for $\alpha = 0$. Applying the Fourier transform to conditions (1.41) and (1.42) and we obtained that

$$\phi'(\alpha, 0) + a\phi(\alpha, 0) = F_+(\alpha) \tag{1.45}$$

$$\phi'(\alpha, 0) + b\phi(\alpha, 0) = G_-(\alpha) + \frac{b-a}{i(\alpha+k)} \tag{1.46}$$

in which the two unknown functions $F_+(\alpha) \equiv \int_0^\infty f(x)e^{i\alpha x}\,dx$ and $G_-(\alpha) \equiv \int_{-\infty}^0 g(x)e^{i\alpha x}$ dx can be shown to be analytic in the overlapping half-planes $\tau > -\delta(\varepsilon)$ and $\tau < \delta(\varepsilon)$, respectively, with $|F_+(\alpha)| = O(|\alpha|^{-1})$ as $|\alpha| \to \infty$ in $\tau > -\delta(\varepsilon)$ and

$$|G_-(\alpha)| = \begin{cases} O(|\alpha|^{-1}) \text{ as } x \to \infty & \text{for finite } |b|, \\ O\left(|\alpha|^{-1/2}\right) \text{ as } x \to \infty & \text{for } |b| = \infty, \end{cases}$$

in $\tau < \delta(\varepsilon)$. Using (1.44) in (1.45) and (1.46) and eliminating $D_1(\alpha)$ we obtain the following two-part Wiener-Hopf functional relation, for the determination of the two functions $F_+(\alpha)$ and $G_-(\alpha)$, as given by

$$\frac{\gamma(\alpha) - b}{\gamma(\alpha) - a} F_+(\alpha) - G_-(\alpha) = \frac{b-a}{i(\alpha+k)} \tag{1.47}$$

valid in the strip $c < \tau < d$ where c and d are chosen such that

$$-\delta(\varepsilon) < -\min(\text{Im } k, \text{Im } \alpha_0) < c < 0 < d < \min(\text{Im } k, \text{Im } \alpha_0) < \delta(\varepsilon) \tag{1.48}$$

with $\alpha_0 = (b^2 + \varepsilon^2)^{1/2}$ that takes the value b for $\varepsilon = 0$.

Now three cases arise according as $b > 0$, $b < 0$ and $|b| = \infty$, and we treat the Wiener-Hopf relation (1.47) in these three cases, in different manners as described below.

Case1: $b > 0$. We note that the coefficient of $F_+(\alpha)$ in (1.47) is $[(\alpha^2 - \alpha_0^2)/(\alpha^2 - k^2)]$ $M(\alpha)$ where $M(\alpha) = (\gamma(\alpha) + a)/(\gamma(\alpha) + b)$. The function $M(\alpha)$ is analytic in the strip $-\delta(\varepsilon) < \tau < \delta(\varepsilon)$ and hence, in the strip $c < \tau < d$, which can be factorized as

$$M(\alpha) = M_+(\alpha)M_-(\alpha) \tag{1.49}$$

where

$M_+(\alpha) = M_-(-\alpha)$, $|M_+(\alpha)| = O(1)$ as $|\alpha| \to \infty$ in $\tau > c$, $M_+(\alpha)$ being analytic in the upper half plane $\tau > c$, and $M_-(\alpha)$ is analytic in the upper half plane $\tau < d$.

Following Noble (1958), $M_+(\alpha)$ is obtained as

$$M_+(\alpha) = \frac{(a - i\varepsilon)^{\frac{1}{2}} \exp\left[\int_0^\alpha \{(\xi + k)/2 + (\xi/a)\Lambda_+(\xi) + (k/a)\Lambda_-(k)\}d\xi/(\xi^2 - k^2)\right]}{(b - i\varepsilon)^{\frac{1}{2}} \exp\left[\int_0^\alpha \{(\xi + \alpha_0)/2 + (\xi/b)\Lambda_+(\xi) + (\alpha_0/b)\Lambda_-(\alpha_0)\}d\xi/(\xi^2 - {\alpha_0}^2)\right]},$$

where

$$\Lambda_+(\xi) = \frac{\mathrm{i}}{\pi\gamma(\xi)} \ln \frac{\gamma(\xi) - \xi + \varepsilon}{\gamma(\xi) + \xi - \varepsilon}, \Lambda_-(\xi) = \Lambda_+(-\xi). \tag{1.50}$$

As is customary in Wiener-Hopf analysis, Eq. (1.47) is written in the form

$$\frac{\alpha + \alpha_0}{\alpha + k} M_+(\alpha) F_+(\alpha) - \frac{2k(b-a)}{i\,(\alpha_0 + k)} \frac{1}{M_{-k}} \frac{1}{\alpha + k} \tag{1.51}$$

$$= \frac{\alpha - k}{\alpha - \alpha_0} \frac{G_-(\alpha)}{M_-(\alpha)} + \frac{b-a}{\mathrm{i}} \left[\frac{\alpha - k}{(\alpha - \alpha_0)(\alpha + k)} \left\{ \frac{1}{M_-(\alpha)} - \frac{1}{M_{-k}} \right\} + \frac{\alpha_0 - k}{\alpha_0 + k} \frac{1}{M_{-k}} \frac{1}{\alpha - \alpha_0} \right].$$

The left hand side of (1.51) is analytic in $\tau > c$ and the right hand side is analytic in $\tau < d$, and as $|\alpha| \to \infty$ in respective half planes, each side tends to zero. Applying the principle of analytic continuation and Liouville's theorem, we find that each side of (1.51) vanishes identically. Thus we find the unknown function $F_+(\alpha)$, as given by

$$F_+(\alpha) = \frac{2k(b-a)}{i\,(\alpha_0 + k) M_+(k)} \frac{1}{(\alpha + \alpha_0) M_+(\alpha)}.$$

Now the use of (1.44) in (1.45) produces $D(\alpha)$ as

$$D(\alpha) = -\frac{F_+(\alpha)}{\gamma(\alpha) - a}$$

so that

$$\varphi(x, y) == -\frac{k(b-a)}{i\,\pi(\alpha_0 + k) M_+(k)} \int_C \frac{e^{-\mathrm{i}\alpha x - \gamma y}}{(\gamma - a)(\alpha + \alpha_0) M_+(\alpha)} d\alpha, \tag{1.52}$$

where C is a line parallel to the real axis in the complex α- plane and lies in the strip $c < \tau < d$.

Case 2: $b < 0$. In this case the coefficient of $F_+(\alpha)$ in (1.47) is written as $L(\alpha)/(\alpha^2 - k^2)$ where $L(\alpha) = (\gamma + a)/(\gamma + |b|)$ which is analytic in the strip $c < \tau < d$. $L(\alpha)$ can be factorised as

$$L(a) = L_+(\alpha) + L_-(\alpha), \tag{3.15}$$

where $L_+(\alpha) = L_-(\alpha)$, $L_+(\alpha)$ is analytic in $\tau > c$, $L_-(\alpha)$, is analytic in $\tau < d$ and that $L_+(\alpha) = O(|\alpha|)$ as $|\alpha| \to \infty$ in $\tau > c$. Following the same procedure as used to obtain $M_+(\alpha)$ above, $L_+(\alpha)$ is found to be

$$L(\alpha) = \{(a - i\epsilon)(-b - i\epsilon)\}^{1/2} \exp\left[\int_0^\alpha \left\{\frac{\xi + k}{2} + \frac{\xi}{a}\,\Lambda_+(\xi) + \frac{k}{a}\,\Lambda_-(k)\right\} \frac{d\xi}{\xi^2 - k^2}\right.$$

$$\left. + \int_0^\alpha \left\{\frac{\xi + \alpha_0}{2} + \frac{\xi}{b}\,\Lambda_+(\xi) + \frac{\alpha_0}{b}\,\Lambda_-(\alpha_0)\right\} \frac{d\xi}{\xi^2 - {\alpha_0}^2}\right] \tag{1.54}$$

where $\Lambda_+(\xi)$ is the same expression as given in relation (1.50).

Using a similar procedure as in Case 1, the function $\varphi(x, y)$ $(y > 0)$ in this case is obtained as

$$\varphi(x,y) == \frac{k(a-b)}{i\,\pi L_+(k)}\int_C \frac{e^{-\mathrm{i}\alpha x-\gamma y}}{(\gamma-a)L_+(\alpha)}d\alpha, \tag{1.55}$$

where C is the same contour as in (1.52).

Case 3: $|b| = \infty$. In this case condition (1.34) assumes the form

$$\varphi = \mathrm{e}^{\mathrm{i}kx} \quad \text{on } y = 0 \text{ for } x > 0$$

so that the functional relation (1.47) is modified as

$$\frac{\gamma(\alpha)+a}{\alpha^2-k^2}F_+(\alpha) - G_-(\alpha) = \frac{1}{i(\alpha+k)} \qquad \text{for } c < \tau < d. \tag{1.56}$$

Here we factorized $P(\alpha) = \gamma(\alpha) + a$ $(a > 0)$ in the form

$$P(\alpha) = P_+(\alpha)\,P_-(\alpha) \tag{1.57}$$

where $P_+(\alpha) = P_-(-\alpha)$, $P_+(\alpha)$ is analytic in $\tau > c$, $P_-(\alpha)$ is analytic in $\tau < d$, $P_+(\alpha) = O(|\alpha|^{1/2})$ as $|\alpha| \to \infty$ in $\tau > c$. $P_+(\alpha)$ is obtained as

$$P_+(\alpha) = (\alpha - \mathrm{i}\varepsilon)^{\frac{1}{2}}\exp\left[\int_0^\alpha \left\{\frac{\xi+k}{2} + \frac{\xi}{a}\,\Lambda_+(\xi) + \frac{k}{a}\,\Lambda_-(k)\right\}\frac{d\xi}{\xi^2-k^2}\right]. \tag{1.58}$$

Following a similar procedure as in Case 1, the function $\varphi(x, y)$ $(y > 0)$ in this case is obtained as

$$\varphi(x,y) = -\frac{k}{i\,\pi P_+(k)}\int_C \frac{e^{-\mathrm{i}\alpha x-\gamma y}}{(\gamma-a)P_+(\alpha)}d\alpha, \tag{1.59}$$

where C is the same contour as in (1.52).

The representations (1.52), (1.55) and (1.59) for $\phi_1(x, y)$ are now analyzed after passing on to the limit $\varepsilon \to 0$ so as to obtain the solution of BVP for $b > 0$, $b < 0$ and $|b| = \infty$ respectively. As $\varepsilon \to 0$, the functions $M_\pm(\alpha)$ in (1.49), $L_\pm(\alpha)$ in (1.53) and $P_\pm(\alpha)$ in (1.57) reduce, respectively, to

$$M_\pm^0(\alpha) = M_-^0(-\alpha) = \frac{(\alpha+a)^{1/2}\exp\left[\frac{1}{i\pi}\int_0^{\alpha/a}\frac{\ln\xi}{\xi^2-1}d\xi\right]}{(\alpha+b)^{1/2}\exp\left[\frac{1}{i\pi}\int_0^{\alpha/b}\frac{\ln\xi}{\xi^2-1}d\xi\right]},$$

$$L_+^0(\alpha) = L_-^0(-\alpha) = (\alpha+a)^{\frac{1}{2}}(\alpha-b)^{\frac{1}{2}}\exp\left[\frac{1}{i\pi}\left\{\int_0^{\alpha/a}\int_0^{\alpha/|b|}\right\}\frac{\ln\xi}{\xi^2-1}d\xi\right], \tag{1.60}$$

$$P_+^0(\alpha) = P_-^0(-\alpha) = (\alpha+a)^{\frac{1}{2}}\exp\left[\frac{1}{i\pi}\int_0^{\alpha/a}\frac{\ln\xi}{\xi^2-1}d\xi\right].$$

Again, as $\varepsilon \to 0$, we note that

$$k \to a,\ \alpha_0 \to b \text{ and } \gamma \to \alpha \operatorname{sgn} \operatorname{Re}\alpha. \tag{1.61}$$

Using (1.61) and (1.60) in (1.52), (1.55) and (1.58), we obtain the solution of the BVP as

$$\phi(x,y)=\begin{cases}-\dfrac{a(b-a)}{i\pi\,(a+b)M_+^0(a)}\displaystyle\int_{-\infty}^{\infty}\dfrac{e^{-\alpha(\text{sgn Re }\alpha)y-i\alpha x}}{(\alpha\text{ sgn Re }\alpha-a)(\alpha+b)M_+^0(\alpha)}d\alpha & \text{for } b>0,\\ -\dfrac{a(b-a)}{i\pi L_+^0(a)}\displaystyle\int_{-\infty}^{\infty}\dfrac{e^{-\alpha(\text{sgn Re }\alpha)y-i\alpha x}}{(\alpha\text{ sgn Re }\alpha-a)L_+^0(\alpha)}d\alpha & \text{for } b<0,\\ -\dfrac{a}{i\pi P_+^0(a)}\displaystyle\int_{-\infty}^{\infty}\dfrac{e^{-\alpha(\text{sgn Re }\alpha)y-i\alpha x}}{(\alpha\text{ sgn Re }\alpha-a)P_+^0(\alpha)}d\alpha & \text{for } b=\infty,\end{cases}\tag{1.62}$$

where the path of the integral is indented above (below), the poles on the negative (positive) real axis.

To evaluate the integrals in (1.62), we introduce the polar coordinates (r,β) where $x=r\cos\beta$, $y=r\sin\beta$ $(0\le\beta\le\pi)$. The poles of the integrands are on the real axis. For $x<0(>0)$ we deform the contour over the bisectors of the first and second (third and fourth) quadrants of the complex α-plane. The integrands decrease exponentially on the bisectors and we retain up to the order of r^{-1} for the integrals on the bisector. Thus we finally obtain the following asymptotic results, valid for large r.

For $x<0$,

$$\phi(x,y)\approx\begin{cases}\dfrac{2(b-a)}{i\pi b(a+b)M_+^0(a)M_+^0(0)}\dfrac{\sin\beta}{r}-\dfrac{2a(b-a)}{\{(a+b)M_+^0(a)\}^2}e^{-ay-iax} & \text{for } b>0,\\ \dfrac{2(b-a)}{i\pi L_+^0(a)L_+^0(0)}\dfrac{\sin\beta}{r}+\dfrac{2a(b-a)}{\{L_+^0(a)\}^2}e^{-ay-iax} & \text{for } b<0,\\ \dfrac{2(b-a)}{i\pi P_+^0(a)P_+^0(0)}\dfrac{\sin\beta}{r}+\dfrac{2a}{\{P_+^0(a)\}^2}e^{-ay-iax} & \text{for } |b|=\infty,\end{cases}\tag{1.63}$$

while for $x>0$

$$\phi(x,y)\approx\begin{cases}-e^{-ay+iax}+\dfrac{2(b-a)}{i\pi b(a+b)M_+^0(a)M_+^0(0)}\dfrac{\sin\beta}{r}+\dfrac{4abM_+^0(b)}{(a+b)^2M_+^0(a)}e^{-by+ibx} & \text{for } b>0,\\ -e^{-ay+iax}-\dfrac{2(b-a)}{i\pi L_+^0(a)L_+^0(0)}\dfrac{\sin\beta}{r} & \text{for } b<0.\\ -e^{-ay+iax}-\dfrac{2(b-a)}{i\pi P_+^0(a)P_+^0(0)}\dfrac{\sin\beta}{r} & \text{for } |b|=\infty.\end{cases}\tag{1.64}$$

Since the total field is given by (cf. Eq. (1.31))

$$\phi^t=e^{-ay+iax}+\varphi(x,y),$$

we observe that the behavior of the total field as $|x|\to\infty$, given by (1.30), is satisfied by (1.63) and (1.67). The complex constants A_1 and B_1 which are the reflection and transmission coefficients (complex), respectively, are now determined explicitly. We note that for $b>0$, there occurs reflection and transmission of the incoming wave train by the discontinuity at (0,0) into the regions $x<0$ and $x>0$, respectively, while for $b<0$ and $|b|=\infty$, there is no transmitted wave in the region $x>0$. This is expected, since in the latter case the inertial surface is too heavy to allow for the propagation of the incoming wave train after it encounters the discontinuity at the origin. We note that the first terms in the right hand side of (1.68) and the second terms in the right hand side of (1.46) arise due to interaction of the incident wave train with the discontinuity at the origin and they die out at large distance from the origin. These do not represent any wave.

Now comparing (1.30) with (1.63) and (1.64) we find that the complex reflection and transmission coefficients are given by

$$A_1, B_1 = \begin{cases} -\dfrac{2a(b-a)}{\{(a+b)M_+^0(a)\}^2}, \dfrac{4abM_+^0(b)}{(a+b)^2 M_+^0(a)} & \text{for } b > 0, \\ \dfrac{2a(b-a)}{\{L_+^0(a)\}^2}, 0 & \text{for } b < 0, \\ \dfrac{2a}{\{P_+^0(a)\}^2}, 0 & \text{for } |b| = \infty. \end{cases} \tag{1.65}$$

Here the reflection and transmission coefficients (real) are obtained as

$$|A_1|, |B_1| = \begin{cases} \left|\dfrac{b-a}{a+b}\right|, \dfrac{2|ab|^{1/2}}{|a+b|} & \text{for } b > 0 \\ 1, 0 & \text{for } b < 0 \\ 1, 0 & \text{for } |b| = \infty. \end{cases} \tag{1.66}$$

In deriving the results in (1.66), we have used from (1.60)

$$|M_+^0(a)| = \left|\frac{2a}{a+b}\right|^{\frac{1}{2}}, \quad |M_+^0(b)| = \left|\frac{a+b}{2b}\right|^{\frac{1}{2}}, |L_+^0(a)| = |2a(a-b)|^{1/2}, |P_+^0(a)| = (2a)^{1/2}. \tag{1.67}$$

The results in (1.66) for $b > 0$ have been obtained by Chakrabarti (2000a) by a different technique. It is also verified from (1.66) that the principle of conservation of energy, viz.

$$|A_1|^2 + |B_1|^2 = 1$$

holds good.

Reflection coefficient

The quantities a and b *are* related and for deep water the relationship can be expressed as

$$b = \frac{a}{1 - l_0 a},$$

where $l_0 \left(= \frac{\sigma}{\rho}\right)$ can be interpreted as the height of a vertical cylinder containing the fluid whose mass is the same as that of the floating matter distributed over the cross-sectional area of the cylinder at the inertial surface. In order that there exist time-harmonic progressive waves at the inertial surface, b must be positive. Hence if the frequency ω of the incident wave in the region $x < 0$ is kept fixed, then $b > 0$ implies that $\sigma < \rho g/\omega^2$, which is usually interpreted as the inertial surface to be light. However, if $b < 0$ or $|b| = \infty$, then $\sigma \geq \frac{\rho g}{\omega^2}$ and the inertial surface is then interpreted as heavy since it does not allow propagation of time harmonic waves, as was pointed out earlier. Again,

$b \lesseqgtr 0$ *also* implies $\omega \lesseqgtr \omega_0$ where $\omega_0 = (\rho g/\sigma)^{1/2}$. This means that ω_0 represents a kind of threshold frequency, since if the frequency of the incident wave train exceeds this frequency, then the inertial surface does not allow propagation of any time-harmonic wave. This phenomenon is well known in the literature cf. Peters (1950), Weitz and Keller (1950), Rhodes-Robinson (1984).

As the inertial surface is in the form of a semi-infinite plane ($y = 0, x > 0, -\infty < z < \infty$), the incident wave field undergoes reflection into the region $x < 0$ by the edge $x = 0$ and transmission into the region $x > 0$ provided the incident wave frequency ω is less than the threshold frequency ω_0. However, when $\omega \geq \omega_0$ the incident wave field from the region $x < 0$ is totally reflected back into the region $x < 0$ by the edge $x = 0$. Of course, there are local excitations by the edge $x = 0$ and these do not propagate as waves and die out away from the edge.

The reflection coefficient $|A_1|$ for $b > 0$ is depicted graphically in Fig. 5.1 against the wave number aL where as mentioned earlier, L is a characteristic length used to nondimensionalize a, b (> 0) and $\sigma/\rho(= l_0)$ for deep water, choosing $\frac{l_0}{L} = 0.01, 0.1$. It is observed from this figure that for fixed l_0/L, $|A_1|$ increases uniformly with the wave number aL.

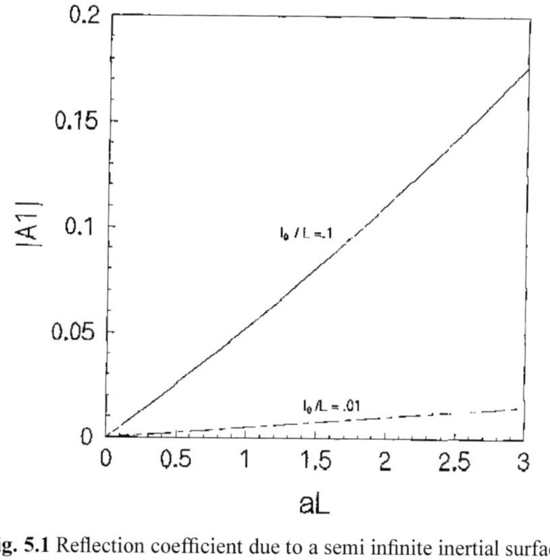

Fig. 5.1 Reflection coefficient due to a semi infinite inertial surface

This is expected since, as the wave number increases, the incident wave remains confined within a thin layer below the free surface in the region $x < 0$ and as it encounters the edge $x = 0$, reflection by the edge becomes more. It is also observed that for fixed wave number, $|A_1|$ increases as l_0/L increases, i.e., as the surface density of the material of the inertial surface increases. This means that as the inertial surface becomes heavier, more energy is reflected by its edge, provided of course the inertial surface is 'light' enough to allow progressive waves to propagate on it.

5.2 Scattering by an inertial surface of finite width

Instead of an inertial surface occupying half the surface of deep water (or half the interface of two superposed fluids) as has been considered in section 5.1, if it occupies an infinite strip of finite width l, say, on the surface (interface), so that it is sandwiched between two surfaces (interfaces), one on the left of $x = 0$ and the other on the right of $x = l$, then we have two points of discontinuity in the surface condition. The mathematical formulation of the corresponding physical problem is described as the following bounded value problem.

Boundary value problem

To solve the Laplace equation

$$\varphi_{xx} + \varphi_{yy} = 0 \;, y > 0, -\infty < x < \infty, \tag{2.1}$$

along with the surface boundary condition, as given by

$$\varphi_y + a\varphi = 0 \quad \text{on} \quad y = 0 \quad \text{for}\, x < 0 \quad \text{and} \quad x > l, \tag{2.2}$$

$$\varphi_y + b\varphi = -(b-a)e^{iax} \quad \text{on} \;\; y = 0 \;\; \text{for} \;\; 0 < x < l \tag{2.3}$$

producing two discontinuities in the surface condition at the points (0,0) and $(l, 0)$, $l(> 0)$ being finite, representing the width of the strip, the edge conditions

$$\varphi = 0(1) \;\; \text{and} \;\; \nabla\varphi = 0(1) \;\; \text{as} \;\; r \to 0 \;\; \text{for finite} \;\; |b|,$$

or

$$\varphi = O(1) \;\; \text{and} \;\; r^{\frac{1}{2}}\nabla\varphi = O(1) \;\; \text{as} \;\; r \to 0 \;\text{for}\; |b| = \infty \tag{2.4}$$

which are given by the physics of the problem and where r represents the distance from the two discontinuities at (0,0) and $(0, l)$, the conditions

$$\varphi, \nabla\varphi \to 0 \;\; \text{as} \;\; y \to \infty, \tag{2.5}$$

and the condition as $|x| \to \infty$, as given by

$$\begin{aligned} \varphi(x,y) + e^{-ay+iax} &\sim A_1 e^{-ay-iax} + e^{-ay+iax} \;\text{as}\; x \to \infty, \\ &= \begin{cases} B_1 e^{-by+ibx} + B_2 e^{-by+ib(x-l)} + \mu_1(x,y) \;, 0 < x < l \,, \text{for}\; b > 0 \\ \mu_2(x,y) \;,\;\; 0 < x < l \,,\;\; \text{for}\;\; b < 0 \;\; \text{or} \;\; |b| = \infty \,, \end{cases} \\ &\sim B_3 e^{-by+ib(x-l)} \;\text{as}\;\; x \to \infty. \end{aligned} \tag{2.6}$$

In (2.6), A_1, B_1, B_2, B_3 are four unknown constants to be determined and the functions $\mu_1(x, y)$, $\mu_2(x, y)$ represent two unknown, ordinary, non-wavy solutions of the Laplace equation. Physically, A_1 represents the reflection coefficient (complex) in the region $x < 0$, B_1 and B_2, respectively, representing the transmission and reflection coefficients (complex) in the region $0 < x < l$ and B_3 represents the transmission coefficient (complex) in the region $x < l$, corresponding to the incident wave field propagating in the region $x < 0$ as in the BVP of the section 5.1.

As in section 5.1, this BVP can be handled for its solution by the aid of Wiener-Hopf technique after generalizing the Laplace equation (2.1) to the Helmholtz equation

$$\varphi_{xx} + \varphi_{yy} + \varepsilon^2\varphi = 0 \ , y > 0, -\infty < x < \infty \tag{2.7}$$

where ε is a complex number with small positive real and imaginary parts as well as by generalizing some conditions and passing on to the limit as $\varepsilon \to 0$. Thus the generalized form of the BVP is to solve the Helmholtz equation (2.7) along with the boundary conditions

$$\varphi_y + a\varphi = 0 \ \text{on} \ \ y = 0 \ \text{for} \ x < 0 \ \text{and} \ x > l, \tag{2.8}$$

$$\varphi_y + b\varphi = -(b-a)e^{ikx} \text{on} \ y = 0 \text{ for } 0 < x < l \tag{2.9}$$

where

$$k = (a^2 + \varepsilon^2)^{\frac{1}{2}} \ \text{with} \ \ k = a \text{ for } \varepsilon = 0, \tag{2.10}$$

the edge conditions

$$\varphi = 0(1) \ \text{and} \ \nabla\varphi = O(1) \text{ as } r \to 0 \ \text{ for finte } |b| \tag{2.11a}$$

or

$$\varphi = O(1) \ \text{and} \ r^{\frac{1}{2}}\nabla \ \varphi = O(1) \text{ as } r \to 0 \ \text{ for } |b| \ = \infty \tag{2.11b}$$

where r denotes the distance from the two discontinuities at (0,0) and (l,0) so that r = $(x^2 + y^2)^{\frac{1}{2}}$ or $\{(x - l)^2 + y^2)\}^{\frac{1}{2}}$, and the conditions at infinity, as given by

$$|\varphi| + |\nabla\varphi| \le \text{const.}\, e^{-\delta(\varepsilon)r} \ \text{ as } r = (x^2 + y^2)^{\frac{1}{2}} \ \to \infty \tag{2.12}$$

where $\delta(\varepsilon)$ is such that

$$\delta(\varepsilon) \to 0+ \ \text{ as } \ \varepsilon \to 0. \tag{2.13}$$

It may be noted that the condition (2.12) is stronger than the condition (2.5) as $y \to \infty$, and is consistent with the behavior of φ as $|x| \to \infty$.

In this case a three-part Wiener-Hopf problem occurs. Its approximate solution has been given by Kanoria et al. (1999).

The three-part Wiener-Hopf technique

Let $\varphi(x, y) \equiv \varphi(x, y; \varepsilon)$ denote the function satisfying the above generalized BVP described by (2.7)–(2.9), (2.11) to (2.12). The Fourier transform of $\varphi(x, y)$ is

$$\Phi(\alpha, y) = \int_{-\infty}^{\infty} \varphi(x, y)e^{i\alpha x} \ dx = \Phi_-(\alpha, y) + \mathrm{e}^{\mathrm{i}\alpha l}\Phi_+(\alpha, y) + \int_0^1 \Phi(\alpha, y)\mathrm{e}^{\mathrm{i}\alpha x} \ dx$$

where

$$\Phi_-(\alpha, y) = \int_{-\infty}^{0} \varphi(x, y)e^{i\alpha x} \ dx \ , \Phi_+(\alpha, y) = \int_{l}^{\infty} \varphi(x, y)e^{i\alpha(x-l)}dx. \tag{2.14}$$

By using condition (2.12) it is observed that $\Phi_+(\alpha,y)$ and $\Phi_-(\alpha,y)$ are analytic functions of α in the half planes $\tau > -\delta(\varepsilon)$ and $\tau < \delta(\varepsilon)$, respectively. Again, by using the edge conditions (2.11) along with the Abelian theorem, it can be shown that

$$\begin{aligned}|\Phi_+(\alpha,y)| &= 0(|\alpha|^{-1}) \text{ as } |\alpha| \to \infty \text{ in } \tau > -\delta(\varepsilon),\\ |\Phi_-(\alpha,y)| &= 0(|\alpha|^{-1}) \text{ as } |\alpha| \to \infty \text{ in } \tau < \delta(\varepsilon).\end{aligned} \tag{2.15}$$

To use the Wiener-Hopf procedure, as in section 5.1, conditions (2.8) and (2.9) are written in the form

$$\varphi_y + a\varphi = \begin{cases} 0 & \text{on } y = 0 \text{ for } x < 0 \text{ and } x > l,\\ f(x) & \text{on } y = 0 \text{ for } 0 < x < l\end{cases} \tag{2.16}$$

and

$$\varphi_y + b\varphi = \begin{cases} u(x) & \text{on } y = 0 \text{ for } x < 0,\\ -(b-a)e^{ikx} & \text{on } y = 0 \text{ for } 0 < x < l,\\ v(x) & \text{on } y = 0 \text{ for } x > l\end{cases} \tag{2.17}$$

where $f(x)$ (for $0 < x < l$), $u(x)$ (for $x < 0$) and $v(x)$ (for $x > l$) are unknown functions having the following behavior at the points $x = 0$ and $x = l$,

$$f(x) = O(1) \text{ as } x \to +0, \text{and } x \to l-0, \tag{2.18}$$

$$u(x) = \begin{cases} O(1) \text{ as } x \to -0 \text{ for finite } |b|,\\ O\left(x^{-1/2}\right) \text{ as } x \to -0 \text{ for } |b| = \infty,\end{cases} \tag{2.19}$$

$$v(x) = \begin{cases} O(1) \text{ as } x \to l+0 \text{ for finite } |b|,\\ O\left(|x-l|^{-1/2}\right) \text{ as } x \to l+0 \text{ for } |b| = \infty,\end{cases} \tag{2.20}$$

obtained from the edge condition (2.11).

Now an appropriate solution for

$$\Phi(\alpha,y) = D(\alpha)e^{-\gamma y}, y > 0, \tag{2.21}$$

where $D(\alpha)$ is an arbitrary function of α, and is determined from the relations obtained by Fourier-transforming conditions (2.16) and (2.17), as given by

$$\begin{aligned}\Phi'(\alpha,0) + a\,\Phi(\alpha,0) &= F_1(\alpha),\\ \Phi'(\alpha,0) + b\Phi(\alpha,0) &= U_-(\alpha) + e^{i\alpha l}V_+(\alpha) - \frac{b-a}{i(\alpha+k)}\left\{e^{i(\alpha+k)l} - 1\right\}.\end{aligned} \tag{2.22}$$

In (2.22) the three unknown functions $U_-(\alpha)$, $V_+(\alpha)$ and $F_1(\alpha)$ are defined by

$$\begin{aligned}U_-(\alpha) = \int_{-\infty}^{0} u(x)e^{i\alpha x}\,dx, \quad V_+(\alpha) &= \int_l^\infty v(x)e^{i\alpha(x-l)}\,dx,\\ F(\alpha) &= \int_0^l f(x)e^{i\alpha x}\,dx.\end{aligned} \tag{2.23}$$

It can be shown that $U_-(\alpha)$ is analytic in the half-plane $\tau < \delta(\varepsilon)$, $V_+(\alpha)$ is analytic in the half-plane $\tau > -\delta(\varepsilon)$ and $F(\alpha)$ is an integral function of α. Use of the edge conditions (4.5)–(4.7) ensures that

$$U_-(\alpha) = \begin{cases} O(|\alpha|^{-1}) \text{ as } |\alpha| \to \infty \text{ in } \tau < \delta(\varepsilon) \text{ for finite } |b|, \\ O(|\alpha|^{-1/2}) \text{ as } |\alpha| \to \infty \text{ in } \tau < \delta(\varepsilon) \text{ for } |b| = \infty, \end{cases}$$

$$V_+(\alpha) = \begin{cases} 0(|\alpha|^{-1}) \text{ as } |\alpha| \to \infty \text{ in } \tau > -\delta(\varepsilon) \text{ for finite } |b|, \\ 0(|\alpha|^{-1/2}) \text{ as } |\alpha| \to \infty \text{ in } \tau > -\delta(\varepsilon) \text{ for } |b| = \infty, \end{cases} \tag{2.24}$$

$$|e^{-i\alpha l}F(\alpha)| = 0(|\alpha|^{-1}) \text{ as } |\alpha| \to \infty \text{ in } \tau < \delta(\varepsilon),$$
$$|F(\alpha)| = 0(|\alpha|^{-1/2}) \text{ as } |\alpha| \to \infty \text{ in } \tau > -\delta(\varepsilon).$$

Using (2.21) and (2.22) and eliminating $D(\alpha)$ we obtain the following three-part Wiener-Hopf functional relation, for the determination of the three unknown functions $U_-(\alpha)$, $V_+(\alpha)$ and $F_1(\alpha)$, as given by

$$\frac{\gamma(\alpha) - b}{\gamma(\alpha) - a} F(\alpha) = U_-(\alpha) + e^{i\alpha l} V_+(\alpha) - \frac{b-a}{i(\alpha + k)} \{e^{i(\alpha+k)l} - 1\} \tag{2.25}$$

valid in the strip $c < \tau < d$ where $c(< 0)$ and $d(> 0)$satisfy inequality

$$-\delta(\varepsilon) < -\min(\text{Im}k, \text{Im}\alpha_0) < c < 0 < d, \min(\text{Im}k, \text{Im}\alpha_0) < \delta(\varepsilon).$$

As in section 5.1, here also three cases arise according as $b > 0$, $b < 0$ and $|b| = \infty$, and we treat the three-part Wiener-Hopf relation (2.25) in three different manners as described below.

Case 1: $b > 0$. Using the same Wiener-Hopf decomposition (1.49), in section 5.1 for $M(\alpha) = (\gamma + a)/(\gamma + b) = M_+(\alpha)\, M_-(\alpha)$, multiplying both sides of (2.15) by $e^{-i\alpha l}/M_+(\alpha)$ and rearranging, we obtain

$$\frac{\alpha + k}{\alpha + \alpha_0} \frac{V_+(\alpha)}{M_+(\alpha)} - \frac{b-a}{i(\alpha + \alpha_0)} \frac{e^{ikl}}{M_+(\alpha)} + \zeta_+(\alpha) + \eta_+(\alpha)$$
$$= \frac{\alpha - \alpha_0}{\alpha - k} M_-(\alpha) e^{-i\alpha l} F(\alpha) - \zeta_-(\alpha) - \eta_-(\alpha) \tag{2.26}$$

where

$$\zeta_+(\alpha) + \zeta_-(\alpha) = \frac{e^{-i\alpha l}(\alpha + k) U_-(\alpha)}{(\alpha + \alpha_0) M_+(\alpha)},$$
$$\eta_+(\alpha) + \eta_-(\alpha) = \frac{(b-a) e^{-i\alpha l}}{(\alpha + \alpha_0) i(\alpha + \alpha_0) M_+(\alpha)}. \tag{2.27}$$

In (2.27), $\zeta_+(\alpha)$, $\eta_+(\alpha)$ are analytic in $\tau > c$ and $\zeta_-(\alpha)$, $\eta_-(\alpha)$ are analytic in $\tau < d$, and their explicit forms can be obtained by employing the additive decomposition theorem, see Noble (1958), p.13. Similarly, multiplying both sides of (2.25) by $1/(M_-(\alpha))$ and rearranging we obtain

$$\frac{\alpha - k}{\alpha - \alpha_0} \frac{U_-(\alpha)}{M_-(\alpha)} + R_-(\alpha) - S_-(\alpha) + \frac{b-a}{i(\alpha - \alpha_0)} \left[\frac{\alpha - k}{\alpha - \alpha_0} \left\{ \frac{1}{M_-(\alpha)} - \frac{1}{M_-(-k)} \right\} + \frac{\alpha_0 - k}{\alpha_0 + k} \frac{1}{M_-(-k)} \right]$$
$$= \frac{\alpha + \alpha_0}{\alpha + k} M_-(\alpha) F(\alpha) - R_+(\alpha) + S_+(\alpha) - \frac{2(b-a)k}{i(\alpha_{0+k}) M_-(-k)} \frac{1}{\alpha + k}, \tag{2.28}$$

where

$$R_+(\alpha) + R_-(\alpha) = \frac{e^{i\alpha l}(\alpha - k)V_+(\alpha)}{(\alpha - \alpha_0)M_-(\alpha)},$$
$$S_+(\alpha) + S_-(\alpha) = \frac{(b-a)(\alpha - k)e^{i(\alpha+k)l}}{i(\alpha + k)(\alpha - \alpha_0)M_-(\alpha)}, \tag{2.29}$$

$R_+(\alpha), S_+(\alpha)$ being analytic in $\tau > c$ and in $R_-(\alpha), S_-(\alpha)$ in $\tau < d$ and their explicit forms can be obtained by employing the additive decomposition theorem mentioned above.

The left-hand side of (2.26) and the right-hand side of (2.28) are analytic in $\tau > c$ while the other sides are analytic in $\tau < d$. Using (2.24), it is seen that each side of (2.26) and (2.28) tends to zero as $|\alpha| \to \infty$ in the appropriate half planes having a common region $c < \tau < d$, so that by Liouville's theorem, each side is identically zero. We are interested in the left-hand sides of (2.26) and (2.28).

For brevity, we introduce the notation

$$\psi_-^*(\alpha) = U_-(\alpha) + \frac{(b-a)}{i(\alpha + k)},$$
$$\psi_+(\alpha) = V_+(\alpha) - \frac{(b-a)}{i(\alpha + k)}e^{ilk} \tag{2.30}$$

where the superscript star is used to indicate that $\psi_-^*(\alpha)$ has a pole at $\alpha = -k$ but apart from this, it is analytic in $\tau < d$, and $\psi_+(\alpha)$ is analytic in $\tau > c$. On equating the left-hand sides of (2.26) and (2.28), and introducing the explicit expressions for $\zeta_+(\alpha)$, $\eta_+(\alpha)$, $R_-(\alpha)$, $S_-(\alpha)$ and using the notation (2.30), we obtain

$$\frac{\alpha + k}{\alpha + \alpha_0}\frac{\psi_+(\alpha)}{M_+(\alpha)} + \frac{1}{2\pi i}\int_{ic_1-\infty}^{ic_1+\infty} \frac{e^{-il\xi}(\xi + k)\psi_-^*(\xi)}{M_+(\xi)(\xi + \alpha_0)(\xi - \alpha)}d\xi = 0, \tau > c \tag{2.31}$$

and

$$\frac{\alpha - k}{\alpha - \alpha_0}\frac{\psi_-^*(\alpha)}{M_-(\alpha)} - \frac{1}{2\pi i}\int_{id_1-\infty}^{id_1+\infty} \frac{e^{il\xi}(\xi - k)\psi_+(\xi)}{M_-(\xi)(\xi - \alpha_0)(\xi - \alpha)}d\xi$$
$$- \frac{2(b-a)k}{iM_-(-k)(\alpha_0 + k)}\frac{1}{\alpha + k} = 0, \qquad \tau < d, \tag{2.32}$$

where $c < c_1 < 0 < d_1 < d$. We choose $c_1 = -h$, $d_1 = h$ where h is positive, then replace ξ by $-\xi$ in (2.31) and α by $-\alpha$ in (2.32). Noting that $M_+(-\alpha) = M_-(\alpha)$, this produces

$$\frac{\alpha + k}{\alpha + \alpha_0}\frac{\psi_+(\alpha)}{M_+(\alpha)} - \frac{1}{2\pi i}\int_{ih-\infty}^{ih+\infty} \frac{e^{il\xi}(\xi - k)\psi_-^*(-\xi)}{M_-(\xi)(\xi - \alpha_0)(\xi + \alpha)}d\xi = 0 \tag{2.33}$$

and

$$\frac{\alpha + k}{\alpha + \alpha_0}\frac{\psi_-^*(-\alpha)}{M_+(\alpha)} - \frac{1}{2\pi i}\int_{ih-\infty}^{ih+\infty} \frac{e^{il\xi}(\xi - k)\psi_+(\xi)}{M_-(\xi)(\xi - \alpha_0)(\xi + \alpha)}d\xi + \frac{2(b-a)k}{iM_-(-k)(\alpha_0 + k)}\frac{1}{\alpha - k} = 0 \tag{2.34}$$

where now $\tau > -h$ in both equations (2.33) and (2.34). We define

$$S_+^*(\alpha) = \psi_+(\alpha) + \psi_-^*(-\alpha), \quad D_+^*(\alpha)= \psi_+(\alpha) + \psi_-^*(\alpha) \tag{2.35}$$

where in this case the star denotes that the expressions are analytic in $\tau > c$ except for simple pole at $\alpha = k$. Then addition and subtraction of Eqs. (2.33) and (2.34) produce

$$\frac{\alpha+k}{\alpha+\alpha_0}\frac{S_+^*(-\alpha)}{M_+(\alpha)} - \frac{1}{2\pi i}\int_{ih-\infty}^{ih+\infty}\frac{e^{il\xi}(\xi-k)S_+^*(\xi)}{M_-(\xi)(\xi-\alpha_0)(\xi+\alpha)}d\xi + \frac{2(b-a)k}{iM_+(k)(\alpha_0+k)}\frac{1}{\alpha-k} = 0, \tau > -h \tag{2.36}$$

and

$$\frac{\alpha+k}{\alpha+\alpha_0}\frac{D_+^*(-\alpha)}{M_+(\alpha)} + \frac{1}{2\pi i}\int_{ih-\infty}^{ih+\infty}\frac{e^{il\xi}(\xi-k)D_+^*(\xi)}{M_-(\xi)(\xi-\alpha_0)(\xi+\alpha)}d\xi - \frac{2(b-a)k}{iM_+(k)(\alpha_0+k)}\frac{1}{\alpha-k} = 0, \tau > -h. \tag{2.37}$$

Eqs. (2.36) and (2.37) are of the same type and can be treated for approximate solution for large l. We write them in a compact form as given by

$$\frac{\alpha+k}{\alpha+\alpha_0}\frac{F_+^*(\alpha;\lambda)}{M_+(\alpha)} + \frac{\lambda}{2\pi i}\int_{ih-\infty}^{ih+\infty}\frac{e^{il\xi}(\xi-k)F_+^*(\xi;\lambda)}{M_-(\xi)(\xi-\alpha_0)(\xi+\alpha)}d\xi = \frac{2\lambda(b-a)k}{iM_+(k)(\alpha_0+k)}\frac{1}{\alpha-k}, \tau > -h \tag{2.38}$$

where $F_+^*(\alpha;\lambda)$ is $S_+^*(\alpha)$ or $D_+^*(\alpha)$ for $\lambda = 1$ or -1, so that from (2.30) and (2.35) to (2.27) we find that

$$F_+^*(\alpha;\lambda) = F_+(\alpha;\lambda) - \frac{(b-a)e^{ikl}}{i(\alpha+k)} + \frac{\lambda(b-a)}{i(\alpha-k)}, \tag{2.39}$$

where $F_+(\alpha;\lambda)$ is analytic in $\tau > c$, it being understood that $F_+(\alpha;1) = V_+(\alpha) - U_-(-\alpha)$ and $F_+(\alpha;-1) = V_+(\alpha) + U_-(-\alpha)$.

Now writing

$$\frac{1}{M_-(\xi)} = M_+(\xi)\frac{\gamma(\xi)+b}{\gamma(\xi)+a} = M_+(\xi)\left[1 - \frac{b-a}{\xi^2-k^2}\{a - (\xi^2-\epsilon^2)^{1/2}\}\right] \tag{2.40}$$

in the integrand of the integral in the left-hand side of relation (2.38) we note that the integrand consists of types of terms. The first type involves simple poles while the second type involves branch points at $\xi = \pm\epsilon$ in the complex ξ-plane. The integrals involving the first type of terms are evaluated by using the residue theorem after completing the contour by a semi circle of large radius in the upper half. To evaluate the integrals involving the branch points, only one branch point, viz. $\xi = \epsilon$ needs to be considered and as such a branch cut in taken parallel to the positive imaginary axis from $\xi = \epsilon$ to infinity. Then the contour is deformed into the two sides of the branch cut and contributions from the poles, if any, are taken into account. The contributions from the two sides of the branch cut involve integrals of the form

$$\int_0^\infty \psi(u)u^{1/2}du, \tag{2.41}$$

where $\psi(u)$ is an analytic function. These are evaluated asymptotically for large l. If $I(l)$ denotes the integral (2.41), then

$$I(l) = \sum_{j=0}^{\infty} \beta_j(l) \frac{\psi^j(0)}{j!},$$

where

$$\beta_j(l) = \int_0^{\infty} z^{j+1/2} e^{-zl}\, dz = \left(\frac{1}{l}\right)^{j+3/2} \Gamma(j+3/2).$$

Thus

$$\begin{aligned} I(l) = \beta_0(l)\left[\psi(0) + \frac{\beta_1}{\beta_0}\psi'(0) + \frac{\beta_2}{2!\,\beta_0}\psi''(0) + \cdots\right] &\approx \beta_0\psi\left(\frac{\beta_1}{\beta_0}\right) + O(l^{-7/2}) \\ &= \frac{\pi^{1/2}}{2}\left(\frac{1}{l}\right)^{3/2} \psi\left(\frac{3}{2l}\right) + O(l^{-7/2}). \end{aligned} \tag{2.42}$$

Incorporating the aforesaid method we find that for large l

$$\int_{ih-\infty}^{ih+\infty} \frac{e^{il\xi}(\xi - k)F_+(\xi;\lambda)}{M_-(\xi)(\xi+\alpha)(\xi-\alpha_0)} d\xi \approx 2\pi i[T(\alpha)F_+(\epsilon';\lambda) + T_1(\alpha)F_+(\alpha_0;\lambda)], \tag{2.43}$$

where

$$T(\alpha) = \frac{(b-a)E_0}{2\pi i(\epsilon' - \alpha_0)(\epsilon' + k)} \frac{1}{\epsilon' + \alpha}$$

with

$$E_0 = -\frac{\pi^{1/2}}{l^{3/2}} e^{i(\epsilon l + (3/4)\pi)} M_+(\epsilon')(\epsilon' + \epsilon)^{1/2}, \quad \epsilon' = \epsilon + \frac{3i}{2l}$$

and

$$T_1(\alpha) = \frac{2b(b-a)M_+(\alpha_0)e^{i\alpha_0 l}}{\alpha_0 + k} \frac{1}{\alpha_0 + \alpha}, \tag{2.44}$$

$$\int_{ih-\infty}^{ih+\infty} \frac{e^{il\xi}}{M_-(\xi)(\xi+\alpha)(\xi-\alpha_0)} d\xi \approx 2\pi i\left[R_1(\alpha) + \frac{T_1(\alpha)}{\alpha_0 - k}\right], \tag{2.45}$$

$$\int_{ih-\infty}^{ih+\infty} \frac{e^{il\xi}(\xi - k)F_+(\xi;\lambda)}{M_-(\xi)(\xi+\alpha)(\xi-\alpha_0)} d\xi \approx 2\pi i\left[R_2(\alpha) + \frac{T_1(\alpha)}{\alpha_0 + k}\right], \tag{2.46}$$

where

$$\begin{aligned} R_1(\alpha) &= \frac{(b-a)E_0}{2\pi i(\epsilon' - \alpha_0)(\epsilon'^2 - k^2)} \frac{1}{\epsilon' + \alpha}, \\ R_2(\alpha) &= \frac{(b-a)E_0}{2\pi i(\epsilon' - \alpha_0)(\epsilon' - k)^2} \frac{1}{\epsilon' + \alpha}. \end{aligned} \tag{2.47}$$

Using results (2.43), (2.45) and (2.46) for large l in (2.38) we obtain an approximate relation between the function $F_+(\alpha;\lambda)$ and the unknown quantities $F_+(\alpha_0;\lambda)$, $F_+(\epsilon';\lambda)$. Setting $\alpha = \alpha_0$ and $\alpha = \epsilon'$ in this we get two quantities involving these two unknowns, which when solved, produce

$$F_+(\alpha_0;\lambda) = -\frac{b-a}{i(A^\lambda D^\lambda - B^\lambda C^\lambda)}\{e^{ikl}(S^\lambda B^\lambda - Q^\lambda D^\lambda) + T^\lambda B^\lambda - R^\lambda D^\lambda\}, \tag{2.48}$$

$$F_+(\epsilon';\lambda) = -\frac{b-a}{i(A^\lambda D^\lambda - B^\lambda C^\lambda)}\{e^{ikl}(S^\lambda A^\lambda - Q^\lambda C^\lambda) + T^\lambda A^\lambda - R^\lambda C^\lambda\}, \tag{2.49}$$

where

$$A^\lambda = \frac{\alpha_0 + k}{2\alpha_0 M_+(\alpha_0)} + \lambda T_+(\alpha_0), B^\lambda = \lambda T(\alpha_0), Q^\lambda = \lambda\left\{R_2(\alpha_0) + \frac{T_1(\alpha_0)}{\alpha_0 + k}\right\},$$

$$R^\lambda = -R_1(\alpha_0) - \frac{T_1(\alpha_0)}{\alpha_0 - k} + \lambda\left\{\frac{2k}{M_+(k)(\alpha_0{}^2 - k^2)} - \frac{\alpha_0 + k}{2M_+(k)\alpha_0(\alpha_0 - k)}\right\}, \tag{2.50}$$

$$C^\lambda = \lambda T_1(\epsilon'), D^\lambda = \frac{\epsilon' + k}{M_+(\epsilon')(\epsilon' + \alpha_0)} + \lambda T(\epsilon'), S^\lambda = \frac{1}{M_+(\epsilon')(\epsilon' + \alpha_0)} + \lambda\left\{R_2(\epsilon') + \frac{T_1(\epsilon')}{\alpha_0 + k}\right\},$$

$$T^\lambda = -R_1(\epsilon') - \frac{T_1(\epsilon')}{\alpha_0 - k} + \lambda\left\{\frac{2k}{M_+(k)(\alpha_0 + k)(\epsilon' - k)} - \frac{\epsilon' + k}{2M_+(\epsilon')(\epsilon' + \alpha_0)(\epsilon' - k)}\right\}.$$

Thus $F_+(\alpha;\lambda)$ is obtained for large l and is given by

$$F_+(\alpha;\lambda) = \frac{(\alpha + \alpha_0)M_+(\alpha)}{\alpha + k}\left[\frac{b-a}{i}\left\{e^{ikl}\left(\frac{1}{(\alpha + \alpha_0)M_+(\alpha)} + \frac{\lambda T_1(\alpha)}{\alpha_0 + k} + \lambda R_2(\alpha)\right)\right.\right.$$

$$\left. -\lambda^2\left(R_1(\alpha) + \frac{T_1(\alpha)}{\alpha_0 - k}\right) - \frac{\lambda(\alpha + k)}{M_+(\alpha)(\alpha - k)(\alpha + \alpha_0)} + \frac{2\lambda k}{M_+(k)(\alpha_0 + k)(\alpha - k)}\right\} \tag{2.51}$$

$$\left. -\lambda\{T(\alpha)F_+(\epsilon';\lambda) + T_1(\alpha)F_+(\alpha_0;\lambda)\}\right]$$

where $F_+(\alpha_0;\lambda)$ and $F_+(\epsilon';\lambda)$ are given in (2.48) and (2.49), respectively. Putting $\lambda = -1$ and 1 in (2.51) we obtain two equation for $V_+(\alpha)$ and $U_-(-\alpha)$ and $V_+(\alpha) - U_-(-\alpha)$. By addition and subtraction we find $V_+(\alpha)$ and $U_-(-\alpha)$ for large l. Replacing α by $-\alpha$ we obtain $U_-(\alpha)$. Thus $V_+(\alpha)$ and $U_-(\alpha)$ are obtained for large l. Now the use of (2.11) in the second equation of (2.12) produces $D(\alpha)$. Thus we obtain $D(\alpha)$. Using this in (2.11) we obtain $\varphi(x, y)$ for large l after taking Fourier inversion, as given by

$$\varphi(x,y) = \varphi_0(x,y) - \frac{b-a}{2\pi i}$$

$$\int_C \left[\frac{(\alpha + \alpha_0)M_+(\alpha)e^{i\alpha l}}{\alpha + k} \times \left\{\frac{i\big(C_1 T(\alpha) + C_2 T_1(\alpha)\big)}{b-a} - R_1(\alpha) - \frac{T_1(\alpha)}{\alpha_0 - k}\right\}\right. \tag{2.52}$$

$$\left. + \frac{(\alpha - \alpha_0)M_-(\alpha)}{\alpha - k}\left\{\frac{i\big(C_3 T(-\alpha) + C_4 T_1(-\alpha)\big)}{b-a} - e^{ikl}\left(R_2(-\alpha) + \frac{T_1(-\alpha)}{\alpha + k}\right)\right\}\right]\frac{e^{-i\alpha x - ry}}{r - b}\, d\alpha$$

with

$$\varphi_0(x,y) = -\frac{k(b-a)}{i\pi(\alpha_0+k)M_+(k)}\int_C \frac{e^{-i\alpha x-\gamma y}}{(\gamma-\alpha)(\gamma+\alpha_0)M_+(\alpha)}d\alpha.$$

Here C is a line parallel to the real axis in the complex α-plane and lies in the strip $c < \tau < d$,

$$\begin{aligned}
C_1 &= \frac{1}{2}\{F_+(\epsilon;-1) - F_+(\epsilon;1)\},\\
C_2 &= \frac{1}{2}\{F_+(\alpha_0;-1) - F_+(\alpha_0;1)\},\\
C_3 &= \frac{1}{2}\{F_+(\epsilon;-1) - F_+(\epsilon;1)\},\\
C_4 &= \frac{1}{2}\{F_+(\alpha_0;-1) - F_+(\alpha_0;1)\}.
\end{aligned} \tag{2.53}$$

The second term in (2.52) may be regarded as due to the presence of the second discontinuity at $(l, 0)$ for large l.

Case 2: $b < 0$. Using the same Wiener–Hopf decomposition (1.53) of section 5.1 for $L(\alpha) = (\gamma + a)(\gamma + |b|)$ and proceeding as in *Case 1,* we obtain in place of (2.38)

$$\begin{aligned}
&\frac{(\alpha+k)\mathrm{F}_+^*(\alpha;\lambda)}{L_+(\alpha)} - \frac{\lambda}{2\pi i}\int_{ih-\infty}^{ih+\infty}\frac{e^{il\xi}(\xi-k)\mathrm{F}_+^*(\xi,\lambda)}{L_-(\xi)(\xi+\alpha)}d\xi\\
&\qquad = -\frac{2\lambda(a-b)k}{iL_+(k)}\frac{1}{\alpha-k}, \quad \tau > -h,
\end{aligned} \tag{2.54}$$

where $F_+^*(\alpha;\lambda)$ has the same meaning as given in Eq. (2.39). We write

$$\frac{1}{L^-(\xi)} = \frac{L_+(\xi)}{(a+b)}\left[\frac{a}{\xi^2-k^2} + \frac{b(\xi^2-k^2)-({\alpha_0}^2-k^2)(\xi^2-\epsilon^2)^{1/2}}{\xi^2-k^2}\right] \tag{2.55}$$

in the integrand of the integral in the left-hand side of (2.54). Following similar arguments as described earlier, we find that, for large l,

$$\int_{ih-\infty}^{ih+\infty}\frac{e^{il\xi}(\xi-k)F_+(\xi;\lambda)}{(\xi+k)L_-(\xi)}d\xi \approx 2\pi i T^1(\alpha)F_+(\epsilon';\lambda)$$

where

$$\begin{aligned}
T^1(\alpha) &= \frac{b-a}{2\pi i}\frac{E_0^1}{(\epsilon'^2-{\alpha_0}^2)(\epsilon'+k)}\frac{1}{\epsilon'+\alpha},\\
E_0^1 &= -\frac{\pi^{1/2}}{l^{3/2}}e^{i(\epsilon l+3\pi/4)}L_+(\epsilon')(\epsilon'+\epsilon)^{1/2},
\end{aligned} \tag{2.56}$$

$$\int_{ih-\infty}^{ih+\infty}\frac{e^{il\xi}}{(\xi+k)L_-(\xi)}d\xi \approx 2\pi i R_1^1(\alpha)$$

where

$$R_1^1(\alpha) = \frac{b-a}{2\pi i}\frac{E_0}{(\epsilon'^2 - k^2)(\epsilon'^2 - \alpha_0{}^2)}\frac{1}{\epsilon' + \alpha'}, \tag{2.57}$$

and

$$\int_{ih-\infty}^{ih+\infty} \frac{(\xi - k)e^{il\xi}}{(\xi + \alpha)(\xi + k)L_-(\xi)} d\xi \approx 2\pi i R_2^1(\alpha),$$

where

$$R_2^1(\alpha) = \frac{b-a}{2\pi i}\frac{E_0}{(\epsilon' + k)^2(\epsilon'^2 - \alpha_0{}^2)}\frac{1}{\epsilon' + \alpha'} \tag{2.58}$$

E_0 being the same as given in (2.44).

Proceeding as in Case 1 we finally obtain $\varphi(x, y)$ for large l in this case, as given by

$$\varphi(x,y) = \varphi_0(x,y) + \frac{a-b}{2\pi i} \tag{2.59}$$

$$\int_C \left[\frac{L_+(\alpha)}{\alpha + k} e^{i\alpha l}\{R_1^1(\alpha) + C_5 T^1(\alpha)\} - \frac{L_-(\alpha)}{\alpha - k}\{e^{ikl} R_2^1(-\alpha) + C_6 T^1(-\alpha)\}\right] \frac{e^{-i\alpha x - \gamma y}}{\gamma - b} d\alpha$$

where now $\varphi_0(x, y)$ is the same expression as given in (2.52) and

$$C_5 = \frac{L_+(\epsilon')}{(\epsilon' + k)^2 - \{T^1(\epsilon')L_+(\epsilon')\}^2}\{(\epsilon' + k)G_2^1(\epsilon') - T^1(\epsilon')L_+(\epsilon')G_2^1(\epsilon')\}$$

with

$$G_1^1(\alpha) = -\frac{e^{ikl}}{L_+(\alpha)} - R_1^1(\alpha),$$

$$G_2^1(\alpha) = \frac{2k}{L_+(k)}\frac{1}{\alpha - k} - \frac{\alpha + k}{L_+(\alpha)(\alpha - k)} - e^{ikl} R_2^1(\alpha) \tag{2.60}$$

and C_6 is the same as C_5 with G_1^1 and G_2^1 interchanged, $R_1^1(\alpha)$ and $R_2^1(\alpha)$ being given by (2.57) and (2.58) respectively. As before, the second term in (2.59) may be regarded as due to the presence of second discontinuity at $(l, 0)$ for large l.

Case 3: $|b| = \infty$. In this case condition (2.9) assumes the from

$$\varphi = -e^{ikx} \text{ on } y = 0 \text{ for } 0 < x < l,$$

so that the modification of relation (2.25) is

$$\frac{\gamma(\alpha) + a}{\alpha^2 - k^2}F(\alpha) = -U_-(\alpha) - e^{i\alpha l}V_-(\alpha) + \frac{1}{i(\alpha + k)}\{e^{i(\alpha+k)l} - 1\}, c < \tau < d. \tag{2.61}$$

Here $P(\alpha) = \gamma + a$ is factorized as $P(\alpha) = P_+(\alpha)\, P_-(\alpha)$ where $P_-(\alpha) = P_-(-\alpha)$ and $P_-(\alpha)$ is given by (1.59) of section 5.1.

Proceeding as in Case 1, we obtain in place of (2.38)

$$\frac{(\alpha+k)G_+^*(\alpha;\lambda)}{P_+(\alpha)} - \frac{\lambda}{2\pi i}\int_{ih-\infty}^{ih+\infty} \frac{e^{il\xi}(\xi-k)G_+^*(\xi,\lambda)}{(\xi+\alpha)P_-(\xi)}d\xi = \frac{2\lambda k}{iP_+(k)}\frac{1}{\alpha-k}, \quad \tau > -h, \tag{2.62}$$

where now

$$G_+^*(\alpha;\lambda) = G_-(\alpha;\lambda) - \frac{e^{ikl}}{i(\alpha+k)} + \frac{\lambda}{i(\alpha-k)} \tag{2.63}$$

with

$$\begin{aligned} G_+(\alpha;1) &= V_+(\alpha) - U_-(-\alpha), \\ G_+(\alpha;-1) &= V_+(\alpha) + U_-(-\alpha). \end{aligned} \tag{2.64}$$

We write

$$\frac{1}{P_-(\xi)} = \frac{-a+(\xi^2-\epsilon^2)^{1/2}}{\xi^2-\epsilon^2}P_+(\xi) \tag{2.65}$$

in the integrand of the integral in the left-hand side of (2.62) and proceeding in a manner similar to Cases 1 and 2 we find that, for large l, $\varphi(x,y)$ is given by

$$\varphi(x,y) = \varphi_0(x,y) + \frac{a-b}{2\pi i} \tag{2.66}$$

$$\int_C \left[\frac{P_+(\alpha)}{\alpha+k}e^{i\alpha l}\{R_1^2(\alpha)+C_7T^2(\alpha)\} - \frac{P_-(\alpha)}{\alpha-k}\{e^{ikl}R_2^2(-\alpha)+C_8T^2(-\alpha)\}\right]e^{-i\alpha x-ry}\,d\alpha$$

where now $\varphi_0(x,y)$ is given by (1.69) of section 5.1.

$$R_1^2(\alpha) = \frac{E_0^{\,2}}{2\pi i}\frac{1}{\left(\epsilon'^2-k^2\right)}\frac{1}{\epsilon'+\alpha}, \quad R_2^2(\alpha) = \frac{E_0^{\,2}}{2\pi i}\frac{1}{(\epsilon'+k)^2}\frac{1}{\epsilon'+\alpha}, T^2(\alpha) = \frac{E_0^{\,2}}{2\pi i}\frac{1}{\epsilon'+k}\frac{1}{\epsilon'+\alpha}$$

with

$$E_0^{\,2} = -\frac{\pi^{1/2}}{l^{3/2}}e^{i(\epsilon l+3\pi/4)}P_+(\epsilon')(\epsilon'+\epsilon)^{1/2},$$

$$C_7 = \frac{P_+(\epsilon')}{(\epsilon'+k)^2-\{T^2(\epsilon')P_+(\epsilon')\}^2}\{(\epsilon'+k)G_2^2(\epsilon') - T^2(\epsilon')P_+(\epsilon')G_1^2(\epsilon') - T^2(\epsilon')P_+(\epsilon')G_1^2(\epsilon')\},$$

where

$$\begin{aligned} G_1^2(\alpha) &= -\frac{e^{ikl}}{P_+(\alpha)} - R_1^2(\alpha) \\ G_2^2(\alpha) &= \frac{2k}{(\alpha-k)P_+(k)} - \frac{\alpha+k}{(\alpha-k)P_+(\alpha)} - R_2^2(\alpha)e^{ikl} \end{aligned} \tag{2.67}$$

and C_8 is the same as C_7 with G_1^2 and G_2^2 interchanged. The second term in (2.66) is due to the presence of the second discontinuity at $(l, 0)$ for large l.

Making $\epsilon \to 0$ in the solution (2.52), (2.59) and (2.66) of the generalized BVP for $b > 0$, $b < 0$ and $|b| = \infty$, respectively, we obtain the solution of the original BVP for $b > 0$, $b < 0$ and $|b| = \infty$. The second terms in (2.52), (2.59) and (2.66), after making $\epsilon \to 0$, involve $e^{-i\alpha(x-l)}$ and $e^{-i\alpha x}$ in the integrands. The integrals involving $e^{-i\alpha x}$ can be evaluated for $x < 0$ ($x > 0$) by deforming the contour along the bisectors of the first and second (third and fourth) quadrants of the complex α-plane as has been done for $\varphi(x, y)$ in section 5.1. The integrals involving $e^{-i\alpha(x-l)}$ can be evaluated similarly for $x < l$ and $x > l$. However, considerable effort is needed in the evaluations of these integrals. Since we are interested only in the wave terms, we need to find only the wave terms of $\varphi(x, y)$ for different cases and different region. If we denote

$$\varphi^t(x,y) = e^{-ay+iax} + \varphi(x,y), \tag{2.68}$$

then asymptotic expressions valid for large l, for the wave terms of $\varphi^t(x, y)$ are obtained, as given below.

For $x < 0$

$$\varphi^t(\mathrm{x},\mathrm{y}) \approx \begin{cases} \left\{ e^{-ay+iax} - \dfrac{2a(b-a)}{\{(a+b)M_+^0(a)\}^2} e^{-ay-iax} \right. \\ + \left[\dfrac{4ab\left(T_1^0(b)\right)^2 M_-^0(a)}{(b^2-a^2)M_+^0(a)} + i\left(\dfrac{3}{2\pi}\right)^{1/2} \dfrac{a(b-a)^2 M_-^0(a) e^{ibl}}{b^2(a+b)M_+^0(a)(la)^2} \right] \\ \times \dfrac{e^{-ay-iax}}{\{M_-^0(b)\}^2 - \{T_1^0(b)\}^2} + \mathrm{O}\left(\dfrac{1}{(la)^3}\right) \quad \text{for } b > 0, \\ e^{-ay+iax} + \dfrac{2a(b-a)e^{-ay-iax}}{\{L_+^0(a)\}^2} + \mathrm{O}\left(\dfrac{1}{(la)^3}\right) \quad \text{for } b < 0, \\ e^{-ay+iax} - \dfrac{2ae^{-ay-iax}}{\{P_+^0(a)\}^2} + \mathrm{O}\left(\dfrac{1}{(la)^3}\right) \quad \text{for } |b| = \infty; \end{cases} \tag{2.69}$$

for $0 < x < l$

$$\varphi^t(\mathrm{x},\mathrm{y}) \approx \begin{cases} \dfrac{4abM_+^0(b)}{(a+b)^2 M_+^0(a)} e^{-by+ibx} - i\left(\dfrac{6}{\pi}\right)^{1/2} \\ \times \dfrac{a^2(b-a)^2 M_+^0(b) e^{ibl} e^{-by+ibx}}{b^2(a+b)^2 M_+^0(a)\left\{\left(M_-^0(b)\right)^2 - \left(T_1^0(b)\right)^2\right\}} \\ +i\left(\dfrac{6}{\pi}\right)^{1/2} \dfrac{1}{(la)^2} \dfrac{a(b-a)^2}{b(a+b)^2} \dfrac{M_+^0(b)}{M_+^0(a)} \\ \times \left[1 - \dfrac{2aT_1^0(b)}{(b-a)\left\{\left(M_-^0(b)\right)^2 - \left(T_1^0(b)\right)^2\right\}}\right] e^{-by-ib(x-l)} \\ +O\left(\dfrac{1}{(la)^3}\right) \quad \text{for } b > 0, \\ 0 \quad \text{for } b < 0, \\ 0 \quad \text{for } |b| = \infty; \end{cases} \tag{2.70}$$

and for $x > l$

$$\varphi^t(x,y) \approx \begin{cases} \left[-\dfrac{2abM_-^0(a)e^{ibl}}{(a+b)M_+^0(a)\left\{\left(M_-^0(b)\right)^2-\left(T_1^0(b)\right)^2\right\}}\right]+i\left(\dfrac{3}{2\pi}\right)^{1/2}\dfrac{1}{(la)^2}\dfrac{(b-a)^2M_-^0(a)}{b(a+b)M_+^0(a)} \\ \times\left\{1-\dfrac{2aT_1^0(b)}{(b-a)\left\{\left(M_-^0(b)\right)^2-\left(T_1^0(b)\right)^2\right\}}\right\}\times e^{-ay+ia(x-l)}+O\left(\dfrac{1}{(la)^3}\right) \text{ for } b>0, \\ -i\left(\dfrac{3}{2\pi}\right)^{1/2}\dfrac{1}{(la)^2}\dfrac{a-b}{b}\dfrac{L_-^0(a)}{L_+^0(a)}e^{-ay+ia(x-l)}+O\left(\dfrac{1}{(la)^3}\right) \text{ for } b<0, \\ \left(\dfrac{3}{2\pi}\right)^{1/2}\dfrac{1}{(la)^2}\dfrac{P_-^0(a)}{P_+^0(a)}e^{-ay+ia(x-l)}+O\left(\dfrac{1}{(la)^3}\right) \quad \text{ for } |b|=\infty, \end{cases} \tag{2.71}$$

where $T_1^0(b) = \lim_{\epsilon\to 0} T_1(\alpha_0) = ((b-a)M_+^0(b)e^{inl})/(a+b)$.

Comparing (2.6) with (2.70) and (2.71) we find that the unknown complex constants A_1, B_1, B_2 and B_3 of (2.6) are now obtained approximately for large l and are given by

$$A_1 = \begin{cases} -\dfrac{2a(b \quad a)}{\{(a+b)M_+^0(a)\}^2}++\left[\begin{array}{l}\dfrac{4ab\left(T_1^0(b)\right)^2M_-^0(a)}{(b^2-a^2)M_+^0(a)}+ \\ i\left(\dfrac{3}{2\pi}\right)^{1/2}\dfrac{a(b-a)^2M_-^0(a)e^{ibl}}{b^2(a+b)M_+^0(a)(la)^2}\end{array}\right] \\ \times\dfrac{1}{\{M_-^0(b)\}^2-\{T_1^0(b)\}^2}+O\left(\dfrac{1}{(la)^3}\right) \text{ for } b>0, \\ \dfrac{2a(b-a)}{\{L_+^0(a)\}^2}+O\left(\dfrac{1}{(la)^3}\right) \text{ for } b<0, \\ \dfrac{2a}{\{P_+^0(a)\}^2}+O\left(\dfrac{1}{(la)^3}\right) \quad \text{ for } |b|=\infty; \end{cases} \tag{2.72}$$

$$B_1 = \frac{4abM_+^0(b)}{(a+b)^2M_+^0(a)} - i\left(\frac{6}{\pi}\right)^{1/2}\frac{a^2(b-a)^2}{b^2(a+b)^2}\frac{M_+^0(b)e^{ibl}}{\{M_-^0(b)\}^2-\{T_1^0(b)\}^2}\frac{1}{(la)^2}+O\left(\frac{1}{(la)^3}\right), \tag{2.73}$$

$$B_2 = i\left(\frac{6}{\pi}\right)^{1/2}\frac{a(b-a)^2}{b(a+b)^2}\frac{M_+^0(b)}{M_+^0(a)}\left[-\frac{2a}{(b-a)}\frac{1}{\{M_-^0(b)\}^2-\{T_1^0(b)\}^2}T_1^0(b)\right]\frac{1}{(la)^2}+O\left(\frac{1}{(la)^3}\right), \tag{2.74}$$

$$B_3 = \begin{cases} -\dfrac{2aM_-^0(a)}{(a+b)M_+^0(a)}\dfrac{e^{ibl}}{\{M_-^0(b)\}^2-\{T_1^0(b)\}^2} \\ +i\left(\dfrac{3}{2\pi}\right)^{1/2}\dfrac{(b-a)^2M_-^0(a)}{b(a+b)M_+^0(a)} \\ \times\left[1-\dfrac{2a}{(b-a)}\dfrac{T_1^0(b)}{\{M_-^0(b)\}^2-\{T_1^0(b)\}^2}\right]\dfrac{1}{(la)^2}+O\left(\dfrac{1}{(la)^3}\right) \text{ for } b>0, \\ -i\left(\dfrac{3}{2\pi}\right)^{1/2}\dfrac{a-b}{b}\dfrac{L_-^0(a)}{L_+^0(a)}\dfrac{1}{(la)^2}+O\left(\dfrac{1}{(la)^3}\right) \text{ for } b>0, \\ i\left(\dfrac{3}{2\pi}\right)^{1/2}\dfrac{P_-^0(a)}{P_+^0(a)}\dfrac{1}{(la)^2}+O\left(\dfrac{1}{(la)^3}\right) \text{ for } |b|=\infty. \end{cases} \tag{2.75}$$

In expressions for A_2 and $B_i (i = 1,2,3)$, $M_\pm^0(a)$, $M_\pm^0(b)$, $P_\pm^0(a)$ are given by (1.67) of *section 5.1*.

It may be noted from (2.6) that B_1 and B_2 exist only for $b > 0$. It is observed from (2.70) that in the region $0 < x < l$, progressive waves exist only for $b > 0$, and these consist of transmitted and reflected waves. In this case the incident wave from the region $x < 0$ undergoes partial transmission below the edge at $x = 0$ which then is partially reflected by the edge at $x = l$. For $b < 0$ or $|b| = \infty$, there is no progressive waves in this region. (2.71) shows that in the region $x > l$, there exist progressive waves due to transmission of the incident wave field through the region below the inertial surface even through there may not any progressive wave in the $0 < x < l$.

Reflection coefficient in the region $x < 0$

As stated in section 5.1, the quantities a and b are related and for deep water the relationship can be expressed as

$$b = \frac{a}{1 - l_0 a},$$

where $l_0 (= \sigma/\rho)$ can be interpreted as the height of a vertical cylinder containing the fluid whose mass is the same as that of the floating matter distributed over the cross-sectional area of the cylinder at the inertial surface. In order that there exist time-harmonic progressive waves at the inertial surface b, must be positive, which is usually interpreted as the inertial surface to be light. However, if $b < 0$ or then the inertial surface is interpreted as heavy since it does not allow propagation of time harmonic waves, as was pointed out earlier. As the inertial surface here is in the form of a strip ($y = 0, 0 \leq x \leq l, -\infty < z < \infty$) there are now two edges, one at $x < 0$ and another at $x = l$. Expressions (2.69) show that the incident wave is reflected into the region $x = 0$ by the left edge of the strip. Expressions (2.70) show that for $b > 0$, there is a transmitted wave and a reflected wave inside the strip, the transmission being through the edge $x = 0$ and the reflection being by the edge $x = l$. For $b < 0$ or $|b| = \infty$, no wave propagates inside the strip apart from some local excitations by the two edges. Finally (2.71) show that in the region right of the strip, progressive waves exist which are due to transmission of the incident wave field through the region below the inertial surface.

The reflection coefficient A_1 for $b > 0$ in the region $x < 0$ is depicted graphically against the wave number aL in Fig. 5.2 taking $l/L = 10$ and $l_0/L = 0.01$ and 0.1 and in Fig. 5.3 taking $l_0/L = 0.01$ and $l/L = 10, 20$. It has been observed from Fig. 5.2 of section 5.1 that in the presence of the semi-infinite inertial surface, the reflection coefficient increases steadily with the wave number. However, when the inertial surface is in the form of a strip, this qualitative behavior of the reflection coefficient ($|A_1|$) is lost. In this case, the behavior of $|A_1|$ changes significantly against the wave number. Each graph of the reflection coefficient has the same basic feature, consisting of a series of concave curves which meet the wave number axis at their ends, so that total transmission occurs for a sequence of discrete values of the wave number. It is observed that larger value of l_0/L leads to higher maxima in $|A_1|$ (cf. Fig. 5.2) and

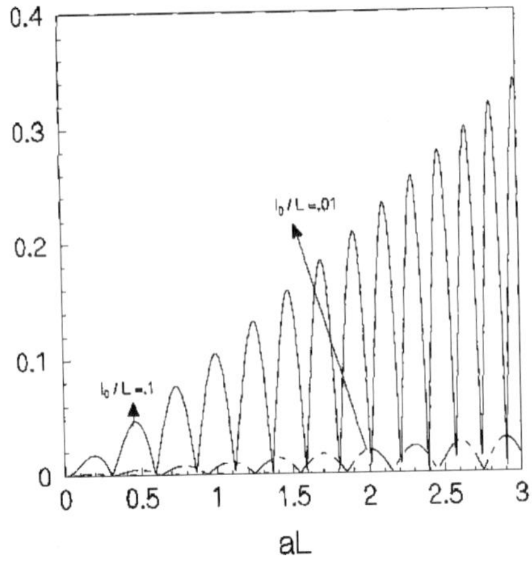

Fig. 5.2 Reflection coefficients $|A_1|$ due to a strip like inertial surface, $\frac{l}{L} = 10$

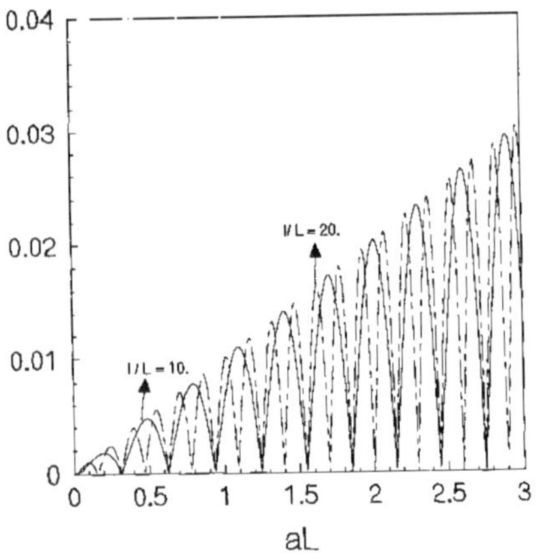

Fig. 5.3 Reflection coefficients $|A_1|$ due to a strip like inertial surface, $\frac{l}{L} = 0$

larger value of l/L leads to more number of zeros of $|A_2|$ (cf. Fig. 5.3). The oscillatory behavior of $|A_1|$ against the wave number may be attributed due to multiple reflections of the wave by two edges of the strip.

5.3 Scattering by two sharp discontinuities

In this section the problem of water wave scattering by two sharp discontinuities in the surface boundary conditions is considered. The sharp discontinuities arise when an infinite strip of inertial surface of surface density $\varepsilon_2\rho$ (ρ being the density of water) and of finite width l lies sandwiched between two inertial surfaces of the same surface density $\varepsilon_1\rho(\varepsilon_2 \neq \varepsilon_1)$ floating on the surface of deep water. ε_1,ε_2 are two constants having dimensions of length. These denote the depths of submergence of the wetted part of the surface. We assume that both ε_1 and ε_2 are less than g/w^2, so that time-harmonic progressive waves can propagate along the two types of inertial surfaces, where g is the acceleration due to gravity and w is the circular frequency of such waves. An inertial surface is a mass-loading model of floating ice where it is regarded as a material of uniform surface density having no elastic property. This is a reasonable model of fragile or pancake ice cf. Fox and Squire (1994) found in Marginal Ice Zone (MIZ) around Antarctica. The MIZ is a region consisting of fragile, pancake ice as well as pack ice and in this region the behaviour of surface waves in quite different from that in open ocean because of the interaction of ice with the waves. This problem has applications in wave propagation through that particular area of MIZ that contains strip of fragile or pancake ice. Gayen et al. (2006) examined this problem by reducing it to two coupled Carleman-type singular integral equations. The method given by Gayen et al. (2006) to solve the problem is given here.

The case $\varepsilon_1 = 0$ (i.e., a strip of inertial surface floating on water while the two surfaces in the left and right sides of the inertial surface are free to the atmosphere) was investigated by Kanoria et al. (1999) by reducing the problem to a Wiener Hopf problem which was solved approximately for large separation length l of the two discontinuities, giving rise to approximate values of the reflection and transmission coefficients associated with the problem.

We have considered both ε_1 and ε_2 to be non-zero, and use Havelock type expansions for the velocity potentials in different regions to reduce the scattering problem to that of solving two coupled Carleman type singular integral equations over a semi-infinite range. Assuming the two discontinuities to be widely separated (i.e., l is large), the two integral equations are decoupled. Solutions of these decoupled integral equations produce the zero-order solutions of the problem under consideration. Then, using these zero-order solutions in the coupled terms of the original equations, and solving the resulting uncoupled equations, the first-order solutions for the unknowns are obtained. Details are given below.

We consider two-dimensional irrotational motion in deep water which is homogeneous and inviscid. A rectangular Cartesian co-ordinate system is chosen in which the y-axis is taken vertically downwards into the water region, the plane $y = 0$ coincides with the undisturbed upper surface of water. This upper surface is covered by an inertial surface of uniform area density $\varepsilon_1\rho$ (ρ being the density of water) except for an infinite strip (along z-direction) of width l (along the x-direction) which is composed of another inertial surface of uniform area density $\varepsilon_2\rho$. The strip occupies the region $y = 0, 0 \le x \le l, -\infty < z < \infty$. Thus there are two discontinuities in the surface boundary conditions. The motion is assumed to be independent of z. If $\Phi(x, y, t) = Re\{\phi(x, y)e^{-i\omega t}\}$ describes the velocity potential for the two-dimensional motion in the fluid region, then the mathematical problem of our interest is to solve the following boundary value problem for ϕ satisfying the Laplace's equation

$$\nabla^2\phi = 0,\ y \ge 0\ \ -\infty < x < \infty, \tag{3.1}$$

the surface boundary conditions

$$K_1\phi + \phi_y = 0 \text{ on } y = 0, -\infty < x < 0 \text{ and } x > l, \tag{3.2}$$

$$K_2\phi + \phi_y = 0 \text{ on } y = 0, 0 < x < l \tag{3.3}$$

producing two discontinuities in the surface boundary conditions at the points (0,0) and (0,l), the bottom condition

$$\nabla\phi \to 0 \text{ as } y \to \infty, \tag{3.4}$$

and the infinity conditions

$$\varphi \to \begin{cases} e^{-K_1 y + iK_1 x} + Re^{-K_1 y - iK_1 x} & \text{as } x \to -\infty, \\ Te^{-K_1 y + iK_1(x-l)} & \text{as } x \to \infty \end{cases} \tag{3.5}$$

where R and T are unknown complex constants to be determined ($|R|$ and $|T|$ being respectively the reflection and transmission coefficients).
In the condition (3.2), (3.3)

$$K_1 = \frac{K}{1-\varepsilon_1 K}, \; K_2 = \frac{K}{1-\varepsilon_2 K} \text{ with } K = w^2/g$$

and $\varepsilon_1, \varepsilon_2 < g/w^2$ so that time-harmonic progressive waves of circular frequency w can propagate along the inertial surfaces. The boundary condition (3.2) or (3.3) is somewhat classical in the water wave literature. It was originally derived by Weitz and Keller (1950) of their paper, although not exactly in the form (3.2), but in an equivalent form. The surface density of the inertial surface floating on water, denoted by m in Weitz and Keller (1950), is taken as $\varepsilon_1\rho$ in (3.2) and $\varepsilon_2\rho$ in (3.3), ρ being the density of water. Use of Archimedes principle shows that if an inertial surface of small but uniform thickness has the density $\varepsilon\rho$, then ε is the depth of submergence of the wetted part of the inertial surface below the mean level of water.

Also in the strip region ϕ is given by

$$\phi = \alpha e^{-K_2 y + iK_2 x} + \beta e^{-K_2 y - iK_2 (x-l)} + \psi(x,y), \; 0 < x < l \tag{3.6}$$

where α and β are unknown complex constants and $\psi(x, y)$ represents an unknown, ordinary non-wavy solution of the Laplace's equation. The condition (3.5) arise due to a train of surface waves represented by $e^{-K_1 y + iK_1 x}$ in the region $x < 0$ being incident on the discontinuity at (0, 0). A part of the wave train is reflected back into the region $x < 0$, the remaining part is transmitted through the strip-like region $0 < x < l$ (where it undergoes multiple reflection and transmission) and eventually transmitted into the region $x > l$.

Let $\phi(x, y)$ be represented by ϕ_1, ϕ_2, ϕ_3 respectively in the regions $x < 0$, $0 < x < l$ and $x > l$ $(y > 0)$. Then using Havelock-type expansions for water wave potentials Ursell (1947), we can write

$$\phi_1(x,y) = e^{-K_1 y + iK_1 x} + Re^{-K_1 y - iK_1 x} + \frac{2}{\pi}\int_0^\infty \frac{A(\xi)}{\xi^2 + K_1^2} L_1(\xi, y) e^{\xi x} d\xi, \; x < 0 \tag{3.7}$$

$$\phi_2(x,y) = \alpha e^{-K_2 y + iK_2 x} + \beta e^{-K_2 y - iK_2 (x-l)} + \frac{2}{\pi}\int_0^\infty \frac{B(\xi)e^{\xi(x-l)} + C(\xi)e^{-\xi x}}{\xi^2 + K_2^2} L_2(\xi, y) d\xi, \; 0 < x < l, \tag{3.8}$$

$$\phi_3(x,y) = Te^{-K_1 y + iK_1 (x-l)} + \frac{2}{\pi}\int_0^\infty \frac{D(\xi)}{\xi^2 + K_1^2} L_1(\xi, y) e^{-\xi(x-l)} d\xi, \; x > l, \tag{3.9}$$

where $A(\xi)$, $B(\xi)$, $C(\xi)$ and $D(\xi)$ are unknown functions to be determined and must be such that the integrals in (3.7)–(3.9) are convergent, and

$$L_i(\xi, y) = \xi \cos \xi y - K_i \sin \xi y, \; i = 1,2. \tag{3.10}$$

Since ϕ and ϕ_x are continuous across the lines $x = 0$ and $x = l$ $(0 < y < \infty)$, we obtain from (3.7)–(3.9) the following four relations

$$(1+R)e^{-K_1 y}+\frac{2}{\pi}\int_0^\infty \frac{A(\xi)}{\xi^2+K_1^2}L_1(\xi,y)d\xi=\left(\alpha+\beta e^{iK_2 l}\right)e^{-K_2 y}+\frac{2}{\pi}\int_0^\infty \frac{B(\xi)e^{-\xi l}+C(\xi)}{\xi^2+K_2^2}L_2(\xi,y)d\xi,\ y>0, \tag{3.11}$$

$$iK_1(1-R)e^{-K_1 y}+\frac{2}{\pi}\int_0^\infty \frac{\xi A(\xi)}{\xi^2+K_1^2}L_1(\xi,y)d\xi$$

$$=iK_2\left(\alpha-\beta e^{iK_2 l}\right)e^{-K_2 y}+\frac{2}{\pi}\int_0^\infty \frac{\xi\left\{B(\xi)e^{-\xi l}-C(\xi)\right\}}{\xi^2+K_2^2}L_2(\xi,y)d\xi,\ y>0, \tag{3.12}$$

$$\left(\alpha e^{iK_2 l}+\beta\right)e^{-K_2 y}+\frac{2}{\pi}\int_0^\infty \frac{B(\xi)+C(\xi)e^{-\xi l}}{\xi^2+K_2^2}L_2(\xi,y)d\xi$$

$$=Te^{-K_1 y}+\frac{2}{\pi}\int_0^\infty \frac{D(\xi)}{\xi^2+K_1^2}L_1(\xi,y)d\xi,\ y>0, \tag{3.13}$$

and

$$iK_2\left(\alpha e^{-iK_2 l}-\beta\right)e^{-K_2 y}+\frac{2}{\pi}\int_0^\infty \frac{\xi\left\{B(\xi)-C(\xi)e^{-\xi l}\right\}}{\xi^2+K_2^2}L_2(\xi,y)d\xi$$

$$=iK_1 Te^{-K_1 y}-\frac{2}{\pi}\int_0^\infty \frac{\xi D(\xi)}{\xi^2+K_1^2}L_1(\xi,y)d\xi,\ y>0. \tag{3.14}$$

Use of Havelock's expansion theorem Ursell (1947) to the relations (3.11) and (3.12) assuming the right side to be known and to (3.13) and (3.14) assuming the left sides to be known, produces

$$A(\xi) = -\frac{(K_1-K_2)\xi}{\xi^2+K_2^2}\left(\alpha+\beta e^{iK_2 l}\right)+\frac{(\xi^2+K_1K_2)\{B(\xi)e^{-\xi l}+C(\xi)\}}{\xi^2+K^2}$$
$$+\frac{2}{\pi}(K_1-K_2)\xi\int_0^\infty \frac{u\{B(u)e^{-ul}+C(u)\}}{(u^2+K_2^2)(u^2-\xi^2)}du, \tag{3.15}$$

$$\frac{1+R}{2K_1}=\frac{\alpha+\beta e^{iK_2 l}}{K_1+K_2}+\frac{2}{\pi}(K_1-K_2)\int_0^\infty \frac{\xi\{B(\xi)e^{-\xi l}+C(\xi)\}}{(\xi^2+K_1^2)(\xi^2+K_2^2)}d\xi; \tag{3.16}$$

$$\xi A(\xi) = -\frac{iK_2(K_1-K_2)(\alpha-\beta e^{iK_2 l})\xi}{\xi^2+K_2^2}+\frac{\xi(\xi^2+K_1K_2)}{\xi^2+K_2^2}\{B(\xi)e^{-\xi l}-C(\xi)\}$$
$$+\frac{2}{\pi}(K_1-K_2)\xi\int_0^\infty \frac{u^2\{B(u)e^{-ul}-C(u)\}}{(u^2+K_2^2)(u^2-\xi^2)}du, \tag{3.17}$$

$$\frac{i(1-R)}{2} = \frac{iK_2(\alpha - \beta e^{iK_2 l})}{K_1 + K_2} + \frac{2}{\pi}(K_1 - K_2)\int_0^\infty \frac{\xi^2\{B(\xi)e^{-\xi l} - C(\xi)\}}{(\xi^2 + K_1^2)(\xi^2 + K_2^2)}d\xi; \tag{3.18}$$

$$D(\xi) = -\frac{(K_1 - K_2)(\alpha e^{iK_2 l} + \beta)\xi}{\xi^2 + K_2^2} + \frac{(\xi^2 + K_1 K_2)\{B(\xi) + C(\xi)e^{-\xi l}\}}{\xi^2 + K_2^2}$$

$$+\frac{2}{\pi}(K_1 - K_2)\xi\int_0^\infty \frac{\{B(u) + C(u)e^{-ul}\}}{(u^2 + K_2^2)(u^2 - \xi^2)}udu, \tag{3.19}$$

$$\frac{T}{2K_1} = \frac{\alpha e^{iK_2 l} + \beta}{K_1 + K_2} + \frac{2}{\pi}(K_1 - K_2)\int_0^\infty \frac{\xi\{B(\xi) + C(\xi)e^{-\xi l}\}}{(\xi^2 + K_1^2)(\xi^2 + K_2^2)}d\xi; \tag{3.20}$$

$$\xi D(\xi) = \frac{iK_2(K_1 - K_2)(\alpha e^{iK_2 l} - \beta)\xi}{\xi^2 + K_2^2} - \frac{\xi(\xi^2 + K_1 K_2)\{B(\xi) - C(\xi)e^{-\xi l}\}}{\xi^2 + K_2^2}$$

$$-\frac{2}{\pi}(K_1 - K_2)\xi\int_0^\infty \frac{u^2\{B(u) - C(u)e^{-ul}\}}{(u^2 + K_2^2)(u^2 - \xi^2)}du, \tag{3.21}$$

$$\frac{iT}{2} = \frac{iK_2(\alpha e^{iK_2 l} - \beta)}{K_1 + K_2} + \frac{2}{\pi}(K_1 - K_2)\int_0^\infty \frac{\xi^2\{B(\xi) - C(\xi)e^{-\xi l}\}}{(\xi^2 + K_1^2)(\xi^2 + K_2^2)}d\xi. \tag{3.22}$$

The integrals in (3.15), (3.17), (3.19) and (3.21) are in the sense of Cauchy principal value.

Elimination of $A(\xi)$ between (3.15) and (3.17), and $D(\xi)$ between (3.19) and (3.21), produces the following two coupled Carleman type singular integral equations for the unknown functions $B(\xi)$ and $C(\xi)$:

$$\lambda(\xi)B_1(\xi) + \frac{1}{\pi}\int_0^\infty \frac{B_1(u)}{u - \xi}du - \frac{1}{\pi}\int_0^\infty \frac{C_1(u)}{u + \xi}e^{-ul}du = F_B(\xi),\ \xi > 0, \tag{3.23}$$

$$\lambda(\xi)C_1(\xi) + \frac{1}{\pi}\int_0^\infty \frac{C_1(u)}{u - \xi}du - \frac{1}{\pi}\int_0^\infty \frac{B_1(u)}{u + \xi}e^{-ul}du = F_C(\xi),\ \xi > 0 \tag{3.24}$$

where

$$B_1(\xi), C_1(\xi) = \frac{\xi}{\xi^2 + K_2^2}\big(B(\xi), C(\xi)\big), \tag{3.25}$$

$$\lambda(\xi) = \frac{\xi^2 + K_1 K_2}{\xi(K_1 - K_2)}, \tag{3.26}$$

$$F_B(\xi) = \frac{\alpha}{2}\frac{e^{iK_2 l}}{\xi - iK_2} + \frac{\beta}{2}\frac{1}{\xi + iK_2}, \tag{3.27}$$

$$F_C(\xi) = \frac{\beta}{2}\frac{e^{iK_2 l}}{\xi - iK_2} + \frac{\alpha}{2}\frac{1}{\xi + iK_2}, \tag{3.28}$$

and the second integrals in (3.23) and (3.24) are in the sense of Cauchy principal value. A method of obtaining approximate solutions for $B_1(\xi)$, $C_1(\xi)$ and approximate expressions for the unknown constants α, β, R, T for large l is described below.

Both the Carleman type singular integral equations (3.23) and (3.24) are similar in nature and can be treated for approximate solutions for large l as described below:

For very large values of l, we can ignore the integrals involving e^{-ul} giving rise to the following uncoupled integral equations:

$$\lambda(\xi)B_1^o(\xi) + \frac{1}{\pi}\int_0^\infty \frac{B_1^o(u)}{u-\xi}du = F_B^o(\xi),\ \xi > 0, \tag{3.29}$$

$$\lambda(\xi)C_1^o(\xi) + \frac{1}{\pi}\int_0^\infty \frac{C_1^o(u)}{u-\xi}du = F_C^o(\xi),\ \xi > 0, \tag{3.30}$$

where the subscript 'o' denotes the *zero-order* approximations so that $F_B^o(\xi)$ and $F_C^o(\xi)$ are the same as the expressions in (3.27) and (3.28) respectively with α, β replaced by α^o, β^o. The two independent equations (3.29) and (3.30) can be solved by converting them into Riemann-Hilbert problems, in a standard manner Gakhov (1966).

The solutions are found to be

$$B_1^o(\xi) = \alpha^o e^{iK_2 l} P_1(\xi) + \beta^o P_2(\xi) \tag{3.31}$$

and

$$C_1^o(\xi) = \alpha^o P_2(\xi) + \beta^o e^{iK_2 l} P_1(\xi) \tag{3.32}$$

where

$$P_1(\xi), P_2(\xi) = \frac{\lambda(\xi)}{2(\xi \mp iK_2)((\lambda(\xi))^2+1)} - \frac{\Lambda_0^+(\xi)}{2\pi(\lambda(\xi)-i)} \int_0^\infty \frac{1}{(\lambda(u)+i)(u \mp iK_2)\Lambda_0^+(u)}\frac{du}{u-\xi} \tag{3.33}$$

with

$$\Lambda_0(\zeta) = \exp\left[\frac{1}{2\pi i}\left\{\int_0^\infty \left(\ln\frac{t-iK_1}{t+iK_1} - 2\pi i\right)\frac{dt}{t-\zeta} - \int_0^\infty \left(\ln\frac{t-iK_2}{t+iK_2} - 2\pi i\right)\frac{dt}{t-\zeta}\right\}\right],$$

$$\zeta \notin (0,\infty) \tag{3.34}$$

and

$$\Lambda^{\pm}(\xi)=\Lambda_0^{\pm}(\xi)\left[\pm\frac{1}{2}\frac{F_B^o(\xi)}{(\lambda(\xi)+i)\Lambda_0^+(\xi)}+\frac{1}{2\pi i}\int_0^{\infty}\frac{F_B^o(u)}{(\lambda(u)+i)\Lambda_0^+(u)}\frac{du}{u-\xi}\right],\xi>0. \tag{3.35}$$

$P_1(\xi)$, $P_2(\xi)$ can be obtained by considering the integrals

$$\int_{C_0}\frac{1}{\Lambda_0(\zeta)(\zeta\mp iK_2)}\frac{d\tau}{\tau-\zeta},\ \zeta\notin C_0, \tag{3.36}$$

where C_0 is a positively oriented closed contour consisting of a loop around the positive real axis and a circle of large radius in the complex τ-plane. Finally these are found as

$$P_1(\xi),P_2(\xi)=\frac{1}{2}\frac{(K_1-K_2)\xi\Lambda_0^+(\xi)}{(\xi-iK_1)(\xi+iK_2)(\xi\mp iK_2)\Lambda_0(\pm iK_2)}. \tag{3.37}$$

To determine α^o, β^o (and also R^o, T^o), we note that these four unknowns satisfy the following relations (obtained by approximating the relations (3.16), (3.18), (3.20) and (3.22) after replacing $B(\xi)$, $C(\xi)$ in terms of $B_1(\xi)$, $C_1(\xi)$ given in (3.25)):

$$\begin{aligned}
\frac{1+R^o}{2K_1} &= \frac{\alpha^o+\beta^o e^{iK_2l}}{K_1+K_2}+\frac{2}{\pi}(K_1-K_2)\int_0^{\infty}\frac{C_1^o(\xi)}{\xi^2+K_1^2}d\xi,\\
\frac{i(1-R^o)}{2} &= \frac{iK_2(\alpha^o-\beta^o e^{iK_2l})}{K_1+K_2}-\frac{2}{\pi}(K_1-K_2)\int_0^{\infty}\frac{\xi C_1^o(\xi)}{\xi^2+K_1^2}d\xi,\\
\frac{T^o}{2K_1} &= \frac{\alpha^o e^{iK_2l}+\beta^o}{K_1+K_2}+\frac{2}{\pi}(K_1-K_2)\int_0^{\infty}\frac{B_1^o(\xi)}{\xi^2+K_1^2}d\xi,\\
\frac{iT^o}{2} &= \frac{iK_2(\alpha^o e^{iK_2l}-\beta^o)}{K_1+K_2}+\frac{2}{\pi}(K_1-K_2)\int_0^{\infty}\frac{\xi B_1^o(\xi)}{\xi^2+K_1^2}d\xi.
\end{aligned} \tag{3.38}$$

Substituting for $B_1^o(\xi)$ and $C_1^o(\xi)$ from (3.31) and (3.32) respectively in the integrals in (3.38), we find that

$$\alpha^o,\beta^o=\frac{1}{a_1^2-a_2^2e^{2iK_2l}}\left(a_1,-a_2e^{iK_2l}\right), \tag{3.39}$$

$$T^o=\frac{(a_1u_1-a_2u_2)e^{iK_2l}}{a_1^2-a_2^2e^{2iK_2l}},\ R^o=\frac{a_1u_2-a_2u_1e^{2ik_2l}}{a_1^2-a_2^2e^{2ik_2l}}-1 \tag{3.40}$$

where

$$a_1=1+\frac{2i}{\pi}(K_1-K_2)\int_0^{\infty}\frac{P_2(\xi)}{\xi+iK_1}d\xi, \tag{3.41}$$

$$a_2=\frac{K_1-K_2}{K_1+K_2}+\frac{2i}{\pi}(K_1-K_2)\int_0^{\infty}\frac{P_1(\xi)}{\xi+iK_1}d\xi, \tag{3.42}$$

$$u_1, u_2 = 2K_1\left[\frac{1}{K_1+K_2} + \frac{2}{\pi}(K_1-K_2)\int_0^\infty \frac{P_1(\xi), P_2(\xi)}{\xi^2+K_1^2}d\xi\right]. \tag{3.43}$$

The integrals in (3.41) and (3.42) can be evaluated by considering the integrals

$$\int_{C_0} \frac{\Lambda_0(\tau)}{(\tau^2+K_1^2)(\tau \pm iK_2)}d\tau \tag{3.44}$$

where C_0 is the same as in (3.36). The results involve $\Lambda_0(\pm iK_1), \Lambda_0(\pm iK_2)$ and these are found to be

$$\Lambda_0(\pm iK_1) = \left(\frac{K_1+K_2}{2K_1}\right)^{1/2}, \ \Lambda_0(\pm iK_2) = \left(\frac{2K_2}{K_1+K_2}\right)^{1/2}. \tag{3.45}$$

Thus $\alpha^o, \beta^o, T^o, R^o$ ('zero-order' approximations to α, β, T, R) are obtained as

$$\alpha^o, \beta^o = \frac{2(K_1K_2)^{1/2}}{(K_1+K_2)^2-(K_1-K_2)^2e^{2iK_2l}}\left(K_1+K_2, -(K_1-K_2)e^{iK_2l}\right), \tag{3.46}$$

$$T^o = \frac{4K_1K_2}{(K_1+K_2)^2-(K_1-K_2)^2e^{2iK_2l}}, R^o = \frac{(K_1-K_2)(1-e^{iK_2l})}{K_1+K_2+(K_1-K_2)e^{iK_2l}}. \tag{3.47}$$

To obtain the next order (first-order) approximate solutions $B_1^1(\xi)$, $C_1^1(\xi)$ of the coupled integral equations (3.23) and (3.24), we decouple it by replacing $C_1(u)$ in (3.23) by the known function $C_1^0(u)$ and $B_1(u)$ in (3.24) by $B_1^0(u)$, and also approximate α, β appearing in the forcing functions by α^1, β^1. This art gives rise to the following pair of Carleman integral equations for $B_1^1(u)$ and $C_1^1(u)$ as

$$\lambda(\xi)B_1^1(\xi) + \frac{1}{\pi}\int_0^\infty \frac{B_1^1(u)}{u-\xi}du = F_B^1(\xi), \xi > 0, \tag{3.48}$$

$$\lambda(\xi)C_1^1(\xi) + \frac{1}{\pi}\int_0^\infty \frac{C_1^1(u)}{u-\xi}du = F_C^1(\xi), \xi > 0, \tag{3.49}$$

where

$$F_B^1(\xi) = \frac{1}{\pi}\int_0^\infty \frac{C_1^o(u)e^{-ul}}{u+\xi}du + \frac{\alpha^1}{2}\frac{e^{iK_2l}}{\xi-iK_2} + \frac{\beta^1}{2}\frac{1}{\xi+iK_2}, \tag{3.50}$$

$$F_C^1(\xi) = \frac{1}{\pi}\int_0^\infty \frac{B_1^o(u)e^{-ul}}{u+\xi}du + \frac{\beta^1}{2}\frac{e^{iK_2l}}{\xi-iK_2} + \frac{\alpha^1}{2}\frac{1}{\xi+iK_2}. \tag{3.51}$$

Note that $F_B^1(\xi)$ and $F_C^1(\xi)$ contain the unknown constants α^1, β^1 (the 'first-order' approximations to α, β).

The equations (3.48) and (3.49) can be solved as before and we find that

$$B_1^1(\xi), C_1^1(\xi) = \frac{(K_1 - K_2)\xi(\xi^2 + K_1K_2)}{(\xi^2 + K_1^2)(\xi^2 + K_2^2)}(F_B^1(\xi), F_C^1(\xi))$$
$$-\frac{\Lambda_0^+(\xi)}{\pi}\frac{\xi(K_1 - K_2)^2}{(\xi - iK_1)(\xi + iK_2)}\int_0^\infty \frac{u(F_B^1(u), F_C^1(u))}{\Lambda_0^+(u)(u + iK_1)(u - iK_2)}du. \quad (3.52)$$

In order to calculate the integral in $B_1^1(\xi)$, we note that

$$\int_{C_0} \frac{1}{\Lambda_0(\tau)}\left[g(\tau) + \frac{\alpha^1 e^{iK_2 l}}{\tau - iK_2} + \frac{\beta^1}{\tau + iK_2}\right]\frac{d\tau}{\tau - \zeta}, \zeta \notin C_0 \quad (3.53)$$
$$= 2\pi i\left[0 - \frac{\alpha^1 e^{iK_2 l}}{(\xi - iK_2)\Lambda_0(iK_2)} - \frac{\beta^1}{(\xi + iK_2)\Lambda_0(-iK_2)}\right]$$

where

$$g(\tau) = \frac{2}{\pi}\int_0^\infty \frac{C_1^o(u)e^{-ul}}{u + \tau}du$$

and the first term in the right side of (3.53) has no contribution since for large l

$$\int_{C_0} \frac{g(\tau)}{\Lambda_0(\tau)(\tau - \varsigma)}d\tau \approx 0, \varsigma \neq C_0.$$

The integral in $C_1^1(\xi)$ can be similarly evaluated.

Thus, ultimately we obtain

$$B_1^1(\xi) = \alpha^1 e^{iK_2 l}P_1(\xi) + \beta^1 P_2(\xi), \quad (3.54)$$

$$C_1^1(\xi) = \alpha^1 P_2(\xi) + \beta^1 e^{iK_2 l}P_1(\xi) \quad (3.55)$$

where $P_1(\xi)$, $P_2(\xi)$ have already been obtained explicitly in (3.37).

To determine the constants α^1, β^1 and R^1, T^1 (the 'first-order' approximations to R, T), we note these satisfy the following four equations (obtained by approximating the relations (3.16), (3.18), (3.20) and (3.22) upto 'first-order'):

$$\frac{1 + R^1}{2K_1} = \frac{\alpha^1 + \beta^1 e^{iK_2 l}}{K_1 + K_2} + \frac{2}{\pi}(K_1 - K_2)\int_0^\infty \frac{B_1^1(\xi)e^{-\xi l} + C_1^1(\xi)}{\xi^2 + K_1^2}d\xi,$$

$$\frac{i(1 - R^1)}{2} = \frac{iK_2(\alpha^1 - \beta^1 e^{iK_2 l})}{K_1 + K_2} - \frac{2}{\pi}(K_1 - K_2)\int_0^\infty \frac{\xi(B_1^1(\xi)e^{-\xi l} - C_1^1(\xi))}{\xi^2 + K_1^2}d\xi,$$

$$\frac{T^1}{2K_1} = \frac{\alpha^1 e^{iK_2 l} + \beta^1}{K_1 + K_2} + \frac{2}{\pi}(K_1 - K_2)\int_0^\infty \frac{B_1^1(\xi) + C_1^1(\xi)e^{-\xi l}}{\xi^2 + K_1^2}d\xi,$$

$$\frac{iT^1}{2} = \frac{iK_2(\alpha^1 e^{iK_2 l} - \beta^1)}{K_1 + K_2} + \frac{2}{\pi}(K_1 - K_2)\int_0^\infty \frac{\xi(B_1^1(\xi) - C_1^1(\xi)e^{-\xi l})}{\xi^2 + K_1^2}d\xi.$$

For the integrals in the above results, expressions for $B_1^1(\xi)$ and $C_1^1(\xi)$ given by (3.54) and (3.55) respectively have to be used. Solving these four equations we obtain $\alpha^1, \beta^1, R^1, T^1$ as

$$\alpha^1, \beta^1 = \frac{1}{b_1^2 - b_2^2}(b_1, -b_2), \tag{3.56}$$

$$T^1 = \frac{b_1 Y_1 - b_2 Y_2}{b_1^2 - b_2^2},\ R^1 = -1 + \frac{b_1 Y_2 - b_2 Y_1}{b_1^2 - b_2^2}, \tag{3.57}$$

$$b_1 = 1 + \frac{2i}{\pi}(K_1 - K_2)\left\{\int_0^\infty \frac{P_2(\xi)}{\xi + iK_1} d\xi - e^{iK_2 l}\int_0^\infty \frac{P_1(\xi)e^{-\xi l}}{\xi - iK_1} d\xi\right\},$$

$$b_2 = \frac{K_1 - K_2}{K_1 + K_2} e^{iK_2 l} + \frac{2i}{\pi}(K_1 - K_2)\left\{e^{iK_2 l}\int_0^\infty \frac{P_1(\xi)}{\xi + iK_1} d\xi - \int_0^\infty \frac{P_2(\xi)e^{-\xi l}}{\xi - iK_1} d\xi\right\},$$

$$Y_1 = 2K_1\left[\frac{e^{iK_2 l}}{K_1 + K_2} + \frac{2}{\pi}(K_1 - K_2)\left\{e^{iK_2 l}\int_0^\infty \frac{P_1(\xi)}{\xi^2 + K_1^2} d\xi - \int_0^\infty \frac{P_2(\xi)e^{-\xi l}}{\xi^2 + K_1^2} d\xi\right\}\right],$$

$$Y_2 = 2K_1\left[\frac{1}{K_1 + K_2} + \frac{2}{\pi}(K_1 - K_2)\left\{\int_0^\infty \frac{P_2(\xi)}{\xi^2 + K_1^2} d\xi + e^{iK_2 l}\int_0^\infty \frac{P_1(\xi)e^{-\xi l}}{\xi^2 + K_1^2} d\xi\right\}\right].$$

The integrals given by the second terms of b_1, b_2, Y_1 and Y_2 can be evaluated by appropriate use of contour integration after substitution of $P_1(\xi)$ and $P_2(\xi)$ from (3.37). However, the integrals given by the third terms of b_1, b_2, Y_1 and Y_2 are evaluated for large l by employing Watson's lemma cf. Jones (1964) and we obtain the following results:

$$b_1 = \frac{\Lambda_0(-iK_1)}{\Lambda_0(-iK_2)} + \frac{i}{\pi} e^{iK_2 l} \frac{(K_1 - K_2)^2}{K_2^2} \frac{1}{\Lambda_0(iK_2)} \frac{1}{(K_1 l)^2} + O\left(\frac{1}{(K_1 l)^3}\right),$$

$$b_2 = \frac{K_1 - K_2}{K_1 + K_2} \frac{\Lambda_0(-iK_1)}{\Lambda_0(iK_2)} e^{iK_2 l} - \frac{i}{\pi} \frac{(K_1 - K_2)^2}{K_2^2} \frac{1}{\Lambda_0(-iK_2)} \frac{1}{(K_1 l)^2} + O\left(\frac{1}{(K_1 l)^3}\right),$$

$$Y_1 = \frac{K_1 - K_2}{K_1 + K_2} \frac{\Lambda_0(-iK_1)}{\Lambda_0(iK_2)} e^{iK_2 l} + \frac{\Lambda_0(iK_1)}{\Lambda_0(iK_2)} e^{iK_2 l}$$

$$-\frac{2i}{\pi} \frac{(K_1 - K_2)^2}{K_2^2} \frac{1}{\Lambda_0(-iK_2)} \frac{1}{(K_1 l)^2} + O\left(\frac{1}{(K_1 l)^3}\right),$$

$$Y_2 = \frac{K_1 - K_2}{K_1 + K_2} \frac{\Lambda_0(iK_1)}{\Lambda_0(-iK_2)} + \frac{\Lambda_0(-iK_1)}{\Lambda_0(-iK_2)} + \frac{2i}{\pi} \frac{(K_1 - K_2)^2}{K_2^2} \frac{e^{iK_2 l}}{\Lambda_0(iK_2)} \frac{1}{(K_1 l)^2} + O\left(\frac{1}{(K_1 l)^3}\right).$$

Substituting the values of $\Lambda_0(\pm iK_j)(j = 1,2)$ from (3.45) in the above results, α^1, β^1, R^1 and T^1 are obtained in terms of computable functions involving K_1, K_2 and l.

Numerical Results

By the use of Green's integral theorem to $\phi(x, y)$ and $\phi^*(x, y)$ (the complex conjugate of ϕ) in the fluid region, the energy identity $|R|^2 + |T|^2 = 1$ can be shown to hold good. The numerical results obtained for $|R^1|$ and $|T^1|$ (the 'first-order' approximations to $|R|$ and $|T|$) computed by using the relations (3.57) for different values of the parameters $\varepsilon_1/L, \varepsilon_2/L, KL, l/L$ where L is a characteristic length introduced to non-dimensionalize $\varepsilon_1, \varepsilon_1, K$ and l, show that this identity is always satisfied approximately (upto five decimal places). Also in the program for numerical computations, if ε_1/L and ε_2/L are made equal, then $|R|(\approx |R^1|)$ is found to be zero for all values of the wave number KL. This means that the incident wave train is fully transmitted. This is due to the fact that since in this case, the regions inside and outside the strip are composed of the same material there is no discontinuity in the surface boundary conditions and as such no reflection occurs.

In Fig. 5.4, $|R|$ is depicted against the wave number KL for $\varepsilon_1/L = 0, \varepsilon_2/L = .01, 0.1$ and $l/L = 10$. This corresponds to water wave scattering by a strip of inertial surface of large width l, floating on clean water. This figure coincides completely with the Fig. 5.2 corresponding figure given in Kanoria et al. (1999) (Fig. 2 there) wherein, as mentioned earlier, the problem was solved approximately employing Wiener-Hopf technique by reducing it to a three-part Wiener-Hopf problem. The Fig. 5.4 shows that $|R|$ is quite small when the surface density of the strip of inertial surface is small, and also the behaviour of $|R|$ as a function of the wave number is oscillatory in nature and there exists a sequence of discrete values of the wave number for which total transmission of the incident wave train occurs. As also mentioned in section 5.2, this oscillatory nature of $|R|$ arises due to multiple reflections of the waves by the two discontinuities.

In Fig. 5.5, $|R|$ is plotted against KL for $\varepsilon_1/L = 0.1, \varepsilon_2/L = 0.2$ and $\varepsilon_1/L = 0.2, \varepsilon_2/L = 0.1$, and $l/L = 10$. As in Fig. 5.4, $|R|$ is also oscillatory in nature and zero reflection occurs

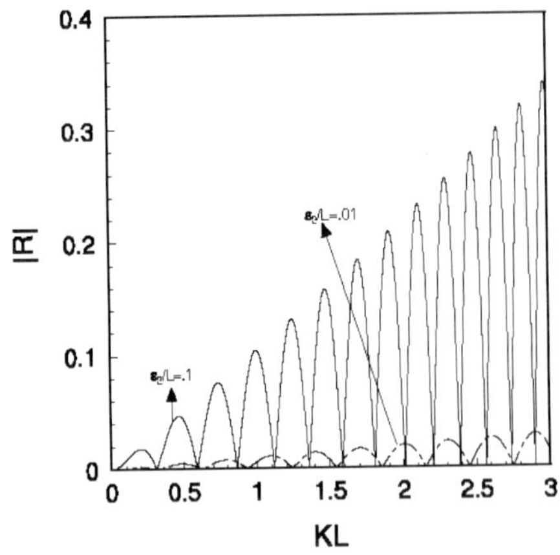

Fig. 5.4 Reflection coefficient, $\varepsilon_1/L = 0$, $l/L = 10$

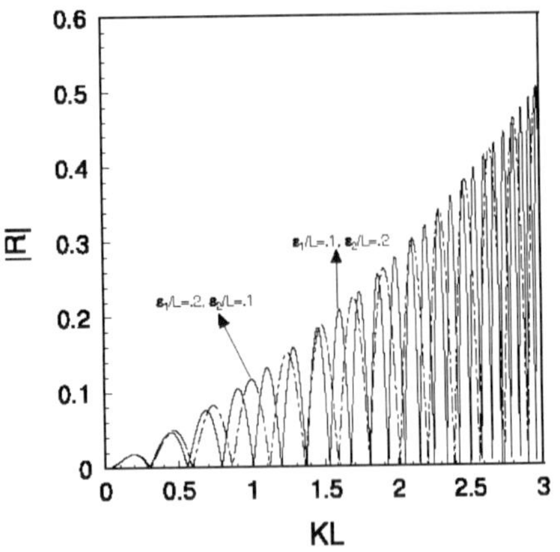

Fig. 5.5 Reflection coefficient, $l/L = 10$

for an increasing sequence of values of KL. Moreover, the number of zero reflection is more when the strip of inertial surface is heavier to the inertial surfaces on its two sides compared to the case when the strip is lighter to the inertial surfaces on its two sides. Also the peak values of $|R|$ steadily increase with the increase in wave number.

The Fig. 5.6 displays $|R|$ for $\varepsilon_1/L = 0$, $\varepsilon_2/L = 0.1$ and $\varepsilon_1/L = 0.1$, $\varepsilon_2/L = 0$ with $l/L = 10$. This corresponds to a strip of inertial surface surrounded by two free surfaces, and a strip of free surface surrounded by two inertial surfaces composed of the same disconnected material. While the oscillatory nature of $|R|$ remains the same for both the situations, the number of zeros of $|R|$ for the first situation is greater than the number of zeros of $|R|$ for the second situation. This may be attributed to more multiple reflections by the two edges of the strip when it is composed of an inertial surface surrounded by free surfaces compared to the case when it consists of a free surface surrounded by inertial surfaces.

Finally, to find the effect of variation of the strip length, $|R|$ is depicted in Fig. 5.7 for $l/L = 10$, 20 and $\varepsilon_1/L = 0.1$ and $\varepsilon_2/L = 0.2$. This figure shows that increase in the strip length results in the increase of the zeros of $|R|$ which was also observed in section 5.2 (Fig. 5.3).

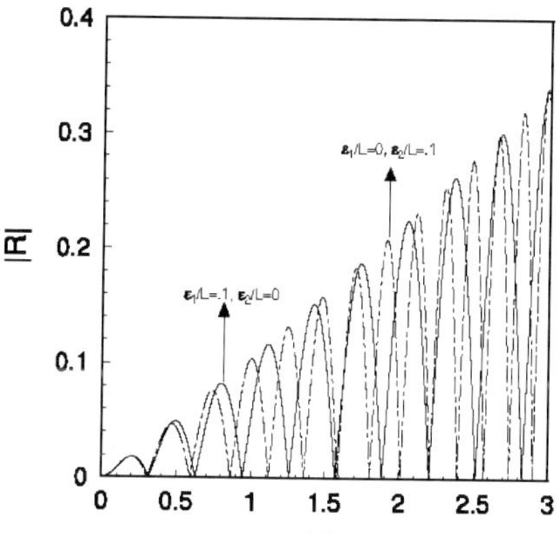

Fig. 5.6 Reflection coefficient, $l/L = 10$

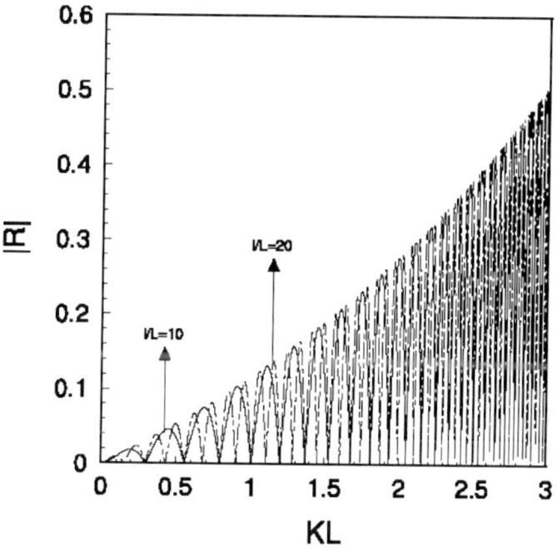

Fig. 5.7 Reflection coefficient, $\varepsilon_1/L = 0.1$, $\varepsilon_2/L = 0.2$

5.4 Scattering by a surface strip

The two dimensional problem of wave scattering by a strip of arbitrary width has been investigated by Gayen et al. (2007) in the context of linearized theory of water waves by reducing it to a pair of Carleman-type singular integral equations. In this section this method is described in some detail. The aforesaid integral equations have been solved in section 5.3 by an iterative process which is valid only for a sufficiently wide strip. A new method is described here by which solutions to these integral equations are determined by solving a set of four Fredholm integral equations of the second kind and the process is valid for a strip of arbitrary width.

As mentioned in section 5.3, there is a considerable amount of interest in the study of surface wave interaction with sea ice. In Antarctica, a region between the ice and the shore-fast sea ice exits, known as Marginal Ice Zone, which consists of continuous sheets of ice as well as broken ice. The latter can be viewed as consisting of non-interacting floating materials having no elasticity, i.e., it can be modeled as an inertial surface. Here we consider wave scattering by a strip of inertial surface of a particular surface density lying sandwiched between other inertial surfaces of a different density. The mathematical formulation is given below:

Mathematical formulation

We consider a strip of inertial surface of surface density $\varepsilon_2\rho$ floating sandwiched between another inertial surface of surface density $\varepsilon_1\rho(\varepsilon_1 \neq \varepsilon_2)$. Here ρ is the density of water. ε_1, ε_2 are two constants which denote the depth of submergence of the immersed parts of the two surfaces and thus have dimension of length. Both the surfaces are infinitely extended in one horizontal direction, say the z-direction so that the problem is two-dimensional in (x, y) coordinates, x-axis being along the width of the strip and y-axis being vertically downwards into the liquid. Thus the intermediate strip-like surface and the surfaces on its two sides occupy the regions $0 < x < l, y = 0$ and $(x < 0) \cup (x > l), y = 0$ respectively. Under the usual assumption of inviscid, incompressible and homogeneous fluid and the motion in it to be time-harmonic and irrotational, there exists a velocity potential $Re\{\phi(x, y)e^{-i\omega t}\}$ where ϕ satisfies

$$\nabla^2\phi = 0,\ y \geq 0 \quad -\infty < x < \infty, \tag{4.1}$$

with the bottom condition

$$\nabla\phi \to 0 \text{ as } y \to \infty, \tag{4.2}$$

and the surface boundary conditions

$$K_1\varphi + \varphi_y = 0 \text{ on } y = 0, (-\infty < x < 0)\cup(x > l), \tag{4.3}$$

$$K_2\varphi + \varphi_y = 0 \text{ on } y = 0, 0 < x < l \tag{4.4}$$

where $K_j = \frac{K}{1-\varepsilon_1 K}$; $(j = 1,2)$, with $K = \frac{\omega^2}{g}$ being the wavenumber. For $0 < \varepsilon_j K < 1$, the form of the boundary conditions (4.4) and (4.3) is merely a modification of the usual free surface condition $K\phi + \phi_y = 0$ on $y = 0$, and as such it allows progressive waves to propagate along the inertial surfaces. However, for waves along the intertial surface. However, for $\varepsilon_j K \geq 1$, the form of (2.3) or (2.4) becomes different and does not allow progressive waves along the inertial surface. This means that the case $\varepsilon_j K \geq 1$ is not important physically. Thus it will be assumed that the inequality $\varepsilon_j < g/\omega^2$ is always satisfied. This is equivalent to the assumption that the inertial surfaces are sufficiently light to support small time-harmonic progressive waves of given angular frequency ω. It may be remarked here that though there are limitations on the use of the present model involving floating ice, the use of the boundary conditions (4.3) and (4.4) provides a useful check of limiting case for more general boundary value problems involving Laplace equation.

The edge conditions at the two discontinuity points (0,0) and $(0,l)$ can be expressed as

$$\left.\begin{aligned}\phi &= O(r)\\ \nabla\phi &= O(1)\end{aligned}\right\} \text{ as } r \to 0 \tag{4.5}$$

where r denotes the distance from either of the edges (0,0) and $(0, l)$. These edge conditions make the solution of the boundary value problem unique.

When a train of surface waves represented by $e^{-K_1y+iK_1x}$ travelling from the direction of $x = -\infty$ is normally incident on the strip at (0,0), a part of it is reflected back into the region $x < 0$, the remaining part is transmitted through the strip $0 < x < l$. In this region it undergoes multiple reflection and transmission and finally it is transmitted into the region $x > l$ through the point $(0, l)$. Thus the far-field behaviour of $\phi(x, y)$ is given by

$$\phi \to \begin{cases} e^{-K_1y+iK_1x} + Re^{-K_1y-iK_1x} & \text{as } x \to -\infty, \\ Te^{-K_1y+iK_1(x-l)} & \text{as } x \to \infty \end{cases} \tag{4.6}$$

while in the strip region it has the form

$$\phi = \alpha\, e^{-K_2y+iK_2x} + \beta\, e^{-K_2y-iK_2(x-l)} + \psi(x, y),\ 0 < x < l. \tag{4.7}$$

In the conditions (4.6) and (4.7), R, T, α, β are unknown constants and $\psi(x, y)$ is the local non-wavy solution of Laplace's equation. R and β are the reflection coefficients due to the discontinuities at the points (0,0) and $(0,l)$ respectively while α and T represent the transmission coefficients due to the same discontinuities. It may be noted that the constants α and β appearing in (4.7) are uniquely defined.

Reduction to Singular Integral Equations

Now the mixed boundary value problem described above is reduced to two Carleman singular integral equations over a semi-infinite range. For this purpose $\phi(x, y)$ is represented in the three regions $x < 0$, $0 < x < l$ and $x > l$ by making use of Havelock's expansions for water wave potentials (Ursell 1947) in the forms

$$\varphi \equiv \varphi_1(x, y) = e^{-K_1y+iK_1x} + Re^{-K_1y-iK_1x} + \frac{2}{\pi}\int_0^\infty \frac{A(\xi)}{\xi^2 + K_1^2} L_1(\xi, y)e^{\xi x}d\xi,\ x < 0, \tag{4.8}$$

$$\varphi \equiv \varphi_2(x, y) = \alpha e^{-K_2y+iK_2x} + \beta e^{-K_2y-iK_2(x-l)} + \frac{2}{\pi}\int_0^\infty \frac{B(\xi)e^{\xi(x-l)} + C(\xi)e^{-\xi x}}{\xi^2 + K_2^2} L_2(\xi, y)d\xi,\ 0 < x < l, \tag{4.9}$$

$$\varphi \equiv \varphi_3(x, y) = Te^{-K_1y+iK_1(x-l)} + \frac{2}{\pi}\int_0^\infty \frac{D(\xi)}{\xi^2 + K_1^2} L_1(\xi, y)e^{-\xi(x-l)}d\xi,\ x > l \tag{4.10}$$

where $A(\xi)$, $B(\xi)$, $C(\xi)$ and $D(\xi)$ are unknown functions and are such that the integrals in (4.8)–(4.10) are convergent, and

$$L_j(\xi, y) = \xi \cos \xi y - K_j \sin \xi y,\ j = 1,2.$$

In order to determine the unknown constants R, T, α, β and the unknown functions $A(\xi)$, $B(\xi)$, $C(\xi)$, $D(\xi)$, the conditions of continuity of $\phi(x, y)$ and $\phi_x(x, y)$ across the lines $x = 0$ and $x = l$ ($y > 0$) are used. These give rise to four equations involving the unknown constants and the functions of the problem valid for $y > 0$.

$$\begin{aligned} &\mathrm{e}^{-K_2 y}(1+R)+\frac{2}{\pi}\int_0^\infty \frac{A(\xi)}{\xi^2+K_1^2}L_1(\xi,y)\mathrm{d}\xi=\mathrm{e}^{-K_2 y}(\alpha+\beta\,\mathrm{e}^{\mathrm{i}K_2 l})+\\ &\qquad\frac{2}{\pi}\int_0^\infty \frac{B(\xi)+C(\xi)}{\xi^2+K_2^2}L_2(\xi,y)\mathrm{e}^{\xi x}d\xi,\ y>0 \end{aligned} \tag{4.11}$$

$$\begin{aligned} &\mathrm{i}K_1\,\mathrm{e}^{-K_2 y}(1-R)+\frac{2}{\pi}\int_0^\infty \frac{A(\xi)}{\xi^2+K_1^2}L_1(\xi,y)\mathrm{d}\xi=\mathrm{i}K_2\,\mathrm{e}^{-K_2 y}(\alpha-\beta\,\mathrm{e}^{\mathrm{i}K_2 l})+\\ &\qquad\frac{2}{\pi}\int_0^\infty \frac{B(\xi)-C(\xi)}{\xi^2+K_2^2}L_2(\xi,y)\mathrm{e}^{\xi x}d\xi,\ y>0, \end{aligned} \tag{4.12}$$

$$\begin{aligned} &\mathrm{e}^{-K_2 y}(\alpha\,\mathrm{e}^{\mathrm{i}K_2 l}+\beta)+\frac{2}{\pi}\int_0^\infty \frac{B(\xi)\mathrm{e}^{\xi l}+C(\xi)\mathrm{e}^{-\xi l}}{\xi^2+K_2^2}L_2(\xi,y)\mathrm{e}^{\xi x}\mathrm{d}\xi\\ &\qquad=T\,\mathrm{e}^{-K_1 y}+\frac{2}{\pi}\int_0^\infty \frac{D(\xi)}{\xi^2+K_1^2}L_1(\xi,y)d\xi,\ y>0, \end{aligned} \tag{4.13}$$

$$\begin{aligned} &\mathrm{i}K_2\,\mathrm{e}^{-K_2 y}(\alpha\,\mathrm{e}^{\mathrm{i}K_2 l}-\beta)+\frac{2}{\pi}\int_0^\infty \frac{\xi\left(B(\xi)\mathrm{e}^{\xi l}+C(\xi)\mathrm{e}^{-\xi l}\right)}{\xi^2+K_2^2}L_2(\xi,y)\mathrm{e}^{\xi x}\mathrm{d}\xi\\ &\qquad=\mathrm{i}K_1 T\,\mathrm{e}^{-K_1 y}-\frac{2}{\pi}\int_0^\infty \frac{\xi D(\xi)}{\xi^2+K_1^2}L_1(\xi,y)d\xi,\ y>0. \end{aligned} \tag{4.14}$$

All the above four equations are basically of the form

$$\chi(y)=\psi_0\,\mathrm{e}^{-Ky}+\int_0^\infty \chi(\xi)\,L(\xi,y)d\xi,\ y>0. \tag{4.15}$$

with

$$L(\xi,y)=\xi\cos\xi y-K\sin\xi y,$$

where the function $\chi(y)$ defined for $y > 0$, is piecewise-continuously differentiable and absolutely integrable on $(0, \infty)$. The constant ψ_0 and the function $\widehat{\chi}(\xi)$ are given by Ursell (1947)

$$\psi_0=2K\int_0^\infty \chi(u)\,\mathrm{e}^{-Ku}\,du, \tag{4.16a}$$

$$\widehat{\chi}(y)=\frac{2}{\pi}\frac{1}{\xi^2+K^2}\int_0^\infty \chi(u)L(\xi,u)\,du, \tag{4.16b}$$

This is generally referred to as the Havelock's inversion theorem in the water wave literature.

Application of this theorem to equations (4.11) to (4.14) (using the result (4.16b) and (4.16a) in each equation) produces the following eight relations (after some simplifications):

$$\begin{aligned}A(\xi) &= -\frac{(K_1-K_2)\xi}{\xi^2+K_2^2}\left(\alpha+\beta e^{iK_2 l}\right)+\frac{(\xi^2+K_1K_2)\{B(\xi)e^{-\xi l}+C(\xi)\}}{\xi^2+K^2}\\ &+\frac{2}{\pi}(K_1-K_2)\xi\int_0^\infty \frac{u\{B(u)e^{-ul}+C(u)\}}{(u^2+K_2^2)(u^2-\xi^2)}du,\end{aligned} \tag{4.17}$$

$$\begin{aligned}\xi A(\xi) &= -\frac{iK_2(K_1-K_2)(\alpha-\beta e^{iK_2 l})\xi}{\xi^2+K_2^2}+\frac{\xi(\xi^2+K_1K_2)}{\xi^2+K_2^2}\{B(\xi)e^{-\xi l}-C(\xi)\}\\ &+\frac{2}{\pi}(K_1-K_2)\xi\int_0^\infty \frac{u^2\{B(u)e^{-ul}-C(u)\}}{(u^2+K_2^2)(u^2-\xi^2)}du,\end{aligned} \tag{4.18}$$

$$\begin{aligned}D(\xi) &= -\frac{(K_1-K_2)(\alpha e^{iK_2 l}+\beta)\xi}{\xi^2+K_2^2}+\frac{(\xi^2+K_1K_2)\{B(\xi)+C(\xi)e^{-\xi l}\}}{\xi^2+K_2^2}\\ &+\frac{2}{\pi}(K_1-K_2)\xi\int_0^\infty \frac{\{B(u)+C(u)e^{-ul}\}}{(u^2+K_2^2)(u^2-\xi^2)}u\,du,\end{aligned} \tag{4.19}$$

$$\begin{aligned}\xi D(\xi) &= \frac{iK_2(K_1-K_2)(\alpha e^{iK_2 l}-\beta)\xi}{\xi^2+K_2^2}-\frac{\xi(\xi^2+K_1K_2)\{B(\xi)-C(\xi)e^{-\xi l}\}}{\xi^2+K_2^2}\\ &-\frac{2}{\pi}(K_1-K_2)\xi\int_0^\infty \frac{u^2\{B(u)-C(u)e^{-ul}\}}{(u^2+K_2^2)(u^2-\xi^2)}du,\end{aligned} \tag{4.20}$$

$$\frac{1+R}{2K_1}=\frac{\alpha+\beta e^{iK_2 l}}{K_1+K_2}+\frac{2}{\pi}(K_1-K_2)\int_0^\infty \frac{\xi\{B(\xi)e^{-\xi l}+C(\xi)\}}{(\xi^2+K_1^2)(\xi^2+K_2^2)}d\xi; \tag{4.21}$$

$$\frac{i(1-R)}{2}=\frac{iK_2(\alpha-\beta e^{iK_2 l})}{K_1+K_2}+\frac{2}{\pi}(K_1-K_2)\int_0^\infty \frac{\xi^2\{B(\xi)e^{-\xi l}-C(\xi)\}}{(\xi^2+K_1^2)(\xi^2+K_2^2)}d\xi; \tag{4.22}$$

$$\frac{T}{2K_1}=\frac{\alpha e^{iK_2 l}+\beta}{K_1+K_2}+\frac{2}{\pi}(K_1-K_2)\int_0^\infty \frac{\xi\{B(\xi)+C(\xi)e^{-\xi l}\}}{(\xi^2+K_1^2)(\xi^2+K_2^2)}d\xi; \tag{4.23}$$

$$\frac{iT}{2}=\frac{iK_2(\alpha e^{iK_2 l}-\beta)}{K_1+K_2}+\frac{2}{\pi}(K_1-K_2)\int_0^\infty \frac{\xi^2\{B(\xi)-C(\xi)e^{-\xi l}\}}{(\xi^2+K_1^2)(\xi^2+K_2^2)}d\xi. \tag{4.24}$$

The integrals in (4.17) to (4.20) are in the sense of Cauchy principal value. It will be seen later that the last four relations will serve the purpose of determination of the four unknown constants.

Elimination of $A(\xi)$ between (3.4) and (3.5), and $D(\xi)$ between (3.6) and (3.7) yields

$$\lambda(\xi)B_1(\xi)+\frac{1}{\pi}\int_0^\infty \frac{B_1(u)}{u-\xi}\mathrm{d}u-\frac{1}{\pi}\int_0^\infty \frac{C_1(u)}{u+\xi}\mathrm{e}^{-ul}du=F_B(\xi),\ \xi>0, \tag{4.25}$$

$$\lambda(\xi)C_1(\xi)+\frac{1}{\pi}\int_0^\infty \frac{C_1(u)}{u-\xi}\mathrm{d}u-\frac{1}{\pi}\int_0^\infty \frac{B_1(u)}{u+\xi}\mathrm{e}^{-ul}du=F_C(\xi),\ \xi>0, \tag{4.26}$$

where

$$B_1(\xi),C_1(\xi)=\frac{\xi}{\xi^2+K_2^2}\left(B(\xi),C(\xi)\right), \tag{4.27}$$

$$\lambda(\xi)=\frac{\xi^2+K_1K_2}{\xi(K_1-K_2)}, \tag{4.28}$$

$$F_B(\xi)=\frac{\alpha}{2}\frac{\mathrm{e}^{\mathrm{i}K_2l}}{\xi-\mathrm{i}K_2}+\frac{\beta}{2}\frac{1}{\xi+\mathrm{i}K_2}, \tag{4.29}$$

$$F_C(\xi)=\frac{\beta}{2}\frac{\mathrm{e}^{\mathrm{i}K_2l}}{\xi-\mathrm{i}K_2}+\frac{\alpha}{2}\frac{1}{\xi+\mathrm{i}K_2}. \tag{4.30}$$

Equations (4.25) and (4.26) are the two coupled Carleman type singular integral equations for determining the unknown functions $B(\xi)$ and $C(\xi)$. It may be noted that these coupled integral equations are decoupled if the strip width is so large that the integrals involving negative exponentials can be neglected. Gayen et al. (2005) utilized this idea to solve similar integral equations arising in the problem of water wave scattering by an ice-strip modeled as a thin elastic plate, by an iterative process and also obtained numerical estimates for the eight unknown constants occurring there. This was valid of course for a sufficiently wide strip. However, if the breadth of the strip is moderate, this iterative process is not applicable, and here a new method given in Gayen et al. (2007) is presented which enables us to solve the coupled singular integral equations for any width of the strip, thereby solving the strip problem for all values of the breadth of the strip.

Solution after reducing to Fredholm integral equations

To solve the two coupled singular integral equations (4.25) and (4.26) for any l we write (4.25) and (4.26) in operator forms as

$$\ell B_1(\xi)+\ell' C_1(\xi)=F_B(\xi),\ \xi>0 \tag{4.31}$$

and

$$\ell C_1(\xi)+\ell' B_1(\xi)=F_C(\xi),\ \xi>0 \tag{4.32}$$

where the operators ℓ and ℓ' are defined by

$$\left.\begin{aligned}\ell f(\xi)&=\lambda(\xi)f(\xi)+\frac{1}{\pi}\int_0^\infty\frac{f(u)}{u-\xi}du,\\ \ell' f(\xi)&=-\frac{1}{\pi}\int_0^\infty\frac{f(u)e^{-ul}}{u+\xi}du\end{aligned}\right\},\ \xi>0. \tag{4.33}$$

We note that the operator ℓ involves CPV integral while ℓ' is a regular integral operator.

It is observed that the Carleman singular integral equation

$$\ell f(\xi)=h(\xi),\ \xi>0 \tag{4.34}$$

can be solved by reducing it to the following Riemann-Hilbert problem (RHP):

$$[\lambda(\xi)+i]\Lambda^+(\xi)-[\lambda(\xi)-i]\Lambda^-(\xi)=h(\xi),\ \xi>0 \tag{4.35}$$

where $\Lambda^\pm(\xi)$ are the limiting values of the sectionally analytic function $\Lambda(\zeta)$ defined by the relation

$$\Lambda(\zeta)=\frac{1}{2\pi i}\int_0^\infty\frac{f(u)}{u-\zeta}du \tag{4.36}$$

in the complex ζ-plane ($\zeta=\xi+i\eta$) cut along the real axis from $\xi=0$ to ∞. Solving the RHP (4.5) in the usual manner see Gakhov (1966) we find that the solution of the Carleman singular integral equation (4.4) is obtained as

$$f(\xi)=\ell^{-1}h(\xi)=\frac{\Lambda_0^+(\xi)}{\lambda(\xi)-i}\hat{\ell}\left[\frac{h(\xi)}{\Lambda_0^+(\xi)(\lambda(\xi)-i)}\right],\xi>0 \tag{4.37}$$

where the operator $\hat{\ell}$ is defined by

$$\hat{\ell} f(\xi)=\lambda(\xi)f(\xi)-\frac{1}{\pi}\int_0^\infty\frac{f(u)}{u-\xi}du,\xi>0 \tag{4.38}$$

and

$$\Lambda_0^+(\xi)=\lim_{\zeta\to\xi+i0}\Lambda_0(\zeta)$$

where

$$\Lambda_0(\zeta)\ =\ \exp\left[\frac{1}{2\pi i}\left\{\int_0^\infty\left(\ln\frac{t-iK_1}{t+iK_1}-2\pi i\right)\frac{dt}{t-\zeta}-\int_0^\infty\left(\ln\frac{t-iK_2}{t+iK_2}-2\pi i\right)\frac{dt}{t-\zeta}\right\}\right],\zeta\notin(0,\infty) \tag{4.39}$$

being the solution of the homogeneous problem corresponding to the RHP (4.35).

We now apply the operator ℓ^{-1} to (4.31) to obtain

$$B_1(\xi)=\ell^{-1}\left[F_B(\xi)-\ell' C_1(\xi)\right],\ \xi>0 \tag{4.40}$$

which when substituted into (4.32) produces

$$\ell C_1(\xi) + \ell'\left[\ell^{-1}(F_B - \ell' C_1)\right](\xi) = F_C(\xi),\ \xi > 0. \tag{4.41}$$

Using the definitions of the integral operators ℓ and ℓ' as given in (4.33), (4.37) it is easy to see that

$$\begin{aligned}\ell m(\xi) &= (\ell^{-1}\ell')m(\xi)\\ &= \frac{\Lambda_0^+(\xi)}{\lambda(\xi)-\mathrm{i}}\left[\frac{\lambda(\xi)}{\Lambda_0^+(\xi)(\lambda(\xi)+\mathrm{i})}\left(-\frac{1}{\pi}\int_0^\infty \frac{m(u)\mathrm{e}^{-ul}}{u+\xi}du\right)\right.\\ &\left.+\frac{1}{\pi^2}\int_0^\infty m(u)\mathrm{e}^{-ul}du\left(\int_0^\infty \frac{\mathrm{d}t}{\Lambda_0^+(t)(\lambda(t)+\mathrm{i})(t+u)(t-\xi)}\right)\right].\end{aligned} \tag{4.42}$$

To evaluate the inner integral in the second term of (4.42), we consider the integral

$$\int_\Gamma \frac{d\tau}{\Lambda_0(\tau)(\tau+u)(\tau-\zeta)},\ \zeta \notin \Gamma, \tag{4.43}$$

where $\Lambda_0(\zeta)$ satisfies the homogeneous RHP

$$[\lambda(\xi)+\mathrm{i}]\Lambda^+(\xi) - [\lambda(\xi)-\mathrm{i}]\Lambda^-(\xi) = 0,\ \xi > 0 \tag{4.44}$$

in the complex ζ-plane cut along the positive real axis and Γ is a positively oriented contour consisting of a loop around the positive real axis having indentations above the point $\zeta = \xi + i0$ and below the point $\zeta = \xi - i0$ and a circle of large radius with centre at the origin in the complex τ-plane.

We observe that

$$\begin{aligned}\int_\Gamma \frac{d\tau}{\Lambda_0(\tau)(\tau+u)(\tau-\zeta)} &= \int_0^\infty \left\{\frac{1}{\Lambda_0^+(t)} - \frac{1}{\Lambda_0^-(t)}\right\}\frac{dt}{(t+u)(t+\zeta)}\\ &= 2i\int_0^\infty \frac{dt}{\Lambda_0^+(t)(\lambda(t)+\mathrm{i})(t+u)(t-\zeta)}\end{aligned} \tag{4.45}$$

after using (4.44).

Also from the residue calculus theorem,

$$\int_\Gamma \frac{d\tau}{\Lambda_0(\tau)(\tau+u)(\tau-\zeta)} = \frac{2\pi\mathrm{i}}{u+\zeta}\left\{\frac{1}{\Lambda_0(\zeta)} - \frac{1}{\Lambda_0(-u)}\right\}. \tag{4.46}$$

Comparing (4.45) and (4.46) we find

$$\frac{1}{u+\zeta}\left\{\frac{1}{\Lambda_0(\zeta)} - \frac{1}{\Lambda_0(-u)}\right\} = \frac{1}{2\pi\mathrm{i}}\int_0^\infty \frac{2\mathrm{i}\,dt}{\Lambda_0^+(t)(\lambda(t)+\mathrm{i})(t+u)(t-\zeta)}. \tag{4.47}$$

Applying Plemelj's formulae to the above relation the inner integral in the second term on the right side of (4.42) is evaluated as

$$\int_0^{\infty} \frac{dt}{\Lambda_0^+(t)(\lambda(t)+\mathrm{i})(t+u)(t-\xi)} = \frac{\pi}{u+\xi}\left\{\frac{\lambda(\xi)}{(\lambda(t)+\mathrm{i})\Lambda_0^+(\xi)} - \frac{1}{\Lambda_0(-u)}\right\}$$

which when substituted into (4.42), produces

$$m(\xi) = -\frac{1}{\pi}\frac{\Lambda_0^+(\xi)}{\lambda(\xi)-\mathrm{i}}\int_0^{\infty}\frac{m(u)\mathrm{e}^{-ul}\,du}{(u+\xi)\Lambda_0(-u)}.$$

Applying the operator ℓ^{-1} to the both sides of (4.11), we find

$$\left[I-\ell^2\right]C_1(\xi) = r(\xi),\ \xi>0 \tag{4.48}$$

where

$$r(\xi) = \ell^{-1}\left[F_C - \ell'\ell^{-1}F_B\right](\xi),\ \xi>0. \tag{4.49}$$

It may be noted that the operator $\ell^{-1}\,\ell'$ is not commutative.

Now $F_B(\xi)$ and $F_C(\xi)$ are substituted from (3.16) and (3.17) into (4.14) to obtain $r(\xi)$ in the form

$$r(\xi) = \alpha r_1(\xi) + \beta r_2(\xi)$$

where

$$r_1(\xi) = \frac{1}{2c}\frac{\Lambda_0^+(\xi)}{\lambda(\xi)-\mathrm{i}}\left[\frac{1}{\xi+\mathrm{i}K_2} + \frac{\mathrm{e}^{\mathrm{i}K_2 l}}{\pi}\int_0^{\infty}\frac{\Lambda_0^+(u)\mathrm{e}^{-ul}\,du}{(\lambda(u)-\mathrm{i})(u+\xi)(u-\mathrm{i}K_2)\Lambda_0(-u)}\right] \tag{4.50}$$

and

$$r_2(\xi) = \frac{1}{2c}\frac{\Lambda_0^+(\xi)}{\lambda(\xi)-\mathrm{i}}\left[\frac{\mathrm{e}^{\mathrm{i}K_2 l}}{\xi-\mathrm{i}K_2} + \frac{1}{\pi}\int_0^{\infty}\frac{\Lambda_0^+(u)\mathrm{e}^{-ul}\,du}{(\lambda(u)-\mathrm{i})(u+\xi)(u+\mathrm{i}K_2)\Lambda_0(-u)}\right] \tag{4.51}$$

with

$$c = \Lambda_0(\pm\mathrm{i}K_2) = \left(\frac{2K_2}{K_1+K_2}\right)^{\frac{1}{2}}.$$

We now define two functions $U(\xi)$ and $V(\xi)$ for $\xi>0$ such that

$$[I+\ell]C_1(\xi) = U(\xi),[I-\ell]C_1(\xi) = V(\xi),\ \xi>0 \tag{4.52}$$

so that

$$C_1(\xi) = \frac{1}{2}[U(\xi)+V(\xi)] \text{ and } \ell\, C_1(\xi) = \frac{1}{2}[U(\xi)-V(\xi)],\ \xi>0. \tag{4.53}$$

Then the integral equation (4.12) can be written either as

$$[I+\ell]V(\xi)=r(\xi),\ \xi>0 \tag{4.54}$$

or as

$$[I-\ell]U(\xi)=r(\xi),\ \xi>0. \tag{4.55}$$

Since

$$r(\xi)=\alpha r_1(\xi)+\beta r_2(\xi),$$

we may express $U(\xi)$, $V(\xi)$ as

$$U(\xi)=[I-\ell]^{-1}r(\xi)=\alpha u_1(\xi)+\beta u_2(\xi) \tag{4.56}$$

and

$$V(\xi)=[I+\ell]^{-1}r(\xi)=\alpha v_1(\xi)+\beta v_2(\xi) \tag{4.57}$$

where $u_j(\xi)$, $v_j(\xi)$, $(j=1,2)$, $\xi>0$ are unknown functions.

The integral equation (4.54) along with the relation (4.56) and the integral equation (4.55) along with the relation (4.58) are satisfied if $u_j(\xi)$, $v_j(\xi)$, $(j=1,2)$ satisfy

$$\begin{aligned}&[I-\ell]u_1(\xi)=r_1(\xi),\ \xi>0,[I-\ell]u_2(\xi)=r_2(\xi),\ \xi>0,\\&[I+\ell]v_1(\xi)=r_1(\xi),\ \xi>0,[I+\ell]v_2(\xi)=r_2(\xi),\ \xi>0.\end{aligned} \tag{4.58}$$

These are in fact Fredholm integral equations with regular kernels, the integral operator ℓ being defined in (4.47). These integral equations are solved numerically by Nystrom's method and then the functions $u_j(\xi)$, $v_j(\xi)$, $(j=1,2)$ are found numerically. It may be noted that some considerable analytical calculations are required to reduce the functions $r_j(\xi)(j=1,2)$ to forms suitable for numerical computation. This is described below.

Simplification of $r_1(\xi)$ and $r_2(\xi)$

The basic step for the evaluation of the integral equations (4.58) and the functions $r_1(\xi)$ and $r_2(\xi)$ is to determine the functions $\Lambda_0(-\xi)$ and $\Lambda_0^+(\xi)$ for $\xi>0$ in computable forms. Here we present an explicit derivation of these functions and also simplify $r_1(\xi)$ and $r_2(\xi)$.

The function $\Lambda_0^+(\xi)$ is given by

$$\Lambda_0^+(\xi)=\left(\frac{\xi-\mathrm{i}K_1}{\xi+\mathrm{i}K_1}\frac{\xi+\mathrm{i}K_2}{\xi-\mathrm{i}K_2}\right)^{\frac{1}{2}}\exp\left(\frac{1}{2\pi\mathrm{i}}\int_0^\infty\frac{\ln\left(\frac{t-\mathrm{i}K_1}{t+\mathrm{i}K_1}\frac{t+\mathrm{i}K_2}{t-\mathrm{i}K_2}\right)}{t-\xi}dt\right),\ \xi>0. \tag{4.59}$$

If we define

$$Y(\xi) = \frac{1}{2\pi \mathrm{i}} \int_0^{\infty} \frac{\ln\left(\dfrac{t-\mathrm{i}K_1}{t+\mathrm{i}K_1}\dfrac{t+\mathrm{i}K_2}{t-\mathrm{i}K_2}\right)}{t-\xi} dt, \xi > 0, \tag{4.60a}$$

$$Y_j(\xi) = \frac{1}{2\pi \mathrm{i}} \int_0^{\infty} \ln\frac{t-\mathrm{i}K_j}{t+\mathrm{i}K_j}\frac{dt}{t-\xi}; (j=1,2), \xi > 0, \tag{4.60b}$$

$$X(\xi) = Y(-\xi) \text{ and } X_j(\xi) = Y_j(-\xi), \xi > 0, \tag{4.60c}$$

then

$$Y(\xi) = Y_1(\xi) - Y_2(\xi), \Lambda_0(-\xi) = \exp(X(\xi)), \tag{4.61a}$$

$$\Lambda_0^+(\xi) = \left(\frac{\xi-\mathrm{i}K_1}{\xi+\mathrm{i}K_1}\frac{\xi+\mathrm{i}K_2}{\xi-\mathrm{i}K_2}\right)^{\frac{1}{2}} \exp(Y(\xi)). \tag{4.61b}$$

Following Varley and Walker (1989) the derivative of $Y_j(\xi)$ is found to be

$$Y_j'(\xi) = -\frac{K_j}{2\pi}\left[\frac{\ln\left(\dfrac{\xi}{-\mathrm{i}K_j}\right)}{\xi(\xi+\mathrm{i}K_j)} + \frac{\ln\left(\dfrac{\xi}{\mathrm{i}K_j}\right)}{\xi(\xi-\mathrm{i}K_j)}\right], j=1,2.$$

It may be observed that $Y_j(\infty) = 0$. Integration of $Y_j'(\xi)$ gives

$$Y_j(\xi) = -\frac{K_j}{2\pi}\int_{\infty}^{\xi}\left[\frac{\ln\left(\dfrac{t}{-\mathrm{i}K_j}\right)}{t(t+\mathrm{i}K_j)} + \frac{\ln\left(\dfrac{t}{\mathrm{i}K_j}\right)}{t(t-\mathrm{i}K_j)}\right] dt$$
$$= -\frac{1}{2\pi \mathrm{i}}\int_{\frac{-\mathrm{i}K_j}{\xi}}^{\frac{\mathrm{i}K_j}{\xi}} \frac{\ln u}{u-1} du. \tag{4.62}$$

After some manipulations $Y(\xi)$ reduces to

$$Y(\xi) = \frac{1}{4}\ln\frac{\xi-\mathrm{i}K_1}{\xi-\mathrm{i}K_2} - \frac{3}{4}\ln\frac{\xi+\mathrm{i}K_1}{\xi+\mathrm{i}K_2} - \frac{1}{\pi}\int_{\frac{K_2}{\xi}}^{\frac{K_1}{\xi}} \frac{\ln v}{v^2+1} dv. \tag{4.63}$$

Hence $\Lambda_0^+(\xi)$ has the alternative form

$$\begin{aligned}\Lambda_0^+(\xi) &= \left(\frac{\xi - \mathrm{i}K_1}{\xi - \mathrm{i}K_2}\right)^{\frac{1}{2}}\left(\frac{\xi + \mathrm{i}K_1}{\xi + \mathrm{i}K_2}\right)^{-\frac{1}{2}} \exp(Y(\xi)) \\ &= \left(\frac{\xi - \mathrm{i}K_1}{\xi - \mathrm{i}K_2}\right)^{\frac{3}{4}}\left(\frac{\xi + \mathrm{i}K_1}{\xi + \mathrm{i}K_2}\right)^{-\frac{5}{4}} E(\xi) \\ &= \left(\frac{\xi^2 + K_1^2}{\xi^2 + K_2^2}\right)^{-\frac{1}{4}} \mathrm{e}^{-2\mathrm{i}(\theta_1 - \theta_2)} E(\xi)\end{aligned} \tag{6.64}$$

where $\theta_j = \tan^{-1}\frac{K_j}{\xi}$, $j = 1,2$, and

$$E(\xi) = \exp\left(-\frac{1}{\pi}\int_{\frac{K_2}{\xi}}^{\frac{K_1}{\xi}} \frac{\ln v}{v^2 + 1}\, dv\right). \tag{4.65}$$

$X(\xi)$ is simplified in a similar manner and we find that

$$X_j(\xi) = Y_j(-\xi) = -Y_j(\xi),\ j = 1,2.$$

Thus $X(\xi) = -\,Y(\xi)$, and

$$\begin{aligned}\Lambda_0(-\xi) &= \exp(X(\xi)) \\ &= \left(\frac{\xi - \mathrm{i}K_1}{\xi - \mathrm{i}K_2}\right)^{-\frac{1}{4}}\left(\frac{\xi + \mathrm{i}K_1}{\xi + \mathrm{i}K_2}\right)^{\frac{3}{4}} (E(\xi))^{-1} \\ &= \left(\frac{\xi^2 + K_1^2}{\xi^2 + K_2^2}\right)^{\frac{1}{4}} \mathrm{e}^{\mathrm{i}(\theta_1 - \theta_2)} (E(\xi))^{-1}.\end{aligned} \tag{4.66}$$

The various complex-valued functions appearing in $r_1(\xi)$ and $r_2(\xi)$ are simplified as follows:

(a) $$\frac{\Lambda_0^+(\xi)}{\lambda(\xi) - \mathrm{i}} = (K_1 - K_2)\xi(\xi^2 + K_1^2)^{-\frac{3}{4}}(\xi^2 + K_2^2)^{-\frac{1}{4}} \mathrm{e}^{-\mathrm{i}(\theta_1 - \theta_2)} E(\xi)(a)$$

where we have used

$$\lambda(\xi) - \mathrm{i} = \frac{(\xi - \mathrm{i}K_1)(\xi + \mathrm{i}K_2)}{\xi(K_1 - K_2)}. \tag{4.67}$$

(b) $$\frac{\Lambda_0^+(\xi)}{\lambda(\xi) - \mathrm{i}} \frac{1}{\Lambda_0(-\xi)} = \frac{(K_1 - K_2)\xi}{\xi^2 + K_1^2} \mathrm{e}^{-2\mathrm{i}(\theta_1 - \theta_2)} (E(\xi))^2,$$

(c) $$\frac{\Lambda_0^+(\xi)}{\lambda(\xi) - \mathrm{i}} \frac{1}{\Lambda_0(-\xi)}\left(\frac{1}{\xi - \mathrm{i}K_2}, \frac{1}{\xi + \mathrm{i}K_2}\right) = \frac{(K_1 - K_2)\xi \mathrm{e}^{-2\mathrm{i}\theta_1}}{(\xi^2 + K_1^2)(\xi^2 + K_2^2)^{\frac{1}{2}}} (E(\xi))^2 (\mathrm{e}^{3\mathrm{i}\theta_2}, \mathrm{e}^{\mathrm{i}\theta_2}).$$

(d) $$\frac{\Lambda_0^+(\xi)}{\lambda(\xi)-\mathrm{i}}\left(\frac{1}{\xi-\mathrm{i}K_2},\frac{1}{\xi+\mathrm{i}K_2}\right)=\frac{(K_1-K_2)\xi\mathrm{e}^{-\mathrm{i}\theta_1}}{\left((\xi^2+K_1^2)(\xi^2+K_2^2)\right)^{\frac{3}{4}}}E(\xi)(\mathrm{e}^{2\mathrm{i}\theta_2},1).$$

Using (a) to (d), $r_1(\xi)$ and $r_2(\xi)$ are simplified as

$$r_1(\xi)=r_0(\xi)\left[\frac{1}{(\xi^2+K_2^2)^{\frac{1}{2}}}+\mathrm{e}^{\mathrm{i}K_2 l}\int_0^\infty M(u,\xi)\mathrm{e}^{3\mathrm{i}\psi_2}du\right], \tag{4.68}$$

$$r_2(\xi)=r_0(\xi)\left[\frac{\mathrm{e}^{\mathrm{i}(K_2 l+2\theta_2)}}{(\xi^2+K_2^2)^{\frac{1}{2}}}+\int_0^\infty M(u,\xi)\mathrm{e}^{\mathrm{i}\psi_2}du\right] \tag{4.69}$$

where

$$r_0(\xi)=\frac{K_1-K_2}{\pi}\frac{\xi E(\xi)\mathrm{e}^{-\mathrm{i}\theta_1}}{(\xi^2+K_1^2)^{\frac{3}{4}}}(\xi^2+K_2^2)^{\frac{1}{4}},$$

$$M(u,\xi)=\frac{K_1-K_2}{2c}\frac{u\mathrm{e}^{-ul}\left(E(\xi)\right)^2\mathrm{e}^{\mathrm{i}(\theta_2-2\psi_1)}}{(u^2+K_1^2)(u^2+K_2^2)^{\frac{1}{2}}(u+\xi)}$$

and

$$\psi_j=\tan^{-1}\frac{K_j}{u};\, j=1,2.$$

The functions $B_1(\xi)$ and $C_1(\xi)$ which satisfy the two coupled singular integral equations (4.25) and (4.26) are now found in a straightforward manner as

$$\begin{aligned}B_1(\xi)&=(\ell^{-1}F_B)(\xi)-\ell C_1(\xi)\\&=(\ell^{-1}F_B)(\xi)-\frac{1}{2}\{U(\xi)-V(\xi)\}\\&=\alpha B_1^\alpha(\xi)+\beta B_1^\beta(\xi)\end{aligned} \tag{4.70}$$

$$C_1(\xi)=\frac{1}{2}\{U(\xi)+V(\xi)\}=\alpha C_1^\alpha(\xi)+\beta C_1^\beta(\xi) \tag{4.71}$$

where

$$B_1^\alpha(\xi)=\frac{1}{2}\left[\frac{\Lambda_0^+(\xi)\mathrm{e}^{\mathrm{i}K_2 l}}{c(\lambda(\xi)-\mathrm{i})(\xi-\mathrm{i}K_2)}-u_1(\xi)+v_1(\xi)\right], \tag{4.72}$$

$$B_1^\beta(\xi)=\frac{1}{2}\left[\frac{\Lambda_0^+(\xi)}{c(\lambda(\xi)-\mathrm{i})(\xi+\mathrm{i}K_2)}-u_2(\xi)+v_2(\xi)\right], \tag{4.73}$$

$$C_1^{\alpha}(\xi) = \frac{1}{2}\{u_1(\xi) + v_1(\xi)\}, \tag{4.74}$$

$$C_1^{\beta}(\xi) = \frac{1}{2}\{u_2(\xi) + v_2(\xi)\}; \tag{4.75}$$

the value of the constant c being given after equation (4.51).

Thus $B_1(\xi)$ and $C_1(\xi)$ are obtained in terms of the unknown constants α and β. We now replace these functions in the equations (4.21)–(4.24) by their expressions in (4.70) and (4.71). This results in a system of four linear equations in α, β, R and T. These equations are solved numerically and the numerical estimates for the reflection and transmission coefficients are computed for different sets of prescribed parameters. Details about how the four linear equations are obtained are given below.

Determination of α, β, R and T

We first replace $B(\xi)$ and $C(\xi)$ in (4.21) to (4.24) in terms of $B_1(\xi)$ and $C_1(\xi)$ using (4.27), then $B_1(\xi)$ and $C_1(\xi)$ in terms of expressions (4.70) and (4.71) involving α and β. These give rise to four equations for the determination of the four unknown constants α, β, R and T. Eliminating R and T from the first two and the last two of these equations respectively, we obtain two equations for α and β which when solved produce

$$\alpha = -\mu_0 \frac{\mu_3\lambda_8 + \lambda_{10}}{\lambda_7\lambda_{10} - \lambda_8\lambda_9}, \; \beta = \mu_0 \frac{\mu_3\lambda_7 + \lambda_9}{\lambda_7\lambda_{10} - \lambda_8\lambda_9} \tag{4.76}$$

where

$$\mu_0 = \left\{ \left(\frac{K_1 - K_2}{K_1 + K_2} \right)^2 e^{2iK_2 l} - 1 \right\}, \; \mu_3 = \frac{K_1 - K_2}{K_1 + K_2} e^{iK_2 l},$$

$$\lambda_7 = 1 - \mu_0(\lambda_{3,l} - \lambda_4), \lambda_8 = \mu_0(\lambda_{5,l} - \lambda_6),$$

$$\lambda_9 = \mu_0(\lambda_{4,l} - \lambda_3), \lambda_{10} = 1 - \mu_0(\lambda_{6,l} - \lambda_5),$$

$$\lambda_j = \int_0^\infty \frac{f_j(\xi)}{\xi + iK_1} d\xi, \; \lambda_{j,l} = \int_0^\infty \frac{f_j(\xi) e^{-\xi l}}{\xi - iK_1} d\xi, j = 3,4,5,6, \tag{4.77}$$

$$f_3(\xi) = \mu_1 B_1^{\alpha}(\xi) - \mu_2 C_1^{\alpha}(\xi), f_4(\xi) = \mu_1 C_1^{\alpha}(\xi) - \mu_2 B_1^{\alpha}(\xi),$$

$$f_5(\xi) = \mu_1 B_1^{\beta}(\xi) - \mu_2 C_1^{\beta}(\xi), f_6(\xi) = \mu_1 C_1^{\beta}(\xi) - \mu_2 B_1^{\beta}(\xi),$$

$$\mu_1 = \frac{2}{\pi i}(K_1 - K_2), \mu_2 = \frac{2}{\pi i} \frac{(K_1 - K_2)^2}{K_1 + K_2} e^{iK_2 l}$$

and the functions $B_1^{\alpha}(\xi)$, $C_1^{\alpha}(\xi)$, $B_1^{\beta}(\xi)$ and $C_1^{\beta}(\xi)$ are given by (4.72) to (4.75). These can be reduced to numerically computable forms by using the calculations of $r_1(\xi)$ and $r_2(\xi)$.

Once α and β are computed from (4.76), R can be determined directly from either (4.21) or (4.22) and T from either (4.23) or (4.24), after writing $B(\xi)$ and $C(\xi)$ in terms of $B_1(\xi)$ and $C_1(\xi)$ and the later functions in terms of α and β.

Numerical results

Owing to the principle of conservation of energy, $|R|^2 + |T|^2 = 1$. Because of this, we present only numerical results, the numerical estimates for the reflection coefficient $|R|$. This identity can be used as a check on the numerical result, which has been done here for all the data points. A characteristic length L, with respect to which the strip width can be regarded as wide or moderate (l/L large or moderate), is introduced to non-dimensionalize the quantities $\varepsilon_j (j = 1,2)$, l and K^{-1}. Figures 5.8(a–c) represent $|R|$ against the wave number KL for a strip of inertial surface floating sandwiched between two semi-infinite free surfaces, i.e., $\varepsilon_1 = 0$ and $\varepsilon_2/L = 0.01$ for Figs. 5.8(a) and 5.8(b) and $\varepsilon_2/L = 0.1$ for Fig. 5.8(c). Figure 5.8(a) depicts $|R|$ for smaller values of the strip width l/L. The overall values of $|R|$ are less than 0.03 in Fig. 5.8(a), it shows that for a strip of sufficiently small width, only a small amount of the incident wave energy is

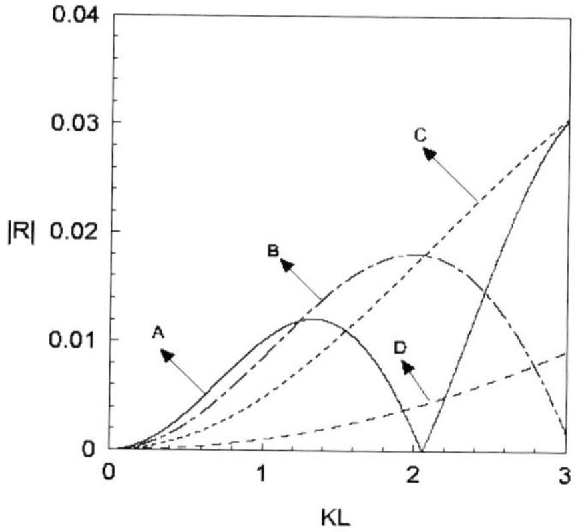

Fig. 5.8a $|R|$ for different values of the strip width $\frac{\varepsilon_1}{L} = 0, \frac{\varepsilon_2}{L} = 0.01, \frac{l}{L} = 0.1(A), 0.5(B), 1.0(C)$

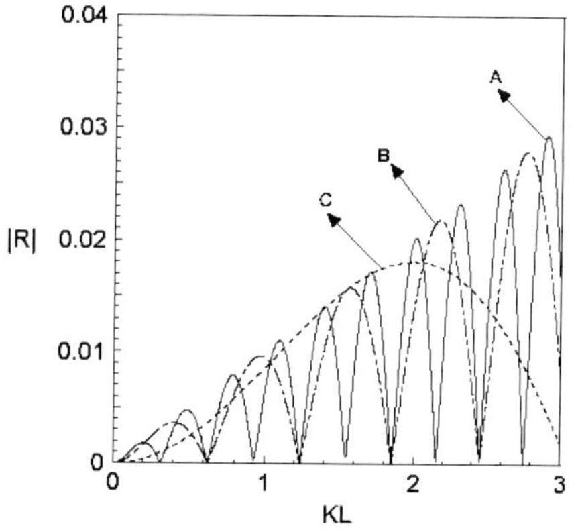

Fig. 5.8b $|R|$ for different values of the strip width $\frac{\varepsilon_1}{L} = 0, \frac{\varepsilon_2}{L} = 0.01, \frac{l}{L} = 1(A), 5(B), 10(C)$

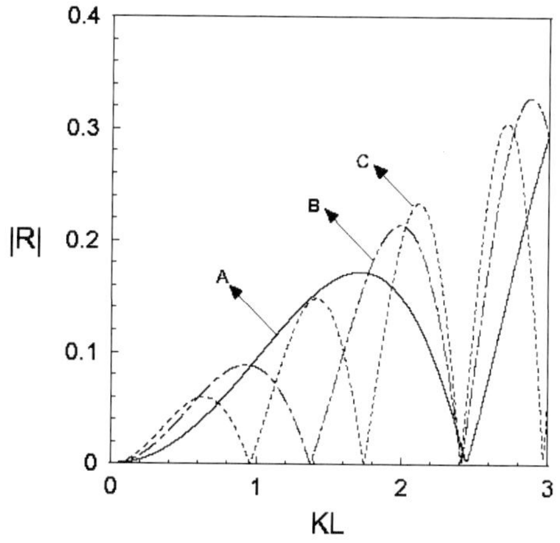

Fig. 5.8c $|R|$ for different values of the strip width $\frac{\varepsilon_1}{L} = 0, \frac{\varepsilon_2}{L} = 0.1, \frac{l}{L} = 1(A), 2(B), 3(C)$

reflected back. As the strip width increases $|R|$ fluctuates and the occurrence of zeros of $|R|$ is observed. This feature is prominent in Fig. 5.8(c). It shows that for larger strip widths ($l/L = 5$, 10) the number of zeros of $|R|$ increases. In Fig. 5.8(c) is considered a heavier strip ($\varepsilon_2/L = 0.1$). Here also there is an increase in the number of zeros of $|R|$ with the increase in strip width occurs. Also the overall values of $|R|$ are increased as the surface density of the material of strip increases.

In Figs. 5.9(a) and 5.9(b) the present results are compared with those given in section 5.2 and also by Kanoria et al. (1999) for a wide strip. The data for $|R|$ computed under the assumption of wide strip are indicated by crosses in Figs. 5.9(a) and 5.9(b). These are seen to lie exactly on the curves drawn from the data obtained following the present method. This provides a good check on the validity of the results obtained here.

If the strip is termed as a scatterer, then the product $(|\varepsilon_1 - \varepsilon_2|K)(Kl)$ is defined as its strength. Since $0 < \varepsilon_1 K$ and $\varepsilon_2 K < 1$, $(|\varepsilon_1 - \varepsilon_2|K)$ is less than unity, so the strength can be increased by increasing Kl. Hence an increase in strip width means an increase in the strength of the scatterer. A wide strip is thus a strong scatterer. Most of the figures shows that as l/L increase, $|R|$ becomes more oscillatory with increasing amplitude as expected.

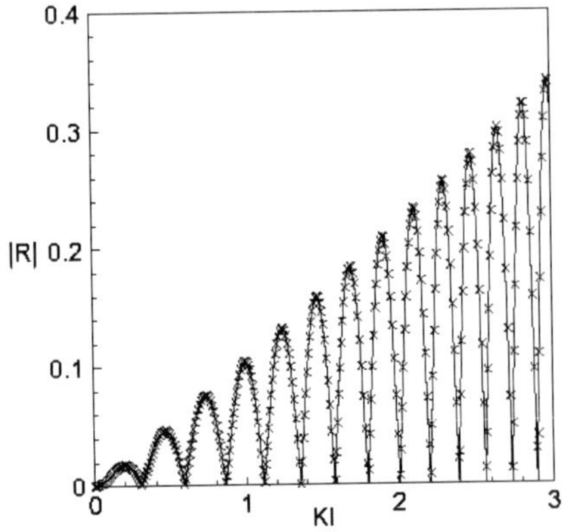

Fig. 5.9a $|R|$ for different values of the strip width $\frac{\varepsilon_1}{L} = 0, \frac{\varepsilon_2}{L} = 0.1, \frac{l}{L} = 10$

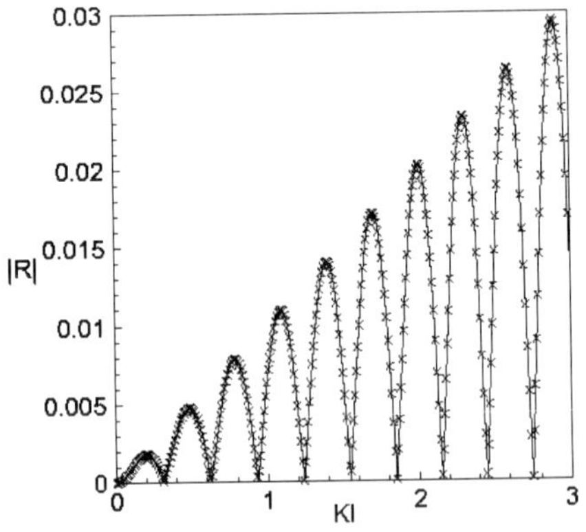

Fig. 5.9b $|R|$ for different values of the strip width $\frac{\varepsilon_1}{L} = 0, \frac{\varepsilon_2}{L} = 0.01, \frac{l}{L} = 10$ (Crosses denote data obtained by Kanoria et al. (1999). The line presents corresponding data from the present analysis)

Figure 5.10 shows $|R|$ for $\varepsilon_1/L = 0.1$, $\varepsilon_2/L = 0$ and $l/L = 1,2,3$, i.e., for the case when there is a gap of finite width between two semi-infinite inertial surfaces of the same surface density. In this case the number of zeros of $|R|$ increases with the increase in the strip width, as was observed in Fig. 5.8.

The two complementary cases, when the intermediate surface is composed of inertial surface surrounded by free surface or a free surface surrounded by inertial surface are compared in Fig. 5.11 for $l/L = 2,3$. The continuous lines IS_2 and IS_3 represent $|R|$ for the case of scattering by a strip of inertial surface floating between free surfaces for $l/L = 2$ and 3 respectively while the dotted lines FS_2 and FS_3 stand for $|R|$ for the complementary case for l/L in the same order. From this figure it is observed that the zeros of $|R|$ for a strip of inertial surface are shifted towards the left of those for a strip of free surface.

For the general case of two non-zero inertial surfaces ($\varepsilon_1/L = 0.02$, $\varepsilon_2/L = 0.05$), the effect of strip width is depicted in Fig. 5.12 which shows a similar feature of the curves as was noticed in Figs. 5.8, 5.10, 5.11.

Figure 5.13 shows $|R|$ associated with a strip which is heavier or lighter compared to the surrounding surfaces. When the strip is heavier, more multiple reflection occurs in comparison to the case when the strip is lighter.

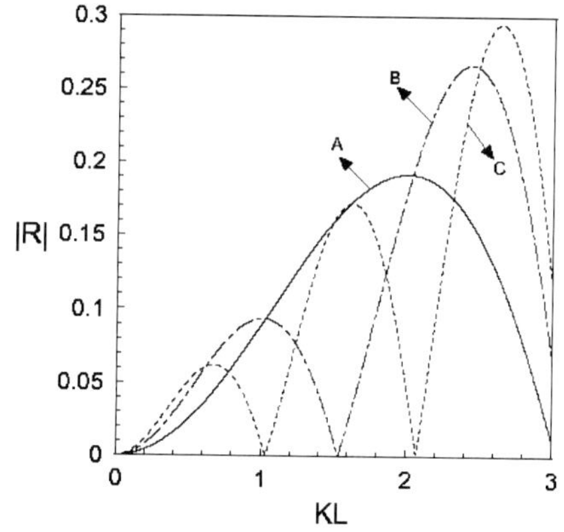

Fig. 5.10 $|R|$ for different values of the strip width $\frac{\varepsilon_1}{L} = 0.1$, $\frac{\varepsilon_2}{L} = 0$, $\frac{l}{L} = 1(A), 2(B), 3(C)$

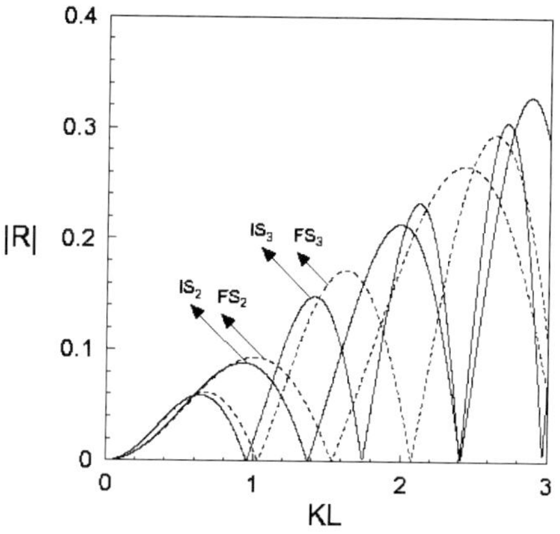

Fig. 5.11 $|R|$ for two complementary case. IS_2: $\frac{\varepsilon_1}{L} = 0$, $\frac{\varepsilon_2}{L} = 0.1$; FS_2: $\frac{\varepsilon_1}{L} = 0.1$, $\frac{\varepsilon_2}{L} = 0$ $\left(\frac{l}{L} = 2\right)$ IS_3: $\frac{\varepsilon_1}{L} = 0$, $\frac{\varepsilon_2}{L} = 0.1$; FS_3: $\frac{\varepsilon_1}{L} = 0.1$, $\frac{\varepsilon_2}{L} = 0$ $\left(\frac{l}{L} = 3\right)$

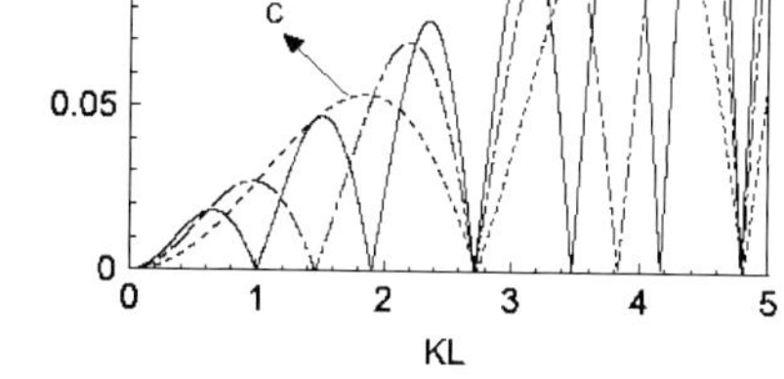

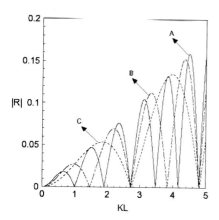

Fig. 5.12 $|R|$ for non-zero surface densities $\frac{\varepsilon_1}{L} = 0.2$, $\frac{\varepsilon_2}{L} = 0.05$, $\frac{l}{L} = 1(A), 2(B), 3(C)$

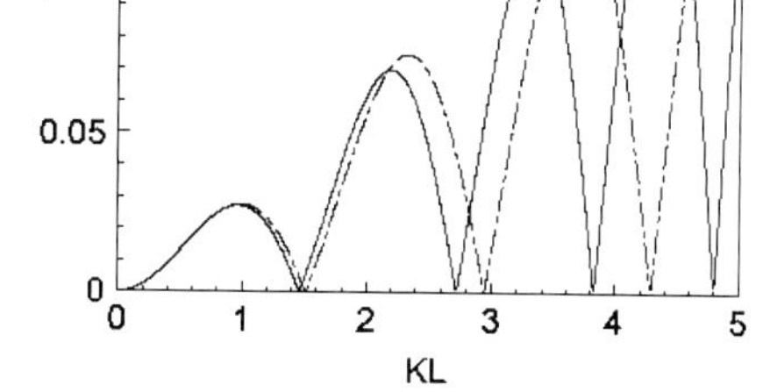

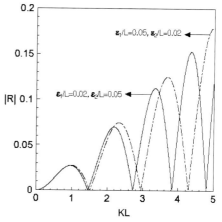

Fig. 5.13 $|R|$ for non-zero surface densities with $\frac{l}{L} = 2$

In Fig. 5.14, $|R|$ is depicted against the surface density parameter ε_2/L of the strip surrounded by free surface ($\varepsilon_1/L = 0$) for two different wave numbers ($KL = 3, 5$) and $l/L = 2$. It is observed that for smaller wave number ($KL = 3$) there is no reflection for $\varepsilon_1/L < 0.02$. However, with the increase in the surface density parameter ε_2/L, the amount of reflection gradually increases. This is also true for $KL = 5$ but in this case the curve for $|R|$ is more oscillatory in nature for various ε_2/L.

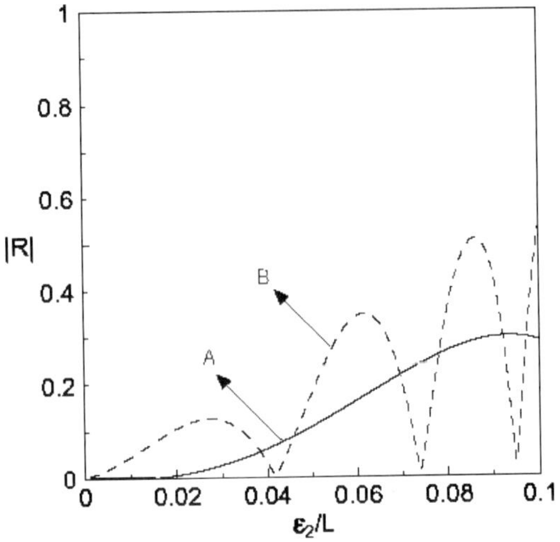

Fig. 5.14 $|R|$ for a strip floating on the water surface vs. its surface density for two different wavelengths $\frac{l}{L} = 2, \frac{\varepsilon_1}{L} = 0, KL = 3(A), 5(B)$

5.5 Scattering by an elastic strip

The method employed in section 5.4 to study the problem of water wave scattering by a surface strip has been further developed by Gayen and Mandal (2009) to study wave scattering by a thin elastic plate of arbitrary width floating on the surface of deep water. An account of this development is given in this section.

The problems of water wave scattering by thin elastic plates of either semi-infinite or of finite width have been investigated by a number of researchers using a variety of mathematical techniques. The interest behind investigating this class of problems arises due to their applications in several practical areas. One of these is associated with understanding the behaviour of the waves while interacting with the sea-ice in Marginal Ice Zone (MIZ) in the Antarctica. Examples of such studies can be found in Fox and Squire (1990, 1994), Meylan and Squire (1994), Squire et al. (1995), Williams and Squire (2006). The effect of surface wave interaction with floating elastic plates is also important in modelling floating breakwaters and very large floating structures (VLFS) like floating runways, offshore pleasure cities, floating oil-storage bases, etc. Problems dealing with wave-VLFS interaction have been considered by Kagemoto et al. (1998), Namba and Okhusu (1999), Kashiwagi (2000), Khapasheva and Korobkin (2002), Okhusu and Namba (2004) and others.

Evans and Davies (1968) derived the explicit solution of the scattering problem involving a semi-infinite thin elastic plate in finite depth water using Wiener-Hopf technique cf. Noble (1958); however no numerical calculation could be carried out due to the complicated nature of the solution. Later this problem was also attacked by Balmforth and Craster (1999) and Chung and Fox (2002) using the Wiener-Hopf technique by incorporating some modifications to determine simpler expressions for the reflection coefficient. Gol'dshtein and Marchenko (1989), Tkacheva (2001, 2003) also employed WH technique to study various problems related to floating elastic plates.

A variety of different techniques can be found in the water wave literature for studying scattering problems involving two-dimensional models of elastic plates floating on surface of either finite or infinite depth water. Newman (1994) presented a methodology for treating the interaction of water waves with arbitrary deformable bodies. His idea was to represent the displacement of the bodies in terms of sets of appropriate model functions and orthogonal polynomials. This theory was generalized by Wu et al. (1995) to a single floating elastic plate. Sahoo et al. (2001) investigated the interaction of surface waves with a semi-infinite elastic plate floating on surface of finite depth water. They used the method of eigenfunction expansions in the mathematical analysis. A mode matching principle was used by Meylan and Squire (1993) to find the reflection and transmission coefficients of ocean waves by a semi-infinite ice-floe (thin elastic plate). In another paper Meylan and Squire (1994) considered a single ice-floe of finite width as well as a pair of ice-floes of the same width and the related problem was reduced to solving a Fredholm integral equation of second kind with logarithmic kernel by the application of Green's function technique. The problem of scattering of water waves by multiple floating plates of variable properties floating on water of uniform finite depth is considered by Kohout et al. (2007) using the principle of matching of eigenfunction expansions at the boundaries of the plates. Also they compared their solution with experimental results. Andrianov and Hermans (2003) and Hermans (2004) considered a single strip or multiple strips of floating elastic platforms employing integro-differential equations along the platforms preceded by application of Green's integral theorem. The diffraction of surface waves by a semi-infinite ice-sheet modelled as a thin elastic plate and by a gap of finite width between two semi-infinite elastic plates floating on water of finite depth were studied by Linton and Chung (2003) and Chung and Linton (2005) respectively by residue calculus technique.

Chakrabarti (2000a) solved explicitly the two-dimensional problem of water wave scattering by a semi-infinite ice-cover floating on the surface of deep water, by reducing the problem to solving a Carleman type singular integral equation. This technique was also employed by Chakrabarti (2000b) to study wave scattering by the discontinuity on the surface of water arising due to the presence of two types of semi-infinite inertial surfaces. A somewhat similar type of problem wherein a single semi-infinite inertial surface is present on the surface of deep water was earlier treated by Kanoria et al. (1999) by Wiener-Hopf technique. They also considered a finite strip of inertial surface floating on the surface of open water. Also see sections 5.1 and 5.2.

In solving problems of different branches of applied mathematics and engineering, Carleman singular integral equations are frequently used. However, as per the knowledge of the authors, no work prior to that of Chakrabarti (2000b) has been done in the field of linear water wave theory, using this method. He showed in his two successive papers (Chakrabarti (2000a, 2000b)) how to reduce two particular scattering problems to a single Carleman singular integral equation. His works are generalized by Gayen et al. (2005, 2006), for finite strip problems. Due to the asymptotic nature of the results, the scheme was not applicable for moderate values of the strip width. So a general method which would work for arbitrary strip widths is required. Gayen et al. (2007) applied it to a strip of an inertial surface floating between another inertial surface and is described in section 5.4. There the effect of a finite strip of inertial surface lying sandwiched between two other semi-infinite inertial surfaces on the wave propagation has been stuided.

Mathematical formulation

Cartesian coordinates are chosen in which the (x, y)-plane corresponds to the undisturbed upper surface and y-axis pointing vertically downwards. We consider the scattering of a normally incident surface wave train by an elastic plate which occupies the position $y=0, 0\leq x\leq l, -\infty<z<\infty$ in deep water. The plate is composed of an elastic material having Young's modulus E and Poisson's ratio ν and is of very small thickness h_0 so that the draft is negligible. Since the plate is infinitely long along z-direction, we can consider the problem to be two-dimensional in (x, y)-coordinates only. We assume that the motion in the fluid is irrotational, time-harmonic with time dependence $e^{-i\sigma t}$, σ being the angular frequency, and the fluid is inviscid and incompressible. Within the framework of linearized theory, the mathematical problem can be described by a velocity potential $\Phi(x, y; t) = Re\{\phi(x, y)e^{-i\sigma t}\}$ where $\phi(x, y)$ is a time-independent complex valued function. It satisfies the Laplace equation

$$\frac{\partial^2\phi}{\partial x^2}+\frac{\partial^2\phi}{\partial y^2}=0 \text{ in the fluid region}, \tag{5.1}$$

the free surface condition

$$K\phi+\phi_y=0 \text{ on } y=0, (-\infty<x<0)\bigcup(l<x<\infty), \tag{5.2}$$

the plate condition

$$D\frac{\partial^5\phi}{\partial^4 x\partial y}+K\phi+\phi_y=0 \text{ on } y=0, 0<x<l, \tag{5.3}$$

D being proportional to flexural rigidity of plate and is given by $D=\frac{Eh_0^3}{12(1-\nu^2)\rho g}$, g being the acceleration due to gravity and $K=\frac{\sigma^2}{g}$, and the bottom condition

$$\nabla\phi\to 0 \text{ as } y\to\infty. \tag{5.4}$$

The conditions of no bending moment and no shearing stress at the two ends of the plate are

$$\begin{aligned}&\phi_{xxy}\to 0 \text{ as } x\to 0^+, l^- \text{ on } y=0,\\ &\phi_{xxxy}\to 0 \text{ as } x\to 0^+, l^- \text{ on } y=0\end{aligned} \tag{5.5}$$

and the requirements at infinity are

$$\phi \to \begin{cases} e^{-Ky+iKx} + Re^{-Ky-iKx} & \text{as } x \to -\infty, \\ Te^{-Ky+iK(x-l)} & \text{as } x \to \infty, \end{cases} \tag{5.6}$$

where R and T are the unknown amplitudes (complex) of the reflected and transmitted waves. Determination of these two quantities is the principal concern here.

It may be noticed that the ice-cover condition given in (5.3) is derived under the assumption that the waves are long compared to the thickness of the ice, i.e., the inertia term is nearly equal to zero. However, if we don't make this assumption then the boundary condition would have been

$$D\frac{\partial^5 \phi}{\partial^4 x \partial y} + (1 - \varepsilon K)\phi + \phi_y = 0 \text{ on } y = 0, 0 < x < l.$$

Unless the frequency of the incident wave is very large, for most of the physical problems we can take $(1 - \varepsilon K) > 0$. Now if we divide the above equation by $(1 - \varepsilon K)$, then it reduces to

$$D'\frac{\partial^5 \phi}{\partial^4 x \partial y} + K'\phi + \phi_y = 0 \text{ on } y = 0, 0 < x < l.$$

where $(D', K') = \frac{(D, K)}{1 - \varepsilon K}$ which has the form similar to equation (5.3).

The above boundary value problem is now reduced to two Carleman type singular integral equations.

Reduction to singular integral equations

We first observe that if we solve the Laplace equation (5.1) subject to the plate condition (5.3) by the method of separation of variable, we obtain the following solutions Chakrabarti et al. (2003):

$$e^{-\lambda Ky \pm i\lambda Kx}, e^{-\lambda_1 Ky \pm i\lambda_1 Kx}, e^{-\bar{\lambda}_1 Ky \pm i\bar{\lambda}_1 Kx}, e^{-\lambda_2 Ky \pm i\lambda_2 Kx}, e^{-\bar{\lambda}_2 Ky \pm i\bar{\lambda}_2 Kx}$$

and $\{\xi(D\xi^4 + 1)\cos \xi y - K \sin \xi y\}e^{\pm \xi x}$, $\xi \in (0, \infty)$, where λK is the real positive root of the equation

$$Dk^5 + k - K = 0 \tag{5.7}$$

whose other roots are $(\lambda_1 K, \bar{\lambda}_1 K)$, $(\lambda_2 K, \bar{\lambda}_2 K)$ with $\text{Re}(\lambda_1) > 0$, $\text{Re}(\lambda_2) < 0$, $\text{Im}(\lambda_1, \lambda_2) > 0$.

Thus in the region below the strip, $\phi(x, y)$ has the form

$$\phi(x, y) = \alpha e^{-\lambda Ky + i\lambda Kx} + \beta e^{-\lambda Ky - i\lambda K(x-l)} + \chi(x, y), 0 < x < l. \tag{5.8}$$

Here it may be observed that we have not taken the solutions involving λ_2 and $\bar{\lambda}_2$ for expressing $\phi(x, y)$ in $0 < x < l$ as these do not satisfy the infinite bottom condition (5.4). The first two terms in equation (5.8) represent the propagating waves with α and β being unknown constants which can be identified with the reflection and transmission coefficients, respectively, through the points $(l,0)$ and $(0,0)$. The function $\chi(x, y)$ is a combination of the solutions $e^{-\lambda_1 Ky \pm i\lambda_1 Kx}$, $e^{-\bar{\lambda}_1 Ky \pm i\bar{\lambda}_1 Kx}$ and $\{\xi(D\xi^4 + 1)\cos \xi y - K \sin \xi y\} e^{\pm \xi x}$, $\xi \in (0, \infty)$. We introduce a reduced potential $\psi(x, y)$ defined by $\phi = \frac{\partial^2 \psi}{\partial x^2}$ and express this function in the three regions $x < 0$, $0 < x < l$, $x > l (y > 0)$. The basic reason to set $\phi = \frac{\partial^2 \psi}{\partial x^2}$ is to ensure the convergence of the various integrals appearing in the mathematical analysis. Working with ψ ensures avoiding divergent integrals altogether, which is not possible if we work with ϕ.

We now employ Havelock's expansion of water wave potential Havelock (1929) to represent $\psi(x, y)$ as

$$\psi(x,y) = -\frac{1}{K^2} e^{-Ky+iKx} - \frac{R}{K^2} e^{-Ky-iKx} + \frac{2}{\pi}\int_0^\infty \frac{A(\xi)}{\xi^2+K^2} L(\xi,y) e^{\xi x} d\xi, \quad x<0,\ y>0, \tag{5.9}$$

$$\psi(x,y) = -\frac{1}{\lambda^2 K^2}\left\{\alpha e^{-\lambda Ky + i\lambda Kx} + \beta e^{-\lambda Ky - i\lambda K(x-l)}\right\} - \frac{1}{\lambda_1^2 K^2}\left\{A_1 e^{-\lambda_1 Ky + i\lambda_1 Kx} + A_2 e^{-\lambda_1 Ky - i\lambda_1 K(x-l)}\right\}$$

$$-\frac{1}{\bar{\lambda}_1^2 K^2}\left\{A_3 e^{-\bar{\lambda}_1 Ky + i\bar{\lambda}_1 K(x-l)} + A_4 e^{-\bar{\lambda}_1 Ky - i\bar{\lambda}_1 Kx}\right\} + \frac{2}{\pi}\int_0^\infty \frac{B(\xi)e^{\xi(x-l)} + C(\xi)e^{-\xi x}}{P(\xi)} \tag{5.10}$$

$$M(\xi,y)d\xi, \ 0<x<l, y>0,$$

$$\psi(x,y) = -\frac{T}{K^2} e^{-Ky+iK(x-l)} + \frac{2}{\pi}\int_0^\infty \frac{G(\xi)}{\xi^2+K^2} L(\xi,y) e^{-\xi(x-l)} d\xi, \ x>l,\ y>0 \tag{5.11}$$

where A_1, A_2, A_3, A_4 are unknown constants and $A(\xi)$, $B(\xi)$, $C(\xi)$ and $G(\xi)$ are unknown functions of ξ and are such that the integrals in (5.9) to (5.11) are convergent, and

$$L(\xi, y) = \xi \cos \xi y - K \sin \xi y,$$

$$M(\xi, y) = D\xi^5 \cos \xi y + L(\xi, y)$$

and

$$P(\xi) = \xi^2 (D\xi^4 + 1)^2 + K^2.$$

From the representation of the function $\psi(x, y)$ it may be noted that the domain of the variable ξ is $(0, \infty)$ and so whenever ξ appears in the rest of the paper it will be assumed that $\xi > 0$.

After obtaining the expansions of $\psi(x, y)$ given in equations (5.9)–(5.11) in the open water region and in the plate covered region, we employ the continuity of ψ and $\frac{\partial \psi}{\partial x}$ across the lines $x = 0$ and $x = l, (y > 0)$. This gives rise to four relations involving the eight unknown constants and the four unknown functions. On these relations we apply the above theorem and we obtain two pairs of representations for the functions $A(\xi)$ and $G(\xi)$ (for details see Gayen et al. (2005)). Elimination of $A(\xi)$ and $G(\xi)$ from their dual representations produces the following Carleman-type singular integral equations for solving the unknown functions $B(\xi)$ and $C(\xi)$:

$$\mu(\xi) B_1(\xi) + \frac{1}{\pi}\int_0^\infty \frac{B_1(u)}{u-\xi} du - \frac{1}{\pi}\int_0^\infty \frac{C_1(u)}{u+\xi} e^{-ul} du = F_B(\xi), \xi > 0, \tag{5.12}$$

and

$$\mu(\xi)C_1(\xi)+\frac{1}{\pi}\int_0^\infty \frac{C_1(u)}{u-\xi}\mathrm{d}u-\frac{1}{\pi}\int_0^\infty \frac{B_1(u)}{u+\xi}\mathrm{e}^{-ul}du=F_C(\xi),\ \xi>0 \tag{5.13}$$

where

$$\left(B_1(\xi),C_1(\xi)\right)=\frac{DK\xi^5}{P(\xi)}\left(B(\xi),C(\xi)\right), \tag{5.14}$$

$$\mu(\xi)=\frac{\xi^2(D\xi^4+1)+K^2}{DK\xi^5} \tag{5.15}$$

and

$$F_B(\xi)=\frac{\lambda-1}{2\lambda^2K}\left\{\frac{\alpha\mathrm{e}^{\mathrm{i}\lambda Kl}}{\xi-\mathrm{i}\lambda K}+\frac{\beta}{\xi+\mathrm{i}\lambda K}\right\}+\frac{\lambda_1-1}{2\lambda_1^2K}\left\{\frac{A_1\mathrm{e}^{\mathrm{i}\lambda_1Kl}}{\xi-\mathrm{i}\lambda_1K}+\frac{A_2}{\xi+\mathrm{i}\lambda_1K}\right\}$$
$$+\frac{\overline{\lambda}_1-1}{2\overline{\lambda}_1^2K}\left\{\frac{A_3}{\xi-\mathrm{i}\overline{\lambda}_1K}+\frac{A_4\mathrm{e}^{-\mathrm{i}\overline{\lambda}_1Kl}}{\xi+\mathrm{i}\overline{\lambda}_1K}\right\}, \tag{5.16}$$

$$F_C(\xi)=\frac{\lambda-1}{2\lambda^2K}\left\{\frac{\alpha}{\xi+\mathrm{i}\lambda K}+\frac{\beta\mathrm{e}^{\mathrm{i}\lambda Kl}}{\xi-\mathrm{i}\lambda K}\right\}+\frac{\lambda_1-1}{2\lambda_1^2K}\left\{\frac{A_1}{\xi+\mathrm{i}\lambda_1K}+\frac{A_2\mathrm{e}^{\mathrm{i}\lambda_1Kl}}{\xi-\mathrm{i}\lambda_1K}\right\}$$
$$+\frac{\overline{\lambda}_1-1}{2\overline{\lambda}_1^2K}\left\{\frac{A_3\mathrm{e}^{-\mathrm{i}\overline{\lambda}_1Kl}}{\xi+\mathrm{i}\overline{\lambda}_1K}+\frac{A_4}{\xi-\mathrm{i}\overline{\lambda}_1K}\right\}. \tag{5.17}$$

Here it may be mentioned that due to the presence of the second integrals involving exp($-ul$) in the equations (5.12) and (5.13), it is not possible to solve them in a straightforward manner. To get rid of these two terms, Gayen et al. (2005) eliminated them by assuming the strip width to be sufficiently large and solved the two equations by a sort of iteration process, after employing the technique of Riemann-Hilbert problem. Here the assumption of largeness of the strip width is not made, and a new method is introduced which is somewhat similar to that in section 5.4 (also see Gayen et al. (2007)). This is explained later.

It will be found in the sequel that the solutions of the integral equations (5.12) and (5.13) can be expressed as linear combinations of some known functions multiplied by six unknown constants $\alpha, \beta, A_1, A_2, A_3, A_4$. Once these expressions are obtained these are substituted into the following eight equations for determining the eight unknown constants $R, T, \alpha, \beta, A_1, A_2, A_3, A_4$:

$$\frac{1+R}{2}=\frac{\alpha+\beta\mathrm{e}^{\mathrm{i}\lambda Kl}}{\lambda^2(\lambda+1)}+\frac{A_1+A_2\mathrm{e}^{\mathrm{i}\lambda_1Kl}}{\lambda_1^2(\lambda_1+1)}+\frac{A_3\mathrm{e}^{-\mathrm{i}\overline{\lambda}_1Kl}+A_4}{\overline{\lambda}_1^2(\overline{\lambda}_1+1)}-\frac{2K^3}{\pi}\int_0^\infty\frac{B_1(\xi)\mathrm{e}^{-\xi l}+C_1(\xi)}{\xi^2+K^2}d\xi, \tag{5.18}$$

$$\frac{1-R}{2}=\frac{\alpha-\beta\mathrm{e}^{\mathrm{i}\lambda Kl}}{\lambda(\lambda+1)}+\frac{A_1-A_2\mathrm{e}^{\mathrm{i}\lambda_1Kl}}{\lambda_1(\lambda_1+1)}+\frac{A_3\mathrm{e}^{-\mathrm{i}\overline{\lambda}_1Kl}-A_4}{\overline{\lambda}_1(\overline{\lambda}_1+1)}+\frac{2K^2\mathrm{i}}{\pi}\int_0^\infty\frac{B_1(\xi)\mathrm{e}^{-\xi l}-C_1(\xi)}{\xi^2+K^2}\xi\, d\xi, \tag{5.19}$$

$$\frac{T}{2}=\frac{\alpha e^{i\lambda Kl}+\beta}{\lambda^2(\lambda+1)}+\frac{A_1e^{i\lambda_1 Kl}+A_2}{\lambda_1^2(\lambda_1+1)}+\frac{A_3+A_4e^{-i\bar{\lambda}_1 Kl}}{\bar{\lambda}_1^2(\bar{\lambda}_1+1)}-\frac{2K^3}{\pi}\int_0^\infty\frac{B_1(\xi)+C_1(\xi)e^{-\xi l}}{\xi^2+K^2}d\xi, \quad (5.20)$$

$$\frac{T}{2}=\frac{\alpha e^{i\lambda Kl}-\beta}{\lambda(\lambda+1)}+\frac{A_1e^{i\lambda_1 Kl}-A_2}{\lambda_1(\lambda_1+1)}+\frac{A_3-A_4e^{-i\bar{\lambda}_1 Kl}}{\bar{\lambda}_1(\bar{\lambda}_1+1)}+\frac{2K^2 i}{\pi}\int_0^\infty\frac{B_1(\xi)-C_1(\xi)e^{-\xi l}}{\xi^2+K^2}\xi d\xi, \quad (5.21)$$

$$\begin{aligned}&(\lambda K)^3(\alpha+\beta e^{i\lambda Kl})+(\lambda_1 K)^3(A_1+A_2e^{i\lambda_1 Kl})+(\bar{\lambda}_1 K)^3(A_3e^{-i\bar{\lambda}_1 Kl}+A_4)\\&\qquad-\frac{2}{D\pi}\int_0^\infty\{B_1(\xi)e^{-\xi l}+C_1(\xi)\}d\xi=0,\end{aligned} \quad (5.22)$$

$$\begin{aligned}&(\lambda K)^4(\alpha-\beta e^{i\lambda Kl})+(\lambda_1 K)^4(A_1-A_2e^{i\lambda_1 Kl})+(\bar{\lambda}_1 K)^4(A_3e^{-i\bar{\lambda}_1 Kl}-A_4)\\&\qquad+\frac{2i}{D\pi}\int_0^\infty\{B_1(\xi)e^{-\xi l}-C_1(\xi)\}\xi d\xi=0,\end{aligned} \quad (5.23)$$

$$\begin{aligned}&(\lambda K)^3(\alpha e^{i\lambda Kl}+\beta)+(\lambda_1 K)^3(A_1e^{i\lambda_1 Kl}+A_2)+(\bar{\lambda}_1 K)^3(A_3+A_4e^{-i\bar{\lambda}_1 Kl})\\&\qquad-\frac{2}{D\pi}\int_0^\infty\{B_1(\xi)+C_1(\xi)e^{-\xi l}\}d\xi=0,\end{aligned} \quad (5.24)$$

$$\begin{aligned}&(\lambda K)^4(\alpha e^{i\lambda Kl}-\beta)+(\lambda_1 K)^4(A_1e^{i\lambda_1 Kl}-A_2)+(\bar{\lambda}_1 K)^4(A_3-A_4e^{-i\bar{\lambda}_1 Kl})\\&\qquad+\frac{2i}{D\pi}\int_0^\infty\{B_1(\xi)-C_1(\xi)e^{-\xi l}\}\xi d\xi=0.\end{aligned} \quad (5.25)$$

The first four equations (5.18) to (5.21) are obtained by application of Havelock's inversion theorem on the relations derived from the continuity of the functions ψ and $\frac{\partial\psi}{\partial x}$ across the lines $x = 0$ and $x = l(y > 0)$, whereas the equations (5.22) to (5.25) are the consequence of the condition (5.4) at the end points of the plate.

Solution for arbitrary width of the plate

Now we solve the two singular integral equations (5.12) and (5.13) for any width of the plate. For this we first introduce a singular integral operator

$$\mathbf{S}: L^2(0,\infty)\to L^2(0,\infty)$$

and a non-singular integral operator

$$\mathbf{S}': C^\infty(0,\infty)\to C^\infty(0,\infty)$$

defined by

$$\mathbf{S}f(\xi)=\mu(\xi)f(\xi)+\frac{1}{\pi}\int_0^\infty\frac{f(u)}{u-\xi}du \quad (5.26)$$

and

$$\boldsymbol{S}'f(\xi) = -\frac{1}{\pi}\int_0^\infty \frac{f(u)e^{-ul}}{u+\xi}du. \tag{5.27}$$

Then the equations (5.12) and (5.13) reduce to the following forms:

$$\boldsymbol{S}B_1(\xi) + \boldsymbol{S}'C_1(\xi) = F_B(\xi) \tag{5.28}$$

and

$$\boldsymbol{S}C_1(\xi) + \boldsymbol{S}'B_1(\xi) = F_C(\xi). \tag{5.29}$$

It may be observed that the analytical form of the inverse operator $\boldsymbol{S}^{-1}$ of $\boldsymbol{S}$ can be determined as follows:

Consider the singular integral equation

$$\boldsymbol{S}f(\xi) = h(\xi). \tag{5.30}$$

Assuming that the right hand side is known, equation (5.30) can be reduced to a Riemann-Hilbert problem (see p.123, Muskhelishvili (1953); p.148, Gakhov (1966))

$$\left(\mu(\xi)+\mathrm{i}\right)\Lambda^+(\xi) - \left(\mu(\xi)-\mathrm{i}\right)\Lambda^-(\xi) = h(\xi) \tag{5.31}$$

after introducing a sectionally analytic function associated with the unknown function $f(\xi)$ satisfying equation (5.30) as

$$\Lambda(\zeta) = \frac{1}{2\pi\mathrm{i}}\int_0^\infty \frac{f(u)}{u-\zeta}du, \quad \zeta = \xi + \mathrm{i}\eta. \tag{5.32}$$

$\Lambda(\zeta)$ is defined in the entire complex ζ-plane cut along the real axis from 0 to ∞. One can solve the Riemann-Hilbert problem given in (5.31) by methods of complex variable theory England (1971), Estrada and Kanwal (2000). The Plemelj's formulae corresponding to equation (5.32) are

$$\Lambda^\pm(\xi) = \pm\frac{1}{2}f(\xi) + \frac{1}{2\pi\mathrm{i}}\int_0^\infty \frac{f(u)}{u-\xi}du$$

so that

$$\Lambda^+(\xi) - \Lambda^-(\xi) = f(\xi) \text{ and } \Lambda^+(\xi) + \Lambda^-(\xi) = \frac{1}{\pi\mathrm{i}}\int_0^\infty \frac{f(u)}{u-\xi}du = 2\Lambda(\xi).$$

Thus the function $f(\xi)$ is found as

$$\begin{aligned} f(\xi) = \boldsymbol{S}^{-1}h(\xi) &= \Lambda^+(\xi) - \Lambda^-(\xi) \\ &= \frac{\Lambda_0^+(\xi)}{\mu(\xi)-\mathrm{i}}\hat{\boldsymbol{S}}\left[\frac{h(\xi)}{\Lambda_0^+(\xi)(\mu(\xi)+\mathrm{i})}\right] \end{aligned} \tag{5.33}$$

where the operator $\hat{S}$ is defined by

$$\hat{\boldsymbol{S}}g(\xi) = \mu(\xi)g(\xi) - \frac{1}{\pi}\int_0^\infty \frac{g(u)}{u-\xi}du \tag{5.34}$$

and

$$\Lambda_0^+(\xi) = \lim_{\eta\to 0^+} \Lambda_0(\zeta); \tag{5.35}$$

$\Lambda_0(\zeta)$ being a solution of the homogeneous problem corresponding to the RHP (5.31) and its explicit form is found to be

$$\Lambda_0(\zeta) = \exp\left[\frac{1}{2\pi\mathrm{i}}\int_0^\infty \frac{\log\left(\dfrac{\mu(t)-\mathrm{i}}{\mu(t)+\mathrm{i}}\right) - \lim\limits_{t\to\infty}\log\left(\dfrac{\mu(t)-\mathrm{i}}{\mu(t)+\mathrm{i}}\right)}{t-\zeta}dt\right], \ (\zeta \notin (0,\infty)). \tag{5.36}$$

It may be noted that the limiting term inside the integral is zero. We now apply the operator $\boldsymbol{S}^{-1}$ to (5.28) to obtain $B_1(\xi)$ in terms of $C_1(\xi)$ as

$$B_1(\xi) = \boldsymbol{S}^{-1}\left[F_B(\xi) - \boldsymbol{S}'C_1(\xi)\right] \tag{5.37}$$

and then substitute $B_1(\xi)$ into (5.29). This yields

$$\boldsymbol{S}C_1(\xi) + \boldsymbol{S}'\left[\boldsymbol{S}^{-1}(F_B - \boldsymbol{S}'C_1)\right](\xi) = F_C(\xi). \tag{5.38}$$

Applying the operator $\boldsymbol{S}^{-1}$ to the above equation we find

$$\left[I - \mathcal{L}^2\right]C_1(\xi) = r(\xi) \tag{5.39}$$

where the operator $\mathcal{L} = \boldsymbol{S}^{-1}\boldsymbol{S}'$ is non-commutative and its analytical form is determined as

$$\mathcal{L}m(\xi) = -\frac{1}{\pi}\frac{\Lambda_0^+(\xi)}{\mu(\xi)-\mathrm{i}}\int_0^\infty \frac{m(u)\mathrm{e}^{-ul}}{(u+\xi)\Lambda_0(-u)}du. \tag{5.40}$$

The explicit derivation of the above expression is outlined in the section 5.4. The right hand side of (5.39) is

$$r(\xi) = \boldsymbol{S}^{-1}\left[F_C - S'S^{-1}F_B\right](\xi) \tag{5.41}$$

and it can be simplified as

$$r(\xi) = \alpha r_\alpha(\xi) + \beta r_\beta(\xi) + A_1 r_{A_1}(\xi) + A_2 r_{A_2}(\xi) + A_3 r_{A_3}(\xi) + A_4 r_{A_4}(\xi) \tag{5.42}$$

where the functions $r_\alpha(\xi)$, $r_\beta(\xi)$, $r_{A_1}(\xi)$, $r_{A_2}(\xi)$, $r_{A_3}(\xi)$, $r_{A_4}(\xi)$ are given by

$$r_\alpha(\xi) = C_\alpha M(\xi)\left[\frac{1}{\Lambda_0(-\mathrm{i}K\lambda)(\xi+\mathrm{i}K\lambda)} + \frac{\mathrm{e}^{\mathrm{i}\lambda Kl}}{\Lambda_0(\mathrm{i}K\lambda)}\int_0^\infty \frac{M_1(\xi,u)}{u-\mathrm{i}K\lambda}du\right], \tag{5.43}$$

$$r_\beta(\xi) = C_\beta M(\xi)\left[\frac{\mathrm{e}^{\mathrm{i}\lambda Kl}}{\Lambda_0(\mathrm{i}K\lambda)(\xi - \mathrm{i}K\lambda)} + \frac{1}{\Lambda_0(-\mathrm{i}K\lambda)}\int_0^\infty \frac{M_1(\xi,u)}{u+\mathrm{i}K\lambda}du\right], \tag{5.44}$$

$$r_{A_1}(\xi) = C_{A_1} M(\xi)\left[\frac{1}{\Lambda_0(-\mathrm{i}K\lambda_1)(\xi + \mathrm{i}K\lambda_1)} + \frac{\mathrm{e}^{\mathrm{i}\lambda_1 Kl}}{\Lambda_0(\mathrm{i}K\lambda_1)}\int_0^\infty \frac{M_1(\xi,u)}{u-\mathrm{i}K\lambda_1}du\right], \tag{5.45}$$

$$r_{A_2}(\xi) = C_{A_2} M(\xi)\left[\frac{\mathrm{e}^{\mathrm{i}\lambda_1 Kl}}{\Lambda_0(\mathrm{i}K\lambda_1)(\xi - \mathrm{i}K\lambda_1)} + \frac{1}{\Lambda_0(-\mathrm{i}K\lambda_1)}\int_0^\infty \frac{M_1(\xi,u)}{u+\mathrm{i}K\lambda_1}du\right], \tag{5.46}$$

$$r_{A_3}(\xi) = C_{A_3} M(\xi)\left[\frac{\mathrm{e}^{-\mathrm{i}\bar{\lambda}_1 Kl}}{\Lambda_0(-\mathrm{i}K\bar{\lambda}_1)(\xi + \mathrm{i}K\bar{\lambda}_1)} + \frac{1}{\Lambda_0(\mathrm{i}K\bar{\lambda}_1)}\int_0^\infty \frac{M_1(\xi,u)}{u-\mathrm{i}K\bar{\lambda}_1}du\right], \tag{5.47}$$

$$r_{A_4}(\xi) = C_{A_4} M(\xi)\left[\frac{1}{\Lambda_0(\mathrm{i}K\bar{\lambda}_1)(\xi - \mathrm{i}K\bar{\lambda}_1)} + \frac{\mathrm{e}^{-\mathrm{i}\bar{\lambda}_1 Kl}}{\Lambda_0(-\mathrm{i}K\bar{\lambda}_1)}\int_0^\infty \frac{M_1(\xi,u)}{u+\mathrm{i}K\bar{\lambda}_1}du\right], \tag{5.48}$$

with

$$M(\xi) = \frac{\Lambda_0^+(\xi)}{\mu(\xi)-\mathrm{i}},\ M_1(\xi,u) = \frac{M(u)\mathrm{e}^{-ul}}{\pi(u+\xi)\Lambda_0(-u)},\ C_\alpha = C_\beta = \frac{\lambda-1}{2\lambda^2 K},$$
$$C_{A_1} = C_{A_2} = \frac{\lambda_1 - 1}{2\lambda_1^2 K},\ C_{A_3} = C_{A_4} = \frac{\bar{\lambda}_1 - 1}{2\bar{\lambda}_1^2 K}. \tag{5.49}$$

In order to determine the functions $r_j(\xi)$ given in (5.43)–(5.48) we have utilized the definition of $\boldsymbol{\mathcal{S}}^{-1}$ given in (5.33) together with (5.34). It may be observed that

$$\boldsymbol{\mathcal{S}}^{-1}\left(\frac{1}{\xi+\xi_0}\right) = \frac{M(\xi)}{(\xi+\xi_0)\Lambda_0(-\xi_0)}$$

where ξ_0 is a positive constant.

Equation (5.34) can be regarded as an ordinary integral equation (involving no singular kernel) for solving $C_1(\xi)$. However, since the forcing function $r(\xi)$ is unknown in the sense that it contains the unknown constants α, β, etc., (5.39) cannot be solved directly. In order to overcome this difficulty we introduce two new functions $U(\xi)$ and $V(\xi)$ in terms of the function $C_1(\xi)$ as

$$[I+\mathcal{L}]C_1(\xi) = U(\xi),\ [I-\mathcal{L}]C_1(\xi) = V(\xi) \tag{5.50}$$

so that

$$C_1(\xi) = \frac{1}{2}[U(\xi)+V(\xi)] \text{ and } \mathcal{L}C_1(\xi) = \frac{1}{2}[U(\xi)-V(\xi)]. \tag{5.51}$$

Then (5.39) can be written in terms of either $U(\xi)$ or $V(\xi)$ as

$$[I-\mathcal{L}]U(\xi)=r(\xi) \tag{5.52}$$

or

$$[I+\mathcal{L}]V(\xi)=r(\xi) \tag{5.53}$$

Because of (5.42) we may express $U(\xi)$ and $V(\xi)$ as

$$U(\xi)=[I-\mathcal{L}]^{-1}r(\xi)=\alpha u_\alpha(\xi)+\beta u_\beta(\xi)+A_1u_{A_1}(\xi)+A_2u_{A_2}(\xi)+A_3u_{A_3}(\xi)+A_4u_{A_4}(\xi) \tag{5.54}$$

and

$$V(\xi)=[I+\mathcal{L}]^{-1}r(\xi)=\alpha v_\alpha(\xi)+\beta v_\beta(\xi)+A_1v_{A_1}(\xi)+A_2v_{A_2}(\xi)+A_3v_{A_3}(\xi)+A_4v_{A_4}(\xi) \tag{5.55}$$

where $u_j(\xi)$, $v_j(\xi)$ (with subscript j denoting $\alpha, \beta, A_1, A_2, A_3, A_4$) are unknown functions. These are determined by incorporating the fact that the integral equation (5.52) along with the relation (5.54) and the integral equation (5.53) along with the relation (5.55) are satisfied simultaneously if $u_j(\xi)$, $v_j(\xi)$ satisfy the following Fredholm integral equations of the second kind:

$$[I-\mathcal{L}]u_j(\xi)=r_j(\xi) \tag{5.56}$$

and

$$[I+\mathcal{L}]v_j(\xi)=r_j(\xi) \tag{5.57}$$

where the subscript j stands for $\alpha, \beta, A_1, A_2, A_3, A_4$. The kernels of the equations and the right hand sides can be computed from the equation (5.40) and the set of relations (5.43) to (5.48) respectively. We solve the integral equations (5.56) and (5.57) by Nystrom's method to derive the functions $u_j(\xi)$, $v_j(\xi)$ numerically.

In order to solve the equations (5.56) and (5.57), we first need to simplify the forms of the operator $\mathcal{L}$ and the functions $\Lambda_0^+(\xi)$, $\Lambda_0(-\xi)$ and $M(\xi)$. The detailed procedure is given below.

Evaluation of the functions $\Lambda_0^+(\xi)$, $\Lambda_0(-\xi)$ and $M(\xi)$

We have

$$\Lambda_0(\zeta)=\exp\left[\frac{1}{2\pi \mathrm{i}}\int_0^\infty \frac{\log\left[\dfrac{\mu(t)-\mathrm{i}}{\mu(t)+\mathrm{i}}\right]}{t-\zeta}dt\right],\ \zeta\notin(0,\infty) \tag{5.58}$$

where

$$\mu(\xi)\mp \mathrm{i}=\frac{1}{K\xi^5}(\xi\mp iK)(\xi\pm \mathrm{i}K\lambda)(\xi\pm \mathrm{i}K\lambda_1)(\xi\pm \mathrm{i}K\bar{\lambda}_1)(\xi\pm \mathrm{i}K\lambda_2)(\xi\pm \mathrm{i}K\bar{\lambda}_2). \tag{5.59}$$

If we define

$$\Gamma_0(\zeta) = \log \Lambda_0(\zeta).$$

then

$$\Gamma_0(\zeta) = \frac{1}{2\pi i}\int_0^\infty \frac{\log\left[\frac{\mu(t)-i}{\mu(t)+i}\right]}{t-\zeta}dt \tag{5.60}$$

and

$$\Gamma_0^+(\xi) = \frac{1}{2}\log\left[\frac{\mu(\xi)-i}{\mu(\xi)+i}\right] + \frac{1}{2\pi i}\int_0^\infty \frac{\log\left[\frac{\mu(t)-i}{\mu(t)+i}\right]}{t-\xi}dt \tag{5.61}$$

so that

$$\Lambda_0^+(\xi) = \left[\frac{\mu(\xi)-i}{\mu(\xi)+i}\right]^{\frac{1}{2}} \exp[Y(\xi)] \tag{5.62}$$

and

$$M(\xi) = \frac{\Lambda_0^+(\xi)}{\mu(\xi)-i} = \frac{1}{\left[1+\mu^2(\xi)\right]^{\frac{1}{2}}}\exp[Y(\xi)] \tag{5.63}$$

with

$$Y(\xi) = \frac{1}{2\pi i}\int_0^\infty \frac{\log\left[\frac{\mu(t)-i}{\mu(t)+i}\right]}{t-\xi}dt. \tag{5.64}$$

Also it can be shown that

$$\Lambda_0(-u) = \exp[Y(-u)]$$

so that

$$[\Lambda_0(-u)]^{-1} = \exp[-Y(-u)]. \tag{5.65}$$

In the following we proceed to simplify the term exp $[Y(\xi)]$ only.

Let

$$Y_1(\xi) = \frac{1}{2\pi i}\int_0^\infty \frac{\log\frac{t-iK}{t+iK}}{t-\xi}dt,\ Y_2(\xi) = \frac{1}{2\pi i}\int_0^\infty \frac{\log\frac{t-iK\lambda}{t+iK\lambda}}{t-\xi}dt,$$

$$Y_3(\xi) = \frac{1}{2\pi i}\int_0^\infty \frac{\log\frac{t-iK\overline{\lambda_1}}{t+iK\lambda_1}}{t-\xi}dt,\ Y_4(\xi) = \frac{1}{2\pi i}\int_0^\infty \frac{\log\frac{t-iK\lambda_1}{t+iK\overline{\lambda_1}}}{t-\xi}dt, \tag{5.66}$$

$$Y_5(\xi) = \frac{1}{2\pi i}\int_0^\infty \frac{\log\frac{t+iK\lambda_2}{t-iK\overline{\lambda_2}}}{t-\xi}dt,\ Y_6(\xi) = \frac{1}{2\pi i}\int_0^\infty \frac{\log\frac{t+iK\overline{\lambda_2}}{t-iK\lambda_2}}{t-\xi}dt.$$

Hence

$$Y(\xi) = Y_1(\xi) - Y_2(\xi) - Y_3(\xi) - Y_4(\xi) + Y_5(\xi) + Y_6(\xi) \tag{5.67}$$

In order to simplify the integrals $Y_j(\xi)\,(j=1,2,...,6)$, we apply the following result of Varley and Walker (1989):

$$V(\xi) = \frac{1}{2\pi \mathrm{i}}\int_0^\infty \frac{\log\dfrac{t-\lambda}{t+\lambda}}{t-\xi}dt$$

$$= -\frac{\sin\theta}{\pi}\int_0^{\frac{\xi}{|\lambda|}} \frac{\ln t}{t^2 - 2t\cos\theta + 1}\mathrm{d}t + \left(1 - \frac{\theta}{2\pi}\right)\log\frac{\xi}{\xi - \bar{\lambda}} - \frac{\theta}{2\pi}\log\frac{\xi}{\xi-\lambda} \tag{5.68}$$

where $\lambda = |\lambda|\,\mathrm{e}^{\mathrm{i}\theta}$.

By virtue of the above result, $Y_j(\xi)\,(j=1,2,...,6)$'s are determined as

$$Y_1(\xi) = -\frac{1}{\pi}\int_0^{\frac{K}{\xi}} \frac{\ln t}{1+t^2}dt - \mathrm{i}\theta_1 + \frac{1}{4}\log\frac{\xi^2}{\xi^2+K^2}, \tag{5.69}$$

$$Y_2(\xi) = -\frac{1}{\pi}\int_0^{\frac{K\lambda}{\xi}} \frac{\ln t}{1+t^2}dt - \mathrm{i}\theta_2 + \frac{1}{4}\log\frac{\xi^2}{\xi^2+\lambda^2K^2}, \tag{5.70}$$

$$Y_3(\xi) = -\frac{\sin\hat{\theta}_3}{\pi}\int_0^{\frac{|\lambda_1|}{\xi}} \frac{\ln t}{t^2 - 2t\cos\hat{\theta}_3 + 1}\mathrm{d}t - \frac{\hat{\theta}_3}{2\pi}\log\frac{\xi}{\xi - K(\omega + \mathrm{i}\nu)} + \left(1 - \frac{\hat{\theta}_3}{2\pi}\right)\log\frac{\xi}{\xi - K(\omega - \mathrm{i}\nu)}, \tag{5.71}$$

$$Y_4(\xi) = -\frac{\sin\hat{\theta}_4}{\pi}\int_0^{\frac{|\lambda_1|}{\xi}} \frac{\ln t}{t^2 - 2t\cos\hat{\theta}_4 + 1}\mathrm{d}t - \frac{\hat{\theta}_4}{2\pi}\log\frac{\xi}{\xi + K(\omega - \mathrm{i}\nu)} + \left(1 - \frac{\hat{\theta}_4}{2\pi}\right)\log\frac{\xi}{\xi + K(\omega + \mathrm{i}\nu)}, \tag{5.72}$$

$$Y_5(\xi) = -\frac{\sin\hat{\theta}_5}{\pi}\int_0^{\frac{|\lambda_2|}{\xi}} \frac{\ln t}{t^2 - 2t\cos\hat{\theta}_5 + 1}\mathrm{d}t - \frac{\hat{\theta}_5}{2\pi}\log\frac{\xi}{\xi - K(\delta + \mathrm{i}\gamma)} + \left(1 - \frac{\hat{\theta}_5}{2\pi}\right)\log\frac{\xi}{\xi - K(\delta - \mathrm{i}\gamma)}, \tag{5.73}$$

$$Y_6(\xi) = -\frac{\sin\hat{\theta}_6}{\pi}\int_0^{\frac{|\lambda_2|}{\xi}} \frac{\ln t}{t^2 - 2t\cos\hat{\theta}_6 + 1}\mathrm{d}t - \frac{\hat{\theta}_6}{2\pi}\log\frac{\xi}{\xi + K(\delta - \mathrm{i}\gamma)} + \left(1 - \frac{\hat{\theta}_6}{2\pi}\right)\log\frac{\xi}{\xi + K(\delta + \mathrm{i}\gamma)}, \tag{5.74}$$

where

$$\theta_1 = \tan^{-1}\frac{K}{\xi}, \theta_2 = \tan^{-1}\frac{K\lambda}{\xi}, \hat{\theta}_3 = \tan^{-1}\frac{\nu}{\omega}, \hat{\theta}_4 = \pi - \hat{\theta}_3, \hat{\theta}_5 = \tan^{-1}\frac{\gamma}{\delta}, \hat{\theta}_6 = \pi - \hat{\theta}_5,$$

$\lambda_1 = \nu + \mathrm{i}\omega$ and $\lambda_2 = -\gamma + \mathrm{i}\delta; \nu, \omega, \gamma, \delta > 0.$

Substituting the explicit forms of $Y_j(\xi)\ (j=1,2,...,6)$ into (5.67) we ultimately find

$$\exp[Y(\xi)] = \exp[V_{12}(\xi) - V_{34}(\xi) + V_{56}(\xi)]e^{-i(\theta_1-\theta_2-\theta_3-\theta_4+\theta_5+\theta_6)}\left|\frac{\xi^2+K^2\lambda^2}{\xi^2+K^2}\right|^{1/4}$$

$$\left|(\xi-K\omega)^2+K^2\lambda^2\right|^{\frac{1}{2}-\frac{\hat{\theta}_3}{2\pi}}$$

$$\left|(\xi+K\omega)^2+K^2\lambda^2\right|^{\frac{\hat{\theta}_3}{2\pi}}\left|(\xi-K\delta)^2+K^2\gamma^2\right|^{-\frac{1}{2}+\frac{\hat{\theta}_5}{2\pi}}\left|(\xi+K\delta)^2+K^2\gamma^2\right|^{-\frac{\hat{\theta}_5}{2\pi}} \quad (5.75)$$

where

$$V_{12} = \frac{1}{\pi}\int_{\frac{K}{\xi}}^{\frac{K\lambda}{\xi}} \frac{\ln t}{1+t^2}dt,\ V_{34} = -\frac{2\sin\hat{\theta}_3}{\pi}\int_0^{\frac{|\lambda_1|}{\xi}} \frac{(t^2+1)\ln t}{(t^2+1)^2-4t^2\cos^2\hat{\theta}_3}dt,$$

$$V_{56} = -\frac{2\sin\hat{\theta}_5}{\pi}\int_0^{\frac{|\lambda_2|}{\xi}} \frac{(t^2+1)\ln t}{(t^2+1)^2-4t^2\cos^2\hat{\theta}_5}dt,$$

$$\theta_3,\theta_4 = \tan^{-1}\frac{K\nu}{\xi\mp K\omega},\ \theta_5,\theta_6 = \tan^{-1}\frac{K\gamma}{\xi\mp K\delta}. \quad (5.76)$$

In deriving (5.74) we have used the following results

$$\xi\mp K\omega + iK\nu = \left\{(\xi\mp K\omega)^2+K^2\nu^2\right\}^{1/2}e^{i(\theta_3,\theta_4)}$$

and

$$\xi\mp K\delta + iK\gamma = \left\{(\xi\mp K\delta)^2+K^2\gamma^2\right\}^{1/2}e^{i(\theta_5,\theta_6)}.$$

Now, the solutions $B_1(\xi)$ and $C_1(\xi)$ of the Carleman integral equations (5.12) and (5.13) are determined in a straightforward manner as

$$B_1(\xi) = (\boldsymbol{S}^{-1}F_B)(\xi) - \boldsymbol{S}C_1(\xi) = (\boldsymbol{S}^{-1}F_B)(\xi) - \frac{1}{2}\{U(\xi)-V(\xi)\}$$

$$= \alpha B_1^{\alpha}(\xi) + \beta B_1^{\beta}(\xi) + A_1B_1^{A_1}(\xi) + A_2B_1^{A_2}(\xi) + A_3B_1^{A_3}(\xi) + A_4B_1^{A_4}(\xi) \quad (5.77)$$

and

$$C_1(\xi) = \frac{1}{2}\{U(\xi)+V(\xi)\} = \alpha C_1^{\alpha}(\xi) + \beta C_1^{\beta}(\xi) + A_1C_1^{A_1}(\xi) + A_2C_1^{A_2}(\xi)$$

$$+ A_3C_1^{A_3}(\xi) + A_4C_1^{A_4}(\xi). \quad (5.78)$$

The functions $B_1^j(\xi), C_1^j(\xi)$ (the superscript j having obvious meanings) can be computed from the following relations involving the functions $u_j(\xi)$ and $v_j(\xi)$:

$$B_1^{\alpha}(\xi)=\frac{C_\alpha M(\xi)\mathrm{e}^{\mathrm{i}\lambda Kl}}{\Lambda_0(\mathrm{i}K\lambda)(\xi-\mathrm{i}K\lambda)}-\frac{1}{2}\{u_\alpha(\xi)-v_\alpha(\xi)\},$$

$$B_1^{\beta}(\xi)=\frac{C_\beta M(\xi)}{\Lambda_0(-\mathrm{i}K\lambda)(\xi+\mathrm{i}K\lambda)}-\frac{1}{2}\{u_\beta(\xi)-v_\beta(\xi)\},$$

$$B_1^{A_1}(\xi)=\frac{C_{A_1}M(\xi)\mathrm{e}^{\mathrm{i}\lambda_1 Kl}}{\Lambda_0(\mathrm{i}K\lambda_1)(\xi-\mathrm{i}K\lambda_1)}-\frac{1}{2}\{u_{A_1}(\xi)-v_{A_1}(\xi)\}$$

$$B_1^{A_2}(\xi)=\frac{C_{A_2}M(\xi)}{\Lambda_0(-\mathrm{i}K\lambda_1)(\xi+\mathrm{i}K\lambda_1)}-\frac{1}{2}\{u_{A_2}(\xi)-v_{A_2}(\xi)\}$$

$$B_1^{A_3}(\xi)=\frac{C_{A_3}M(\xi)}{\Lambda_0(\mathrm{i}K\bar{\lambda}_1)(\xi-\mathrm{i}K\bar{\lambda}_1)}-\frac{1}{2}\{u_{A_3}(\xi)-v_{A_3}(\xi)\}$$

$$B_1^{A_4}(\xi)=\frac{C_{A_4}M(\xi)\mathrm{e}^{-\mathrm{i}\bar{\lambda}_1 Kl}}{\Lambda_0(-\mathrm{i}K\bar{\lambda}_1)(\xi+\mathrm{i}K\bar{\lambda}_1)}-\frac{1}{2}\{u_{A_3}(\xi)-v_{A_3}(\xi)\} \quad (5.79)$$

and

$$C_1^j(\xi)=\frac{1}{2}\{u_j(\xi)+v_j(\xi)\}. \quad (5.80)$$

Thus the functions $B_1(\xi)$ and $C_1(\xi)$ are now derived as linear combinations of unknown constants $\alpha,\beta,A_1,A_2,A_3,A_4$. We then replace $B_1(\xi)$ and $C_1(\xi)$ appearing in the equations (5.18) to (5.25) by their forms in (5.79) and (5.80). This yields a set of eight linear equations for determining the eight unknown constants including R,T. These equations are solved numerically to compute the numerical estimates for the unknown constants. The numerical results for the reflection coefficient for different parameters are discussed below.

Numerical results

In order to establish the correctness of the numerical results obtained by the present analysis, we have compared $|R|$ with the results given in Meylan and Squire (1994). Choosing the same values of various physical quantities such as Young's modulus (E = 6 GPa), Poisson's ratio ($\nu = 0.3$), densities of water (1025 kgm^{-3}) and ice (922.5 kgm^{-3}), wavelength 100 m and g = 9.81 ms^{-2} as given in Meylan and Squire (1994), $|R|$ is depicted in Fig. 5.15 for different values of thickness of the ice-sheet (h_0 = 1 m, 2 m, 5 m) against its width (floe-length in meter). If this Fig. 5.15 is compared with the corresponding figure (Fig. 2) of Meylan and Squire (1994), it is obvious that these are almost identical.

Figure 5.16 shows $|R|$ for plate width $l/L = 10$, ice cover parameter $D/L^4 = 0.001$. Here we notice that when $KL < 0.7$, $|R|$ is almost zero implying that there occurs total transmission for incident waves with sufficiently smaller frequencies. The continuous line in this figure is drawn on the basis of our present approach while the triangles represent corresponding data in Gayen et al. (2005), wherein the mathematical analysis was based on the assumption of largeness of the plate width. It is evident that the present results completely match with the previous ones which establishes the validity of the theory presented in this paper.

In Figs. 5.17 and 5.18 the reflection coefficient is plotted against D/L^4 for two different wave numbers $KL = 2$ and $KL = 4$ taking the strip width $l/L = 10$. It is observed that overall amount of reflection increases with an increase in the wave number. Also the number of zeros of $|R|$ increases for larger frequency.

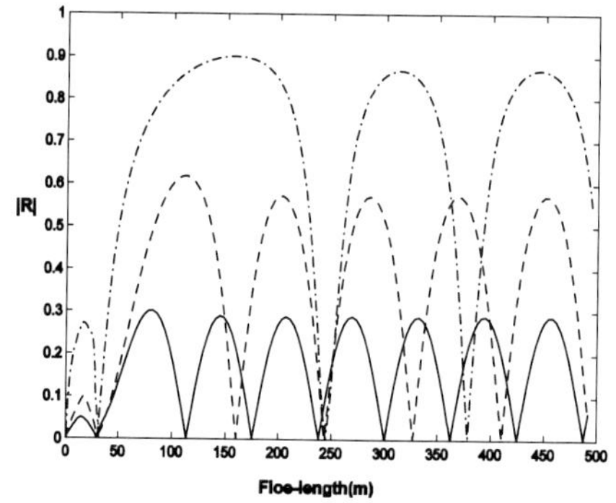

Fig. 5.15 $|R|$ for different thickness: $h_0 = 1$ m (solid curve), 2 m (dashed curve), 5 m (dash-dot curve)

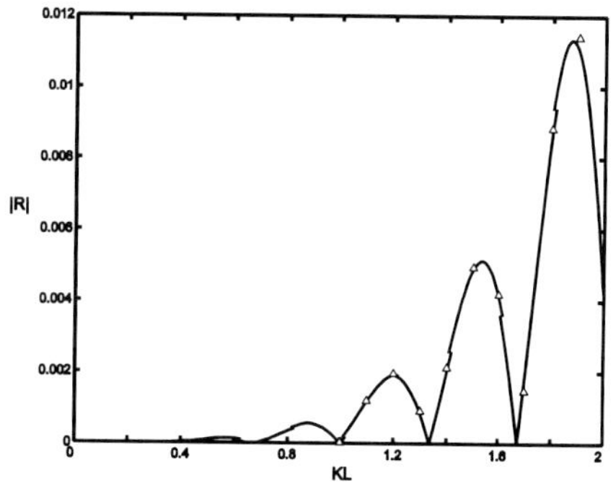

Fig. 5.16 $|R|$ for plate width $\frac{l}{L} = 10$, $\frac{D}{L^4} = 0.0001$. Triangles denote data obtained by Gayen, Mandal and Chakrabarti (2005). The line denotes $|R|$ computed by the present method

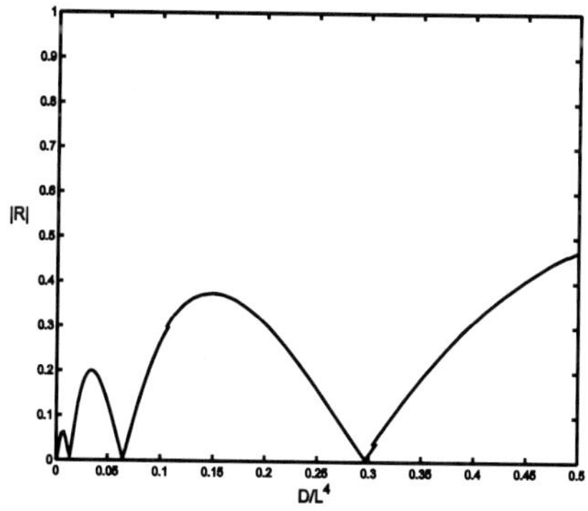

Fig. 5.17 $|R|$ for plate width $\frac{l}{L} = 10$, $KL = 2$

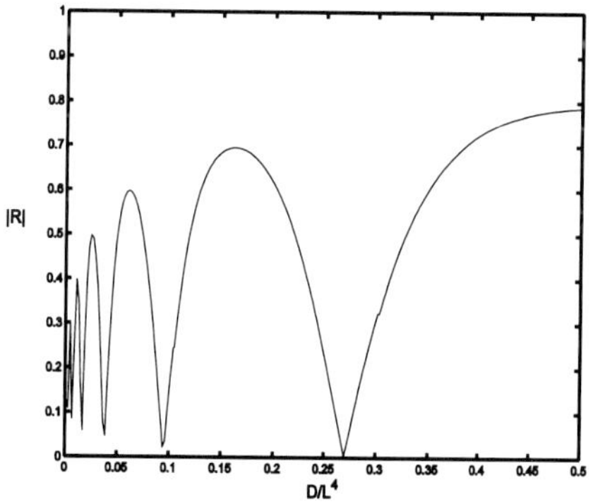

Fig. 5.18 $|R|$ for plate width $\frac{l}{L} = 10$, $KL = 4$

The effect of frequency is again compared in Figs. 5.19 and 5.20 by choosing a larger strip width, i.e., $l/L = 100$ for $KL = 2, 4$. From these two figures it is obvious that maximum values of $|R|$ as well as number of zeros of $|R|$ increase with increase of wave number.

Now comparing the Figs. 5.17 and 5.19 or, the Figs. 5.18 and 5.20, it is observed that $|R|$ becomes more oscillatory in nature for a wider strip. A similar feature was observed by Chung and Linton (2005) and Williams and Squire (2006) for a gap of finite width between two floating plates where the number of zeros of $|R|$ was found to be greater with larger gap width.

The effect of the thickness of the plate can be examined by varying the parameter D if we assume that the Young's modulus and the Poisson's ratio are kept fixed. This has been shown in Fig. 5.21 by taking $l/L = 5$ and $D/L^4 = 0.1, 0.4, 0.7$. The curves in Fig. 5.21 reveal that $|R|$ increases with an increase of D/L^4. Thus for plates with same elastic parameters, amount of reflected wave energy is increased for thicker plates. This is also evident from Fig. 1.

In Fig. 5.22 we have again considered the effect of the plate width on $|R|$ for fixed value 0.1 of D/L^4. Here we have taken moderate plate widths viz. $l/L = 1,2,3$. Here also $|R|$ increases with an increase in the plate width but the phenomenon of multiple

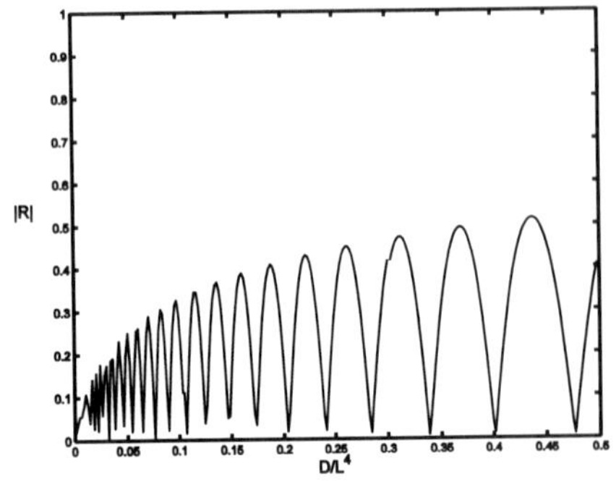

Fig. 5.19 $|R|$ for plate width $\frac{l}{L} = 100$, $KL = 2$

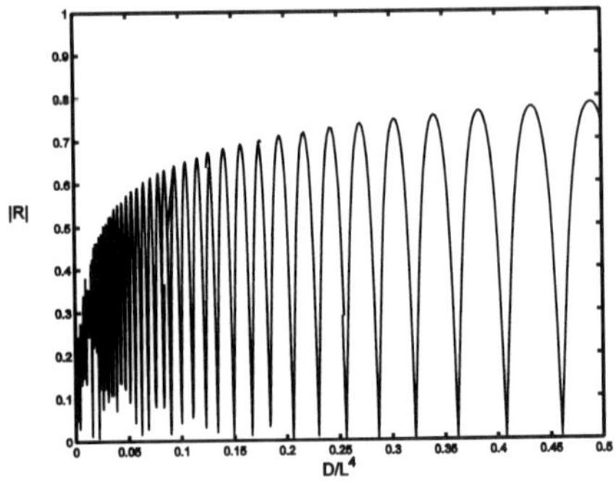

Fig. 5.20 $|R|$ for plate width $\frac{l}{L} = 100$, $KL = 4$

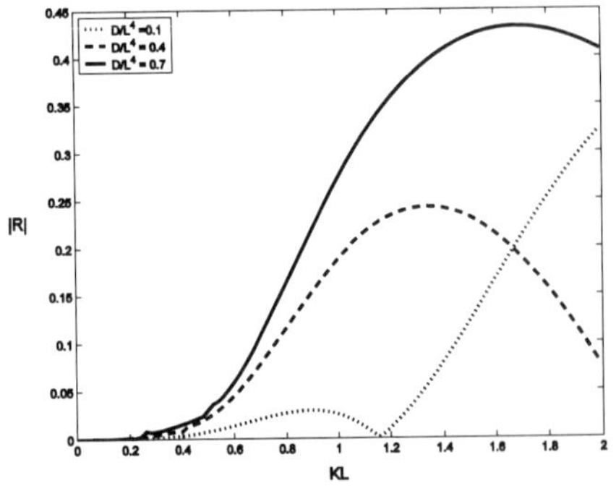

Fig. 5.21 $|R|$ for plate width $\frac{l}{L} = 5$

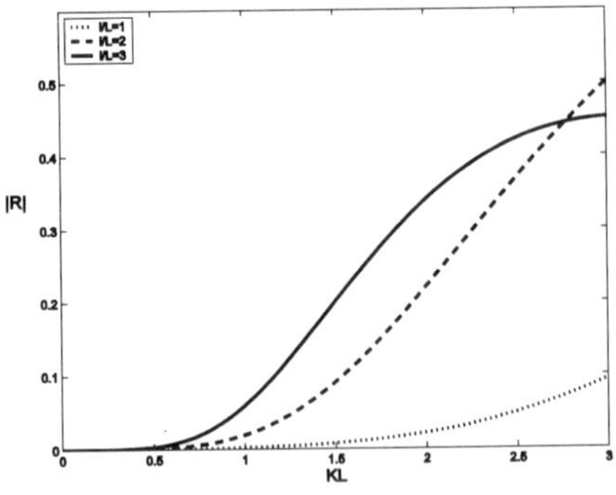

Fig. 5.22 $|R|$ for plate width $\frac{D}{L^4} = 0.1$

reflection does not occur for moderate widths unlike the Figs. 5.17 to 5.20 where the plate width was taken to be sufficiently large. It may also be noticed from Figs. 5.21 and 5.22 that total transmission occurs for smaller frequencies as was found in Fig. 5.16.

CHAPTER VI

Long Horizontal Cylinder

6.1 Scattering by a half-immersed circular cylinder in water with a free surface

In this section the problem of scattering of surface waves obliquely incident on a fixed half immersed circular cylinder in deep water is considered. This has been solved by Mandal and Goswami (1984) by reducing it to the solution of an integral equation involving the unknown scattered velocity potential on the cylinder by a simple use of Green's theorem in the fluid region. The kernel of the integral equation is expanded in a series involving different orders of the sine of the angle of incidence, the expression being valid for all angles of incidence from 0° to 90° and small wave number. This expression of the kernel suggests the corresponding form of the expression of the unknown function in the integral equation and thereby we obtain a set of integral equations of the second order. Here the integral equations is solved approximately for long wave only, as it is already mentioned above that the expression of the kernel of the original integral equation is valid only when the wave number is small, i.e., for the long wave case. The same problem is solved again by forming the scattered velocity potential by the method of multipoles using the general expansion theorem of Ursell (1968) for the scattered potential.

Statement of the Problem

We consider the scattering of surface water waves obliquely incident on a fixed half immersed circular cylinder of radius a and use a co-ordinate system in which the y-axis is vertically downwards, and the z-axis is the axis of the cylinder which is horizontal and is on the mean free surface. Polar co-ordinates are defined by $x = r\sin\theta$, $y = r\cos\theta$. Assuming the fluid to be inviscid and incompressible and the motion to be irrotational and simple harmonic in time with circular frequency σ and small amplitude, a velocity potential exists and it may be taken to be $\mathrm{Re}\{\chi(x, y, z)e^{-i\sigma t}\}$ satisfying the equations

$$\nabla^2\chi = 0 \qquad \text{in the fluid;}$$

$$\frac{\partial \chi}{\partial y} + K\chi = 0 \quad \text{on} \quad y = 0, \ |x| > a;$$

$$\frac{\partial \chi}{\partial r} = 0 \quad \text{on} \quad r = a, \quad -\frac{\pi}{2} < \theta < \frac{\pi}{2},$$

where $K = \sigma^2 / g$.

A train of waves travelling from the negative infinity and represented by

$$\chi^{inc} = \exp(-Ky + i\mu x + ivz),$$

where $\mu = K\cos\beta$, $v = K\sin\beta$, is assumed to be incident at an angle β to the x-axis. Such a wave will be partially reflected and transmitted by the cylinder, and in view of the geometry of the cylinder it is reasonable to assume $\chi(x, y, z) = \Phi(x, y)\exp(ivz)$. Then Φ must satisfy

$$\frac{\partial^2 \Phi}{\partial^2 x} + \frac{\partial^2 \Phi}{\partial^2 y} - v^2\Phi = 0 \quad \text{in the fluid;} \tag{1.1}$$

$$\frac{\partial \Phi}{\partial y} + K\Phi = 0 \quad \text{on} \quad y = 0, \ |x| > a; \tag{1.2}$$

$$\frac{\partial \Phi}{\partial r} = 0 \quad \text{on} \quad r = a, \quad -\frac{\pi}{2} < \theta < \frac{\pi}{2}. \tag{1.3}$$

Now let us write $\Phi(x, y) = \phi^{\text{inc}}(x, y) + \phi(x, y)$, where $\phi^{\text{inc}}(x, y) = \exp(-Ky + i\mu x)$ and $\phi(x, y)\exp(ivz)$ is the scattered velocity potential. Hence $\phi(x, y)$ satisfies (1.1) and (1.2). Further $\phi(x, y)$ should satisfy the following asymptotic forms at infinity:

$$\begin{aligned} \phi(x,y) &\sim (T-1)\exp(-Ky + \mathrm{i}\mu x) \quad (x \to +\infty), \\ \phi(x,y) &\sim R \ \exp(-Ky - \mathrm{i}\mu x) \quad (x \to -\infty), \end{aligned} \tag{1.4}$$

T and *R* being the (complex) transmission and reflection coefficients respectively. From the conditions (1.4) it is evident that $\phi(x, y)$ represents an outgoing wave at infinity.

Reduction to an integral equation

Following Levine (1965) the generalized Green's function G, satisfying (2.1), (2.2) and representing an outgoing wave at infinity, is obtained as

$$G(x, y; \xi, \eta) = K_0(v\rho) - K_0(v\rho^*) + \frac{2\pi i K}{(K^2 - v^2)^{\frac{1}{2}}} \exp\left\{-K(y+\eta) + i(K^2 - v^2)^{\frac{1}{2}}|x - \xi|\right\}$$

$$+2\int_v^\infty \frac{(K^2 - v^2)^{\frac{1}{2}}\cos\left[(k^2 - v^2)^{\frac{1}{2}}(y+\eta)\right] - K\sin\left[(k^2 - v^2)^{\frac{1}{2}}(y+\eta)\right]}{K^2 + k^2 - v^2} \exp\{-k|x - \xi|\} dk \tag{1.5}$$

where $\rho^2, \rho^{*2} = (x - \xi)^2 + (y \mp \eta)^2$.

Now taking (ξ, η) on the semicircle $\xi = a \sin \alpha$, $\eta = a \cos \alpha$, and applying Green's integral theorem to $\phi(x, y)\, G(x, y; \xi, \eta)$ and in the fluid region with a small indentation at (ξ, η) on the semicircle, we obtain

$$-\pi\phi(\alpha) + \int_{-\frac{1}{2}\pi}^{\frac{1}{2}\pi} \phi(\theta)\,\langle a\frac{\partial G}{\partial r}(\theta,\alpha)\rangle\, d\theta = -\int_{-\frac{1}{2}\pi}^{\frac{1}{2}\pi} \langle a\frac{\partial \phi^{\text{inc}}}{\partial r}\rangle \langle G(\theta,\alpha)\rangle\, d\theta, -\tfrac{1}{2}\pi < \alpha < \tfrac{1}{2}\pi, \tag{1.6}$$

where $G(\theta, \alpha) = G(r \sin \theta, r \cos \theta; \alpha \sin \alpha, \alpha \cos \alpha)$ and we use angular brackets $\langle\rangle$ to indicate that r is to be put equal to a.

Next let us write $\epsilon = \sin \beta$ for $0^\circ \le \beta \le 90^\circ$ and Ka not being large, we obtain Mandal and Goswami (1984)

$$G(x,y;\xi,\eta) = G_0(x,y;\xi,\eta) + \epsilon^2 \ln\epsilon\, G_1(x,y;\xi,\eta) + \epsilon^2 G_2(x,y;\xi,\eta) + O(\epsilon^4 \ln\epsilon, \epsilon^4),$$
$$0 \le \epsilon < 1, 0 < \rho, \rho^* < 2a, \tag{1.7}$$

where

$$G_0(x,y;\xi,\eta) = -\ln\frac{\rho}{\rho^*} + 2\pi i \exp\{-K(y+\eta) + iK|x-\xi|\}$$
$$-2\int_0^\infty \frac{K\sin k(y+\eta) - k\cos k(y+\eta)}{K^2+k^2} \exp\{-k|x-\xi|\}\; dk, \tag{1.8}$$

$$G_1(x,y;\xi,\eta) = (1-Ky)(1-K\eta)\,, \tag{1.9}$$

$$G_2(x,y;\xi,\eta) = \big(\sqrt{2} - 1 + \ln(2-\sqrt{2})\big)\big(1 - K(y+\eta)\big) - \big(\ln 2 + \psi(2)\big)K^2 y\eta$$
$$+\frac{1}{4}(K^2\rho^{*2}\ln K\rho^* - K^2\rho^2 \ln K\rho) + \pi i(1 - iK|x-\xi| \exp\{-K(y+\eta) + iK|x-\xi|\}$$
$$+K^2\int_K^\infty \frac{K\sin k(y+\eta) - k\cos k(y+\eta)}{k^2(K^2+k^2)} \times (1 + k|x-\xi|)\exp\{-k|x-\xi|\}\; dk\,. \tag{1.10}$$

By (1.7) and (1.6) we write

$$\phi(\alpha) = \phi_0(\alpha) + \epsilon^2 \ln\epsilon\, \phi_1(\alpha) + \epsilon^2\phi_2(\alpha) + O(\epsilon^4\ln\epsilon, \epsilon^4), -\tfrac{1}{2}\pi < \alpha < \tfrac{1}{2}\pi\,, \tag{1.11}$$

And by (1.6), (1.7) and (1.11), we obtain

$$-\pi\phi_0(a) + \int_{-\frac{1}{2}\pi}^{\frac{1}{2}\pi} \phi_0(\theta)\,\langle a\frac{\partial G_0}{\partial r}(\theta,\alpha)\rangle\, d\theta = -\int_{-\frac{1}{2}\pi}^{\frac{1}{2}\pi} \langle a\frac{\partial \phi_0{}^{\text{inc}}}{\partial r}\rangle \langle G_0(\theta,\alpha)\rangle\, d\theta, -\tfrac{1}{2}\pi < \alpha < \tfrac{1}{2}\pi, \tag{1.12}$$

$$-\pi\phi_1(a) + \int_{-\frac{1}{2}\pi}^{\frac{1}{2}\pi} \phi_1(\theta)\,\langle a\frac{\partial G_0}{\partial r}(\theta,\alpha)\rangle\, d\theta = -\int_{-\frac{1}{2}\pi}^{\frac{1}{2}\pi} \phi_0(\theta)\;\langle a\frac{\partial G_1}{\partial r}(\theta,\alpha)\rangle\, d\theta$$
$$-\int_{-\frac{1}{2}\pi}^{\frac{1}{2}\pi} \langle a\frac{\partial \phi_0{}^{\text{inc}}}{\partial r}\rangle \langle G_1(\theta,\alpha)\rangle\, d\theta, -\tfrac{1}{2}\pi < \alpha < \tfrac{1}{2}\pi, \tag{1.13}$$

$$\pi\phi_2(a) + \int_{-\frac{1}{2}\pi}^{\frac{1}{2}\pi} \phi_2(\theta)\,\langle a\frac{\partial G_0}{\partial r}(\theta,\alpha)\rangle\, d\theta = -\int_{-\frac{1}{2}\pi}^{\frac{1}{2}\pi} \phi_0(\theta)\,\langle a\frac{\partial G_2}{\partial r}(\theta,\alpha)\rangle\, d\theta$$

$$-\tfrac{1}{2}\pi\,\langle a\frac{\partial {\phi_0}^{\text{inc}}}{\partial r}\rangle\,\langle G_2(\theta,\alpha)\rangle d\theta - \int_{-\frac{1}{2}\pi}^{\frac{1}{2}\pi} \langle a\frac{\partial {\phi_1}^{\text{inc}}}{\partial r}\rangle\,\langle G_2(\theta,\alpha)\rangle\, d\theta, \quad \tfrac{1}{2}\pi < \alpha < \tfrac{1}{2}\pi, \tag{1.14}$$

where

$$\phi^{\text{inc}}(x,y) = {\phi_0}^{\text{inc}}(x,y) + \epsilon^2 {\phi_1}^{\text{inc}}(x,y) + O(\epsilon^4).$$

The integral equations (1.12), (1.13), (1.14), etc. cannot be solved analytically for any *Ka*. Since the expansion (1.7) valid for *Ka* is not large, we try to solve these integral equations approximately for small *Ka* only. Following Yu and Ursell (1961) we can show that

$$\langle G_0(\theta,\alpha)\rangle = -\ln\left|\frac{\sin\frac{1}{2}(\theta-\alpha)}{\cos\frac{1}{2}(\theta+\alpha)}\right| - 2(\gamma + \ln 2Ka - \pi i)\langle \exp\{-K(y+\eta)\}\cos K(x-\xi)\rangle$$

$$+(\alpha-\theta)\langle \exp\{-K(y+\eta)\}\sin K(x-\xi)\rangle$$

$$-2\sum_{m=1}^{\infty}(-1)^{m-1}\frac{1+\frac{1}{2}+\cdots+\left(\frac{1}{m}\right)}{m!}\left(2Ka\cos\frac{1}{2}(\theta+\alpha)\right)^m \cos\frac{1}{2}m(a-\theta), \tag{1.15}$$

and

$$\langle a\frac{\partial}{\partial r}G_0(\theta,\alpha)\rangle = -\langle a\exp\{-K(y+\eta)\}\cos K(x-\xi)\rangle - 2\left(\gamma + \ln 2Ka - \pi i + \ln\cos\frac{1}{2}(\theta+\alpha)\right)$$

$$\times\langle a\frac{\partial}{\partial r}[\exp\{-K(y+\eta)\}\cos K(x-\xi)]\rangle + 2\,\langle \zeta a\frac{\partial}{\partial r}[\exp\{-K(y+\eta)\}\sin K(x-\xi)]\rangle$$

$$-2\sum_{m=1}^{\infty}(-1)^{m-1}\frac{1+\frac{1}{2}+\cdots+\left(\frac{1}{m}\right)}{m!}\langle a\frac{\partial}{\partial r}[(K\rho)^m\cos m\zeta]\rangle, \tag{1.16}$$

where

$$\xi = a\sin\alpha,\quad \eta = a\cos\alpha,\quad \langle\zeta\rangle = \frac{1}{2}(a-\theta),\langle\rho\rangle = 2a\cos\frac{1}{2}(\theta+\alpha)\ ,-\tfrac{1}{2}\pi < \theta,\alpha < \tfrac{1}{2}\pi.$$

The integral equation (1.12) can be written as

$$\pi\phi_0(\alpha) + \int_{-\frac{1}{2}\pi}^{\frac{1}{2}\pi} M(\theta,\alpha)\phi_0(\theta)d\theta = L(\alpha,Ka) - A(Ka)\langle \exp(-K\eta)\cos K\xi\rangle$$

$$-B(Ka)\langle \exp(-K\eta)\sin K\xi\rangle = \pi q(\alpha)\ (\text{say}),\ -\tfrac{1}{2}\pi < \alpha < \tfrac{1}{2}\pi. \tag{1.17}$$

where

$$M(\theta,\alpha) = 2\ln\cos\frac{1}{2}(\theta+\alpha)\,\langle a\frac{\partial}{\partial r}[\exp\{-K(y+\eta)\}\cos K(x-\xi)]\rangle$$

$$-2\,\langle \zeta a\frac{\partial}{\partial r}[\exp\{-K(y+\eta)\}\sin K(x-\xi)]\rangle$$

$$+2\sum_{m=1}^{\infty}(-1)^{m-1}\frac{1+\frac{1}{2}+\cdots+\left(\frac{1}{m}\right)}{m!}\langle a\frac{\partial}{\partial r}[(K\rho)^m\cos m\zeta]\rangle, \tag{1.18}$$

$$L(\alpha,Ka) = -Ka\int_{-\frac{1}{2}\pi}^{\frac{1}{2}\pi}\exp\{-Ka(\cos\theta - i\sin\theta) - i\theta\}\langle G_0(\theta,\alpha)\rangle\,d\theta,$$

and

$$\begin{matrix}A\\B\end{matrix} = \int_{-\frac{1}{2}\pi}^{\frac{1}{2}\pi}\phi_0(\theta)\,\langle\exp(-Ky)\begin{matrix}\cos\\\sin\end{matrix}Kx\rangle\,d\theta + 2(\gamma+\ln 2Ka-\pi i)$$

$$\times\int_{-\frac{1}{2}\pi}^{\frac{1}{2}\pi}\phi_0(\theta)\,\langle a\frac{\partial}{\partial r}\exp(-Ky)\begin{matrix}\cos\\\sin\end{matrix}Kx\rangle\,d\theta. \tag{1.20}$$

Neglecting terms of order $O((Ka)^2 \text{ In } Ka, (Ka)^2)$ we obtain

$$M(\theta,\alpha) = Ka\,M_0(\theta,\alpha) + O((Ka)^2\ln Ka,(Ka)^2),$$

where

$$M_0(\theta,\alpha) = -2\ln\cos\frac{1}{2}(\theta+\alpha)\cos\theta + (\alpha-\theta)\sin\theta + \cos\theta + \cos\alpha,$$

so that $M_0(\theta, \alpha)$ is bounded for all $-\frac{1}{2}\pi \le \theta, \alpha \le \frac{1}{2}\pi$. Also

$$L(\alpha,Ka) - Ka\,L_0(\alpha) + O((Ka)^2\ln Ka,(Ka)^2),$$

where

$$L_0(\alpha) = 4(\gamma+\ln 2Ka-\pi i) - 2\ln 2 - 2 + 2\ln\cos\alpha + 2\sin\alpha\ln\tan\left(\frac{1}{4}\pi+\frac{1}{2}\alpha\right) + i\pi\sin\alpha \tag{1.21}$$

Thus $L_0(\alpha)$ is bounded for all $-\frac{1}{2}\pi \le \alpha \le \frac{1}{2}\pi$. Hence the kernel of the integral equation is small for small Ka, so that the following theorem due to Ursell (1961) is applicable.

Theorem. If (i) $q(\alpha)$ is bounded in $-\frac{1}{2}\pi \le \alpha \le \frac{1}{2}\pi$ and

(ii) $\int_{-\frac{1}{2}\pi}^{\frac{1}{2}\pi}|M(\theta,\alpha)|d\theta \le m < 1, -\frac{1}{2}\pi \le \alpha \le \frac{1}{2}\pi,$

then the integral equation (1.17) may be solved by iteration, and

$$\phi_0(\alpha) = q(\alpha) + \sum_{s=1}^{\infty} q_s(\alpha)$$

where

$$q_1(\alpha) = -\frac{1}{\pi}\int_{-\frac{1}{2}\pi}^{-\frac{1}{2}\pi} M(\theta,\alpha)q(\theta)d\theta$$

and

$$q_s(\alpha) = -\frac{1}{\pi}\int_{-\frac{1}{2}\pi}^{-\frac{1}{2}\pi} M(\theta,\alpha)q_{s-1}(\theta)d\theta\,, \qquad s > 1,$$

the domain of validity of Ka being determined by the second condition. A rough upper bound of Ka is found to be approximately 0.93. Thus we obtain

$$\pi\phi_0(\alpha) = KaL_0(\alpha) - \frac{2\pi(\gamma + \ln Ka - \pi i)Ka}{\pi - (1 + \gamma + \ln 4Ka - \pi i)Ka} + O((Ka)^2 \ln Ka\,, (Ka)^2(Ka)^2),$$
$$-\tfrac{1}{2}\pi \le \alpha \le \tfrac{1}{2}\pi, \tag{1.22}$$

$L_0(\alpha)$ being given by (1.21). Similarly by (1.13), (1.14), (1.22) and the above theorem, we obtain

$$\phi_1(\alpha) = \frac{-Ka}{\pi - (1 + \gamma + \ln 4Ka - \pi i)Ka} + O((Ka)^2 \ln Ka\,, (Ka)^2(Ka)^2), -\tfrac{1}{2}\pi \le \alpha \le \tfrac{1}{2}\pi$$

$$\phi_2(\alpha) = \left(\sqrt{2} - 1 + \ln\left(\sqrt{2} - 1\right) + \pi i\right)\phi_1(\alpha) - \frac{i}{2}Ka \sin\alpha\,, -\tfrac{1}{2}\pi \le \alpha \le \tfrac{1}{2}\pi, \tag{1.23}$$

etc.

Reflection and transmission coefficients

Taking (ξ, η) in the fluid and applying Green's integral theorem on $\phi(x, y)$ and $G(x, y; \xi, \eta)$, we obtain

$$2\pi\phi(\xi,\eta) = \int_{-\frac{1}{2}\pi}^{\frac{1}{2}\pi} \phi(\theta)\,\langle a\frac{\partial G}{\partial r}(x,y;\xi,\eta)\rangle\, d\theta + \int_{-\frac{1}{2}\pi}^{\frac{1}{2}\pi} \langle a\frac{\partial \phi^{inc}}{\partial r}\rangle\,\langle G(x,y;\xi,\eta)\rangle d\theta,$$

and using (1.4) and (1.5) as $|\xi| \to \infty$, we obtain

$$R = \frac{iK}{\mu}\int_{-\frac{1}{2}\pi}^{\frac{1}{2}\pi} \phi(\theta)\,\langle a\frac{\partial}{\partial r}\exp(-Ky + i\mu x)\rangle\, d\theta$$
$$+\frac{iK}{\mu}\int_{-\frac{1}{2}\pi}^{\frac{1}{2}\pi} \langle a\frac{\partial}{\partial r}\exp(-Ky + i\mu x)\rangle\,\langle\exp(-Ky + i\mu x)\rangle d\theta, \tag{1.24}$$

and

$$T = 1 + \frac{\mathrm{i}K}{\mu}\int_{-\frac{1}{2}\pi}^{\frac{1}{2}\pi}\varphi(\theta)\left\langle a\frac{\partial}{\partial r}\exp\left(-Ky - \mathrm{i}\,\mu x\right)\right\rangle d\theta$$

$$\frac{\mathrm{i}K}{\mu}\int_{-\frac{1}{2}\pi}^{\frac{1}{2}\pi}\left\langle a\frac{\partial}{\partial r}\exp\left(-Ky + \mathrm{i}\,\mu x\right)\right\rangle\left\langle\exp\left(-Ky - \mathrm{i}\,\mu x\right)\right\rangle d\theta. \tag{1.25}$$

In view of (1.11), by (1.24) and (1.25) we may write

$$R = R_0 + \varepsilon^2 \ln\varepsilon R_1 + \varepsilon^2 R_2 + O\left(\varepsilon^4 \ln\varepsilon, \varepsilon^4\right),$$

$$T = T_0 + \varepsilon^2 \ln\varepsilon T_1 + \varepsilon^2 T_2 + O\left(\varepsilon^4 \ln\varepsilon, \varepsilon^4\right). \tag{1.26}$$

Thus by (1.26) and (1.24), we obtain

$$R_0 = \mathrm{i}\int_{-\frac{1}{2}\pi}^{\frac{1}{2}\pi}\varphi_0(\theta)\left\langle a\frac{\partial}{\partial r}\exp\left(-Ky + \mathrm{i}\,Kx\right)\right\rangle d\theta + \mathrm{i}\int_{-\frac{1}{2}\pi}^{\frac{1}{2}\pi}\left\langle a\frac{\partial}{\partial r}\exp\left(-Ky + \mathrm{i}\,Kx\right)\right\rangle\left\langle\exp\left(-Ky + \mathrm{i}\,Kx\right)\right\rangle d\theta,$$

$$R_1 = \mathrm{i}\int_{-\frac{1}{2}\pi}^{\frac{1}{2}\pi}\varphi_1(\theta)\left\langle a\frac{\partial}{\partial r}\exp\left(-Ky + \mathrm{i}\,Kx\right)\right\rangle d\theta, \tag{1.27}$$

$$R_2 = \mathrm{i}\int_{-\frac{1}{2}\pi}^{\frac{1}{2}\pi}\varphi_2(\theta)\left\langle a\frac{\partial}{\partial r}\exp\left(-Ky + \mathrm{i}\,Kx\right)\right\rangle d\theta - \int_{-\frac{1}{2}\pi}^{\frac{1}{2}\pi}\varphi_0(\theta)\left\langle\frac{a}{2}\frac{\partial}{\partial r}\left\{a\exp\left(-Ky + \mathrm{i}\,Kx\right)\right\}\right\rangle d\theta + \frac{R_0}{2} + \pi\,\mathrm{i}\left(Ka\right)^2$$

etc.

Therefore, by (1.22), (1.23) and (1.27), we obtain

$$R_0 = -4\left(Ka\right)^2 - \mathrm{i}\left[2Ka + \left(\frac{4\gamma}{\pi} + \frac{4}{\pi}\ln 4Ka - \frac{6}{\pi} + \frac{1}{2}\pi\right)\left(Ka\right)^2\right] + O\left(\left(Ka\right)^3 \ln Ka, \left(Ka\right)^3\right),$$

$$R_1 = \mathrm{i}\frac{2}{\pi}\left(Ka\right)^2 + O\left(\left(Ka\right)^3 \ln Ka, \left(Ka\right)^3\right), \tag{1.28}$$

$$R_2 = \left(\sqrt{2} - 1 - \ln\left(\sqrt{2} + 1\right) + \mathrm{i}\,\pi\right)R_1 + \frac{1}{2}\left(R_0\right) + \mathrm{i}\frac{3\pi}{2}\left(Ka\right)^2$$

etc.

Similarly, by (1.22), (1.23), (1.26) and (1.25) we obtain

$$T_0 = 1 + R_0 + \mathrm{i}\,2\pi\left(Ka\right)^2; T_1 = 1 + R_1; T_2 = R_2 - \mathrm{i}\,\pi\left(Ka\right)^2; \quad \text{etc.} \tag{1.29}$$

Solution by the method of multipoles

Following Ursell (1968), the scattered potential $\varphi(x,y)\equiv\varphi(vr,\theta)$ can be written as

$$\phi(vr,\theta)=-\frac{\mathrm{i}C}{\pi}\cos\beta\,\psi_0(vr,\theta)+\frac{D}{\pi K}\frac{\partial}{\partial x}\psi_0(vr,\theta)+A\exp\left(-Ky+\mathrm{i}\,Kx\cos\beta\right)$$

$$+B\exp\left(-Ky-\mathrm{i}\,Kx\cos\beta\right) \tag{1.30}$$

$$+2\sum_{m=1}^{\infty}p_{2m}\frac{(va/2)^{2m}}{(2m-1)!}\psi_{2m}(vr,\theta)+2\sum_{m=1}^{\infty}p_{2m+1}\frac{(va/2)^{2m+1}}{(2m-1)!}\psi_{2m+1}(vr,\theta),-\frac{1}{2}\pi\le\theta\le\frac{1}{2}\pi,$$

where

$$\psi_0(vr,\theta)=(\pi i-\lambda)\coth\lambda\left\{I_0(vr)+2\sum_{\infty}^{m=1}(-1)^m\cosh m\lambda\,I_m(vr)\cos m\theta\right\}$$

$$K_0(vr)-2\coth\lambda\sum_{m=1}^{\infty}(-1)^m\sinh m\lambda\,I_m(vr)\left[\frac{\partial}{\partial r}\left(I_r(vr)\cos\tau\theta\right)\right]_{r=m},$$

$$\psi_{2m}(vr,\theta)=K_{2m}(vr)\cos 2m\theta+(2/\sin\beta)K_{2m-1}(vr)\cos(2m-1)\theta$$
$$+K_{2m-2}(vr)\cos(2m-2)\theta\qquad(m=1,2,3,\ldots\ldots\ldots..),$$

$$\psi_{2m+1}(vr,\theta)=K_{2m+1}(vr)\sin(2m+1)\theta+(2/\sin\beta)K_{2m-1}(vr)\sin 2m\theta$$
$$+K_{2m-1}(vr)\sin(2m-1)\theta\qquad(m=1,2,3,\ldots\ldots\ldots..),$$

$I_m(vr)$ and $K_{2m}(vr)$ being modified Bessel functions, and by (1.4)

$$C-D=T-1, \tag{1.31}$$

$$C+D=R, \tag{1.32}$$

$$A=0=B, \tag{1.33}$$

C, D and p_{2m}, p_{2m+1} being complex constants to be determined to find the complex transmission and reflection coefficients T and R. Also, λ is defined by

$$\cosh\lambda=K/v=\operatorname{cosec}\beta,\ \sinh\lambda=\cot\beta.$$

By the condition (1.3)

$$-\frac{\mathrm{i}C}{\pi}\cos\beta\left\langle\frac{\partial\psi_0}{\partial r}\right\rangle+\frac{D}{\pi K}\left\langle\frac{\partial}{\partial r}\frac{\partial}{\partial x}\psi_0\right\rangle+2\sum_{m=1}^{\infty}p_{2m}\frac{(va/2)^{2m}}{(2m-1)!}\left\langle\frac{\partial\psi_{2m}}{\partial r}\right\rangle$$

$$+2\sum_{m=1}^{\infty}p_{2m+1}\frac{(va/2)^{2m+1}}{(2m-1)!}\left\langle\frac{\partial\psi_{2m+1}}{\partial r}\right\rangle=-\left\langle\frac{\partial}{\partial r}\exp\left(-Ky+\mathrm{i}\,Kx\cos\beta\right)\right\rangle,-\frac{1}{2}\pi\le\theta\le\frac{1}{2}\pi. \tag{1.34}$$

Multiplying both sides of (1.34) by $\cos 2n\theta$ or $\sin(2n+1)\theta$ $(n = 1,2,3,.........)$, and integrating between $-\pi/2$ and $\pi/2$, we obtain two infinite system of simultaneous linear equations involving $C, p_2, p_4,..........$, etc. and $D, p_3, p_5,..........$, etc. respectively.

For any given value of Ka and any given value of β the systems of infinite simultaneous equations mentioned above are solved approximately by retaining only a finite number of unknowns and the same number of equations.

Here we obtained the following approximate expressions for C and D by neglecting terms of order

$$O\left((Ka)^3 \ln Ka;\quad (\nu a)^3 \ln \nu a;\quad (Ka)^3;\quad (\nu a)^3\right):$$

$$C = -4(Ka)^2 \sec^2 \beta$$

$$-\mathrm{i}\left[2(Ka)\sec\beta - \left(\frac{4}{3} - \gamma - \ln\frac{1}{2}Ka - \ln\sin\beta\sec\beta\ln\tan\frac{1}{2}\beta + \frac{1}{8}\pi^2\sin^2\beta\right)\frac{4}{\pi}(Ka)^2\sec\beta\right],$$

$$D = -\mathrm{i}\pi(Ka)^2 \sec\beta, \tag{1.35}$$

where $\gamma = 0.577...$ is Euler's constant, R and T being given by (1.31) and (1.32) respectively.

Discussion

(a) The reflection and transmission coefficients obtained by the IE method here are valid for all angles of incidence $\beta\,(0^o \le \beta \le 90^o)$ and for Ka not large, as is obvious from the expansion of $G(x, y; \xi, \eta)$, on the cylinder, in terms of $\varepsilon = \sin\beta$,

The domain of Ka is determined by the conditions (ii) in the Theorem. Since in our calculations we have retained only three terms in the expansion of $G(x, y; \xi, \eta)$ and to keep the errors in our numerical calculations within some preassigned limit, we have confined ourselves within $\beta \le 30^\circ$. However, for other values of $\beta(<90^\circ)$, the calculations could be performed taking more appropriate terms in the different series involving $\varepsilon^n \ln\varepsilon\,(n \ge 4)$, to minimize the errors of approximations. Taking $Ka = 0.05$ and 0.1 respectively, $|R|$ and $|T|$ are calculated for $\beta = 0^\circ, 5^\circ, 10^\circ, 15^\circ, 20^\circ, 25^\circ$ and 30°, correct up to four decimal places. If we are to take larger values of Ka within the domain of validity of Ka determined by the condition (ii) in the theorem, then the approximate expressions for $\phi_0(\theta), \phi_1(\theta), \phi_2(\theta)$, etc. which were evaluated up to certain orders of Ka (here first order), are to be modified by taking more appropriate terms involving higher orders of Ka. The condition (ii) of the Theorem provides us with a rough upper bound of Ka as 0.93 so the IE method will obviously fail if Ka exceeds this value. In spite of this we use the IE method since it is based on the results of the integral equation for the case of a normally incident wave, whose solution is supposed to be known. The method of multipoles can be applied to solve the problem for larger values of Ka.

(b) Again $|R|$ and $|T|$ obtained by the method of multipoles are calculated for $\beta = 0^\circ, 5^\circ, 10^\circ, 15^\circ, 20^\circ, 25^\circ$ and 30° and $Ka = 0.05$ and 0.1 respectively, correct up to four decimal places. Both the (a) sets (b) and are presented in tabular form (Tables 6.1 and 6.2).

Table 6.1 Values of $|R|$

Ka/β		0°	5°	10°	15°	20°	25°	30°
0.05	(a)	0.0964	0.0967	0.0976	0.0991	0.1010	0.1034	0.1061
	(b)	0.0964	0.0967	0.0977	0.0992	0.1015	0.1046	0.1087
0.1	(a)	0.1964	0.1970	0.1985	0.2009	0.2041	0.2080	0.2124
	(b)	0.1966	0.1971	0.1987	0.2013	0.2051	0.2103	0.2173

(a) IE method; (b) Multipole method.

Table 6.2 Values of $|T|$

Ka/β		0°	5°	10°	15°	20°	25°	30°
0.05	(a)	0.9932	0.9932	0.9931	0.9928	0.9925	0.9922	0.9917
	(b)	0.9932	0.9932	0.9931	0.9928	0.9924	0.9919	0.9912
0.1	(a)	0.9687	0.9685	0.9679	0.9669	0.9655	0.9639	0.9620
	(b)	0.9687	0.9685	0.9679	0.9667	0.9651	0.9627	0.9595

(a) IE method; (b) Multipole method.

From the two Tables (6.1) and (6.2) it is observed that, for a fixed β, as Ka increases the reflection coefficient increases while transmission coefficient decreases, which is plausible from the physical considerations of the problem. Again, for a fixed Ka, as β increases, the reflection coefficient increases and transmission coefficient decreases. Levine (1965) studied the effect of a submerged circular cylinder on oblique surface waves and found that the reflection coefficient $|R|$ increases from 0 to 1 as the angle of incidence β increases from 0° to 90°. Garrison (1969) considered the interaction of a shallow draft cylinder at the free surface with a train of oblique waves and similar conclusions were drawn by him. The results obtained here therefore appear to be consistent. Finally, the results obtained by the method of multipoles are found to be in good agreement with those obtained by the IE method.

6.2 Scattering by a circular cylinder half-immersed in water with an ice-cover

In section 6.1 we have considered wave scattering by a circular-cylinder half-immersed in water with a free surface. In recent times there has been considerable interest in the investigation of ice-wave interaction problems due to an increase in scientific activities in polar regions. The ice-cover is modeled as a thin ice-sheet of which still a smaller part is immersed in water, and is composed of materials having elastic properties.

In this section the problem of scattering of surface waves obliquely incident on a fixed half-immersed circular cylinder in deep water with an ice-cover is considered. It has been solved by Das and Mandal (2009) by using the method of multipoles, an account of which is given here. From a practical point of view, this particular problem considered here may be thought of to model a floating storage tank in the form of a long cylinder of circular cross section in polar ocean. It can also be thought of as a long ship at rest in polar ocean subjected to incident wave train.

Mathematical formulation

Let a train of surface waves be obliquely incident on a long horizontal circular cylinder of radius a half-immersed in deep water with an ice-cover. A Cartesian co-ordinate system is chosen with y-axis directed vertically downwards, and z-axis as the axis of the cylinder. The plane $y = 0$ $(r = (x^2 + z^2)^{1/2} > a, -\infty < z < \infty)$ is the mean position of the ice-cover modeled as a floating elastic plate of small thickness h_0. Polar co-ordinates are defined by $x = r \sin \theta$, $y = r \cos \theta$. Under the assumption of irrotational motion and linear theory, the obliquely incident wave field can be described by the potential function $\text{Re}\{\phi_0(x, y)e^{-i\sigma t + i\gamma z}\}$. Here $\gamma = \lambda \sin \alpha$, α being the angle of incidence of the wave train, σ is the angular frequency and λ is the unique real positive root of the dispersion equation

$$k(Dk^4 + 1 - \varepsilon K) - K = 0 \tag{2.1}$$

The other roots of (2.1) are $\lambda_1, \overline{\lambda_1}, \lambda_2$ and $\overline{\lambda_2}$ where $\text{Re}(\lambda_1) > 0$ and $\text{Re}(\lambda_2) < 0$. In (2.1) $K = \sigma^2/g$, g is the acceleration due to gravity, $D = L/\rho_1 g$, L is the flexural rigidity of the ice-cover and $\varepsilon = \frac{\rho_0}{\rho_1} h_0$, ρ_0 and ρ_1 being the densities of ice and water respectively, and

$$\phi_0(x, y) = e^{-\lambda y + i\mu x} \tag{2.2}$$

with $\mu = \lambda \cos \alpha$.

Let $\Phi(x, y, z, t) = \text{Re}\{\phi(x, y)e^{-i\sigma t + i\gamma z}\}$ describe the velocity potential for the ensuing fluid motion. Then $\phi(x, y)$ satisfies the modified Helmholtz's equation

$$\left(\nabla^2 - \gamma^2\right)\phi(x, y) = 0, \quad y > 0, \quad r > a, \tag{2.3}$$

the linearized boundary condition at the ice-cover

$$\left(D\left(\frac{\partial^2}{\partial x^2} - \gamma^2\right)^2 + 1 - \varepsilon K\right)\varphi_y + K\varphi = 0 \text{ on } y = 0, \quad |x| > |a|. \tag{2.4}$$

At the free edges of the cylinder, conditions of zero bending moments and zero shear stresses are expressed as

$$\mathcal{L}\varphi(x, 0) \to 0 \text{ as } x \to a^+, \quad x \to -a^-, \tag{2.5a}$$

$$\mathcal{M}\varphi(x, 0) \to 0 \text{ as } x \to a^+, \; x \to -a^- \tag{2.5b}$$

where the operators $\mathcal{L}$ and $\mathcal{M}$ are defined by

$$\mathcal{L} = \left(\frac{\partial^2}{\partial x^2} + \nu\gamma^2\right)\frac{\partial}{\partial y},$$

$$\mathcal{M} = \frac{\partial}{\partial x}\left(\frac{\partial^2}{\partial x^2} + (2 - \nu)\gamma^2\right)\frac{\partial}{\partial y},$$

ν being the Poisson's ratio of the material of the ice-cover. Other forms of boundary conditions representing fixed edges to the ice-cover can also be taken. However, this has not been considered here. The boundary condition on the half-immersed cylinder is

$$\frac{\partial \phi}{\partial r} = 0 \text{ on } r = a, \quad -\pi/2 < \theta < \pi/2. \tag{2.6}$$

Also condition at the large depth is

$$\nabla \phi \to 0 \text{ as } y \to \infty, \tag{2.7}$$

and at infinite distance from the cylinder,

$$\phi(x, y) \sim \begin{cases} T\phi_0(x, y) & \text{as } x \to \infty, \\ \phi_0(x, y) + R\phi_0(-x, y) & \text{as } x \to -\infty, \end{cases} \tag{2.8}$$

where T and R respectively denote the transmission and reflection coefficients and are unknown complex constants to be determined.

Solution

Let $\psi(x, y)$ denote the source function satisfying the Helmholtz's equation (2.3) except at the origin with boundary conditions (2.4) on $y = 0, -\infty < x < \infty$, and the additional condition that it behaves as an outgoing wave as $|x| \to \infty$. Following Thorne (1953) one can construct $\psi(x, y)$ as

$$\psi(x, y) = \int_{\gamma}^{\infty} A(k) e^{-ky} \cos\left((k^2 - \gamma^2)^{\frac{1}{2}} x \right) dk, \tag{2.9}$$

where $A(k)$ is a function of k to be found such that the integral exists in some sense. The ice-cover condition is satisfied if $A(k)$ is chosen as

$$A(k) = 2 \frac{k(Dk^4 + 1 - \varepsilon K)}{(k^2 - \gamma^2)^{\frac{1}{2}} \{k(Dk^4 + 1 - \varepsilon K) - K\}}. \tag{2.10}$$

It is noted that $A(k)$ has a pole on the real k-axis at $k = \lambda$, and thus we take

$$\psi(x, y) = 2 \int_{\gamma}^{\infty} \frac{k(Dk^4 + 1 - \varepsilon K)}{(k^2 - \gamma^2)^{\frac{1}{2}} \{k(Dk^4 + 1 - \varepsilon K) - K\}} e^{-ky} \cos\left((k^2 - \gamma^2)^{\frac{1}{2}} x \right) dk, \tag{2.11}$$

where the contour is indented below the pole at $k = \lambda$ on the k-axis to account for the outgoing nature of $\psi(x, y)$ as $|x| \to \infty$.

Using the well-known result

$$e^{\frac{1}{2} X (P + P^{-1})} = \sum_{m=0}^{\infty} \frac{1}{2} \varepsilon_m (P^m + P^{-m}) I_m(X) \tag{2.12}$$

where $\varepsilon_0 = 0, \varepsilon_m = 2, m \geq 1$ and $I_m(X)$ is the modified Bessel function of first kind, (2.11) can be expanded in terms of polar co-ordinates as

$$\psi(x, y) = \sum_{m=0}^{\infty} A_m I_m(\gamma r) \cos m\theta \tag{2.13}$$

where

$$A_m = 2\int_{\gamma}^{\infty} (-1)^m \varepsilon_m \frac{k(Dk^4 + 1 - \varepsilon K)}{(k^2 - \gamma^2)^{\frac{1}{2}} \{k(Dk^4 + 1 - \varepsilon K) - K\}} \cosh m\eta dk, \tag{2.14}$$

with $k = \gamma \cosh \eta$. The wave dipole is defined as

$$\frac{\partial \psi}{\partial x} = -2\int_{\gamma}^{\infty} \frac{k(Dk^4 + 1 - \varepsilon K)}{k(Dk^4 + 1 - \varepsilon K) - K} e^{-ky} \sin\left((k^2 - \gamma^2)^{\frac{1}{2}} x \right) dk. \tag{2.15}$$

Using (2.12), it can be expanded in terms of polar co-ordinates as

$$\frac{\partial \psi}{\partial x} = \sum_{m=0}^{\infty} B_m I_m(\gamma r) \sin m\theta \tag{2.16}$$

where

$$B_m = 2i(-1)^m \varepsilon_m \int_{\gamma}^{\infty} \frac{k(Dk^4 + 1 - \varepsilon K)}{k(Dk^4 + 1 - \varepsilon K) - K} \sinh m\eta dk. \tag{2.17}$$

The incident wave potential (2.2) has the expansion

$$\phi_0(x, y) = \sum_{m=0}^{\infty} (-1)^m \varepsilon_m I_m(\gamma r)[\cosh m\eta_\lambda \cos m\theta - i \sinh m\eta_\lambda \sin m\theta] \tag{2.18}$$

with $\cosh \eta_\lambda = \lambda/\gamma = \dfrac{1}{\sin \alpha}$.

To solve the scattering problem we express the potential function describing the ensuing fluid motion as Ursell (1968)

$$\varphi(x, y) = \varphi_0(x, y) - \frac{ic_1}{2\pi\lambda S(\lambda)} \cos \alpha \psi(x, y) + \frac{c_2}{2\pi\lambda^2 S(\lambda)} \psi_x(x, y) + Ae^{-\lambda y + i\mu x} + Be^{-\lambda y - i\mu x}$$
$$+2\sum_{m=1}^{\infty} p_m \frac{(\gamma a/2)^m}{(m-1)!} \chi_m^{(s)}(x, y) + 2\sum_{m=1}^{\infty} q_m \frac{(\gamma a/2)^m}{(m-1)!} \chi_m^{(a)}(x, y) + \begin{cases} a_1\varphi_1(x, y) + a_2\varphi_2(x, y), & x > 0 \\ a_3\varphi_3(x, y) + a_4\varphi_4(x, y), & x < 0, \end{cases} \tag{2.19}$$

where $c_1, c_2 A, B, p_m, q_m$ $(m = 1,2,3,...)$ and a_i's $(i = 1,2,3,4)$ are unknown constants to be determined, and

$$\chi_m^{(s)}(x, y) = K_{m-2}(\gamma r)\cos(m-2)\theta + K_m(\gamma r)\cos m\theta + \frac{2}{\sin \alpha} K_{m-1}(\gamma r)\cos(m-1)\theta$$
$$+\frac{2(-1)^m}{\lambda \sin \alpha} \int_0^{\infty} \frac{\left(K + \upsilon(D\upsilon^4 + 1 - \varepsilon K)\right)\cosh(m-1)k}{D(\upsilon^4 + \lambda\upsilon^3 + \lambda^2\upsilon^2 + \lambda^3\upsilon + \lambda^4) + 1 - \varepsilon K} \cos(\gamma x \sinh k) e^{-\upsilon y} dk, \tag{2.20}$$

$$\chi_m^{(a)}(x,y) = K_{m-1}(\gamma r)\sin(m-1)\theta + K_{m+1}(\gamma r)\sin(m+1)\theta + \frac{2}{\sin\alpha}K_m(\gamma r)\sin m\theta + \frac{2(-1)^{(m-1)}}{\lambda\sin\alpha}$$

$$\int_0^\infty \frac{\left(K + \upsilon(D\upsilon^4 + 1 - \varepsilon K)\right)\sinh mk}{D(\upsilon^4 + \lambda\upsilon^3 + \lambda^2\upsilon^2 + \lambda^3\upsilon + \lambda^4) + 1 - \varepsilon K}\sin(\gamma x \sinh k)e^{-\upsilon y}dk, \tag{2.21}$$

where $\upsilon = \gamma\cosh k$, and

$$S(\lambda) = \frac{D\lambda^4 + 1 - \varepsilon K}{5D\lambda^4 + 1 - \varepsilon K}, \tag{2.22}$$

$$\phi_1(x,y) = e^{-\lambda_1 y + i(\lambda_1^2 - \gamma^2)^{\frac{1}{2}}x}, \quad \phi_3(x,y) = e^{-\lambda_1 y - i(\lambda_1^2 - \gamma^2)^{\frac{1}{2}}x}, \tag{2.23}$$

$$\phi_2(x,y) = e^{-\overline{\lambda_1}y - i(\overline{\lambda_1^2} - \gamma^2)^{\frac{1}{2}}x}, \quad \phi_4(x,y) = e^{-\overline{\lambda_1}y + i(\overline{\lambda_1^2} - \gamma^2)^{\frac{1}{2}}x}, \tag{2.24}$$

and $K_n(z)$ is the modified Bessel function of second kind. Here $\chi_m^{(s),(a)}(x,y)$ are wave-free potentials and have been constructed so as to satisfy the ice-cover condition (2.4) on $y = 0, \ -\infty < x < \infty$.

Comparing (2.19) with (2.8) as $|x| \to \infty$ we find that

$$T = 1 + c_1 - c_2, \tag{2.25}$$

$$R = c_1 + c_2, \tag{2.26}$$

$$A = B = 0. \tag{2.27}$$

Also using the conditions (2.5a) and (2.5b) we get

$$[a_1\mathcal{L}\varphi_1(x,y) + a_2\mathcal{L}\varphi_2(x,y) + \mathcal{L}(E(x,y))]_{x=a,y=0} = 0, \tag{2.28}$$

$$[a_1\mathcal{M}\varphi_1(x,y) + a_2\mathcal{M}\varphi_2(x,y) + \mathcal{M}(E(x,y))]_{x=a,y=0} = 0, \tag{2.29}$$

$$[a_3\mathcal{L}\varphi_3(x,y) + a_4\mathcal{L}\varphi_4(x,y) + \mathcal{L}(E(x,y))]_{x=-a,y=0} = 0, \tag{2.30}$$

$$[a_3\mathcal{M}\varphi_3(x,y) + a_4\mathcal{M}\varphi_4(x,y) + \mathcal{M}(E(x,y))]_{x=-a,y=0} = 0, \tag{2.31}$$

where

$$E(x,y) = \varphi_0(x,y) - \frac{ic_1}{2\pi\lambda S(\lambda)}\cos\alpha\psi(x,y) + \frac{c_2}{2\pi\lambda^2 S(\lambda)}\psi_x(x,y)$$

$$+2\sum_{m=1}^{\infty} p_m \frac{(\gamma a/2)^m}{(m-1)!}\chi_m^{(s)}(x,y) + 2\sum_{m=1}^{\infty} q_m \frac{(\gamma a/2)^m}{(m-1)!}\chi_m^{(a)}(x,y). \tag{2.32}$$

Solving the equations (2.28), (2.29) we obtain the unknowns a_1 and a_2 in terms of the unknowns c_1, c_2, p_m, q_m and similarly solving the equations (2.30) and (2.31), we find that the unknowns a_3 and a_4 in terms of the unknowns c_1, c_2, p_m, q_m.

The polar expansions of $\chi_m^{(s)}, \chi_m^{(a)}, \phi_i(x,y)'s\ (i=1,2,3,4)$ are given by

$$\chi_m^{(s)}(x,y) = K_{m-2}(\gamma r)\cos(m-2)\theta + K_m(\gamma r)\cos m\theta$$
$$+\frac{2}{\sin\alpha}K_{m-1}(\gamma r)\cos(m-1)\theta + \frac{2(-1)^m}{\lambda\sin\alpha}\sum_{s=0}^{\infty}\Theta_{ms}^{(1)}I_s(\gamma r)\cos s\theta, \tag{2.33}$$

$$\chi_m^{(a)}(x,y) = K_{m-1}(\gamma r)\sin(m-1)\theta + K_{m+1}(\gamma r)\sin(m+1)\theta$$
$$+\frac{2}{\sin\alpha}K_m(\gamma r)\sin m\theta + \frac{2(-1)^{(m-1)}}{\lambda\sin\alpha}\sum_{s=0}^{\infty}\Theta_{ms}^{(2)}I_s(\gamma r)\sin s\theta, \tag{2.34}$$

$$\varphi_1(x,y),\ \varphi_3(x,y) = \sum_{m=0}^{\infty}(-1)^m\varepsilon_m I_m(\gamma r)\left[\cosh m\eta_{\lambda_1}\cos m\theta \mp i\sinh m\eta_{\lambda_1}\sin m\theta\right], \tag{2.35}$$

$$\varphi_2(x,y),\ \varphi_4(x,y) = \sum_{m=0}^{\infty}(-1)^m\varepsilon_m I_m(\gamma r)\left[\cosh m\eta_{\overline{\lambda}_1}\cos m\theta \pm i\sinh m\eta_{\overline{\lambda}_1}\sin m\theta\right], \tag{2.36}$$

where

$$\Theta_{ms}^{(1)} = (-1)^s\varepsilon_s\int_0^{\infty}\frac{\left(K+\upsilon(D\upsilon^4+1-\varepsilon K)\right)}{D(\upsilon^4+\lambda\upsilon^3+\lambda^2\upsilon^2+\lambda^3\upsilon+\lambda^4)+1-\varepsilon K}\times\cosh(m-1)k\cosh skdk, \tag{2.37}$$

$$\Theta_{ms}^{(2)} = i(-1)^s\varepsilon_s\int_0^{\infty}\frac{\left(K+\upsilon(D\upsilon^4+1-\varepsilon K)\right)}{D(\upsilon^4+\lambda\upsilon^3+\lambda^2\upsilon^2+\lambda^3\upsilon+\lambda^4)+1-\varepsilon K}\sinh mk\sinh skdk, \tag{2.38}$$

and $\cosh\eta_{\lambda_1} = \dfrac{\lambda_1}{\gamma}$, $\cosh\eta_{\overline{\lambda}_1} = \dfrac{\overline{\lambda}_1}{\gamma}$.

Using the polar expansions (2.13), (2.16), (2.18), (2.33) to (2.36) in the body boundary condition $\dfrac{\partial\phi}{\partial r} = 0$ on $r = a$, we obtain

$$\left[\left\langle\frac{\partial}{\partial r}E(x,y)\right\rangle\right]_{-\frac{\pi}{2}\le\theta\le\frac{\pi}{2}} + \left\{\begin{array}{l}\left[a_1\left\langle\frac{\partial}{\partial r}\phi_1(x,y)\right\rangle + a_2\left\langle\frac{\partial}{\partial r}\phi_2(x,y)\right\rangle\right]_{0\le\theta\le\frac{\pi}{2}}\\ \left[a_3\left\langle\frac{\partial}{\partial r}\phi_3(x,y)\right\rangle + a_4\left\langle\frac{\partial}{\partial r}\phi_4(x,y)\right\rangle\right]_{-\frac{\pi}{2}\le\theta\le 0}\end{array}\right. = 0 \tag{2.39}$$

where the angular brackets ⟨⟩ indicate that r is to be put equal to a. Multiplying both sides of (2.39) by $\cos n\theta$ or $\sin n\theta$ $(n = 0,1,2,...)$ and integrating with respect to θ between $-\pi/2$ to $\pi/2$, we obtain two infinite systems of linear equations involving the unknowns and these are given by

$$
\begin{aligned}
&-c_1 \frac{i\cos\alpha}{2\pi\lambda S(\lambda)} \sum_{m=0}^{\infty} A_m I'_m(\gamma a) M_{m,n} \\
&+2\sum_{m=1}^{\infty} p_m \frac{(\gamma a/2)^m}{(m-1)!}\left[K'_{m-2}(\gamma a)M_{m-2,n} + K'_m(\gamma a)M_{m,n} + \frac{2}{\sin\alpha}K'_{m-1}(\gamma a)M_{m-1,n} + \frac{2(-1)^m}{\lambda\sin\alpha}\sum_{s=0}^{\infty}\Theta^{(1)}_{ms} I'_s(\gamma a) M_{s,n}\right] \\
&= -\sum_{m=0}^{\infty}(-1)^m \varepsilon_m I'_m(\gamma a)\left[\cosh m\eta_\lambda M_{m,n} + (a_1\cosh m\eta_{\lambda_1} + a_2\cosh m_{\bar{\lambda}_1}) I^{(1)}_{m,n} + (a_3 \cosh m\eta_{\lambda_1} + a_4\cosh m\eta_{\bar{\lambda}_1}) I^{(2)}_{m,n}\right],
\end{aligned}
$$

$$n = 0,1,2,..., \qquad (2.40)$$

and

$$
\begin{aligned}
&c_2 \frac{1}{2\pi\lambda^2 S(\lambda)} \sum_{m=0}^{\infty} B_m I'_m(\gamma a) N_{m,n} \\
&+2\sum_{m=1}^{\infty} q_m \frac{(\gamma a/2)^m}{(m-1)!}\left[K'_{m+1}(\gamma a)N_{m+1,n} + K'_{m-1}(\gamma a)N_{m-1,n} + \frac{2}{\sin\alpha}K'_m(\gamma a)N_{m,n}\right. \\
&\qquad\qquad\left. + \frac{2(-1)^{(m-1)}}{\lambda\sin\alpha}\sum_{s=0}^{\infty}\Theta^{(2)}_{ms} I'_s(\gamma a) N_{s,n}\right] \\
&= i\sum_{m=0}^{\infty}(-1)^m \varepsilon_m I'_m(\gamma a)\left[\sinh(m)\eta_\lambda N_{m,n} + (a_1\sinh(m)\eta_{\lambda_1} - a_2\sinh(m)\eta_{\bar{\lambda}_1}) I^{(3)}_{m,n}\right. \\
&\qquad\qquad\left. + (-a_3\sinh(m)\eta_{\lambda_1} + a_4\sinh(m)\eta_{\bar{\lambda}_1}) I^{(4)}_{m,n}\right],
\end{aligned}
$$

$$n = 0,1,2,..., \qquad (2.41)$$

where

$$M_{m,n} = \int_{-\pi/2}^{\pi/2} \cos m\theta \cos n\theta d\theta, \qquad N_{m,n} = \int_{-\pi/2}^{\pi/2} \sin m\theta \sin n\theta d\theta,$$

$$I^{(1)}_{m,n} = \int_0^{\pi/2} \cos m\theta \cos n\theta d\theta, \qquad I^{(2)}_{m,n} = \int_{-\pi/2}^{0} \cos m\theta \cos n\theta d\theta,$$

$$I^{(3)}_{m,n} = \int_0^{\pi/2} \sin m\theta \sin n\theta d\theta, \qquad I^{(4)}_{m,n} = \int_{-\pi/2}^{0} \sin m\theta \sin n\theta d\theta,$$

Now substituting the constants $a_i's$ $(i = 1,2,3,4)$ in (2.40) and (2.41) and using the relations

$$[\mathcal{L}\,\varphi_{1,2}(x,y)]_{\theta=\pi/2} = [\mathcal{L}\,\varphi_{3,4}(x,y)]_{\theta=-\pi/2}, \; [\mathcal{M}\,\varphi_{1,2}(x,y)]_{\theta=\pi/2} = -[\mathcal{M}\,\varphi_{3,4}(x,y)]_{\theta=-\pi/2},$$

$$[\mathcal{L}\,\psi(x,y)]_{\theta=\pi/2} = [\mathcal{L}\,\psi(x,y)]_{\theta=-\pi/2}, \; [\mathcal{M}\,\psi(x,y)]_{\theta=\pi/2} = -[\mathcal{M}\,\psi(x,y)]_{\theta=-\pi/2},$$

$$[\mathcal{L}\,\psi_x(x,y)]_{\theta=\pi/2} = -[\mathcal{L}\,\psi_x(x,y)]_{\theta=-\pi/2}, [\mathcal{M}\,\psi_x(x,y)]_{\theta=\pi/2} = [\mathcal{M}\,\psi_x(x,y)]_{\theta=-\pi/2},$$

we obtain two infinite systems of linear equations involving $c_1, p_1, p_2, \ldots$ and $c_2, q_1, q_2, \ldots$, etc. respectively and these are given by

$$-c_1 \frac{i\cos\alpha}{2\pi\lambda S(\lambda)} \sum_{m=0}^{\infty} \left[A_m + (-)^m \varepsilon_m \frac{1}{\Delta} F_m^{(1)} \right] I_m'(\gamma a) M_{m,n} + \frac{1}{\Delta} \sum_{m=0}^{\infty} (-1)^m \varepsilon_m I_m'(\gamma a) F_m^{(2)} M_{m,n}$$

$$+ 2\sum_{m=1}^{\infty} p_m \frac{(\gamma a/2)^m}{(m-1)!} \left(K_{m-2}'(\gamma a) M_{m-2,n} + K_m'(\gamma a) M_{m,n} + \frac{2}{\sin\alpha} K_{m-1}'(\gamma a) M_{m-1,n} \right.$$

$$\left. + \frac{2(-1)^m}{\lambda \sin\alpha} \sum_{s=0}^{\infty} \Theta_{ms}^{(1)} I_s'(\gamma a) M_{s,n} \right)$$

$$= -\sum_{m=0}^{\infty} \left[\cosh m\eta_\lambda + \frac{1}{2\Delta} F_m^3 \right] (-1)^m \varepsilon_m I_m'(\gamma a) M_{m,n}, n = 0,1,2,\ldots, \quad (2.42)$$

and

$$c_2 \frac{1}{2\pi\lambda^2 S(\lambda)} \sum_{m=0}^{\infty} \left[B_m - i(-1)^m \varepsilon_m \frac{1}{\Delta} F_m^{(4)} \right] I_m'(\gamma a) N_{m,n} - \frac{i}{\Delta} \sum_{m=0}^{\infty} (-1)^m \varepsilon_m I_m'(\gamma a) F_m^{(5)} N_{m,n}$$

$$+ 2\sum_{m=1}^{\infty} q_m \frac{(\gamma a/2)^m}{(m-1)!} \left(K_{m+1}'(\gamma a) N_{m+1,n} + K_{m-1}'(\gamma a) N_{m-1,n} + \frac{2}{\sin\alpha} K_m'(\gamma a) N_{m,n} \right.$$

$$\left. + \frac{2(-1)^{(m-1)}}{\lambda \sin\alpha} \sum_{s=0}^{\infty} \Theta_{ms}^{(2)} I_s'(\gamma a) N_{s,n} \right)$$

$$= i\sum_{m=0}^{\infty} \left[\sinh(m)\eta_\lambda + \frac{1}{2\Delta} F_m^{(6)} \right] (-1)^m \varepsilon_m I_m'(\gamma a) N_{m,n}, n = 0,1,2,\ldots, \quad (2.43)$$

where

$$\Delta = [\mathcal{L}(\varphi_2)\mathcal{M}(\varphi_2) - \mathcal{L}(\varphi_1)\mathcal{M}(\varphi_1)]_{\theta=\pi/2},\ F_m^{(1)} = \mathcal{G}_1 \cosh m\eta_{\lambda_1} + \mathcal{G}_2 \cosh m\eta_{\bar{\lambda}_1},$$

$$F_m^{(2)} = \mathcal{H}_1 \cosh m\eta_{\lambda_1} + \mathcal{H}_2 \cosh m\eta_{\bar{\lambda}_1},\ F_m^{(3)} = \mathcal{V}_1 \cosh m\eta_{\lambda_1} + \mathcal{V}_2 \cosh m\eta_{\bar{\lambda}_1},$$

$$F_m^{(4)} = \mathcal{G}_3 \sinh(m)\eta_{\lambda_1} + \mathcal{G}_4 \sinh(m)\eta_{\bar{\lambda}_1},\ F_m^{(5)} = \mathcal{H}_3 \sinh(m)\eta_{\lambda_1} + \mathcal{H}_4 \sinh(m)\eta_{\bar{\lambda}_1},$$

$$F_m^{(6)} = \mathcal{V}_3 \sinh(m)\eta_{\lambda_1} + \mathcal{V}_4 \sinh(m)\eta_{\bar{\lambda}_1},$$

with

$$\mathcal{G}_1 = [\mathcal{L}(\varphi_2)\mathcal{M}(\psi) - \mathcal{L}(\varphi_2)\mathcal{M}(\psi)]_{\theta=\pi/2},\ \mathcal{G}_2 = [-\mathcal{L}(\varphi_1)\mathcal{M}(\psi) + \mathcal{L}(\varphi_1)\mathcal{M}(\psi)]_{\theta=\pi/2},$$

$$\mathcal{G}_3 = [\mathcal{L}(\varphi_2)\mathcal{M}(\psi_x) - \mathcal{M}(\varphi_2)\mathcal{L}(\psi_x)]_{\theta=\pi/2},\ \mathcal{G}_4 = [\mathcal{L}(\varphi_1)\mathcal{M}(\psi_x) - \mathcal{M}(\varphi_1)\mathcal{L}(\psi_x)]_{\theta=\pi/2},$$

$$\mathcal{H}_1 = 2\left[\mathcal{L}(\varphi_2) \sum_{k=1}^{\infty} p_k \frac{(\gamma a/2)^k}{(k-1)!} \mathcal{M}(\chi_k^{(s)}) - \mathcal{M}(\varphi_2) \sum_{k=1}^{\infty} p_k \frac{(\gamma a/2)^k}{(k-1)!} \mathcal{L}(\chi_k^{(s)}) \right]_{\theta=\pi/2}$$

$$\mathcal{H}_2 = 2\left[\mathcal{M}(\varphi_1)\sum_{k=1}^{\infty} p_k \frac{(\gamma a/2)^k}{(k-1)!}\mathcal{L}(\chi_k^{(s)}) - \mathcal{L}(\varphi_1)\sum_{k=1}^{\infty} p_k \frac{(\gamma a/2)^k}{(k-1)!}\mathcal{M}(\chi_k^{(s)})\right]_{\theta=\pi/2},$$

$$\mathcal{H}_3 = 2\left[\mathcal{L}(\varphi_2)\sum_{k=1}^{\infty} q_k \frac{(\gamma a/2)^k}{(k-1)!}\mathcal{M}(\chi_k^{(a)}) - \mathcal{M}(\varphi_2)\sum_{k=1}^{\infty} q_k \frac{(\gamma a/2)^k}{(k-1)!}\mathcal{L}(\chi_k^{(a)})\right]_{\theta=\pi/2},$$

$$\mathcal{H}_4 = 2\left[\mathcal{L}(\varphi_1)\sum_{k=1}^{\infty} q_k \frac{(\gamma a/2)^k}{(k-1)!}\mathcal{M}(\chi_k^{(a)}) - \mathcal{M}(\varphi_1)\sum_{k=1}^{\infty} q_k \frac{(\gamma a/2)^k}{(k-1)!}\mathcal{L}(\chi_k^{(a)})\right]_{\theta=\pi/2},$$

$$\mathcal{V}_1 = \{([\mathcal{M}(\varphi_0)]_{\theta=\pi/2} - [\mathcal{M}(\varphi_0)]_{\theta=-\pi/2})[\mathcal{L}(\varphi_2)]_{\theta=\pi/2}$$
$$-([\mathcal{L}(\varphi_0)]_{\theta=\pi/2} + [\mathcal{L}(\varphi_0)]_{\theta=-\pi/2})[\mathcal{M}(\varphi_2)]_{\theta=\pi/2}\},$$
$$\mathcal{V}_2 = \{([\mathcal{L}(\varphi_0)]_{\theta=\pi/2} + [\mathcal{L}(\varphi_0)]_{\theta=-\pi/2})[\mathcal{M}(\varphi_1)]_{\theta=\pi/2}$$
$$-([\mathcal{M}(\varphi_0)]_{\theta=\pi/2} - [\mathcal{M}(\varphi_0)]_{\theta=-\pi/2})[\mathcal{L}(\varphi_1)]_{\theta=\pi/2}\},$$
$$\mathcal{V}_3 = \{([\mathcal{M}(\varphi_0)]_{\theta=\pi/2} + [\mathcal{M}(\varphi_0)]_{\theta=-\pi/2})[\mathcal{L}(\varphi_2)]_{\theta=\pi/2}$$
$$-([\mathcal{L}(\varphi_0)]_{\theta=\pi/2} - [\mathcal{L}(\varphi_0)]_{\theta=-\pi/2})[\mathcal{M}(\varphi_2)]_{\theta=\pi/2}\},$$
$$\mathcal{V}_4 = \{([\mathcal{M}(\varphi_0)]_{\theta=\pi/2} + [\mathcal{M}(\varphi_0)]_{\theta=-\pi/2})[\mathcal{L}(\varphi_1)]_{\theta=\pi/2}$$
$$-([\mathcal{L}(\varphi_0)]_{\theta=\pi/2} - [\mathcal{L}(\varphi_0)]_{\theta=-\pi/2})[\mathcal{M}(\varphi_1)]_{\theta=\pi/2}\},$$

The systems (2.42) and (2.43) can be solved numerically by truncating these to an $N \times N$ systems and increasing N until the solution converges to some required degree of accuracy.

Thus the constants c_1 and c_2 appearing in (2.25) and (2.26) for obtaining the transmission and reflection coefficients can be obtained numerically by solving the linear systems (2.42) and (2.43) after truncation. Here these systems are truncated up to five terms only. It has been verified that the errors in the numerical results are of the order of 10^{-6}.

Numerical results

In Figs. 6.1–6.10 the reflection and transmission coefficients $|R|$ and $|T|$ are plotted against the wave number Ka. In all the plots, $\varepsilon/a = 0.01$ and $\nu = 0.3$. The incident wave angle α is taken as 15°, 30°, 45°, 60°, 75° in the different figures. The different curves in each figure correspond to different values of D/a^4, which are taken as 0.05, 0.1, 0.5, and 1.5.

From these figures it is observed that as the flexural rigidity parameter (D/a^4) increases, $|R|$ increases while $|T|$ decreases. Also the Figs. 6.1, 6.3, 6.5, 6.7, and 6.9 show that $|R|$ first increases with Ka, then oscillates in an irregular manner and finally decreases slowly as Ka further increases for any D/a^4. Similarly Figs. 6.2, 6.4, 6.6, 6.8 and 6.10 describe the behaviour of $|T|$ which is complementary to the behaviour of $|R|$. This is obvious due to the energy identity $|R|^2 + |T|^2 = 1$.

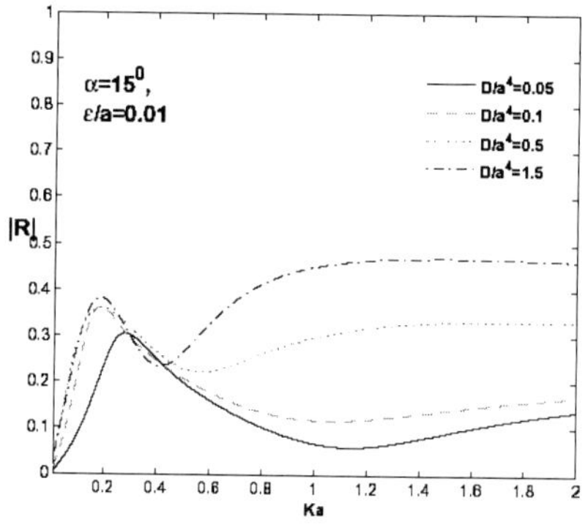

Fig. 6.1 Reflection coefficient against the wave number *Ka*

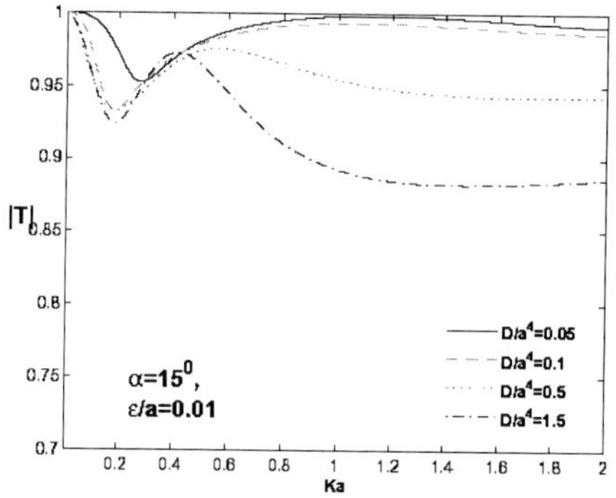

Fig. 6.2 Transmission coefficient against the wave number *Ka*

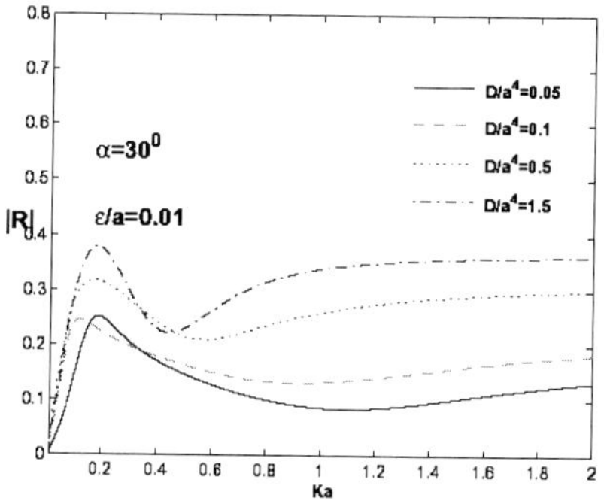

Fig. 6.3 Reflection coefficient against the wave number *Ka*

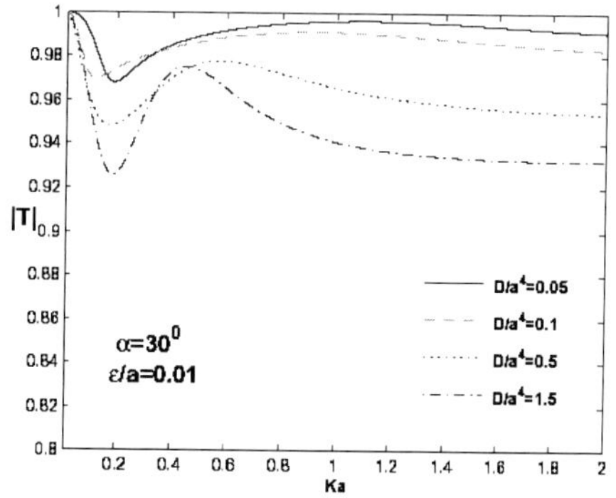

Fig. 6.4 Transmission coefficient against the wave number *Ka*

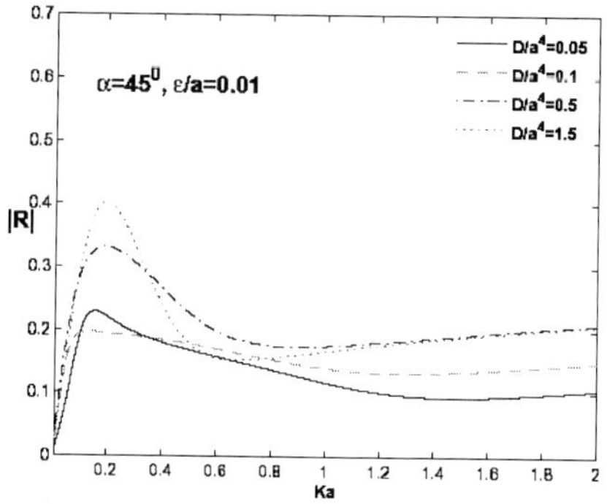

Fig. 6.5 Reflection coefficient against the wave number *Ka*

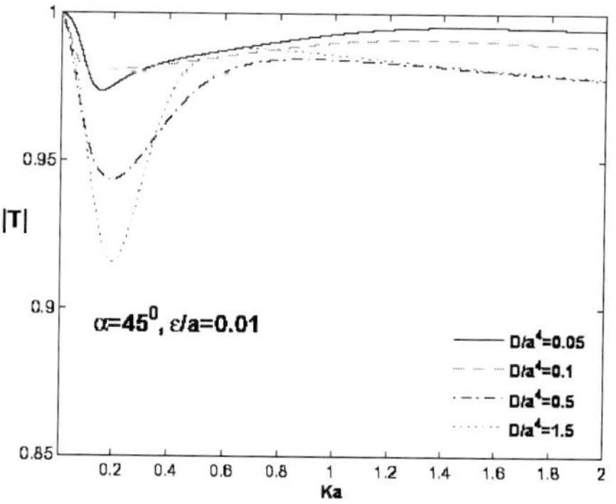

Fig. 6.6 Transmission coefficient against the wave number *Ka*

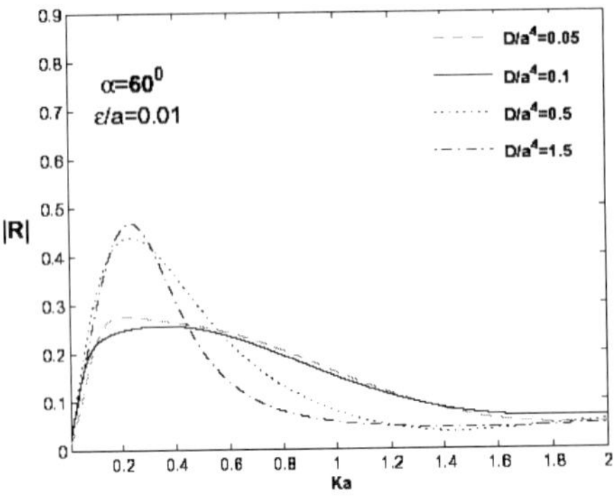

Fig. 6.7 Reflection coefficient against the wave number *Ka*

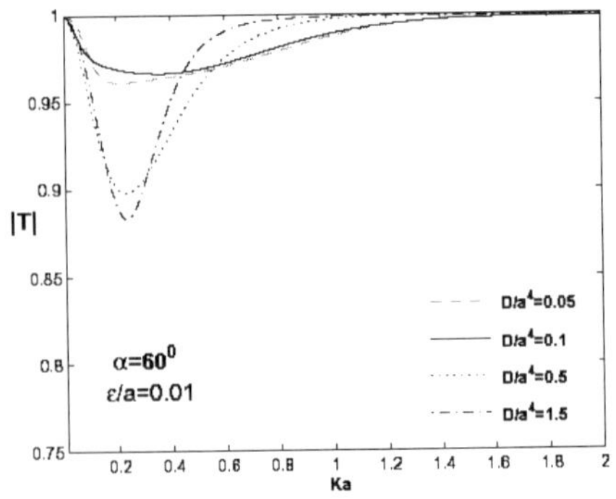

Fig. 6.8 Transmission coefficient against the wave number *Ka*

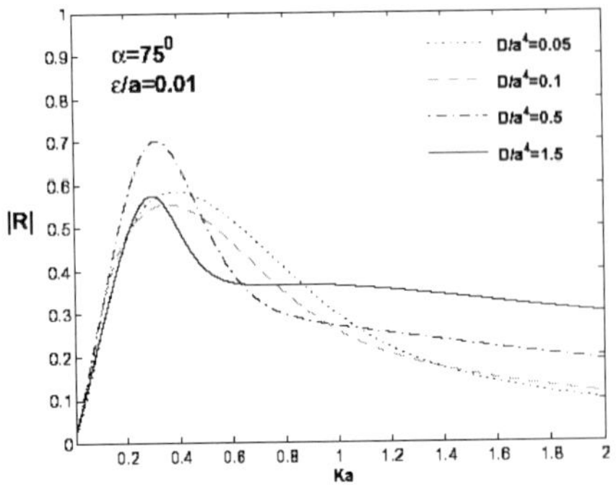

Fig. 6.9 Reflection coefficient against the wave number *Ka*

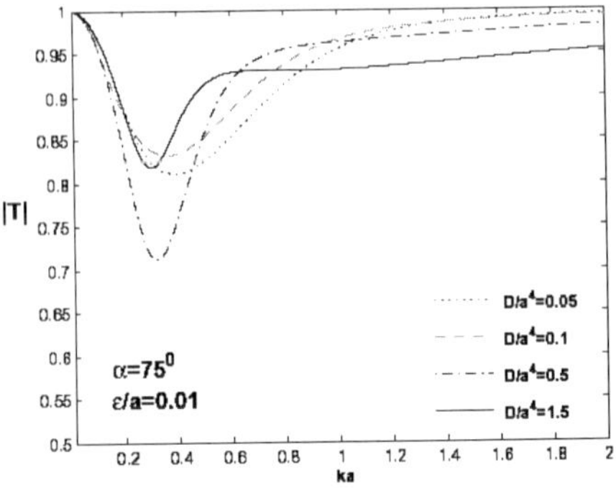

Fig. 6.10 Transmission coefficient against the wave number *Ka*

Figures 6.11, 6.12 depict $|R|$ and $|T|$ against Ka for $\varepsilon/a = 0.01, D/a^4 = 1, \nu = 0.3$. The different curves correspond to different values of the angle of incidence α, viz, α = 5°, 15°, 30°, 45°. Figures 6.13, 6.14 depict $|R|$ and $|T|$ against Ka for

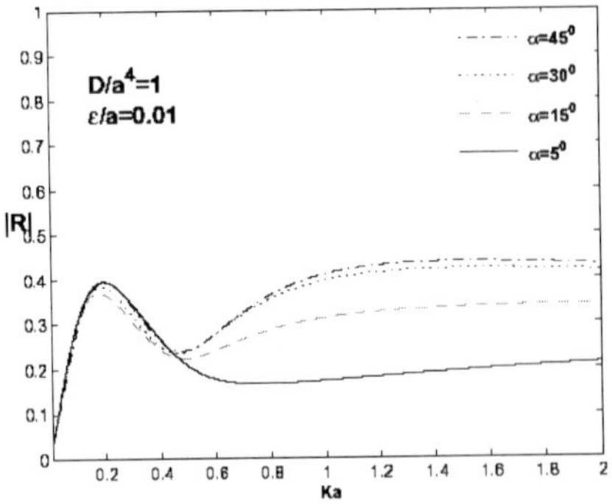

Fig. 6.11 Reflection coefficient against the wave number *Ka*

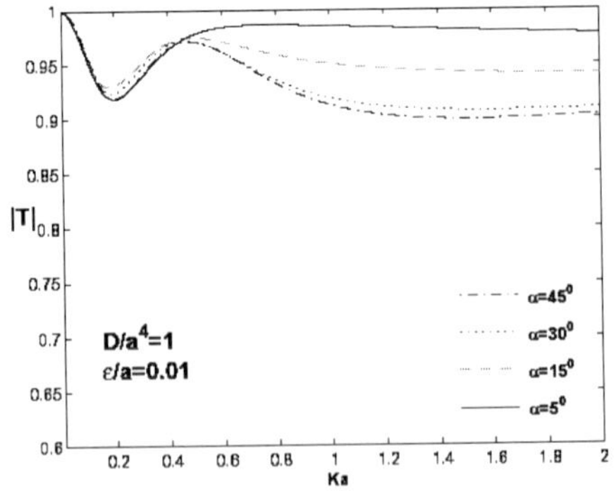

Fig. 6.12 Transmission coefficient against the wave number *Ka*

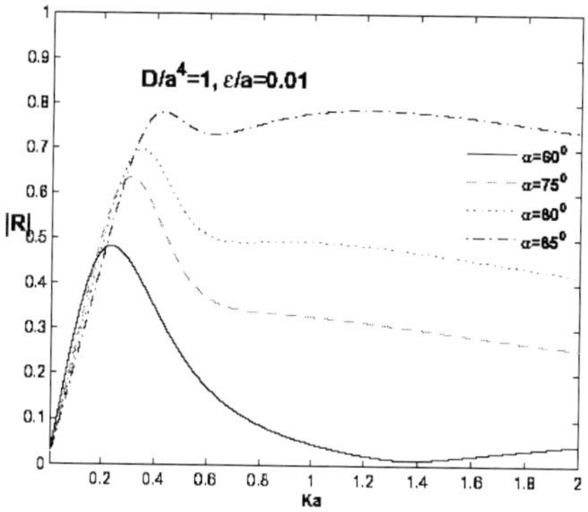

Fig. 6.13 Reflection coefficient against the wave number *Ka*

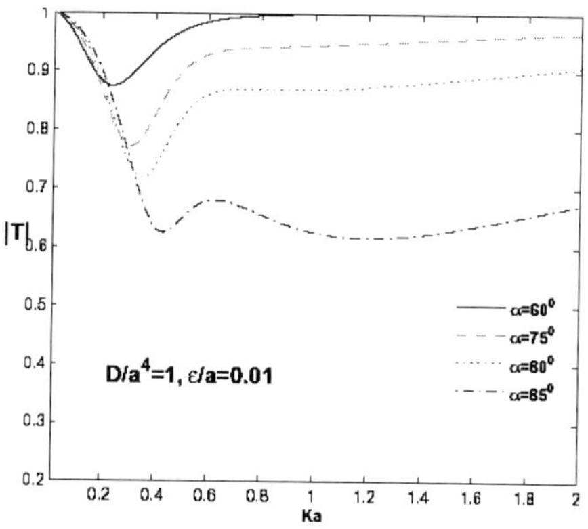

Fig. 6.14 Transmission coefficient against the wave number *Ka*

$\varepsilon/a = 0.01, D/a^4 = 1, \nu = 0.3$. The different curves correspond to different values of the angle of incidence α, viz, α = 60°, 75°, 80°, 85°. It is observed that as the angle of incidence increases, $|R|$ increases while $|T|$ decreases. Also due to the presence of ice-cover $|R|$ gets enlarged and $|T|$ gets reduced. Also it is observed that $|R|$ and $|T|$ are oscillatory in nature. This may be attributed due to interaction of the incident wave train between the boundary of the circular cylinder and the ice-cover. All the numerical results for $|R|$ and $|T|$ have been checked for their correctness by the energy identity.

6.3 Scattering by a circular cylinder submerged beneath an ice-cover

There is a classical result in the linearized theory of water waves which states that a long circular cylinder with horizontal axis completely submerged in deep water with a free surface, experiences no reflection Dean (1948), Ursell (1950), Ogilive (1963) when a train of surface water waves is normally incident on it. However, for oblique incidence of the wave train, the cylinder does experience reflection Levine (1965). It is shown in this section that the same phenomenon also holds good when the deep water has an ice-cover instead of a free surface, the ice-cover being modelled as a thin elastic plate, as has been shown by Das and Mandal (2006).

Mathematical formulation

Let a circular cylinder represented by $r = (x^2 + y^2)^{\frac{1}{2}} = a$ be submerged fully in an infinitely deep water occupying the region $y \geq -h,\ -\infty < x < \infty,\ -\infty < z < \infty,\ r \geq a$, where $h(> a)$ is the depth of submergence of the centre of a cross-section of the cylinder, a its radius, and y-axis is taken vertically downwards. On the surface of water floats a very thin ice-sheet which is modeled as an elastic plate. Let a harmonically time-dependent train of waves described by the potential function $\mathrm{Re}\left\{\phi_0(x, y)e^{-i\sigma t + i\gamma z}\right\}$ be obliquely incident on the fixed circular cylinder. Here $\gamma = \lambda \sin \alpha$, where α is the angle of incidence of the wave train and σ is the angular frequency, and λ is the unique positive real root of the dispersion equation

$$k(Dk^4 + 1 - \varepsilon K) = K. \tag{3.1}$$

In (3.1) $K = \sigma^2/g$, g being the acceleration due to gravity, D is the flexural rigidity of the ice-cover and $\varepsilon = \frac{\rho_0}{\rho_1} h_0$, ρ_0 and ρ_1 are the densities of ice and water, respectively and h_0 is the very small thickness of the ice-cover. The other roots of (3.1) are $\lambda_1, \overline{\lambda_1}$ and $\lambda_2, \overline{\lambda_2}$ where $\mathrm{Re}(\lambda_1) > 0$ and $\mathrm{Re}(\lambda_2) < 0$, and

$$\varphi_0(x, y) = e^{-\lambda y + i\mu x} \tag{3.2}$$

with $\mu = \lambda \cos\alpha$.

The velocity potential function describing the resulting motion in the water can be represented by $\mathrm{Re}\{\phi(x, y)e^{-i\sigma t + i\gamma z}\}$, where the complex valued potential function $\phi(x, y)$ satisfies

$$\left(\nabla^2 - \gamma^2\right)\varphi = 0, \qquad r > a, \quad y > -h, \quad -\infty < x < \infty, \tag{3.3}$$

$$\left(D\left(\frac{\partial^2}{\partial x^2} - \gamma^2\right)^2 + 1 - \varepsilon K\right)\phi_y + K\phi = 0 \quad \text{on} \quad y = -h, \tag{3.4}$$

$$\frac{\partial \varphi}{\partial r} = 0 \qquad \text{on} \quad r = a, \tag{3.5}$$

where $x = r\sin\theta$, $y = r\cos\theta$ $(-\pi \le \theta \le \pi)$,

$$\nabla\phi \to 0 \quad \text{as} \quad y \to \infty, \tag{3.6}$$

and

$$\phi(x, y) \sim \begin{cases} T\phi_0(x, y) & \text{as} \quad x \to \infty, \\ \phi_0(x, y) + R\phi_0(-x, y) & \text{as} \quad x \to -\infty. \end{cases} \tag{3.7}$$

In (3.7) T and R denote respectively the transmission and reflection coefficients and are unknown complex constants to be determined.

Solution of the problem

Let $G(x, y; \xi, \eta)$ denote the Green's function satisfying (3.3) except at (ξ, η) $(\eta > -h)$ with boundary conditions (3.4), (3.6) and represent an outgoing wave at infinity. We can represent $G(x, y; \xi, \eta)$ as

$$G(x, y; \xi, \eta) = K_0(\gamma r_1) - K_0(\gamma r_2) + \int_\gamma^\infty A(k)e^{-k(y+h)} \cos\left((k^2 - \gamma^2)^{\frac{1}{2}}(x - \xi)\right)dk, \tag{3.8}$$

where $K_0(x)$ denotes the modified Bessel function of second kind, and

$$r_1 = ((x-\xi)^2 + (y-\eta)^2)^{\frac{1}{2}}, \quad r_2 = ((x-\xi)^2 + (y+\eta+2h)^2)^{\frac{1}{2}}$$

and $A(k)$ is a function of k to be found such that the integral exits in some sense. The ice-cover condition (2.4) is satisfied if $A(k)$ is chosen as

$$A(k) = 2\frac{k(Dk^4+1-\varepsilon K)}{(k^2-\gamma^2)^{\frac{1}{2}}\left\{k(Dk^4+1-\varepsilon K)-K\right\}}. \tag{3.9}$$

Thus we can take

$$G(x,y;\xi,\eta) = K_0(\gamma r_1) - K_0(\gamma r_2) + 2\int_0^\infty \frac{k(Dk^4+1-\varepsilon K)}{(k^2-\gamma^2)^{\frac{1}{2}}\{k(Dk^4+1-\varepsilon K)-K\}} e^{-k(y+\eta+2h)} \cos\left((k^2-\gamma^2)^{\frac{1}{2}}(x-\xi)\right) dk \tag{3.10}$$

where the contour is indented below the pole $k = \lambda$ on the real k-axis to take care of its outgoing nature at infinity. An alternative representation in which its behaviour as $(x-\xi) \to \pm\infty$ is evident is given by

$$G(x,y;\xi,\eta) = K_0(\gamma r_1) - K_0(\gamma r_2) + 2\pi i\left[\frac{\lambda g(\lambda)}{\cos\alpha} e^{-\lambda(y+\eta+2h)+i(\lambda^2-\gamma^2)^{\frac{1}{2}}|x-\xi|} + \frac{\lambda_1 g(\lambda_1)}{(\lambda_1^2-\gamma^2)^{\frac{1}{2}}} e^{-\lambda_1(y+\eta+2h)+i(\lambda_1^2-\gamma^2)^{\frac{1}{2}}|x-\xi|} + \frac{\overline{\lambda_1} g(\overline{\lambda_1})}{(\overline{\lambda_1^2}-\gamma^2)^{\frac{1}{2}}} e^{-\overline{\lambda_1}(y+\eta+2h)+i(\overline{\lambda_1^2}-\gamma^2)^{\frac{1}{2}}|x-\xi|}\right] \tag{3.11}$$
$$+2\int_0^\infty \frac{k(Dk^4+1-\varepsilon K)[k(Dk^4+1-\varepsilon K)\cos k(y+\eta+2h) - K\sin k(y+\eta+2h)]}{(k^2-\gamma^2)^{\frac{1}{2}}\{k^2(Dk^4+1-\varepsilon K)^2+K^2\}} e^{-(k^2+\gamma^2)^{\frac{1}{2}}|x-\xi|} dk$$

where

$$g(\lambda) = \frac{D\lambda^4+1-\varepsilon K}{5D\lambda^4+1-\varepsilon K}. \tag{3.12}$$

To obtain a representation of $\phi(x,y)$ at a point (ξ,η), we apply Green's integral theorem to the functions $\psi(x,y) = \phi(x,y) - \phi_0(x,y)$ and the Green's function $G(x,y;\xi,\eta)$ within the entire fluid domain surrounding the rigid cylinder. This gives

$$2\pi\psi(\xi,\eta) = -\int_{-\pi}^{\pi}\psi(\theta)\left\langle a\frac{\partial}{\partial r}G(x,y;\xi,\eta)\right\rangle d\theta - \int_{-\pi}^{\pi} G(x,y;\xi,\eta)\left\langle a\frac{\partial}{\partial r}\phi_0(x,y)\right\rangle d\theta, \tag{3.13}$$

where $\psi(\theta)$ is the unknown scattered potential function on the contour of the cylinder and the angular bracket denotes the values at $r = a$.

By another use of Green's integral theorem to $\psi(x,y)$ and $G(x,y;a\sin\beta,a\cos\beta)$ in the fluid region with a small indentation at the point $(a\sin\beta,a\cos\beta)$ on the circle $r=a$, $\psi(\theta)$ can be shown to satisfy an integral equation of the second kind given by

$$\pi\psi(\beta)=-\int_{-\pi}^{\pi}\psi(\theta)\left\langle a\frac{\partial G}{\partial r}\right\rangle d\theta-\int_{-\pi}^{\pi}G(x,y;a\sin\beta,a\cos\beta)\left\langle a\frac{\partial}{\partial r}\varphi_0(x,y)\right\rangle d\theta \cdot \quad (3.14)$$

Making $\xi\to\mp\infty$ in (3.13), we get

$$R,T-1=-i\frac{g(\lambda)}{\cos\alpha}e^{-2\lambda h}\left[\int_{-\pi}^{\pi}\psi(\theta)\left\langle a\frac{\partial}{\partial r}e^{-\lambda y\pm i\mu x}\right\rangle d\theta+\int_{-\pi}^{\pi}e^{-\lambda y\pm i\mu x}\left\langle a\frac{\partial}{\partial r}\phi_0(x,y)\right\rangle d\theta\right]. \quad (3.15)$$

Now we expand $\psi(\theta)$ as

$$\psi(\theta)=a_0+\sum_{n=1}^{\infty}(a_n\cos n\theta+b_n\sin n\theta),\quad -\pi<\theta<\pi, \quad (3.16)$$

where $a_0,a_n,b_n\ (n=1,2,\ldots)$ are unknown Fourier coefficients. Using (3.16) and multiplying both sides of (3.14) by $\cos s\beta,\sin s\beta$ respectively and integrating with respect to β form $-\pi$ to π, the following two infinite linear systems for a_n,b_n are obtained:

$$\pi^2a_s+\sum_{n=0}^{\infty}a_nP_{ns}^{(1)}+\sum_{n=1}^{\infty}b_nP_{ns}^{(2)}+\sum_{n=0}^{\infty}(-1)^n\frac{(\lambda a)^n}{n!}B_{ns}^{(1)}=V_s^{(1)},s=0,1,2,\ldots \quad (3.17)$$

$$\pi^2b_s+\sum_{n=0}^{\infty}a_nP_{ns}^{(3)}+\sum_{n=1}^{\infty}b_nP_{ns}^{(4)}+\sum_{n=0}^{\infty}(-1)^n\frac{(\lambda a)^n}{n!}B_{ns}^{(2)}=V_s^{(2)},s=1,2,\ldots \quad (3.18)$$

where

$$P_{ns}^{(1),(2)}=\int_{-\pi}^{\pi}\int_{-\pi}^{\pi}\left\langle a\frac{\partial}{\partial r}G(x,y;a\sin\beta,a\cos\beta)\right\rangle\cos s\beta\,d\theta\,d\beta, \quad (3.19)$$

$$P_{ns}^{(3),(4)}=\int_{-\pi}^{\pi}\int_{-\pi}^{\pi}\left\langle a\frac{\partial}{\partial r}G(x,y;a\sin\beta,a\cos\beta)\right\rangle\sin s\beta\,d\theta\,d\beta, \quad (3.20)$$

$$B_{ns}^{(1),(2)}=\int_{-\pi}^{\pi}\int_{-\pi}^{\pi}\left\langle a\frac{\partial}{\partial r}G(x,y;a\sin\beta,a\cos\beta)\right\rangle(\cos\theta-i\cos\alpha\sin\theta)^n d\theta\,d\beta, \quad (3.21)$$

$$V_s^{(1)(2)}=-\pi\int_{-\pi}^{\pi}\phi_0(\xi,\eta)\,d\beta. \quad (3.22)$$

Since

$$\int_{-\pi}^{\pi}\left\langle a\frac{\partial}{\partial r}G(x,y;\xi,\eta)\right\rangle d\theta=0,$$

it is obvious that there is no effect of a_0 on the function $\psi(\xi,\eta)$ in (3.13).

Then R and T reduce to

$$R,T-1=-i\frac{g(\lambda)}{\cos\alpha}e^{-2\lambda h}\left[\sum_{n=1}^{\infty}a_n(E_n\pm ip_n)+\sum_{n=1}^{\infty}b_n(q_n\pm iH_n)\right]$$
$$+\sum_{n=0}^{\infty}(-1)^n\frac{(\lambda a)^n}{n!}\int_{-\pi}^{\pi}\left\langle a\frac{\partial\varphi_0}{\partial r}\right\rangle(\cos\theta\mp i\cos\alpha\sin\theta)^n\,d\theta, \tag{3.23}$$

where

$$E_n,p_n=\int_{-\pi}^{\pi}\left\langle a\frac{\partial}{\partial r}e^{-\lambda y}\right\rangle\cos n\theta d\theta,$$

$$q_n,H_n=\int_{-\pi}^{\pi}\left\langle a\frac{\partial}{\partial r}e^{-\lambda y}\right\rangle\sin n\theta d\theta.$$

To evaluate these integrals, we use the following result given by Levine (1965):

Let $u(x,y)$ satisfy

$$\left(\frac{\partial^2}{\partial x^2}+\frac{\partial^2}{\partial y^2}-\lambda^2\sin^2\alpha\right)u(x,y)=0,\quad \text{for } 0\le r\le a \tag{3.24}$$

and be regular for $0\le r\le a$, then

$$\frac{1}{2\pi}\int_{-\pi}^{\pi}\left[\frac{\partial}{\partial r}u(r,\theta)\right]_{r=a}\begin{matrix}\cos n\theta\\ \sin n\theta\end{matrix}\,d\theta=(-1)^n\left[\frac{d}{dr}I_n(\lambda r\sin\alpha)\right]_{r=a}\begin{matrix}\cos nD\\ \sin nD\end{matrix}\,u(0), \tag{3.25}$$

where

$$\cos D=-\frac{1}{\lambda\sin\alpha}\frac{\partial}{\partial y},\qquad \sin D=-\frac{1}{\lambda\sin\alpha}\frac{\partial}{\partial x}$$

and $x=r\sin\theta, y=r\cos\theta$ $(-\pi\le\theta\le\pi)$, $I_n(z)$ being the modified Bessel function of first kind.

Using this result, it can be shown that

$$p_n=q_n=0, \tag{3.26}$$

$$P_{ns}^{(2)}=P_{ns}^{(3)}=0. \tag{3.27}$$

Thus the equations (3.17) and (3.18) reduce to

$$\pi^2a_s+\sum_{n=1}^{\infty}a_nP_{ns}^{(1)}+\sum_{n=1}^{\infty}(-1)^n\frac{(\lambda a)^n}{n!}B_{ns}^{(1)}=V_s^{(1)},\quad s=1,2,... \tag{3.28}$$

$$\pi^2b_s+\sum_{n=1}^{\infty}b_nP_{ns}^{(4)}+\sum_{n=1}^{\infty}(-1)^n\frac{(\lambda a)^n}{n!}B_{ns}^{(2)}=V_s^{(2)},\quad s=1,2,... \tag{3.29}$$

as the constants $B_{01}^{(1)}, B_{01}^{(2)}$ can be shown to be zero. It is now shown that the infinite linear systems (3.28) and (3.29) possess unique solutions.

The equations (3.28) and (3.29) can be written in the forms

$$(\mathrm{I}+\mathrm{C}^{(1)})\mathrm{a}+\mathrm{B}^{(1)}=\mathrm{w}^{(1)}, \tag{3.30}$$

$$(\mathrm{I}+\mathrm{C}^{(2)})\mathrm{b}+\mathrm{B}^{(2)}=\mathrm{w}^{(2)}, \tag{3.31}$$

where I is the infinite unit matrix, $\mathrm{a}, \mathrm{b}, \mathrm{w}^{(1)}, w^{(2)}$ are the column vectors $(a_s), (b_s), (w_s^{(1)}), (w_s^{(2)})$ respectively, where

$$w_s^{(1),(2)}=\frac{1}{\pi^2}V_s^{(1),(2)}, \tag{3.32}$$

$$C_{s,n}^{(1)}=\frac{1}{\pi^2}P_{ns}^{(1)}, \tag{3.33}$$

$$C_{s,n}^{(2)}=\frac{1}{\pi^2}P_{ns}^{(4)}, \tag{3.34}$$

$$B_{s,n}^{(1)}=(-1)^n\frac{1}{\pi^2}\frac{(\lambda a)^n}{n!}B_{ns}^{(1)}, \tag{3.35}$$

$$B_{s,n}^{(2)}=(-1)^n\frac{1}{\pi^2}\frac{(\lambda a)^n}{n!}B_{ns}^{(2)}. \tag{3.36}$$

Equations of the forms (3.30) and (3.31) have been studied by Ursell (1950).

The linear system (3.30) possesses unique solution under the following conditions:

$$(a1)\quad \sum_{s=1}^{\infty}|w_s^{(1)}|<\infty, \tag{3.37}$$

$$(a2)\quad \sum_{s=1}^{\infty}\sum_{n=1}^{\infty}|C_{s,n}^{(1)}|<\infty, \tag{3.38}$$

$$(a3)\quad \sum_{s=1}^{\infty}\sum_{n=1}^{\infty}|B_{s,n}^{(1)}|<\infty, \tag{3.39}$$

Similarly for the system (3.31):

$$(b1)\quad \sum_{s=1}^{\infty}|w_s^{(2)}|<\infty, \tag{3.40}$$

$$(b2)\quad \sum_{s=1}^{\infty}\sum_{n=1}^{\infty}|C_{s,n}^{(2)}|<\infty, \tag{3.41}$$

$$(b3)\quad \sum_{s=1}^{\infty}\sum_{n=1}^{\infty}|B_{s,n}^{(2)}|<\infty. \tag{3.42}$$

Now

$$|w_s^{(1)}|=\frac{1}{\pi^2}|V_s^{(1)}|=2\sum_{k=0}^{\infty}\frac{(\frac{1}{2}a)^{2k+s}(\gamma)^{2k}\lambda^s}{k!(k+s)!}. \tag{3.43}$$

Since the above series is convergent, the condition (a1) is satisfied. Similarly the condition (b1) is satisfied, where

$$|w_s^{(2)}| = \left| 2i\sum_{k=0}^{\infty} \frac{(\frac{1}{2}a)^{2k+s}(\gamma)^{2k}\lambda^{s-1}\mu}{k!(k+s)!} \right|. \tag{3.44}$$

Again

$$|C_{s,n}^{(1)}| = \frac{1}{\pi^2}|P_{ns}^{(1)}| = \left| \frac{2^{n+s}aI_n'(a\gamma)I_s(a\gamma)}{\gamma^{n+s-2}} \frac{\partial^{n+s}}{\partial y^n \partial \eta^s} G(x,y;\xi,\eta)|_{x=\xi=0,y=\eta=0} \right| \leq |N_1||M_1|, \tag{3.45}$$

where

$$N_1 = 2\left(\frac{a}{h}\right)^n \left(\frac{a}{h}\right)^s \left(\frac{1}{2(n-1)!}+1\right)\left(\sum_{i=1}^{\infty} \frac{\left(\frac{1}{4}a^2\gamma^2\right)^i}{i!(i+n)!}\right)\left(\sum_{j=0}^{\infty} \frac{\left(\frac{1}{4}a^2\gamma^2\right)^j}{j!(j+n)!}\right), \tag{3.46}$$

$$M_1 = \int_\gamma^\infty \frac{u(D_1u^4+1-\varepsilon K)+Kh}{\left(u^2-(\gamma h)^2\right)^{\frac{1}{2}}\{u(D_1u^4+1-\varepsilon K)-Kh\}} u^{n+s}e^{-2u}du + 2\pi i \frac{g(\lambda h)}{\cos\alpha}(\lambda h)^{n+s}e^{-2\lambda h} \tag{3.47}$$

with $D_1 = \frac{D}{h^4}$. Also

$$|C_{s,n}^{(2)}| \leq |N_1||M_1|, \tag{3.48}$$

where

$$M_2 = \int_\gamma^\infty \frac{\left(u^2-(\gamma h)^2\right)^{\frac{1}{2}}\{u(D_1u^4+1-\varepsilon K)+Kh\}}{\{u(D_1u^4+1-\varepsilon K)-Kh\}} u^{n+s-2}e^{-2u}du + 2\pi i g(\lambda h)h^2$$

$$(\lambda h)^{n+s-2}\cos\alpha e^{-2\lambda h}. \tag{3.49}$$

Since $a < h$ and $|M_1|$ and $|M_2|$ are bounded. Thus the conditions ($a2$) and ($b2$) are satisfied. Again

$$\left|A_{s,n}^{(1),(2)}\right| = \frac{1}{\pi^2}\left|\frac{(\lambda a)^n}{n!} B_{ns}^{(1),(2)}\right|.$$

We see that $B_{ns}^{(1),(2)}$ is the finite series of $P_{ns}^{(1),(4)}$. Hence the conditions ($a3$) and ($b3$) are satisfied. Thus the infinite linear systems (3.30) and (3.31) possess unique solution.

Thus R and T in (3.16) can be written in the forms

$$R, T-1 = -i\frac{g(\lambda)}{\cos\alpha}e^{-2\lambda h}\left[\sum_{n=1}^{\infty}(a_nE_n \pm ib_nH_n) + \sum_{n=0}^{\infty}(-1)^n \frac{(\lambda a)^n}{n!}\int_{-\pi}^{\pi}\left\langle a\frac{\partial\varphi_0}{\partial r}\right\rangle(\cos\theta \mp i\cos\alpha\sin\theta)^n d\theta\right]. \tag{3.50}$$

Now for the case of normal incidence, $\alpha = 0$ and in this case, it can be shown that

$$E_n = -H_n, \qquad n = 0, 1, 2, \ldots,$$

$$P_{ns}^{(1)} = P_{ns}^{(4)}, \qquad n, s = 1, 2, \ldots,$$

and

$$V_s^{(1)} = iV_s^{(2)}, \qquad s = 1, 2, \ldots$$

Thus for $\alpha = 0$, the equations (3.28) and (3.29) reduce to

$$\pi^2 A_s^{(1)} + \sum_{n=1}^{\infty} P_{ns}^{(1)} A_n^{(1)} = 2(-1)^s \frac{\pi^2 (a\lambda)^s}{s!}, \qquad s = 1, 2, \ldots \tag{3.51}$$

$$\pi^2 A_s^{(2)} + \sum_{n=1}^{\infty} P_{ns}^{(1)} A_n^{(2)} = 2(-1)^s \frac{\pi^2 (a\lambda)^s}{s!}, \qquad s = 1, 2, \ldots \tag{3.52}$$

where

$$\begin{pmatrix} A_n^{(1)} \\ A_n^{(2)} \end{pmatrix} = \begin{pmatrix} a_n \\ ib_n \end{pmatrix} + (-1)^n \frac{(a\lambda)^n}{n!}. \tag{3.53}$$

It is noted that the two linear systems (3.51) and (3.52) are exactly the same, and hence $A_n^{(1)} = A_n^{(2)}$ so that

$$a_n = ib_n, \qquad n = 1, 2, \ldots \tag{3.54}$$

Thus R and $T - 1$ reduce to

$$R, T - 1 = -ig(\lambda) e^{-2\lambda h} \sum_{n=1}^{\infty} (A_n^{(1)} E_n \pm A_n^{(2)} H_n)$$

and this produces for normal incidence of the wave train, the classical result

$$R \equiv 0.$$

This result holds for all frequencies and radius of the submerged cylinder. Hence, as in the case of a horizontal circular cylinder submerged beneath a free surface, here also the circular cylinder experiences no reflection for normally incident wave train.

When $\alpha \neq 0$, the infinite systems (3.28) and (3.29) have to be solved by truncation. A first approximation to R and T can be obtained by assuming a_1, b_1 to be the only non vanishing coefficients in the systems (3.28) and (3.29), so that we find, after truncation

$$a_1 = \lambda a + \frac{V_1^{(1)} - \pi^2 \lambda a}{\pi^2 + P_{11}^{(1)}}, \tag{3.55}$$

$$b_1 = i\left(-\lambda a \cos\alpha + \frac{\pi^2 \lambda a \cos\alpha - iV_1^{(2)}}{\pi^2 + P_{11}^{(4)}} \right). \tag{3.56}$$

The constants $P_{11}^{(1)}$, $P_{11}^{(4)}$, $B_{11}^{(1)}$, $B_{11}^{(2)}$, $V_1^{(1)}$ and $V_1^{(2)}$ can be evaluated by using the result (3.25) and these are given by

$$P_{11}^{(1)} = -\frac{2\pi E_1 I_1(a\lambda)}{\sin\alpha}\left[\left\{K_0(2\gamma h)+\frac{K_1(2\gamma h)}{2\gamma h}-\frac{1}{2}\frac{K_1'(a\gamma)}{I_1'(a\gamma)}\right\}\sin^2\alpha-\frac{d^2}{dl^2}F(l,\gamma)\big|_{l=2h}\right], \tag{3.57}$$

where $I_1(x), K_n(x)$ $(n = 0,1)$ denote the modified Bessel functions of first kind and second kind respectively and dash denotes its derivative, and

$$E_1 = -2\pi\lambda a I_1'(a\gamma), \tag{3.58}$$

$$F(l,\gamma) = 2\int_0^\infty \frac{D(z^2+\gamma^2)^2+1-\varepsilon K}{(z^2+\gamma^2)^{\frac{1}{2}}\left\{D(z^2+\gamma^2)^2+1-\varepsilon K\right\}-K}e^{-l\sqrt{z^2+\gamma^2}}\,dz, \tag{3.59}$$

the contour being indented below the pole at $z = \mu$,

$$P_{11}^{(4)} = -\frac{2\pi E_1 I_1(a\lambda)}{\sin\alpha}\left[\left\{\frac{K_1(2\gamma h)}{2\lambda h}\sin\alpha-\frac{1}{2}\frac{K_1'(a\gamma)}{I_1'(a\gamma)}\sin^2\alpha\right\}+(\gamma^2-\frac{d^2}{dl^2})F(l,\gamma)\big|_{l=2h}\right], \tag{3.60}$$

$$B_{11}^{(1)} = P_{11}^{(1)}, \qquad B_{11}^{(2)} = -i\cos\alpha P_{11}^{(4)}, \qquad V_1^{(1)} = \frac{2\pi^2 I_1(a\gamma)}{\sin\alpha}, \tag{3.61}$$

$$V_1^{(2)} = -i\frac{2\pi^2 I_1(a\gamma)}{\tan\alpha}. \tag{3.62}$$

Now $F(l,\gamma)$ can be simplified as

$$F(l,\gamma) = 2\pi i\left[\frac{e^{-l\lambda}}{\cos\alpha}g(\lambda)+\frac{\lambda_1 e^{-l\lambda_1}}{(\lambda_1^2-\gamma^2)^{\frac{1}{2}}}g(\lambda_1)+\frac{\overline{\lambda_1}e^{-l\overline{\lambda_1}}}{(\overline{\lambda_1}^2-\gamma^2)^{\frac{1}{2}}}g(\overline{\lambda_1})\right]+2K_0(l\gamma) \tag{3.63}$$
$$-2[g(\lambda)M(\lambda)-g(\lambda_1)M(\lambda_1)+g(\overline{\lambda_1})M(\overline{\lambda_1})-g(\lambda_2)M(\lambda_2)+g(\overline{\lambda_2})M(\overline{\lambda_2})],$$

with

$$M(u) = \left[\frac{u}{(u^2-\gamma^2)^{\frac{1}{2}}}\ln\left\{\frac{u+(u^2-\gamma^2)^{\frac{1}{2}}}{\gamma}\right\}+\int_0^{lu} e^x K_0\left(\frac{\gamma x}{u}\right)dx\right]e^{-lu}. \tag{3.64}$$

Thus a first order approximation to reflection and transmission coefficients is given by

$$R^{(1)} = -i\frac{g(\lambda)}{\cos\alpha}e^{-2\lambda h}\left[E_0+\left(\frac{V_1^{(1)}-\pi^2\lambda a}{\pi^2+P_{11}^{(1)}}\right)E_1-\left(\frac{\pi^2\lambda a\cos\alpha-iV_1^{(2)}}{\pi^2+P_{11}^{(4)}}\right)H_1\right], \tag{3.65}$$

where

$$E_0 = 2\pi\gamma a I_1(a\gamma),$$

$$H_1 = 2\pi\mu a I_1'(a\gamma),$$

$$T^{(1)} = 1 - i\frac{g(\lambda)}{\cos\alpha}e^{-2\lambda h}\left[E_0 + \left(\frac{V_1^{(1)} - \pi^2\lambda a}{\pi^2 + P_{11}^{(1)}}\right)E_1 + \left(\frac{\pi^2\lambda a\cos\alpha - iV_1^{(2)}}{\pi^2 + P_{11}^{(4)}}\right)H_1\right], \quad (3.66)$$

A second approximation to the reflection co-efficient $R^{(2)}$ and transmission co-efficient $T^{(2)}$ follow from the assumption that a_1, a_2 and b_1, b_2 be the only non vanishing coefficients in the linear systems (3.28) and (3.29). Hence we find from (3.28) and (3.29) that

$$\pi^2 a_s + a_1 P_{1s}^{(1)} + a_2 P_{2s}^{(1)} - \lambda a B_{1s}^{(1)} + \frac{(\lambda a)^2}{2}B_{2s}^{(1)} = V_s^{(1)}, \quad s = 1, 2, \ldots \quad (3.67)$$

$$\pi^2 b_s + b_1 P_{1s}^{(4)} + b_2 P_{2s}^{(4)} - \lambda a B_{1s}^{(2)} + \frac{(\lambda a)^2}{2}B_{2s}^{(2)} = V_s^{(2)}, \quad s = 1, 2, \ldots \quad (3.68)$$

From two equations (3.67) and (3.68), we get

$$a_1 = \lambda a + \frac{\left(V_1^{(1)} - \pi^2\lambda a\right)\left(P_{22}^{(1)} + \pi^2\right) - \left(V_2^{(1)} + \frac{\pi^2(\lambda a)^2}{4}(1+\cos^2\alpha)\right)P_{12}^{(1)}}{\left(\pi^2 + P_{11}^{(1)}\right)\left(\pi^2 + P_{22}^{(1)}\right) - \left(P_{12}^{(1)}\right)^2}, \quad (3.69)$$

$$a_2 = -\frac{(\lambda a)^2}{4}(1+\cos^2\alpha) + \frac{\left(V_2^{(1)} + \frac{\pi^2(\lambda a)^2}{4}(1+\cos^2\alpha)\right)\left(\pi^2 + P_{11}^{(1)}\right) - \left(V_1^{(1)} - \pi^2\lambda a\right)P_{12}^{(1)}}{\left(\pi^2 + P_{11}^{(1)}\right)\left(\pi^2 + P_{22}^{(1)}\right) - \left(P_{12}^{(1)}\right)^2} \quad (3.70)$$

and

$$b_1 = i\left\{-\lambda a\cos\alpha + \frac{\left(\pi^2\lambda a\cos\alpha - iV_1^{(2)}\right)\left(\pi^2 + P_{22}^{(4)}\right) + \left(\frac{\pi^2(\lambda a)^2}{2}\cos\alpha + iV_2^{(2)}\right)P_{12}^{(4)}}{\left(\pi^2 + P_{11}^{(4)}\right)\left(\pi^2 + P_{22}^{(4)}\right) - \left(P_{12}^{(4)}\right)^2}\right\}, \quad (3.71)$$

$$b_2 = i\left\{\frac{(\lambda a)^2}{2}\cos\alpha - \frac{\left(\pi^2\lambda a\cos\alpha - iV_1^{(2)}\right)P_{12}^{(4)} + \left(\frac{\pi^2(\lambda a)^2}{2}\cos\alpha + iV_2^{(2)}\right)\left(\pi^2 + P_{11}^{(4)}\right)}{\left(\pi^2 + P_{11}^{(4)}\right)\left(\pi^2 + P_{22}^{(4)}\right) - \left(P_{12}^{(4)}\right)^2}\right\}, \quad (3.72)$$

The constants can be calculated by using the result (3.25) and these are given as follows:

$$P_{12}^{(1)} = P_{21}^{(1)} = -4\pi \frac{E_1 I_2(a\gamma)}{\sin^2\alpha}\left[-\frac{1}{\lambda^3}\int_\gamma^\infty \frac{z^3}{(z^2-\gamma^2)^{\frac{1}{2}}}dz + K_1''(2\gamma h)\sin^3\alpha + \frac{d^3}{dl^3}F(l,\gamma)|_{l=2h}\right], \tag{3.73}$$

where $I_2(x)$ denotes the modified Bessel function of first kind,

$$P_{22}^{(1)} = 2\pi \frac{E_1 I_2(a\gamma)}{(2-\sin^2\alpha)\sin^2\alpha} \tag{3.74}$$

$$\left[4\left\{\frac{1}{\lambda^4}\int_\gamma^\infty \frac{z^4}{(z^2-\gamma^2)^{\frac{1}{2}}}dz + K_1'''(2\gamma h)\sin^4\alpha + \frac{d^4}{dl^4}F(l,\gamma)|_{l=2h}\right\} + \frac{2\sin^3\alpha}{\pi E_1 I_1(a\gamma)}P_{11}^{(1)}\right]$$

with

$$E_2 = \frac{2\pi\lambda a}{\sin\alpha}(2-\sin^2\alpha)I_2'(a\gamma), \tag{3.75}$$

$$B_{12}^{(1)} = P_{12}^{(1)}, \qquad B_{21}^{(1)} = \frac{1}{2}(1+\cos^2\alpha)P_{12}^{(1)}, \tag{3.76}$$

$$P_{12}^{(4)} = P_{21}^{(4)} = -4\pi \frac{E_1 I_2(a\gamma)}{\sin^2\alpha}\left[\frac{1}{\lambda^3}\int_\gamma^\infty z(z^2-\gamma^2)^{\frac{1}{2}}dz + K_1''(2\gamma)\sin^3\alpha \right.$$
$$\left. + \left(\frac{d^3}{dl^3} - \sin^2\alpha\frac{d}{dl}\right)F(l,\gamma)|_{l=2h}\right], \tag{3.77}$$

$$P_{22}^{(4)} = 8\pi \frac{E_1 I_2(a\gamma)}{(2-\sin^2\alpha)\sin^2\alpha}\left[-\frac{1}{\lambda^4}\int_\gamma^\infty z^2(z^2-\gamma^2)^{\frac{1}{2}}dz + K_1'''(2\gamma h)\sin^4\alpha \right.$$

$$\left. + \left(\frac{d^4}{dl^4} - \sin^2\alpha\frac{d^2}{dl^2}\right)F(l,\gamma)|_{l=2h}\right], \tag{3.78}$$

$$B_{12}^{(2)} = B_{21}^{(2)} = -i\cos\alpha P_{12}^{(4)}, \tag{3.79}$$

$$B_{22}^{(2)} = -i\cos\alpha P_{22}^{(4)}, \tag{3.80}$$

$$V_2^{(1)} = -2\pi^2\frac{I_2(a\gamma)}{\sin^2\alpha}(2-\sin^2\alpha), \tag{3.81}$$

$$V_2^{(2)} = 4\pi^2 i\frac{I_2(a\gamma)}{\sin^2\alpha}\cos\alpha. \tag{3.82}$$

Thus the second order approximation to reflection and transmission coefficients is given by

$$R^{(2)} = -i\frac{g(\lambda)}{\cos\alpha}e^{-2\lambda h}\left[\left(1+\frac{(\lambda a)^2}{4}\sin^2\alpha\right)E_0 + \left\{\frac{\left\{V_2^{(1)} + \pi^2\frac{(\lambda a)^2}{4}(1+\cos^2\alpha)\right\}\left[\left(\pi^2 + P_{11}^{(1)}\right)E_2 - P_{12}^{(1)}E_1\right]}{\left(\pi^2 + P_{11}^{(1)}\right)\left(\pi^2 + P_{22}^{(1)}\right) - \left(P_{12}^{(1)}\right)^2}\right.\right.$$
$$\left. + \frac{\left(V_1^{(1)} - \pi^2\lambda a\right)\left[\left(P_{22}^{(1)} + \pi^2\right)E_1 - P_{12}^{(1)}E_2\right]}{\left(\pi^2 + P_{11}^{(1)}\right)\left(\pi^2 + P_{22}^{(1)}\right) - \left(P_{12}^{(1)}\right)^2}\right\} - \left\{\frac{\left(\pi^2\lambda a\cos\alpha - iV_1^{(2)}\right)\left[\left(\pi^2 + P_{22}^{(4)}\right)H_1 - P_{12}^{(4)}H_2\right]}{\left(\pi^2 + P_{11}^{(4)}\right)\left(\pi^2 + P_{22}^{(4)}\right) - \left(P_{12}^{(4)}\right)^2}\right.$$
$$\left.\left. + \frac{\left\{\pi^2\frac{(\lambda a)^2}{2}\cos\alpha + iV_2^{(2)}\right\}\left[P_{12}^{(4)}H_1 - \left(\pi^2 + P_{11}^{(4)}\right)H_2\right]}{\left(\pi^2 + P_{11}^{(4)}\right)\left(\pi^2 + P_{22}^{(4)}\right) - \left(P_{12}^{(4)}\right)^2}\right\}\right] \tag{3.83}$$

and

$$T^{(2)} = 1 - i\frac{g(\lambda)}{\cos\alpha}e^{-2\lambda h}\left[\left(1+\frac{(\lambda a)^2}{4}\sin^2\alpha\right)E_0 + \left\{\frac{\left\{V_2^{(1)} + \pi^2\frac{(\lambda a)^2}{4}(1+\cos^2\alpha)\right\}\left[\left(\pi^2 + P_{11}^{(1)}\right)E_2 - P_{12}^{(1)}E_1\right]}{\left(\pi^2 + P_{11}^{(1)}\right)\left(\pi^2 + P_{22}^{(1)}\right) - \left(P_{12}^{(1)}\right)^2}\right.\right.$$
$$\left. + \frac{\left(V_1^{(1)} - \pi^2\lambda a\right)\left[\left(P_{22}^{(1)} + \pi^2\right)E_1 - P_{12}^{(1)}E_2\right]}{\left(\pi^2 + P_{11}^{(1)}\right)\left(\pi^2 + P_{22}^{(1)}\right) - \left(P_{12}^{(1)}\right)^2}\right\} + \left\{\frac{\left(\pi^2\lambda a\cos\alpha - iV_1^{(2)}\right)\left[\left(\pi^2 + P_{22}^{(4)}\right)H_1 - P_{12}^{(4)}H_2\right]}{\left(\pi^2 + P_{11}^{(4)}\right)\left(\pi^2 + P_{22}^{(4)}\right) - \left(P_{12}^{(4)}\right)^2}\right.$$
$$\left.\left. + \frac{\left\{\pi^2\frac{(\lambda a)^2}{2}\cos\alpha + iV_2^{(2)}\right\}\left[P_{12}^{(4)}H_1 - \left(\pi^2 + P_{11}^{(4)}\right)H_2\right]}{\left(\pi^2 + P_{11}^{(4)}\right)\left(\pi^2 + P_{22}^{(4)}\right) - \left(P_{12}^{(4)}\right)^2}\right\}\right] \tag{3.84}$$

with

$$H_2 = -\frac{4\pi\lambda a}{\sin\alpha}\cos\alpha I_2'(a\gamma). \tag{3.85}$$

Also higher order approximations to R and T can be calculated, but this is not pursued here.

Numerical results

Figures 6.15–6.18 depict the reflection and transmission coefficients against the wave number Kh for $a/h = 0.6, \varepsilon/h = 0.01, D/h^4 = 0.1$, $|R^{(1)}|$, $|T^{(1)}|$ denoting first order approximations while $|R^{(2)}|$, $|T^{(2)}|$ denoting the second order approximations. The different curves correspond to $\alpha = 15°, 30°, 45°, 60°, 75°$. It is remarkable that the first and second approximations are somewhat close and possess the same features.

Figures 6.19–6.22 show $|R^{(1)}|$, $|T^{(1)}|$ and $|R^{(2)}|$, $|T^{(2)}|$ against Kh for $a/h = 0.6, \varepsilon/h = 0.01, \alpha = 60°$. The different curves correspond to different values of the flexural rigidity, i.e., $D/h^4 = 0.01, 0.05, 0.1, 0.5, 1$.

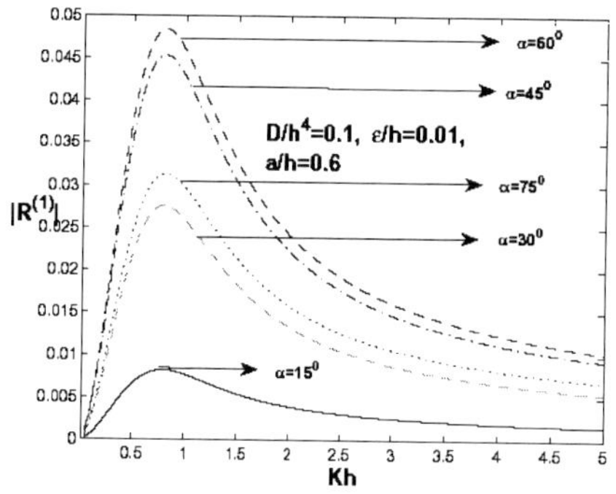

Fig. 6.15 Reflection coefficient against wave number

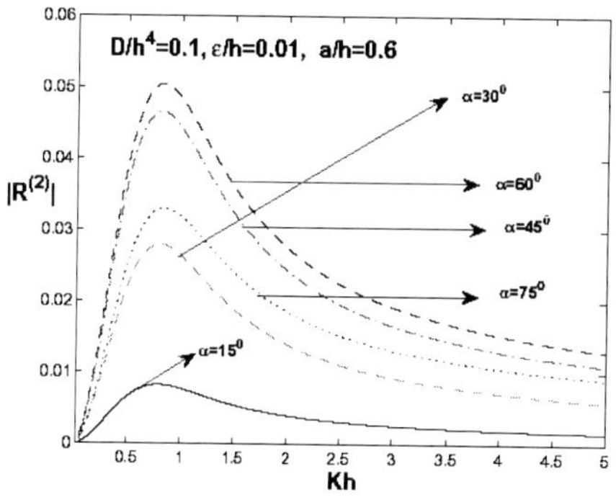

Fig. 6.16 Reflection coefficient against wave number

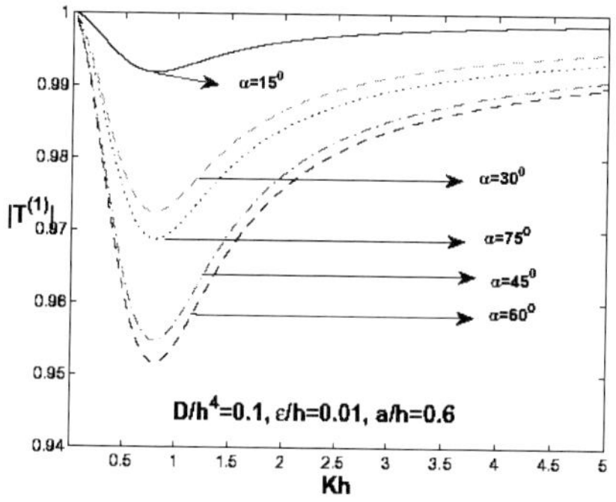

Fig. 6.17 Transmission coefficient against wave number

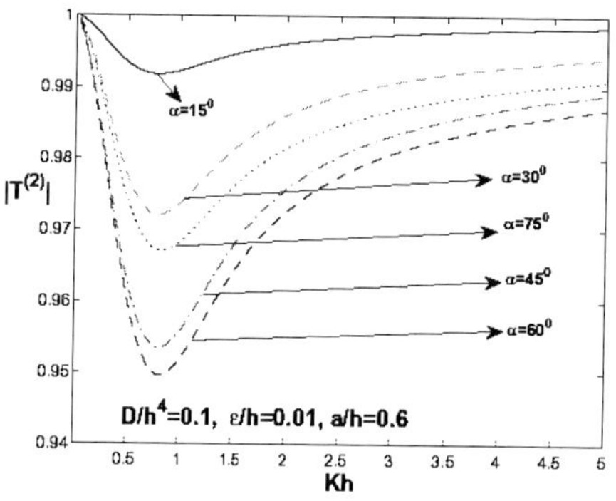

Fig. 6.18 Transmission coefficient against wave number

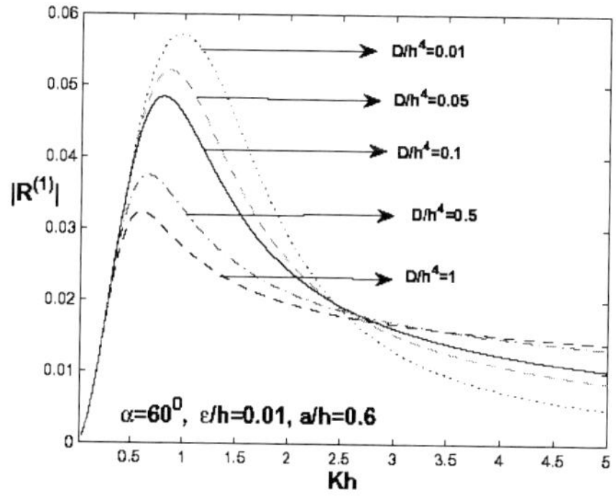

Fig. 6.19 Reflection coefficient against wave number

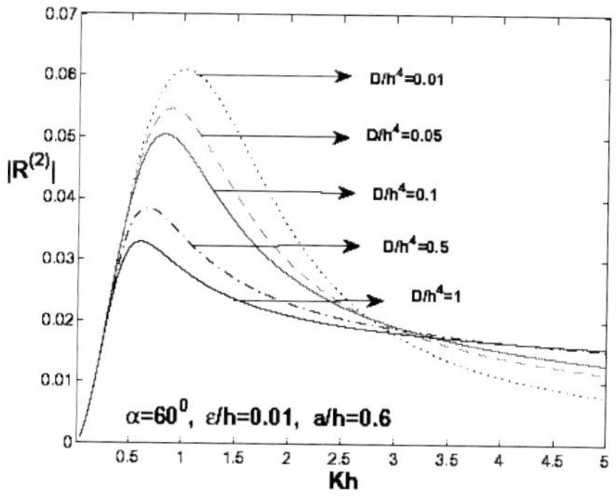

Fig. 6.20 Reflection coefficient against wave number

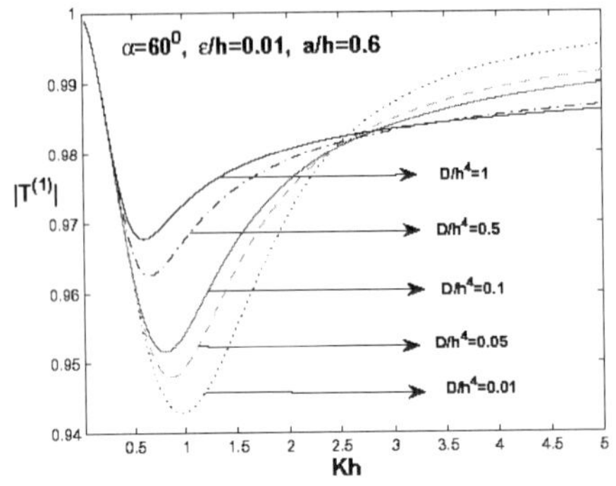

Fig. 6.21 Transmission coefficient against wave number

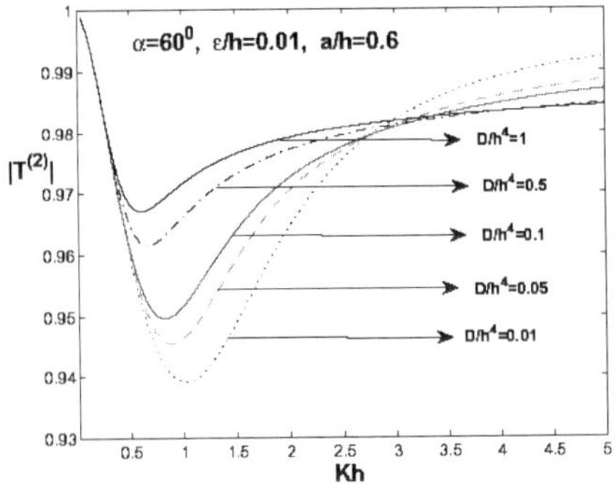

Fig. 6.22 Transmission coefficient against wave number

Figure 6.15 and Fig. 6.16 show that the reflection coefficient $|R^{(1)}|$ or $|R^{(2)}|$ regarded as a function of the wave number Kh, first increases as Kh increases, attains a maximum value and then decreases as Kh increases. Also $|R|$ (i.e., $|R^{(1)}|$ or $|R^{(2)}|$) first increases as α increases, attains a maximum value and then decreases as α further increases. Similarly Figs. 3 and 4 show that $|T|$ (i.e., $|T^{(1)}|$ or $|T^{(2)}|$) first decreases as Kh increases, attains a minimum value then increases as Kh further increases. It first decreases as α increases, attains a minimum value and then increases as α further increases. Figures 6.19 and 6.20 show that $|R|$ first decreases as D/h^4 increases for low to moderate values of Kh but it increases as D/h^4 further increases for somewhat large value of Kh. Figures 7 and 8 describe the behaviour of $|T|$ which is complimentarily to the behaviour of $|R|$ described by Figs. 6.19 and 6.20. This is obvious due to the energy identity $|R|^2 + |T|^2 = 1$.

CHAPTER VII

Energy Identities

Energy identities for free-surface boundary condition with higher order derivatives

In the linearized theory of water waves, the reflection and transmission coefficients R, T in any wave scattering problem involving a finite number of bodies present in a single-layer fluid with a free-surface satisfy the energy identity $|R|^2 + |T|^2 = 1$ which has been derived by a simple use of Green's integral theorem in the fluid region in section 1.7. For a two-layer fluid with a free-surface there exist two different modes at which time-harmonic progressive waves can propagate, and as such two energy identities involving reflection and transmission coefficients of two different modes corresponding to incident waves of again two different modes, have also been derived in section 1.7 by using Green's integral theorem in both the fluid regions Linton and McIver (1995). The various hydrodynamic relations such as the energy-conservation principle and Haskind-Hanaoka relation in any radiation problem involving a free-surface can be derived by a simple use of Green's integral theorem in the fluid region Mei (1982). For a two-layer fluid with a free surface, the energy-conservation principle and Haskind-Hanaoka relations can also be derived by using Green's integral theorem in both the fluid regions Newman (1976), Yeung and Nguyen (2000), Ten and Kashiwagi (2004), Kashiwagi et al. (2006). In these problems the free-surface condition involves first-order partial derivative. However, if the free-surface involves higher-order partial derivatives Manam et al. (2006), then the derivation of the energy identity cannot be achieved by a straightforward application of the standard form of Green's integral theorem. The same is true also for a two-layer fluid wherein the free-surface of the upper fluid involves higher-order partial derivatives. Such higher-order derivatives in the free-surface condition arise for a large class of problems in the area of ocean structure interaction, e.g., ice sheet covering a vast area of the ocean surface in the cold regions (Antarctic region), the ice sheet being regarded as thin elastic plate, very large floating structure constructed for the purpose of using it as a large floating air port, etc. Kashiwagi (1998), Gayen et al. (2005), Manam et al. (2006), Fox and Squire (1994), Lawrie and Abrahams (1999), Evans and Porter (2003) and others. Although some part of the free-surface may be covered by a floating elastic plate, the remaining part of the free surface is of ordinary gravity waves. Thus there may co-exist two different free-surface conditions. For such a case, the energy identity for a single-layer fluid of uniform finite depth has been derived by Balmforth and Craster (1999) (cf. the relation (7.10) in their paper).

A modified form of Green's integral theorem designed appropriately to take care of the free-surface condition with higher-order derivatives throughout the entire free-surface, has been employed by Das et al. (2008) to derive the energy identity for any wave scattering problem, the energy-conservation principle and the *modified* Haskind-Hanaoka relation for any radiation problem in a single-layer fluid. For a two-layer fluid again with higher-order derivatives in the free-surface condition, two forms of energy identities satisfied by the reflection and transmission coefficients of different wave modes exist and these are also derived by Das et al. (2008) by appropriate uses of the modified form of Green's integral theorem in the upper and lower fluid layers. Also the energy-conservation principle and the *modified* Haskind-Hanaoka relations are derived in such a two-layer fluid. An account of this is described in this chapter.

Formulation of a general boundary value problem

Single-layer fluid

A general boundary value problem describing wave propagation in the presence of a finite number of bodies is here considered assuming linear theory. The usual assumptions of incompressible, homogeneous and inviscid fluid, irrotational and simple harmonic motion with angular frequency ω under gravity only, are made. The y-axis is chosen vertically upwards and the plane $y = 0$ is taken as the mean horizontal position of the upper surface of the fluid. Two-dimensional motion depending on x, y only is considered. The fluid occupies the region $y < 0$ if it is infinitely deep, or $-h < y < 0$ if it is of uniform finite depth h. If $Re\{\phi(x, y)e^{-i\omega t}\}$ denotes the velocity potential describing the motion in the fluid, then ϕ satisfies

$$\nabla^2\phi = 0 \text{ in the fluid region}, \tag{1}$$

where ∇^2 denotes the two-dimensional Laplace operator. On the upper surface having the mean position $y = 0$, ϕ satisfies the free-surface condition with higher-order derivatives of the form

$$\left(D\frac{\partial^4}{\partial x^4} + (1-\varepsilon K)\right)\frac{\partial\phi}{\partial y} - K\phi = 0 \quad \text{on } y = 0 \tag{2}$$

if the free-surface has an ice-cover modelled as a thin elastic plate, where $D = Eh_0^3/12(1-\nu^2)\rho g$, $\varepsilon = \rho_0 h_0/\rho$, ρ_0 is the density of ice, ρ is density of water, h_0 is the small thickness of ice-cover, E, ν are the Young's modulus and Poisson's ratio of the ice and $K = \omega^2/g$, g being the acceleration due to gravity. A generalization of (2) for more higher-order derivatives has been introduced by Manam et al. (2006) and has the form

$$\ell\phi_y - K\phi = 0 \text{ on } y = 0, \tag{2'}$$

where ℓ is a linear differential operator of the form

$$\ell = \sum_{m=0}^{m_0} c_m \frac{\partial^{2m}}{\partial x^{2m}}. \tag{3}$$

In (3) $c_m (m = 0, 1, \cdots, m_0)$ are known constants. Keeping in mind various physical problems involving fluid structure interaction, only the even order partial derivatives in x are considered in the differential operator ℓ. The bottom condition is given by

$$\nabla\phi \to 0 \quad \text{as} \quad y \to -\infty \tag{4a}$$

for infinitely deep water, or by

$$\phi_y = 0 \quad \text{on} \quad y = -h \tag{4b}$$

for water of uniform finite depth h. Finally, the body boundary conditions are given by

$$\phi_n \text{ is prescribed on } B = B_F \bigcup B_S, \tag{5}$$

where B_F denotes the wetted parts of the floating bodies while B_S denotes the submerged body boundaries and ϕ_n denotes the derivative normal to B (see Fig. 7.1).

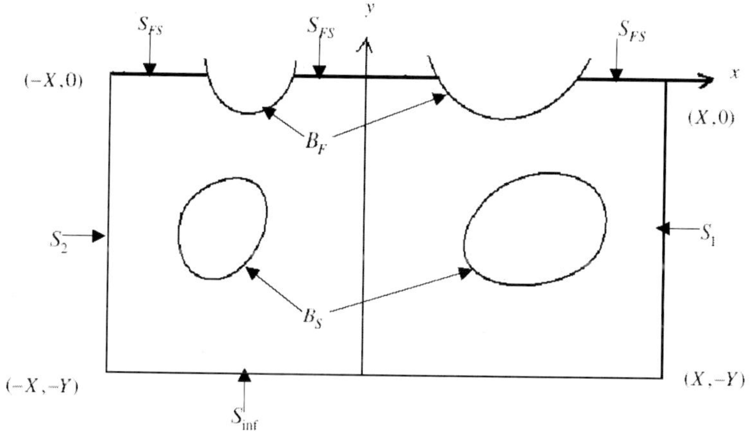

Fig. 7.1 Boundaries of a single-layer fluid

The forms of the far-field on the two sides of B are given by

$$\phi(x,y) \to \begin{cases} (A^{\pm}e^{\pm ip_0x} + B^{\pm}e^{\pm ip_0x})e^{p_0y} \text{ as } x \to \pm\infty \text{ for deep water,} \\ (A^{\pm}e^{\pm ip_0x} + B^{\pm}e^{\pm ip_0x})g(y) \text{ as } x \to \pm\infty \text{ for finite depth water,} \end{cases} \tag{6}$$

where

$$g(y) = \frac{\cosh p_0(y+h)}{\cosh p_0 h}, \tag{7}$$

and p_0 satisfies the transcendental equation

$$\sum_{m=0}^{m_0}(-1)^m c_m p^{2m} = K \text{ for deep water}, \tag{8a}$$

$$\left(\sum_{m=0}^{m_0}(-1)^m c_m p^{2m}\right)\tanh ph = K \text{ for finite depth water.} \tag{8b}$$

Under specific assumptions involving the constants $c_m (m = 0,1,\cdots m_0)$, the equation (8a) or (8b) is assumed to possess only one real positive root. This is also physically realistic since wave of only one wavenumber can propagate on the upper surface.

A convenient short notation for (6) is Linton and McIver (1995),

$$\phi \sim \left(A^-, B^-; A^+, B^+\right) \tag{9}$$

where $A^\pm, B^\pm$ denote the amplitudes as $x \to \pm\infty$ of the outgoing and incoming waves set up at either infinities.

Two-layer fluid

In a two-layer fluid, both the upper and lower fluids are assumed to be homogeneous, incompressible and inviscid. Let ρ^I be the density of the upper fluid and $\rho^{II}(> \rho^I)$ be the same for the lower fluid. Let the lower fluid extend infinitely downwards while the upper one has a finite height h above the mean interface. Let y-axis point vertically upwards from the undisturbed interface $y = 0$. Thus the upper layer occupies the region $0 < y < h$ while the lower layer occupies the region $y < 0$. Under the usual assumption of linear theory and irrotational two-dimensional motion, velocity potentials $Re\{\phi^{I,II}(x, y)e^{-i\omega t}\}$ describing the fluid motion in the upper and lower layers exist. For a general boundary value problem, $\phi^{I,II}$ satisfy

$$\nabla^2\phi^I = 0,\ 0 < y < h, \tag{10a}$$

$$\nabla^2\phi^{II} = 0,\ y < 0. \tag{10b}$$

The linearized boundary conditions at the interface $y = 0$ are

$$\phi_y^I = \phi_y^{II} \text{ on } y = 0, \tag{11a}$$

$$\rho\left(\phi_y^I - K\phi^I\right) = \phi_y^{II} - K\phi^{II} \text{ on } y = 0 \tag{11b}$$

where $\rho = \rho^I/\rho^{II} (< 1)$, while the free-surface condition with higher-order derivatives at $y = h$ is

$$\ell\phi_y^I - K\phi^I = 0 \text{ on } y = h \tag{12}$$

where the differential operator ℓ has the same form as given in (3). The bottom condition is given by

$$\nabla \phi^{II} \to 0 \quad \text{as } y \to -\infty. \tag{13}$$

The conditions on the body boundary are given by

$$\phi_n^I \text{ is prescribed on } B_I = B_{IF} \bigcup B_{IS} \tag{14a}$$

where B_{IF} denotes the wetted parts of the floating bodies, while B_{IS} denotes the body boundaries submerged in the upper fluid (see Fig. 7.2) and

$$\phi_n^{II} \text{ is prescribed on } B_{II} \tag{14b}$$

where B_{II} represents the body boundaries submerged in the lower fluid.

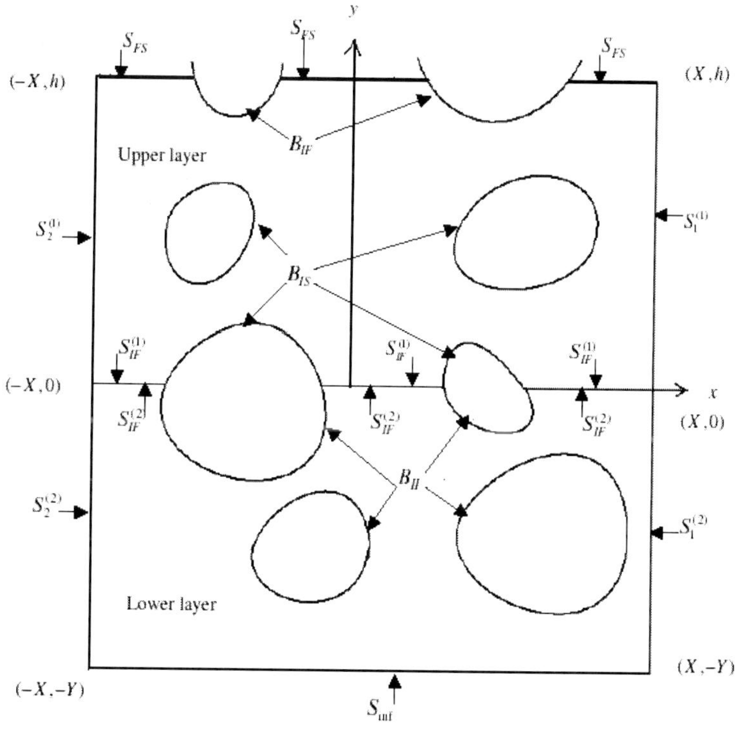

Fig. 7.2 Boundaries of a two-layer fluid

It is well known that there exists two distinct values of the wavenumbers for time-harmonic progressive waves of a particular frequency, one propagating at the upper surface and the other at the interface. Thus for any wave propagation problem, the far-field consists of outgoing and incoming waves at each of the two wavenumbers λ_1, λ_2, say, where λ_1, λ_2 are defined in (17) below. Thus it is given by

$$\phi^I \to \left(A^\pm e^{\pm i\lambda_1 x}+C^\pm e^{\mp i\lambda_1 x}\right) g_1(y)+\left(B^\pm e^{\pm i\lambda_2 x}+D^\pm e^{\mp i\lambda_2 x}\right)g_2(y) \text{ as } x\to\pm\infty, \quad (15a)$$

$$\phi^{II} \to \left(A^\pm e^{\pm i\lambda_1 x}+C^\pm e^{\mp i\lambda_1 x}\right) e^{\lambda_1 y}+\left(B^\pm e^{\pm i\lambda_2 x}+D^\pm e^{\mp i\lambda_2 x}\right)e^{\lambda_2 y} \text{ as } x\to\pm\infty, \quad (15b)$$

where $g_1(y), g_2(y)$ are defined in (18) below.

In the notation of Linton and McIver (1995), a short-hand version of (15) is

$$\phi \sim \left(A^-,B^-,C^-,D^-;A^+,B^+,C^+,D^+\right). \quad (16)$$

In (15), the real positive number $\lambda_1, \lambda_2 (> \lambda_1)$ are the only two real positive roots of the transcendental equation

$$\left(\sum_{m=0}^{m_0}(-1)^m c_m \lambda^{2m+1} \sinh \lambda h - K\cosh \lambda h\right)(\lambda-\rho\lambda-K)$$
$$-\rho K\left(\sum_{m=0}^{m_0}(-1)^m c_m \lambda^{2m+1} \cosh \lambda h - K\sinh \lambda h\right) = 0, \quad (17)$$

and the functions $g_j(y)(j=1,2)$ are given by

$$g_j(y) = \frac{\{(1-\rho)\lambda_j - K\}}{\rho K\left[\sum_{m=0}^{m_0}(-1)^m c_m \lambda_j^{2m+1}\cosh \lambda_j h - K\sinh \lambda_j h\right]}$$
$$\times\left[\left\{\sum_{m=0}^{m_0}(-1)^m c_m \lambda_j^{2m+1} + K\right\}e^{\lambda_j(y-h)} + \left\{\sum_{m=0}^{m_0}(-1)^m c_m \lambda_j^{2m+1} - K\right\}e^{-\lambda_j(y-h)}\right]. \quad (18)$$

The constants $c_m (m = 0,1,\cdots m_0)$ are assumed to be such that the equation (17) possesses only two real positive roots, which correspond to the two different wavenumbers (modes) at which progressive waves propagate at the upper surface and the interface of the two-layer fluid. It is emphasized that the physical constants $c_m (m = 0,1,\cdots m_0)$ are such that only two real positive roots of the equation (17) exist due to physical reasons.

Derivation of energy identity (identities)

Modified form of Green's integral theorem

In the case of a single-layer fluid having free-surface condition with first-order derivative, a relation exists between the various hydrodynamics quantities that arise in a general boundary value problem involving a finite number of body boundaries, and can be determined by a judicious application of the standard Green's integral theorem for harmonic functions in the form

$$\int_S (\phi\psi_n - \psi\phi_n)ds = 0. \quad (19)$$

In (19), S denotes the boundary of the fluid region and ϕ_n, ψ_n denote partial derivatives along the normal to S. A similar relation can also be determined in the case of a two-layer fluid having free-surface condition with first-order derivative.

For the determination of a similar relation for a general boundary value problem in a single-layer fluid or a two-layer fluid with higher-order derivatives in the free-surface condition, Green's theorem (19) has to be modified taking into account this higher-order derivatives. For boundary condition of the form as given by (2) for a single-layer fluid or by (12) for a two-layer fluid, the modified form of (19) is given by

$$\int_S (\phi L_n \psi - \psi L_n \phi) ds = 0 \tag{20}$$

where the operator L_n is of the form

$$L_n = \sum_{m=0}^{m_0} (-1)^m c_m \frac{\partial^{2m+1}}{\partial n^{2m+1}}, \tag{21}$$

$\frac{\partial}{\partial n}$ being the derivative normal to S.

The proof of the generalization (20) of the standard Green's theorem (19) is given by Das et al. (2008).

The modified form of Green's theorem given by (20) is now employed to obtain the desired relations between different hydrodynamic quantities for a general boundary value problem in a single-layer or a two-layer fluid with higher-order derivatives in the free-surface.

Single-layer fluid

Let ϕ, ψ be the solutions of the two different problems with ϕ_n, ψ_n being prescribed on the submerged body boundary B. Let the far-field form of ϕ be given by (2.9) while that for ψ be given by

$$\psi \sim \left(P^-, Q^-; P^+, Q^+\right). \tag{22}$$

Let S be chosen as the boundary of the fluid region as given in Fig. 7.1 where X and Y are arbitrarily large, S_{FS} denotes the portions of free surface between $-X$ to X. Then application of (19) to ϕ, ψ produces

$$\left(\int_{S_{FS}} + \int_{S_2} + \int_{S_{inf}} + \int_{S_1} + \int_B \right) (\phi L_n \psi - \psi L_n \phi) ds = 0. \tag{23}$$

The first integral in (23) is

$$\int_{S_{FS}} (\phi L_y \psi - \psi L_y \phi)(x,0) dx = \sum_{m=0}^{m_0} c_m \int_{S_{FS}} \left(\phi \frac{\partial^{2m+1} \psi}{\partial x^{2m} \partial y} - \psi \frac{\partial^{2m+1} \phi}{\partial x^{2m} \partial y} \right)(x,0) dx.$$

Use of the free-surface condition with higher-order derivatives (2′) on $y = 0$ makes this integral identically equal to zero for any X.

The second integral in (23) is

$$\int_{S_2}(\psi L_x\phi-\phi L_x\psi)(-X,y)dy.$$

Making $X\to\infty, Y\to\infty$, this reduces to, after using the far-field condition (6a),

$$i\sum_{m=0}^{m_0}c_m p_0^{2m}\left(Q^-A^- - P^-B^-\right). \tag{24}$$

Similarly, the fourth integral in (23) reduces to, as $X, Y\to\infty$,

$$-i\sum_{m=0}^{m_0}c_m p_0^{2m}\left(P^+B^+ - Q^+A^+\right). \tag{25}$$

Again, the third integral in (23) tends to 0 as $Y\to\infty$ by using the bottom condition (4a). Finally, the last integral in (23) is

$$\int_B\left[\phi\sum_{m=0}^{m_0}(-1)^m c_m\frac{\partial^{2m+1}\psi}{\partial n^{2m+1}}-\psi\sum_{m=0}^{m_0}(-1)^m c_m\frac{\partial^{2m+1}\phi}{\partial n^{2m+1}}\right]ds.$$

Let the cross-section B of the body boundaries be described parametrically by $x=X(\theta), y=Y(\theta), (0\le\theta\le 2\pi$ for a submerged body and $\alpha\le\theta\le\beta$ for wetted portion of floating body, α,β may be negative) where $\theta=0$ is chosen to be coincident with the line $x=0$. Defining (s,n) as rectangular co-ordinates along the normal and tangent to B at any point of B, and using the relation (2.14) of Porter (2002), the harmonic functions ϕ,ψ satisfy $\nabla_1^2\phi=0, \nabla_1^2\psi=0$ where $\nabla_1^2=\frac{\partial^2}{\partial s^2}+\frac{\partial^2}{\partial n^2}+\kappa(s)\frac{\partial}{\partial n}$, $\kappa(s)$ being the curvature as a function of the arc length s. Now

$$\frac{\partial^{2m}}{\partial n^{2m}}=(-1)^m\left(\frac{\partial^2}{\partial s^2}+\kappa(s)Q(s)\frac{\partial}{\partial s}\right)^{2m}$$

where $Q(s)=\frac{(Y'(\theta))^2-(X'(\theta))^2}{X'(\theta)Y'(\theta)}$, and is a function of s. Using these we find the last integral in (23) to be

$$\int_B\left[\varphi\mathcal{M}_s\frac{\partial\psi}{\partial n}-\psi\mathcal{M}_s\frac{\partial\varphi}{\partial n}\right]ds, \tag{26}$$

where

$$\mathcal{M}_s\equiv\sum_{m=0}^{m_0}c_m\left(\frac{\partial^2}{\partial s^2}+\kappa(s)Q(s)\frac{\partial}{\partial s}\right)^{2m}.$$

If $\frac{\partial\phi}{\partial n}$ and $\frac{\partial\psi}{\partial n}$ vanish on B (for scattering problems), then this integral vanishes identically.

Collecting all the terms in (23) we obtain the relation

$$i\sum_{m=0}^{m_0}c_m p_0^{2m}\left(Q^-A^- - P^-B^-\right)+\int_B\cdots ds=i\sum_{m=0}^{m_0}c_m p_0^{2m}\left(P^+B^+ - Q^+A^+\right) \tag{27}$$

where $\int_B \cdots ds$ is the same as given by (26) above.

For a *scattering problem*, $\frac{\partial \phi}{\partial n} = 0$ on B and the far-field form of ϕ is given by

$$\phi \sim (R, 1; T, 0) \tag{28}$$

where R and T are the reflection and transmission coefficients respectively due to an incident field propagating from the direction of $x = -\infty$. Let $\overline{\phi}$ denote the complex conjugate of ϕ. Then $\frac{\partial \overline{\phi}}{\partial n} = 0$ on B, and the far-field form of $\overline{\phi}$ is

$$\overline{\phi} \sim \left(1, \overline{R}; 0, \overline{T}\right). \tag{29}$$

Writing $\psi = \overline{\phi}$ in (27), we obtain

$$\left(|T|^2 + |R|^2 - 1\right)\sum_{m=0}^{m_0} c_m p_0^{2m} = 0,$$

giving

$$|T|^2 + |R|^2 = 1 \tag{30}$$

which is the desired *energy identity*.

For a *scattering* potential function ϕ and a *radiation* potential function ψ, $\frac{\partial \phi}{\partial n} = 0$ on B and the far-field form of ϕ is given by (28) and $\frac{\partial \psi}{\partial n} = n_j$, where n_j is the component of the normal to the body in the jth mode of motion. The far-field form of ψ is

$$\psi \sim \left(\Theta_j^-, 0;\ \Theta_j^+, 0\right). \tag{31}$$

Then we obtain from (27)

$$i\rho\omega \int_B \varphi \mathcal{M}_s n_j \, ds = -\rho\omega \sum_{m-0}^{m_0} c_m p_0^{2m} \Theta_j^- \tag{32}$$

where ρ is the density of the fluid. For $m_0 = 0$ and $c_0 = 1$, i.e., when the free surface condition has the usual form $K\phi + \phi_y = 0$, the left side of (32) produces $i\rho\omega \int_B \varphi n_j \, ds$, which is the hydrodynamic force on the body B in the jth mode of motion. Thus we may call (32) as the modified form of Haskind-Hanaoka relation. However, it should be noted that, for $m_0 > 0$, the left side of (32) cannot be termed as the actual hydrodynamic force on the body since it cannot be obtained by integrating over the body surface the expression of pressure from Bernoulli's equation.

Now we consider the case of two *radiation* potential functions. Let $\phi = \phi_j$ and $\psi = \phi_k$ be two *radiation* potentials whose behaviour in the far-field is given by

$$\phi_j \sim \left(\Theta_j^-, 0;\ \Theta_j^+, 0\right),$$

$$\phi_k \sim \left(\Theta_k^-, 0;\ \Theta_k^+, 0\right)$$

and which satisfy the body boundary condition

$$\frac{\partial \phi_j}{\partial n} = n_j, \quad \frac{\partial \phi_k}{\partial n} = n_k \quad \text{on } B.$$

Then we obtain from (27)

$$\int_B \left(\varphi_j \mathcal{M}_s n_k - \varphi_k \mathcal{M}_s n_j\right) ds = 0. \tag{33}$$

If we now use $\psi = \overline{\phi}_k$, the complex conjugate of ϕ_k, then we find from (27) that

$$\int_B \left(\varphi_j \mathcal{M}_s n_k - \overline{\varphi}_k \mathcal{M}_s n_j\right) ds = -i \sum_{m=0}^{m_0} c_m p_0^{2m} \left(\Theta_j^- \overline{\Theta}_k^- + \Theta_j^+ \overline{\Theta}_k^+\right). \tag{34}$$

In particular, for the case when $j = k$ equation (27) becomes

$$\int_B \left(\varphi_j \mathcal{M}_s n_j - \overline{\varphi}_j \mathcal{M}_s n_j\right) ds = -i \sum_{m=0}^{m_0} c_m p_0^{2m} \left(\left|\Theta_j^-\right|^2 + \left|\Theta_j^+\right|^2\right). \tag{35}$$

This is an extension of the modified energy-conservation principle to the case of a wave propagation problem in a single-layer fluid having free-surface condition with higher-order derivatives.

For the case of water of uniform finite depth h, we replace Y by h in (23). Then as $X \to \infty$, the second integral reduces to

$$i \sum_{m=0}^{m_0} c_m p_0^{2m} \left(1 - e^{-2p_0 h}\right)\left(Q^- A^- - P^- B^-\right) \tag{36}$$

while the fourth integral reduces to

$$-i \sum_{m=0}^{m_0} c_m p_0^{2m} \left(1 - e^{-2p_0 h}\right)\left(P^+ B^+ - Q^+ A^+\right) \tag{37}$$

Thus, in this case, we obtain the relation

$$i \sum_{m=0}^{m_0} c_m p_0^{2m} \left(1 - e^{-2p_0 h}\right)\left\{\left(Q^- A^- - P^- B^-\right) - \left(P^+ B^+ - Q^+ A^+\right)\right\} + \int_B \cdots ds = 0 \tag{38}$$

where the integral $\int_B \cdots ds$ is the same as given by (26).

For a scattering problem $\phi_n = 0$ on B and choosing $\psi = \overline{\phi}$ so that $\psi_n = 0$ on B, we finally obtain, after using (8b), the same identity (30).

For a scattering potential function ϕ and a radiation potential function ψ, we obtain from (38) the relation

$$i\rho\omega \int_B \varphi\, \mathcal{M}_s n_j\, ds = -\rho\omega \sum_{m=0}^{m_0} c_m p_0^{2m} \left(1 - e^{-2p_0 h}\right)\Theta_j^- \tag{39}$$

which is the *modified* form of Haskind-Hanaoka relation.

Again if ϕ_j and ψ_k denote solutions of two radiation problems satisfying on the body boundary B, $\dfrac{\partial \phi_j}{\partial n} = n_j$, $\dfrac{\partial \psi_k}{\partial n} = n_k$, then we find from (38) the relation for $j = k$

$$\int_B \left(\varphi_j \mathcal{M}_s n_j - \overline{\varphi}_j \mathcal{M}_s n_j\right) ds = i \sum_{m=0}^{m_0} c_m p_0^{2m} \left(1 - e^{-2p_0 h}\right)\left(\left|\Theta_j^-\right|^2 + \left|\Theta_j^+\right|^2\right). \tag{40}$$

This is the modified form of the energy-conservation principle.

Thus the modified forms of the energy identity for any diffraction problem, the Haskind-Hanaoka relation for any radiation and diffraction problems, the energy-conservation principle for any radiation problem are established for a single-layer fluid having free-surface boundary condition with higher-order derivatives. When we put $c_0 = 1,\ c_m = 0\,(m = 1, 2, 3, \cdots, m_0)$ in these relations, the energy identity, the Haskind-Hanaoka relation and the energy-conservation principle for a single-layer fluid with the usual free-surface condition as given by Mei (1982) are obtained.

Two-layer fluid

Let there be situated a finite number of bodies in a two-layer fluid, some in the upper layer and some in the lower layer (cf. Fig. 7.2). There may be some bodies which are present in both the layers. Let the boundaries of the bodies lying in the upper layer be denoted by B_I and those in the lower layer by B_{II}. Let ϕ, ψ be the solutions of two different boundary value problems with ϕ_n, ψ_n being prescribed on the boundaries B_I and B_{II}, and the far-field form of ϕ being given by (16) and that of ψ is given by

$$\psi \sim \left(M^-, N^-, P^-, Q^-; M^+, N^+, P^+, Q^+\right). \tag{41}$$

In this case we choose S in (20) first to be the boundary of the region in the upper fluid as shown in Fig. 7.2 and ultimately make $X \to \infty$, and next to be the boundary of the region in the lower fluid as shown in Fig. 7.2 and ultimately make both $X, Y \to \infty$.

For the upper layer, (20) produces

$$\left(\int_{S_{FS}} + \int_{S_2^{(1)}} + \int_{S_{IF}^{(1)}} + \int_{S_1^{(1)}} + \int_{B_I}\right)\left(\phi^I L_n \psi^I - \psi^I L_n \phi^I\right) ds = 0. \tag{42}$$

The first integral in (42) vanishes identically due to the boundary condition (12) satisfied by both ϕ^I and ψ^I. The second integral in (42) is

$$\int_{S_2^{(1)}} \left(\phi^I L_x \psi^I - \psi^I L_x \phi^I\right)(-X, y)\, dy.$$

Making use of the far-field behavior of ϕ^I, ψ^I for large X, this produces

$$\begin{aligned}
&2i\left[\sum_{m=0}^{m_0} c_m \left\{\lambda_1^{2m+1}\left(P^-A^- - M^-C^-\right)\int_0^h \left(g_1(y)\right)^2 dy + \lambda_2^{2m+1}\left(Q^-B^- - N^-D^-\right)\int_0^h \left(g_2(y)\right)^2 dy\right\}\right] \\
&\quad + i\left[\sum_{m=0}^{m_0} c_m \left\{\left(\lambda_1^{2m+1} - \lambda_2^{2m+1}\right)\left(\left(N^-A^- - M^-B^-\right)e^{-i(\lambda_1+\lambda_2)X} + \left(Q^-C^- - D^-P^-\right)e^{i(\lambda_1+\lambda_2)X}\right)\right.\right. \\
&\quad \left.\left. + \left(\lambda_1^{2m+1} + \lambda_2^{2m+1}\right)\left(\left(Q^-A^- - M^-D^-\right)e^{-i(\lambda_1-\lambda_2)X} + \left(B^-P^- - N^-C^-\right)e^{i(\lambda_1-\lambda_2)X}\right)\right\} . \times \int_0^h g_1(y) g_2(y)\, dy\right].
\end{aligned} \tag{43}$$

Similarly the fourth integral in (42) produces an expression which is similar to (43) with the subscripts minus (–) replaced by plus (+). The third integral in (42) is

$$\int_{S_F^{(1)}} \left(\phi^I L_y \psi^I - \psi^I L_y \phi^I\right)(x,0)dx. \tag{44}$$

Finally, the last integral in (42) becomes, after using the same reasoning as used in obtaining (26),

$$\int_{B_I} \left[\varphi^I \mathcal{M}_s\left(\frac{\partial \psi^I}{\partial n}\right) - \psi^I \mathcal{M}_s\left(\frac{\partial \varphi^I}{\partial n}\right)\right] ds. \tag{45}$$

For the lower layer, (20) produces

$$\left(\int_{S_{IF}^{(2)}} + \int_{S_2^{(2)}} + \int_{S_{inf}} + \int_{S_1^{(2)}} + \int_{B_{II}}\right)\left(\phi^{II} L_n \psi^{II} - \psi^{II} L_n \phi^{II}\right) ds = 0. \tag{46}$$

The first integral in (46) is

$$\int_{S_{IF}^{(2)}} \left(\phi^{II} L_y \psi^{II} - \psi^{II} L_y \phi^{II}\right)(x,0)dx. \tag{47}$$

The second integral in (46) reduces to, after using (15b) and making $Y \to \infty$,

$$\begin{aligned}
&i\left[\sum_{m=0}^{m_0} c_m \left\{\lambda_1^{2m}\left(P^- A^- - M^- C^-\right) + \lambda_2^{2m}\left(Q^- B^- - N^- D^-\right)\right\}\right] \\
&+i\left[\sum_{m=0}^{m_0} c_m \left\{\left(\lambda_1^{2m+1} - \lambda_2^{2m+1}\right)\left(\left(N^- A^- - M^- B^-\right)e^{-i(\lambda_1+\lambda_2)X} + \left(Q^- C^- - D^- P^-\right)e^{i(\lambda_1+\lambda_2)X}\right)\right.\right. \\
&\left.\left. + \left(\lambda_1^{2m+1} + \lambda_2^{2m+1}\right)\left(\left(Q^- A^- - M^- D^-\right)e^{-i(\lambda_1-\lambda_2)X} + \left(B^- P^- - N^- C^-\right)e^{i(\lambda_1-\lambda_2)X}\right)\right\}\right. \\
&\left. \times \int_{-\infty}^{0} e^{(\lambda_1+\lambda_2)y} dy\right].
\end{aligned} \tag{48}$$

Similarly, the fourth integral in (46) reduces to the same expression (48) with the subscripts minus (–) replaced by plus (+).

Again, the third integral in (46) tends to 0 as $Y \to -\infty$, after using the conditions at infinite depth. Finally, the last integral in (46) becomes,

$$\int_{B_{II}} \left[\varphi^{II} \mathcal{M}_s\left(\frac{\partial \psi^{II}}{\partial n}\right) - \psi^{II} \mathcal{M}_s\left(\frac{\partial \varphi^{II}}{\partial n}\right)\right] ds. \tag{49}$$

Substituting all these results in (42) and (46), and using the condition at the interface given by

$$\rho\left(\phi^I L_y \psi^I - \psi^I L_y \phi^I\right) = \phi^{II} L_y \psi^{II} - \psi^{II} L_y \phi^{II} \quad \text{on } y = 0 \tag{50}$$

and the result

$$\rho \int_0^h g_1(y) g_2(y) dy + \int_{-\infty}^{0} e^{(\lambda_1+\lambda_2)y} dy = 0, \tag{51}$$

and making $X \to \infty$, we obtain

$$\rho\int_{B_I}[\ ]ds+\int_{B_{II}}[\ ]ds+iJ_{\lambda_1}\left(P^+A^+-M^+C^++P^-A^--M^-C^-\right)$$
$$+iJ_{\lambda_2}\left(Q^+B^+-N^+D^++Q^-B^--N^-D^-\right)=0 \tag{52}$$

where $\int_{B_I}[\ ]ds$ is given in (45), $\int_{B_{II}}[\ ]ds$ is given in (49), and

$$J_{\lambda_j}=\sum_{m=0}^{m_0}c_m\lambda_j^{2m}\left\{1+2\rho\lambda_j\int_0^h(g_j(y))^2dy\right\},\ j=1,2. \tag{53}$$

The equation (52) gives a relation between the wave amplitudes of the two boundary value problems described by ϕ,ψ in terms of their values together with their normal derivatives on the body boundaries B_I and B_{II}.

If we now consider wave scattering by a fixed set of bodies, then in general there are two problems to be considered. Scattering of an incident wave at mode λ_1 is referred to as Problem 1. Problem 2 refers to scattering of an incident wave at mode λ_2.

The notations R_{λ_j} and T_{λ_j} $(j=1,2)$ are used to denote the reflection and transmission coefficients respectively corresponding to waves of wave number λ_1 due to an incident wave of wave number λ_j $(j=1,2)$ while r_{λ_j} and t_{λ_j} $(j=1,2)$ are used to denote reflection and transmission coefficients corresponding to waves of wave number λ_2 due to an incident wave of wave number λ_j $(j=1,2)$. Thus the two scattering problems are characterized by

$$\phi_1\sim\left(R_{\lambda_1},r_{\lambda_1},1,0;T_{\lambda_1},t_{\lambda_1},0,0\right), \tag{54}$$

$$\phi_2\sim\left(R_{\lambda_2},r_{\lambda_2},0,1;T_{\lambda_2},t_{\lambda_2},0,0\right). \tag{55}$$

Also $\frac{\partial\phi^I}{\partial n}=0$ on B_I, $\frac{\partial\phi^{II}}{\partial n}=0$ on B_{II}. Applying (52) to $\phi=\phi_1$ and $\psi=\overline{\phi_1}$, the complex conjugate of ϕ_1, we obtain the identity relating the reflection coefficients $R_{\lambda_1},r_{\lambda_1}$ and the transmission coefficients $T_{\lambda_1}\ t_{\lambda_1}$ as given by

$$|R_{\lambda_1}|^2+|T_{\lambda_1}|^2+J\left(|r_{\lambda_1}|^2+|t_{\lambda_1}|^2\right)=1 \tag{56}$$

where

$$J=J_{\lambda_2}/J_{\lambda_1}. \tag{57}$$

Similarly we obtain for the scattering problem 2

$$|R_{\lambda_2}|^2+|T_{\lambda_2}|^2+J\left(|r_{\lambda_2}|^2+|t_{\lambda_2}|^2\right)=J. \tag{58}$$

The relations (56) and (57) are the two desired identities.

Let ψ be a radiation potential function with far-field behaviour given by

$$\psi \sim \left(\Theta_{\lambda_1}^-, \Theta_{\lambda_2}^-, 0, 0;\ \Theta_{\lambda_1}^+, \Theta_{\lambda_2}^+, 0, 0\right) \tag{59}$$

and on the body boundaries $\frac{\partial \psi}{\partial n} = n_j$. Applying (52) to $\phi = \phi_1$ and ψ, we obtain the relation

$$i\rho_{II}\omega\left[\rho\int_{B_I}\varphi^I \mathcal{M}_s n_j\,ds + \int_{B_{II}}\varphi^{II}\mathcal{M}_s n_j\,ds\right] = -\rho_{II}\omega J_{\lambda_1}\Theta_{\lambda_1}^-. \tag{60}$$

Similarly, for $\phi = \phi_2$ we obtain

$$i\rho_{II}\omega\left[\rho\int_{B_I}\varphi^I \mathcal{M}_s n_j\,ds + \int_{B_{II}}\varphi^{II}\mathcal{M}_s n_j\,ds\right] = -\rho_{II}\omega J_{\lambda_2}\Theta_{\lambda_2}^-. \tag{61}$$

The relations (60) and (61) represent the modified Haskind-Hanaoka relations in a two-layer fluid having free-surface boundary condition with higher-order derivatives.

Let ϕ and ψ denote the radiation potential functions whose far-field behaviours are given by

$$\phi \sim \left(A_{\lambda_1}^-, A_{\lambda_2}^-, 0, 0;\ A_{\lambda_1}^+, A_{\lambda_2}^+, 0, 0\right),$$

$$\psi \sim \left(\Theta_{\lambda_1}^-, \Theta_{\lambda_2}^-, 0, 0;\ \Theta_{\lambda_1}^+, \Theta_{\lambda_2}^+, 0, 0\right),$$

and on the body boundaries $\frac{\partial \phi}{\partial n} = n_j$ and $\frac{\partial \psi}{\partial n} = n_k$.

Applying (52) to ϕ and ψ we obtain

$$\rho\int_{B_I}\left(\varphi^I \mathcal{M}_s n_k - \psi^I \mathcal{M}_s n_j\right)ds + \int_{B_{II}}\left(\varphi^{II}\mathcal{M}_s n_k - \psi^{II}\mathcal{M}_s n_j\right)ds = 0. \tag{62}$$

Now applying (52) to ϕ and $\overline{\psi}$, the complex conjugate of ψ, we obtain

$$\rho\int_{B_I}\left(\varphi^I \mathcal{M}_s n_k - \overline{\psi}^I \mathcal{M}_s n_j\right)ds + \int_{B_{II}}\left(\varphi^{II}\mathcal{M}_s n_k - \overline{\psi}^{II}\mathcal{M}_s n_j\right)ds$$

$$= -iJ_{\lambda_1}\left(A_{\lambda_1}^+\overline{\Theta}_{\lambda_1}^+ + A_{\lambda_1}^-\overline{\Theta}_{\lambda_1}^-\right) - iJ_{\lambda_2}\left(A_{\lambda_2}^+\overline{\Theta}_{\lambda_2}^+ + A_{\lambda_2}^-\overline{\Theta}_{\lambda_2}^-\right). \tag{63}$$

Thus in particular, for $\psi = \overline{\phi}$, the above relation reduces to

$$\rho\int_{B_I}\left(\varphi^I \mathcal{M}_s n_j - \overline{\varphi}^I \mathcal{M}_s n_j\right)ds + \int_{B_{II}}\left(\varphi^{II}\mathcal{M}_s n_j - \overline{\varphi}^{II}\mathcal{M}_s n_j\right)ds$$

$$= -iJ_{\lambda_1}\left(\left|A_{\lambda_1}^+\right|^2 + \left|A_{\lambda_1}^-\right|^2\right) - iJ_{\lambda_2}\left(\left|A_{\lambda_2}^+\right|^2 + \left|A_{\lambda_2}^-\right|^2\right). \tag{64}$$

This relation (3.46) is the modified form of the energy-conservation principle for a wave propagation problem in a two-layer fluid having free-surface boundary condition with higher-order derivatives.

Thus we have obtained above various hydrodynamic relations such as the energy identities, the modified Haskind-Hanaoka relations and the energy-conservation principle for scattering and radiation problems in a two-layer fluid having free-surface boundary condition with higher-order derivatives. If we put $c_0 = 1$, $c_m = 0\,(m = 1, 2, 3, \cdots, m_0)$, then these reduce to the corresponding relations in a two-layer fluid with the usual free-surface as in Newman (1976), Linton and McIver (1995), Yeung and Nguyen (2000), Ten and Kashiwagi (2004), Kashiwagi et al. (2006).

CHAPTER VIII

Two-Layer Fluid

8.1 Scattering by a thin vertical plate in a two layer fluid

Stokes (1847) first investigated propagation of waves in a two-layer fluid assuming linear theory. In the classical treatise by Lamb (1932), it was shown that in a two-layer fluid with a free surface there exist two possible linear wave systems at a given frequency each with a different wave number, the waves with lower wave number (or mode) propagates along the free surface while those with higher wave number propagates along the interface. When a wave train of a particular mode encounters an obstacle, it is partially reflected into waves of both modes and also partially transmitted similarly. Thus there is a transfer of energy from surface wave to interface wave and vice-versa. This makes the study of wave scattering problems in a two-layer fluid interesting. Linton and McIver (1995) developed the general theory for two-dimensional motion in a two-layer fluid in which the lower layer of heavier density extends infinitely downwards and the upper layer of lower density has a free surface. Dhillon et al. (2014) investigated the problem of scattering of surface and interface waves by a thin vertical plate partially immersed in the upper layer of a two-layer fluid consisting of a layer of finite depth bounded above by a free surface and below by an infinite layer of fluid of density greater than the upper one. They formulated the problem in terms of a hypersingular integral equation for the difference of velocity potential across the plate. The hypersingular integral equation was then solved numerically by using a collocation method based on Chebyshev polynomial approximation Parsons and Martin (1992, 1994). The solution of this hypersingular integral equation was then utilized to compute the reflection and transmission coefficients. The energy identities were used to check the correctness of the numerical results presented here. An account of this is presented in this section.

We consider two-dimensional irrotational, time-harmonic motion in two-layer fluid of which the upper layer is of density ρ_1 and the lower layer is of density $\rho_2(>\rho_1)$. A thin rigid vertical plate described by $x = 0, 0 < y < a$, is partially immersed in the upper layer which occupies the region $0 < y < h (h > a)$ with $y = 0$ as the mean free surface, y-axis being taken vertically downwards. The lower fluid occupies the region $h < y < \infty$ where $y = h$ is the undisturbed mean interface of the two fluids.

Under the usual assumption of linear theory, the velocity potentials describing fluid motions in the upper and lower fluid regions are

$$\Phi(x,y;t) = Re\{\phi(x,y)e^{-i\omega t}\},$$

$$\Psi(x,y;t) = Re\{\psi(x,y)e^{-i\omega t}\},$$

respectively, where ω is the circular frequency. Here, ϕ and ψ satisfy the Laplace equation

$$\nabla^2\phi = 0, \quad 0 < y < h, \tag{1.1}$$

$$\nabla^2\psi = 0, \quad h < y < \infty. \tag{1.2}$$

Linearized boundary conditions at the free surface, plate and at the interface are

$$K\phi + \phi_y = 0, \quad \text{on} \quad y = 0, \tag{1.3}$$

$$\phi_x = 0, \quad \text{on} \quad 0 < y < a, \tag{1.4}$$

$$\phi_y = \psi_y, \quad \text{on} \quad y = h, \tag{1.5}$$

$$s(K\phi + \phi_y) = K\psi + \psi_y \quad \text{on} \quad y = h, \tag{1.6}$$

where $s = \frac{\rho_1}{\rho_2}$ and $K = \frac{\omega^2}{g}$, g being the acceleration due to gravity.

Also, the bottom condition is

$$\nabla\psi \to 0, \quad \text{as} \quad y \to \infty. \tag{1.7}$$

In a two-layer fluid, progressive waves are described by

$$\phi = Ae^{\pm iux}[(K\sigma - u)e^{2uh}e^{-uy} + (K - u)e^{uy}], \tag{1.8}$$

$$\psi = Ae^{\pm iux}K(\sigma - 1)e^{2uh}e^{-uy}, \tag{1.9}$$

where u satisfies the dispersion relation

$$(u - K)[K(\sigma + e^{-2uh}) - u(1 - e^{-2uh})] = 0 \tag{1.10}$$

with $\sigma = \frac{1+s}{1-s}$. It follows that $u = K$ and $u = v$ where

$$(K + v)e^{-vh} + (K\sigma - v)e^{vh} = 0. \tag{1.11}$$

The relation (1.11) has exactly one positive root v Linton and McIver (1995) which lies in the range

$$K\sigma < v < \frac{K(\sigma+1)}{1-e^{-2Kh\sigma}}. \tag{1.12}$$

In any wave scattering problem, the far-field will take the form of incoming and outgoing waves at each of the wave numbers K and v. It is given by

$$\phi \backsim A^{\pm}e^{\pm iKx}e^{-Ky} + B^{\pm}e^{\pm ivx}g(y) + C^{\pm}e^{\mp iKx}e^{-Ky} + D^{\pm}e^{\mp ivx}g(y) \quad (1.13)$$

and

$$\psi \backsim A^{\pm}e^{\pm iKx}e^{-Ky} + B^{\pm}e^{\pm ivx}e^{-v(y-h)} + C^{\pm}e^{\mp iKx}e^{-Ky} + D^{\pm}e^{\mp ivx}e^{-v(y-h)} \quad (1.14)$$

as $x \longrightarrow \pm\infty$, where

$$g(y) = \frac{K\sigma - v}{K(\sigma-1)}e^{-v(y-h)} + \frac{K-v}{K(\sigma-1)}e^{v(y-h)}, 0 < y < h. \quad (1.15)$$

Here, two problems are being considered:

Problem-I : The scattering of an incident wave of wave number K from the direction of $x \longrightarrow -\infty$.

Problem-II : The scattering of an incident wave of wave number v from the direction of $x \longrightarrow -\infty$.

In each case there may be reflected and transmitted waves of wavenumbers K and v. As in Linton and McIver (1995), the notations R and T are used to represent reflection and transmission coefficients corresponding to waves of wavenumber K while r and t are used for waves of wavenumber v. Incident plane wave of wavenumber K has the form

$$\phi^{inc}(x,y) = e^{iKx-Ky}, \quad (1.16)$$

$$\psi^{inc}(x,y) = e^{iKx-Ky},$$

and an incident plane wave of wave number v is given by

$$\phi^{inc}(x,y) = e^{ivx}g(y),$$

$$\psi^{inc}(x,y) = e^{ivx}e^{-v(y-h)}, \quad (1.17)$$

Therefore, the far-field forms of ϕ and ψ for incident wave of wavenumber K can be written as

$$\phi \rightarrow \begin{cases} e^{iKx-Ky} + R_1 e^{-iKx-Ky} + r_1 e^{-ivx}g(y), & \text{as} \quad x \rightarrow -\infty, \\ T_1 e^{iKx-Ky} + t_1 e^{ivx}g(y), & \text{as} \quad x \rightarrow \infty \end{cases} \quad (1.18)$$

$$\psi \rightarrow \begin{cases} e^{iKx-Ky} + R_1 e^{-iKx-Ky} + r_1 e^{-ivx}e^{-v(y-h)}, & \text{as} \quad x \rightarrow -\infty, \\ T_1 e^{iKx-Ky} + t_1 e^{ivx}e^{-v(y-h)}, & \text{as} \quad x \rightarrow \infty \end{cases} \quad (1.19)$$

and the far-field forms of ϕ and ψ for incident wave of wavenumber v can be written as

$$\phi \rightarrow \begin{cases} e^{ivx}g(y) + r_2 e^{-ivx}g(y) + R_2 e^{-iKx-Ky}, & \text{as} \quad x \rightarrow -\infty, \\ t_2 e^{ivx}g(y) + T_2 e^{iKx-Ky}, & \text{as} \quad x \rightarrow \infty, \end{cases} \quad (1.20)$$

$$\psi \to \begin{cases} e^{ivx}e^{-v(y-h)} + r_2 e^{-ivx}e^{-v(y-h)} + R_2 e^{-iKx-Ky}, & \text{as} \quad x \to -\infty, \\ t_2 e^{ivx}e^{-v(y-h)} + T_2 e^{iKx-Ky}, & \text{as} \quad x \to \infty. \end{cases} \tag{1.21}$$

Here, R_1, r_1 are the reflection coefficients corresponding to the waves of wavenumbers K and v, respectively, due to an incident wave of wave number K and T_1 and t_1 are the transmission coefficients corresponding to the wave of wavenumber K and v, respectively, due to an incident wave of wavenumber K. Similarly, R_2, r_2 are the reflection coefficients corresponding to the waves of wave numbers K and v, respectively, due to an incident wave of wavenumber v and T_2 and t_2 are the transmission coefficients corresponding to the wave of wavenumber K and v, respectively, due to an incident wave of wave number v.

In case of a single layer fluid, for any scattering problem, the reflection and transmission coefficients satisfy the energy identity, which is generally used as a partial check on the correctness of the analytic or computed values of these coefficients. For a two-layer fluid with a free surface, there exists two energy identities corresponding to scattering of incident waves of two different wave numbers Linton and McIver (1995) as in Chapter 7 and are given by

$$|R_1|^2 + |T_1|^2 + J(|r_1|^2 + |t_1|^2) = 1, \tag{1.22}$$

$$|R_2|^2 + |T_2|^2 + J(|r_2|^2 + |t_2|^2) = J, \tag{1.23}$$

with

$$J = \frac{J_v}{J_K},$$

where

$$J_K = i\left[e^{-2Kh} + 2Ks\int_0^h e^{-2Ky}\,dy\right], \tag{1.24}$$

$$J_v = i\left[1 + 2vs\int_0^h (g(y))^2\,dy\right]. \tag{1.25}$$

These identities are used here as a partial numerical checks.

Solution of the Problem-I

Let the functions $\phi_1(x, y)$ and $\psi_1(x, y)$ be defined as

$$\phi_1 = \phi(x,y) - \phi^{inc}(x,y), \tag{1.26}$$

$$\psi_1 = \psi(x,y) - \psi^{inc}(x,y), \tag{1.27}$$

then ϕ_1 and ψ_1 satisfies

$$\nabla^2\phi_1 = 0, \quad 0 < y < h, \tag{1.28}$$

$$\nabla^2\psi_1 = 0, \quad h < y < \infty. \tag{1.29}$$

$$K\phi_1 + \phi_{1y} = 0, \quad \text{on} \quad y = 0, \ -\infty < x < \infty, \tag{1.30}$$

$$\phi_{1x} = -iKe^{-Ky}, \quad \text{on} \quad x = 0, \ 0 < y < a, \tag{1.31}$$

$$\phi_{1y} = \psi_{1y}, \quad \text{on} \quad y = h, \tag{1.32}$$

$$s(K\phi_1 + \phi_{1y}) = K\psi_1 + \psi_{1y}, \quad \text{on} \quad y = h, \tag{1.33}$$

$$\nabla\psi_1 \to 0, \quad \text{as} \quad y \to \infty, \tag{1.34}$$

$$\phi_1 \to \begin{cases} R_1 e^{-iKx-Ky} + r_1 e^{-ivx} g(y), & \text{as} \quad x \to -\infty, \\ (T_1 - 1)e^{iKx-Ky} + t_1 e^{ivx} g(y), & \text{as} \quad x \to \infty \end{cases} \tag{1.35}$$

$$\psi_1 \to \begin{cases} R_1 e^{-iKx-Ky} + r_1 e^{-ivx} e^{-v(y-h)}, & \text{as} \quad x \to -\infty, \\ (T_1 - 1)e^{iKx-Ky} + t_1 e^{ivx} e^{-v(y-h)}, & \text{as} \quad x \to \infty. \end{cases} \tag{1.36}$$

Let $G(x, y; \xi, \eta)$ be the source potential in the upper region due to a line source of unit strength situated at (ξ, η) $(0 < \eta < h)$ in the upper fluid and $H(x, y; \xi, \eta)$ be the source potential in the lower region due to source at (ξ, η) in the upper fluid. Expressions for G and H can be obtained from the results given in Mandal and Chakrabarti (1983) and Rhodes-Robinson (1994). By appropriate uses of Green's integral theorem, we obtain

$$2\pi\phi_1(\xi,\eta) = \int_0^a f(y) G_x(0, y; \xi, \eta)\, dy, \tag{1.37}$$

where

$$f(y) = \phi_1(+0, y) - \phi_1(-0, y) = \phi(-0, y) - \phi(+0, y), \quad 0 < y < a \tag{1.38}$$

so that

$$f(a) = 0$$

Now, by the condition (1.31), we have

$$\phi_{1\xi} = -iKe^{-K\eta}, \quad 0 < \eta < a. \tag{1.39}$$

Using this result in (1.37), we obtain

$$\int_0^a f(y) \frac{\partial^2}{\partial x \partial \xi} G(0, y; 0, \eta)\, dy = -2\pi i K e^{-K\eta}, \quad 0 < \eta < a, \tag{1.40}$$

where the integral is interpreted as Hadamrd finite part integral of order 2.

Now, from Chakrabarti and Mandal (1983) we find,

$$G_{x\xi}(0, y; 0, \eta) = -\frac{1}{(y-\eta)^2} + \frac{1}{(y+\eta)^2} - L_1(y, \eta), \tag{1.41}$$

where

$$L_1(y,\eta) = -\frac{1}{(y+\eta)^2} + 4\mu \int_0^\infty k \ \sin ky \ \sin k\eta \left(\frac{\mu+\cos 2kh}{1+\mu^2+2\mu\cos 2kh}\right) dk$$
$$-\int_0^\infty \frac{2k^2}{K} \frac{e^{-kh}\left(e^{k(h-\eta)}+\mu e^{-k(h-\eta)}\right)}{1+\mu e^{-2kh}} \cosh ky \, dk - \frac{2\pi is K^2 e^{-K(y+\eta-h)}}{((1-2s)\sinh Kh - \cosh Kh)}$$
$$-\frac{2\pi i v^2(1-s)(v\cosh vy - K\sinh vy)\sinh vh}{KD}\left(s\mu e^{-v(h-\eta)} - \frac{e^{-vh}\left(e^{v(h-\eta)}+\mu e^{-v(h-\eta)}\right)}{1+\mu e^{-2vh}}\right. \tag{1.42}$$
$$\left.\left(s\mu e^{-vh} + \frac{((K-v(1-s))\sinh vh + sK\cosh vh)}{K}\right)\right),$$

$$\mu = \frac{1-s}{1+s} \quad \left(=\frac{1}{\sigma}\right)$$

and

$$D = ((1-2s)\,sinh\,vh - cosh\,vh)((1-s-Kh)\,sinh\,vh + (v(1-s)-sK)h\,cosh\,vh). \tag{1.43}$$

The hypersingular integral equation (1.40) becomes

$$\int_0^a f(y)\left[\frac{1}{(y-\eta)^2} + L_1(y,\eta)\right] dy = H_1(\eta), \tag{1.44}$$

where

$$H_1(\eta) = 2\pi i K e^{-K\eta} \tag{1.45}$$

and $f(y)$ is bounded at $y = 0$ while $f(a) = 0$.

To solve the above equation, we substitute $y = ap$ and $\eta = aq$ in (1.44) to get

$$\int_0^1 F(p)\left[\frac{1}{(p-q)^2} + L_2(p,q)\right] dp = H_2(q), \quad 0 < q < 1, \tag{1.46}$$

where

$$F(p) = f(ap) \tag{1.47}$$

so that $F(1) = 0$ and $F(0)$ is bounded.

$$H_2(q) = aH_1(aq) = 2\pi i K_1 e^{K_1 q}, \quad 0 < q < 1 \tag{1.48}$$

and

$$L_2(p,q) = a^2 L_1(ap,aq) = -\frac{1}{(p+q)^2} + 4\mu \int_0^\infty k_1 \ \sin k_1 p \ \sin k_1 q \left(\frac{\mu+\cos 2k_1h_1}{1+\mu^2+2\mu\cos 2k_1h_1}\right) dk_1 -$$
$$\int_0^\infty \frac{2k_1^2}{K_1} \frac{e^{-k_1h_1}\left(e^{k_1(h_1-q)}+\mu e^{-k_1(h_1-q)}\right)}{1+\mu e^{-2k_1h_1}} \cosh k_1 p \, dk_1 - \frac{2\pi is K_1^2 e^{-K_1(p+q-h_1)}}{((1-2s)\sinh K_1h_1 - \cosh K_1h_1)} - \left(s\mu e^{-v_1(h_1-q)} -\right.$$
$$\left.\frac{e^{-v_1h_1}\left(e^{v_1(h_1-q)}+\mu e^{-v_1(h_1-q)}\right)}{1+\mu e^{-2v_1h_1}}\left(s\mu e^{-v_1h_1} + \frac{((K_1-v_1(1-s))\sinh v_1h_1 + sK_1\cosh v_1h_1)}{K_1}\right)\right). \tag{1.49}$$

Here, $h_1 = h/a$, $v_1 = va$, $K_1 = Ka$ are all in the non-dimensionalized form.

Following the methodology used by Parsons and Martin (1992, 1994), we assume

$$F(p) \simeq \sqrt{1-p^2}\sum_{n=0}^{N} a_n U_n(p), \tag{1.50}$$

where $U_n(p)$ is the Chebyshev polynomial of second kind given by

$$U_n(\cos\theta) = \frac{\sin(n+1)\theta}{\sin\theta} \tag{1.51}$$

and a_n $(n = 0,1, \dots . N)$ are unknowns to be determined.

Substituting (1.50) in (1.46), we find

$$\sum_{n=0}^{N} a_n c_n(q) = H_2(q), \quad 0 < q < 1, \tag{1.52}$$

where

$$c_n(q) = \int_0^1 \frac{\sqrt{1-p^2}\, U_n(p)}{(p-q)^2}\, dp + \int_0^1 \sqrt{1-p^2}\, U_n(p) L_2(p,q) dp. \tag{1.53}$$

The first integral in (1.53) is evaluated by a recurrence relation given in Parsons and Martin (1994). To find the unknown constants a_n $(n = 0,1, \dots .N)$ we put $q = q_j$, $(j = 0,1, \dots . .N)$ in (1.52) to obtain the linear system

$$\sum_{n=0}^{N} a_n c_n(q_j) = H_2(q_j), \quad j = 0,1, \dots . N, \tag{1.54}$$

which can be solved by standard methods. The collocation points $q = q_j$ are chosen as (Parsons and Martin (1994))

$$q_j = \frac{1}{2}\left[\cos\left(\frac{(2j+1)\pi}{2N+2}\right)\right], \quad j = 0,1, \dots \dots N. \tag{1.55}$$

Knowing a_n, one can obtain $F(p)$, $f(p)$ can be obtained from (1.47). T_1, R_1, t_1 and r_1 can then be obtained from (1.37) by making $\xi \to \pm\infty$, and noting the behavior of $\phi_1(\xi, \eta)$ from (1.35) as $\xi \to \pm\infty$ we get

$$T_1 = 1 - sKA_0 e^{Kh} \int_0^a f(y) e^{-Ky}\, dy, \tag{1.56}$$

$$R_1 = sKA_0 e^{Kh} \int_0^a f(y) e^{-Ky}\, dy, \tag{1.57}$$

$$t_1 = -\nu s^3 K^2 A_1 e^{\nu h} \int_0^a f(y) g(y)\, dy, \tag{1.58}$$

$$r_1 = \nu s^3 K^2 A_1 e^{\nu h} \int_0^a f(y) g(y)\, dy, \tag{1.59}$$

where

$$A_0 = \frac{1}{2s \sinh Kh + e^{-Kh}} \tag{1.60}$$

and

$$A_1 = \frac{1}{(1-s)\nu(\nu-K)(1-s-Kh)\sinh \nu h + (\nu(1-s)-sK)h \cosh \nu h}. \tag{1.61}$$

T_1, R_1, t_1 and r_1 and are now evaluated numerically after finding the values of a_n.

Solution of the Problem-II

Let the functions $\phi_2(x, y)$ and $\psi_2(x, y)$ be defined as

$$\phi_2 = \phi(x,y) - \phi^{inc}(x,y), \tag{1.62}$$

$$\psi_2 = \psi(x,y) - \psi^{inc}(x,y), \tag{1.63}$$

then ϕ_2 and ψ_2 satisfies

$$\nabla^2\phi_2 = 0, \quad 0 < y < h, \tag{1.64}$$

$$\nabla^2\psi_2 = 0, \quad h < y < \infty, \tag{1.65}$$

$$K\phi_2 + \phi_{2y} = 0, \quad \text{on} \quad y = 0, \ -\infty < x < \infty, \tag{1.66}$$

$$\phi_{2x} = -ivg(y), \quad \text{on} \quad x = 0, \ 0 < y < a, \tag{1.67}$$

$$\phi_{2y} = \psi_{2y}, \quad \text{on} \quad y = h, \tag{1.68}$$

$$s(K\phi_2 + \phi_{2y}) = K\psi_2 + \psi_{2y}, \quad \text{on} \quad y = h, \tag{1.69}$$

$$\nabla\psi_2 \to 0, \quad \text{as} \quad y \to \infty, \tag{1.70}$$

and $\phi^{inc}(x, y)$ and $\psi^{inc}(x, y)$ are given by (1.17).

Following a similar procedure as before we obtain

$$\int_0^a f(y)\left[\frac{1}{(y-\eta)^2} + L_1(y,\eta)\right] dy = 2\pi ivg(y), \tag{1.71}$$

where $f(y)$ and $L_1(y, \eta)$ are defined earlier.

In a similar manner we get

$$T_2 = -sKA_0e^{Kh} \int_0^a f(y)e^{-Ky}\, dy, \tag{1.72}$$

$$R_2 = sKA_0e^{Kh} \int_0^a f(y)e^{-Ky}\, dy, \tag{1.73}$$

$$t_2 = 1 - vs^3K^2A_1e^{vh} \int_0^a f(y)g(y)\, dy, \tag{1.74}$$

$$r_2 = vs^3K^2A_1e^{vh} \int_0^a f(y)g(y)\, dy. \tag{1.75}$$

Similarly, T_2, R_2, t_2 and r_2 are evaluated numerically once $f(y)$ is known using the methodology given by Parson and Martin (1992, 1994).

Numerical results

For numerical computation we have used $N = 15$ in (1.50) and the collocation points given by (1.55). $|T_1|$, $|R_1|$, $|t_1|$ ($|t_1| = |r_1|$), $|t_2|$, $|r_2|$, $|T_2|$, ($|T_2| = |R_2|$), are now computed numerically against Ka for various values of a/h and s.

By increasing N to $N = 20$, we get almost the same numerical results for these quantities. So, we fix $N = 15$ for all computations. Figures 8.1–8.3 for $s = 0.1$, 0.25 and 0.5 respectively and $a/h = 0.2$, 0.5, 0.8, 0.99 represent the reflection and transmission coefficients corresponding to incident waves of wave number K. Figures 8.4–8.6 for $s = 0.1$, 0.25 and 0.5 respectively and $a/h = 0.2$, 0.5, 0.8, 0.99 represent the reflection and transmission coefficients corresponding to incident waves of wave number ν.

In Figs. 8.1(a), 8.2(a) and 8.3(a), $|T_1|$ is depicted against Ka for $s = 0.1$, 0.25 and 0.5 respectively. $|T_1|$ shows an oscillatory behaviour which is more pronounced when $a/h = 0.5$. For s = 0.1 and 0.25, $|T_1|$ is almost equal to one when $a/h = 0.2$, 0.8, 0.99 showing that the wave energy is almost totally transmitted. For $s = 0.5$, $|T_1|$ shows an oscillatory behaviour for all values of a/h. This shows that when the density of the

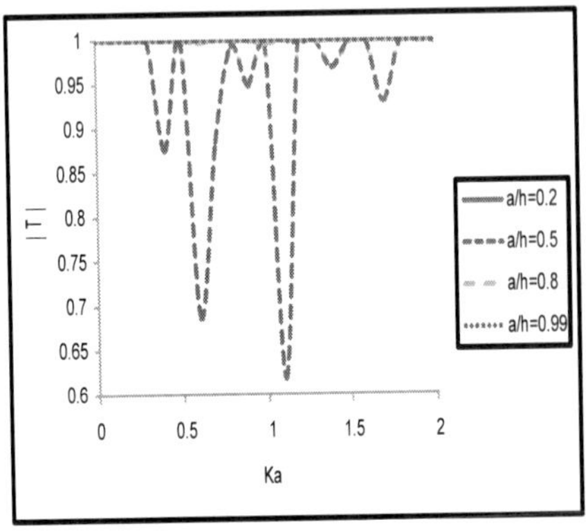

Fig. 8.1a Transmission coefficient $|T_1|$ for a wave of wave number K due to an incident wave of wave number K ($s = 0.1$)

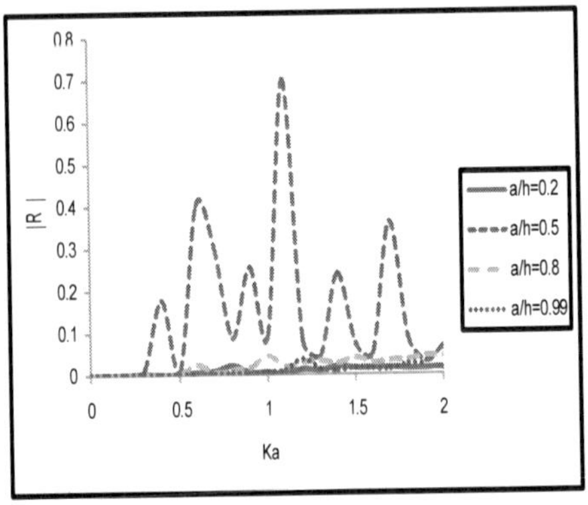

Fig. 8.1b Reflection coefficient $|R_1|$ for a wave of wave number K due to an incident wave of wave number K ($s = 0.1$)

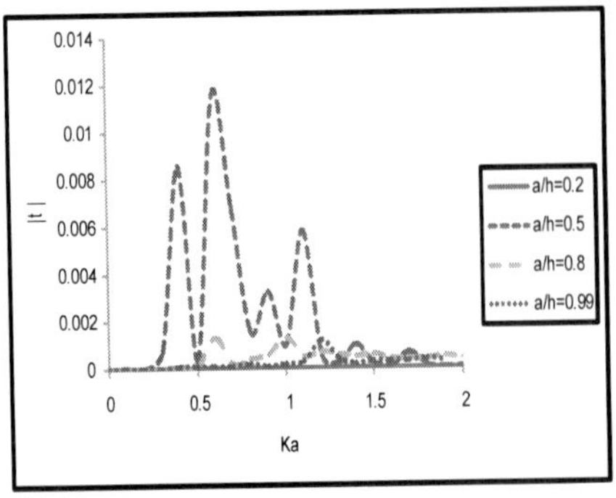

Fig. 8.1c Transmission coefficient $|t_1|$ for a wave of wave number ν due to an incident wave of wave number K ($s = 0.1$)

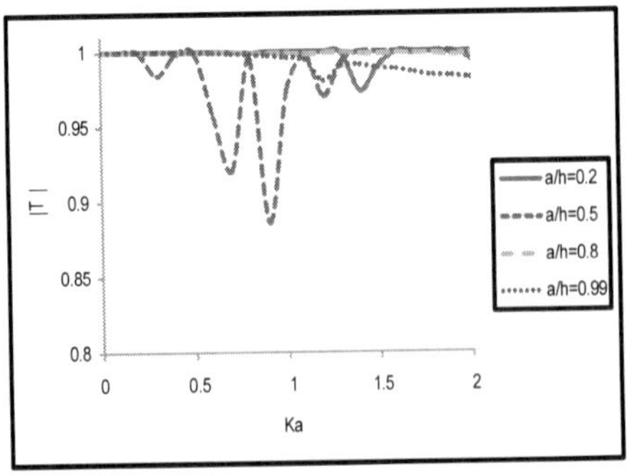

Fig. 8.2a Transmission coefficient $|T_1|$ for a wave of wave number K due to an incident wave of wave number K ($s = 0.25$)

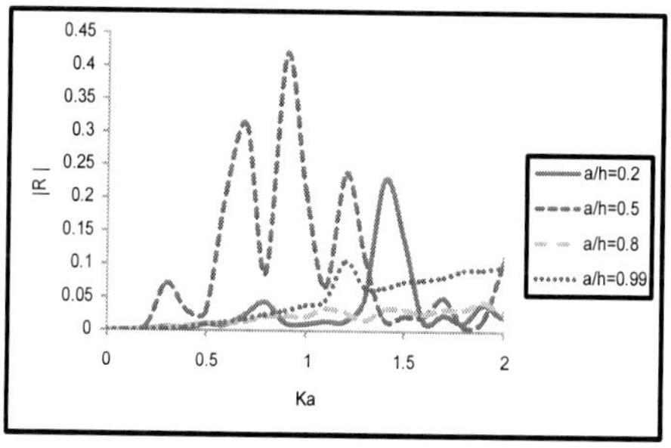

Fig. 8.2b Reflection coefficient $|R_1|$ for a wave of wave number K due to an incident wave of wave number K (s = 0.25)

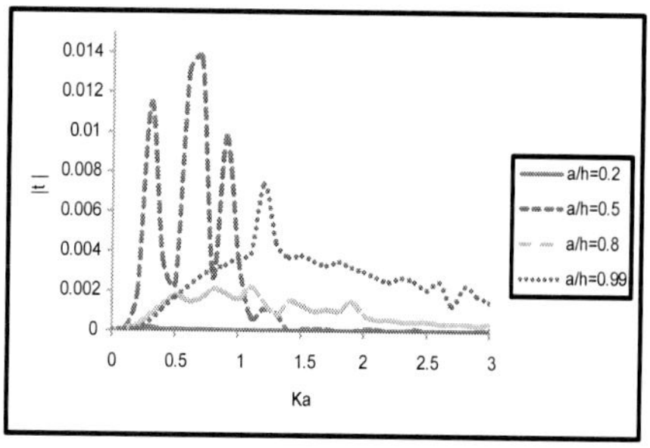

Fig. 8.2c Transmission coefficient $|t_1|$ for a wave of wave number ν due to an incident wave of wave number K ($s = 0.25$)

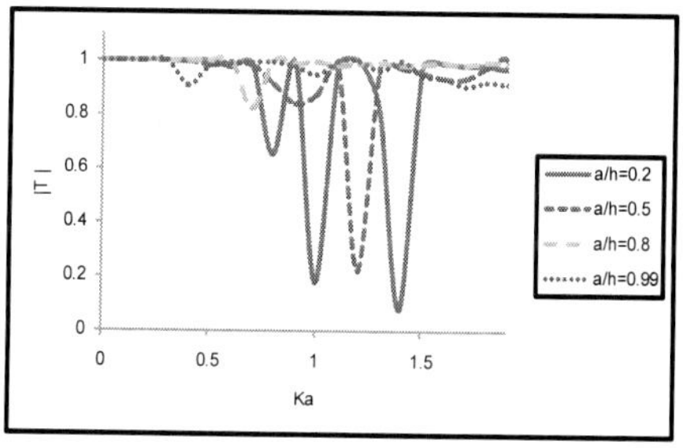

Fig. 8.3a Transmission coefficient $|T_1|$ for a wave of wave number K due to an incident wave of wave number K ($s = 0.5$)

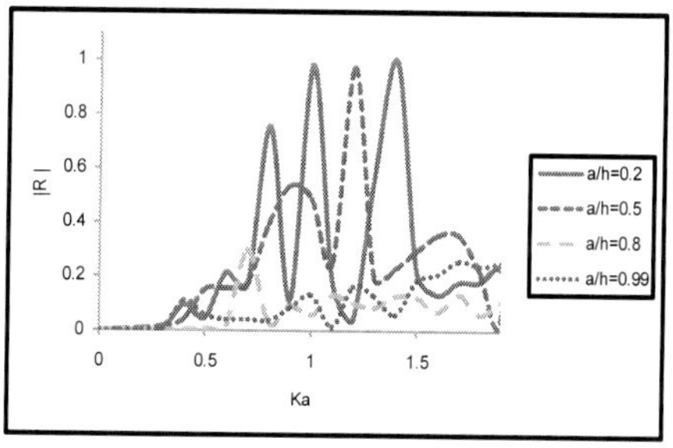

Fig. 8.3b Reflection coefficient $|R_1|$ for a wave of wave number K due to an incident wave of wave number K ($s = 0.5$)

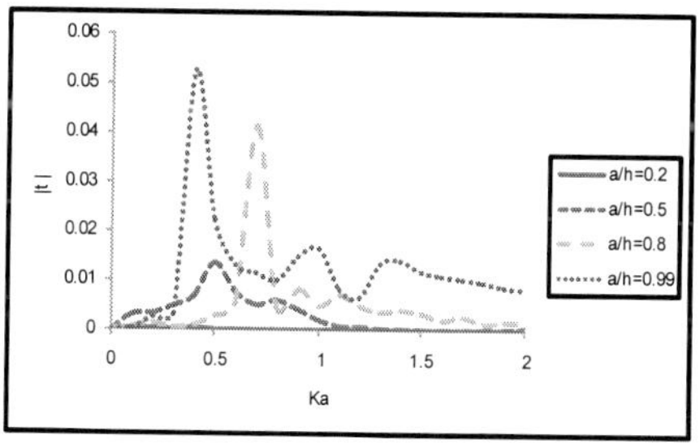

Fig. 8.3c Transmission coefficient $|t_1|$ for a wave of wave number ν due to an incident wave of wave number K ($s = 0.5$)

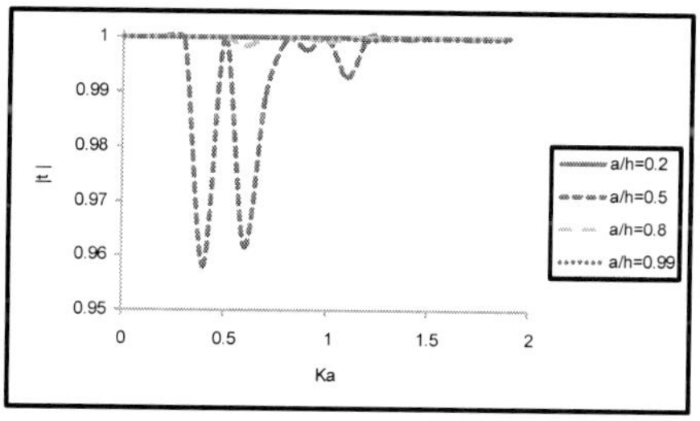

Fig. 8.4a Transmission coefficient $|t_2|$ for a wave of wave number ν due to an incident wave of wave number ν ($s = 0.1$)

upper fluid is half the density of the lower fluid there occurs a significant interaction between transmitted wave and the barrier.

In Figs. 8.1(b), 8.2(b) and 8.3(b), is depicted against Ka for $s = 0.1$, 0.25 and 0.5 respectively. $|R_1|$ also shows an oscillatory behaviour showing peaks at the points where $|T_1|$ shows dip. The reflection is more for $a/h = 0.5$ compared to the cases when $a/h = 0.2$, 0.8, 0.99.

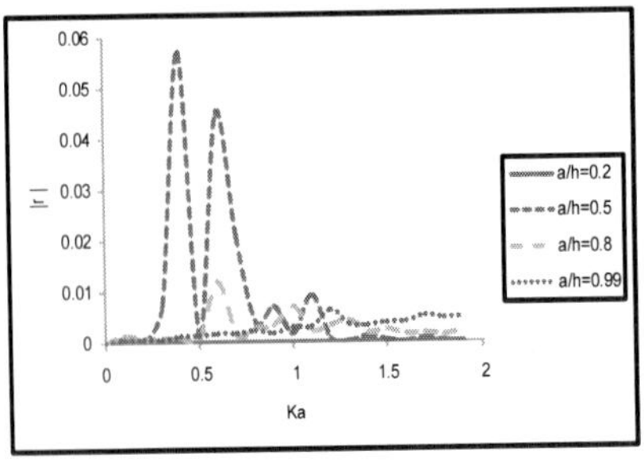

Fig. 8.4b Reflection coefficient $|r_2|$ for a wave of wave number ν due to an incident wave of wave number ν ($s = 0.1$)

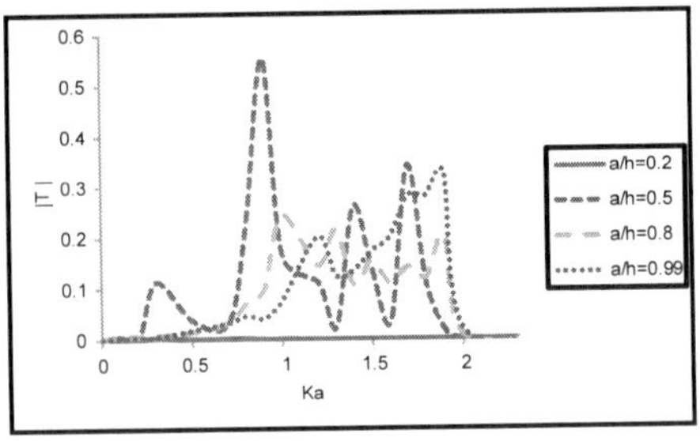

Fig. 8.4c Transmission coefficient $|T_2|$ for a wave of wave number K due to an incident wave of wave number ν ($s = 0.1$)

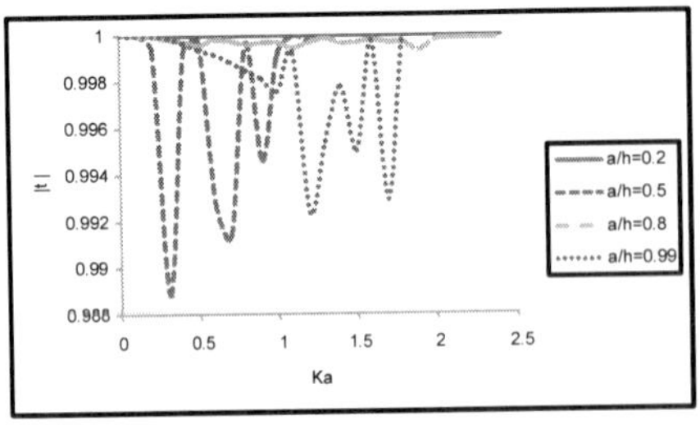

Fig. 8.5a Transmission coefficient $|t_2|$ for a wave of wave number ν due to an incident wave of wave number ν ($s = 0.25$)

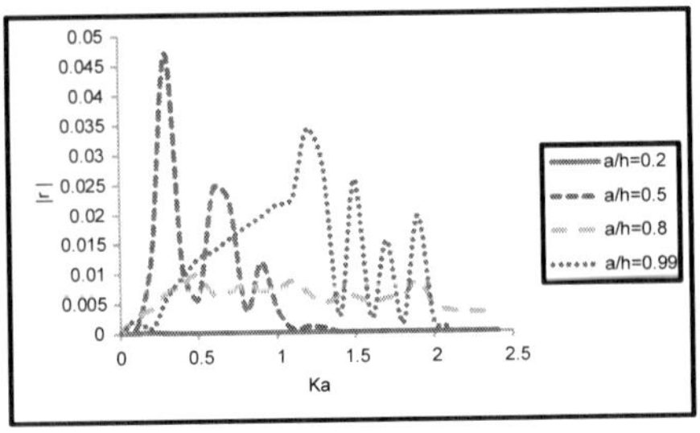

Fig. 8.5b Reflection coefficient $|r_2|$ for a wave of wave number ν due to an incident wave of wave number ν ($s = 0.25$)

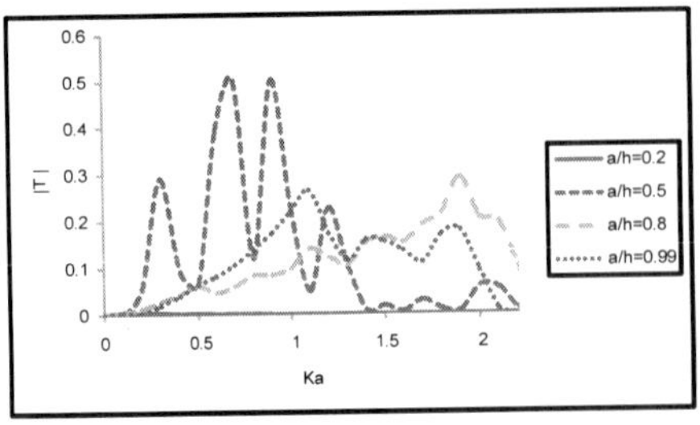

Fig. 8.5c Transmission coefficient $|T_2|$ for a wave of wave number K due to an incident wave of wave number ν ($s = 0.25$)

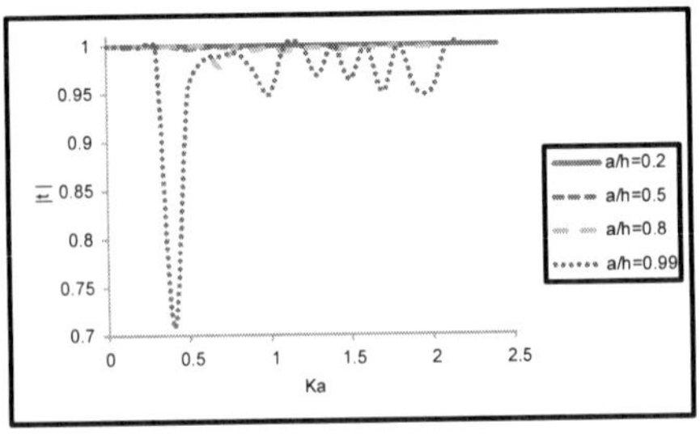

Fig. 8.6a Transmission coefficient $|t_2|$ for a wave of wave number ν due to an incident wave of wave number ν ($s = 0.5$)

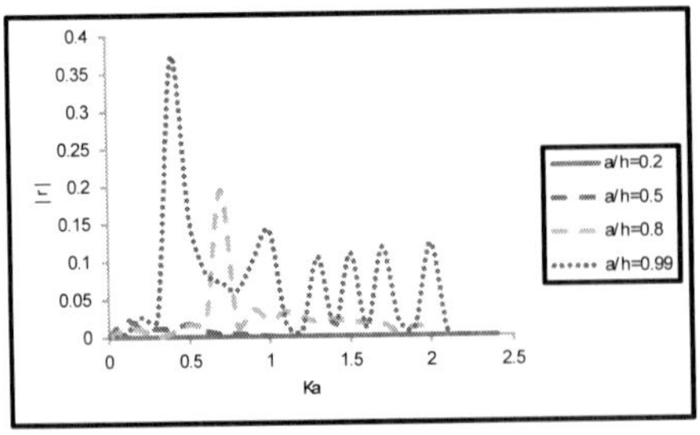

Fig. 8.6b Reflection coefficient $|r_2|$ for a wave of wave number ν due to an incident wave of wave number ν ($s = 0.5$)

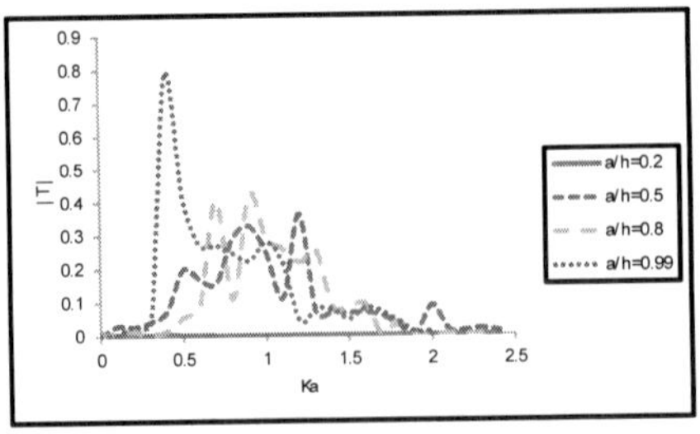

Fig. 8.6c Transmission coefficient $|T_2|$ for a wave of wave number K due to an incident wave of wave number ν ($s = 0.5$)

In Figs. 8.1(c), 8.2(c) and 8.3(c), $|t_1|$ is depicted against Ka for $s = 0.1, 0.25$ and 0.5 respectively. It is observed that the transmission coefficient corresponding to the waves of wave number v due to an incident wave of wave number K exhibits an oscillatory behaviour when $0.1 < Ka < 1.5$. It eventually dies out as Ka increases further. $|t_1|$ is almost negligible when $a/h = 0.2$ as compared to the cases when $a/h = 0.5, 0.8, 0.99$.

In Figs. 8.4(a), 8.5(a) and 8.6(a), $|t_2|$ is depicted against Ka for $s = 0.1, 0.25$ and 0.5 respectively for incident wave of wave number v. In all these figures, it is observed that $|t_2|$ is almost 1 for $a/h = 0.2$ but for $a/h = 0.5, 0.8, 0.99$, $|t_2|$ shows an oscillatory behaviour first and then becomes almost equal to 1 as Ka further increases. For $s = 0.5$ and $a/h = 0.99$ there occurs a significant interaction between transmitted wave and the barrier.

Figures 8.4(b), 8.5(b) and 8.6(b), show the behaviour of $|r_2|$ against Ka for $s = 0.1, 0.25$ and 0.5 respectively. It is observed that there is almost no reflection for $a/h = 0.2$. For $a/h = 0.5, 0.8, 0.99$, $|r_2|$ shows an oscillatory behaviour and then decreases to zero as Ka increases.

In Figs. 8.4(c), 8.5(c) and 8.6(c), $|T_2|$ is plotted against Ka for $s = 0.1, 0.25$ and 0.5 respectively. $|T_2|$ shows an oscillatory behaviour which goes on decreasing with the increase of Ka. $|T_2|$ is nearly zero for $a/h = 0.2$ in all the figures. The peak of oscillation decreases for $a/h = 0.5$ as s increases from 0.1 to 0.5. Thus, the interaction of waves with the barrier is quite significant when the density of the upper fluid is half the density of the lower fluid.

For all the computations, the energy identities are checked and found to be satisfied.

8.2 Scattering by a circular cylinder in a two layer fluid

As mentioned in section 8.1 Linton and McIver (1995) developed a general theory for two-dimensional wave propagation in a two-layer fluid. Due to the presence of an obstacle, an incident wave of a particular wave number gets reflected and transmitted into waves of both the wave numbers, so that on scattering by an obstacle, transfer of energy from one mode to another takes place. When the obstacle is in the form of a horizontal circular cylinder, situated in either the lower or the upper layer, Linton and McIver (1995) calculated reflection and transmission coefficients associated with both the wave numbers for a wave train of again both the wave numbers normally incident on the cylinder. They observed that when the cylinder is in the lower layer then the reflection coefficients are identically zero.

This is in conformity with the classical result Ursell (1950) that a horizontal circular cylinder submerged in deep water is transparent to a normally incident wave train. However, the cylinder does experience reflection if it is submerged in the upper layer. This problem arose in connection with modelling an underwater pipe bridge across Norwegian fjords consisting of a layer of fresh water on the top of a deep layer of salt water. Linton and Cadby (2002) also investigated the problem of scattering of obliquely incident waves by a long circular cylinder in a two-layer fluid and observed that for an incident angle above a critical angle defined by a relation involving the density ratio between the two fluids, there is no transfer of energy from the waves of higher wave number to the waves of lower wave number while for incident angles less than the critical angle, energy transfer only occurs at somewhat higher frequencies, and the phenomenon of zero transmission occurs at some particular frequencies.

There is a considerable interest in the study of various types of water wave problems in the presence of a thin ice-sheet floating on water, termed as water with an ice-cover, the ice-sheet being modeled as a thin elastic plate. In section 6.3 wave scattering by a long circular cylinder in a single-layer fluid of infinite depth with an ice-cover was considered, Das and Mandal (2006). It was observed that the cylinder does not experience reflection for normal incidence, but does experience reflection for oblique incidence.

Wave scattering by a horizontal circular cylinder situated in either of the two layers of an ice-covered two-layer fluid is considered in this section. This problem has been investigated by Das and Mandal (2007). An account of this investigation is presented below.

Oblique waves scattering in a two-layer fluid

We are here concerned with irrotational motion in two superposed non-viscous incompressible fluids under the action of gravity, neglecting any effect due to surface tension at the interface of the two fluids, the upper being of finite depth h and covered by a thin uniform ice sheet modelled as a thin elastic plate, while the lower layer being infinitely deep. The upper and lower layer fluids have densities ρ_1 and $\rho_2 (> \rho_1)$ respectively. Cartesian co-ordinates are chosen such that (x, z)-plane coincides with the undisturbed interface between the two fluids. The y-axis points vertically upwards with $y = 0$ as the mean position of the interface and $y = h (> 0)$ as the mean position of the thin ice-cover. Under the usual assumptions of linear water wave theory a velocity potential can be defined for oblique waves in the form

$$\Phi(x, y, z, t) = Re\{\phi(x, y) e^{-i\sigma t + i\gamma z}\},$$

where $\phi(x, y)$ is a complex valued potential function, γ is the wavenumber component along the z-direction and σ is defined earlier.

The upper fluid, $0 < y < h$, will be referred to as region I, while the lower fluid, $y < 0$, will be referred to as region II. The potential in the upper fluid will be denoted by ϕ^I and that in the lower fluid by ϕ^{II}. ϕ^I and ϕ^{II} satisfied Helmholtz equation

$$\left(\nabla^2 - \gamma^2\right)\phi^I = 0, \quad \text{for } 0 < y < h, \tag{2.1}$$

$$\left(\nabla^2 - \gamma^2\right)\phi^{II} = 0, \quad \text{for } -\infty < y < 0. \tag{2.2}$$

Linearized boundary conditions at the interface and at the ice-cover are

$$\phi_y^I = \phi_y^{II} \quad \text{on } y = 0, \tag{2.3}$$

$$\rho\left(\phi_y^I - K\phi^I\right) = \phi_y^{II} - K\phi^{II} \quad \text{on } y = 0, \tag{2.4}$$

where $\rho=\frac{\rho_1}{\rho_2}(<1)$,

$$\left(D\left(\frac{\partial^2}{\partial x^2}-\gamma^2\right)^2+1-\varepsilon K\right)\phi_y^I-K\phi^I=0,\ \text{on}\ \ y=h, \tag{2.5}$$

where $K=\sigma^2/\text{g}$, g being the gravity, $D=\frac{L}{\rho_1 g}$ where L is the flexural rigidity of the elastic ice-cover and $\varepsilon=\frac{\rho_0}{\rho_1}h_0$, ρ_0 is the density of the ice and h_0 is the very small thickness of the ice-cover. The boundary conditions (2.3) and (2.4) are obtained from the continuity of normal velocity and pressure at the interface respectively.

Also the condition at large depth is

$$\nabla\phi^{II}\to 0 \qquad \text{as}\ \ y\to-\infty. \tag{2.6}$$

In a two-layer fluid progressive waves have the form (except for a multiplicative constant)

$$\phi^I=e^{\pm ix(k^2-\gamma^2)^{\frac{1}{2}}}\left[\{k(Dk^4+1-\varepsilon K)+K\}e^{k(y-h)}+\{k(Dk^4+1-\varepsilon K)-K\}e^{-k(y-h)}\right], \tag{2.7}$$

and

$$\phi^{II}=e^{\pm ix(k^2-\gamma^2)^{\frac{1}{2}}}e^{ky}\left[\{k(Dk^4+1-\varepsilon K)+K\}e^{-kh}+\{k(Dk^4+1-\varepsilon K)-K\}e^{kh}\right], \tag{2.8}$$

where k satisfies the dispersion equation

$$\begin{aligned}H(k)\equiv K\rho\{k(Dk^4+1-\varepsilon K)\cosh kh-K\sinh kh\}\\ -\{k(1-\rho)-K\}\{k(Dk^4+1-\varepsilon K)\sinh kh-K\cosh kh\}=0.\end{aligned} \tag{2.9}$$

The equation (2.9) has exactly two positive real roots λ_1 and $\lambda_2(\lambda_1<\lambda_2)$ (say). Also, it has one negative real root and four complex roots in the four quadrants of the complex k-plane.

For the case $k=\lambda_j(j=1,2)$ progressive waves are thus of the form

$$\phi^I(x,y)=e^{\pm i\beta_j x}g_j(y),\quad j=1,2, \tag{2.10}$$

$$\phi^{II}(x,y)=e^{\pm i\beta_j x+\lambda_j y},\quad j=1,2, \tag{2.11}$$

where $\beta_j=(\lambda_j^2-\gamma^2)^{\frac{1}{2}}$ in which that branch of the square root is chosen for which $\beta_j=\lambda_j$ for $\gamma=0$,

$$g_j(y)=\frac{\{\lambda_j(1-\rho)-K\}}{K\rho\{\lambda_j(D\lambda_j^4+1-\varepsilon K)\cosh\lambda_j h-K\sinh\lambda_j h\}}$$

$$\times\left[\{\lambda_j(D\lambda_j^4+1-\varepsilon K)+K\}e^{\lambda_j(y-h)}+\{\lambda_j(D\lambda_j^4+1-\varepsilon K)-K\}e^{-\lambda_j(y-h)}\right], j=1,2. \tag{2.12}$$

We require $\gamma<\lambda_1$ for $j=1$ and $\gamma<\lambda_2$ for $j=2$, for the progressive waves to exist.

In any wave scattering problem therefore, the far-field will take the form of incoming and outgoing waves at each of the wave numbers $\lambda_j (j = 1,2)$. It is given by

$$\phi^I \sim \left(A^{\pm} e^{\pm i\beta_1 x} + C^{\pm} e^{\mp i\beta_1 x}\right) g_1(y) + \left(B^{\pm} e^{\pm i\beta_2 x} + D^{\pm} e^{\mp i\beta_2 x}\right) g_2(y), \tag{2.13}$$

$$\phi^{II} \sim \left(A^{\pm} e^{\pm i\beta_1 x} + C^{\pm} e^{\mp i\beta_1 x}\right) e^{\lambda_1 y} + \left(B^{\pm} e^{\pm i\beta_2 x} + D^{\pm} e^{\mp i\beta_2 x}\right) e^{\lambda_2 y}, \tag{2.14}$$

as $x \to \pm\infty$, for which, in the notation of Linton and McIver (1995),

$$\phi \sim \left(A^-, B^-, C^-, D^-; A^+, B^+, C^+, D^+\right). \tag{2.15}$$

Incident plane wave ϕ_{inc} of wave number λ_1 making an angle $\alpha (0 \le \alpha \le \frac{\pi}{2})$ with the positive x-axis has the form

$$\phi^I_{inc} = e^{i\lambda_1 x \cos\alpha} g_1(y), \tag{2.16}$$

$$\phi^{II}_{inc} = e^{i\lambda_1 x \cos\alpha + \lambda_1 y}. \tag{2.17}$$

In this case

$$\gamma = \lambda_1 \sin\alpha,\ \beta_1 = \lambda_1 \cos\alpha,\ \beta_2 = \left(\lambda_2^2 - \lambda_1^2 \sin^2\alpha\right)^{\frac{1}{2}}. \tag{2.18}$$

It is obvious that β_2 is real since $\lambda_1 < \lambda_2$ and so scattered waves of wave number λ_2 will exist for all values of λ_1 (i.e., for all values of K, since for different values of K, we get different λ_1 and λ_2 $(> \lambda_1)$) and for all incident angles α. The angle α_{λ_2} for the scattered waves of wave number λ_2 is given by

$$\tan\alpha_{\lambda_2} = \frac{\gamma}{\beta_2} = \frac{\lambda_1 \sin\alpha}{\left(\lambda_2^2 - \lambda_1^2 \sin^2\alpha\right)^{\frac{1}{2}}}. \tag{2.19}$$

Since $\beta_2 > \beta_1$ we know that $\tan\alpha_{\lambda_2} < \tan\alpha$ and hence $\alpha_{\lambda_2} < \alpha$.

An incident plane wave of wave number λ_2 making an angle $\alpha (0 \le \alpha \le \frac{\pi}{2})$ with the positive x-axis is given by

$$\phi^I_{inc} = e^{i\lambda_2 x \cos\alpha} g_2(y), \tag{2.20}$$

$$\phi^{II}_{inc} = e^{i\lambda_2 x \cos\alpha + \lambda_2 y}. \tag{2.21}$$

In this case

$$\gamma = \lambda_2 \sin\alpha,\ \beta_1 = \left(\lambda_1^2 - \lambda_2^2 \sin^2\alpha\right)^{\frac{1}{2}},\ \beta_2 = \lambda_2 \cos\alpha. \tag{2.22}$$

For a given angle α there may be a value of K for which $\lambda_1 = \lambda_2 \sin\alpha$ and thus $\beta_1 = 0$. We will call this K as the *cut-off frequency* and denote it by K_c. For a value of K for which $\lambda_1 > \lambda_2 \sin\alpha$ (for fixed α) we get real β_1 and so waves of wave number λ_1 will propagate. For a value of K for which $\lambda_1 < \lambda_2 \sin\alpha$ (for fixed α), β_1 becomes imaginary and in that case there exists no propagating wave of wave number λ_1. Figure 8.7 shows the cut-off frequency $K_c a$, plotted against incident wave angle

$$\alpha = \sin^{-1}\left(\frac{\lambda_1}{\lambda_2}\right), \tag{2.23}$$

for density ratio $\rho = 0.5$, $h/a = 2$, and different values of D/a^4 and ε/a, a being the radius of the circular cylinder considered in the next section. Instead of using a, we could have used h to non-dimensionalize ε and D, but that would not have changed this discussion.

The different curves in Fig. 8.7 correspond to $D/a^4 = 2, 1.5, 1, 0.5, 0.1$ and $\varepsilon/a = 0.01$ (except one curve for which $\varepsilon / a = 0.0001$, $D / a^4 = 0.0001$). It is observed from this figure that for any angle α for which the point (α, Ka) is situated on the right side of the curve there are no propagating waves of wave number λ_1 for this value of Ka. It may be noted that for very small value of D/a^4, i.e., 0.0001 with small $\varepsilon/a = 0.0001$, the curve almost coincides with the curve for the case of upper fluid with a free surface (Fig. 1 in Linton and Cadby (2002)). Due to the presence of ice-cover, we observe from this figure that for some values of α for which the point (Ka, α) is situated on the left side of the curve there are two cut-off frequencies and only for frequencies lying between these two cut-off frequencies will there be conversion of wave of wave number λ_1 from wave of wave number λ_2. In Fig. 8.8 the critical angle α_c is plotted against D/a^4, for $K_c a = 0.3, 0.5, 0.9$. These curves show that α_c decreases as D/a^4 increases for higher values of $K_c a$.

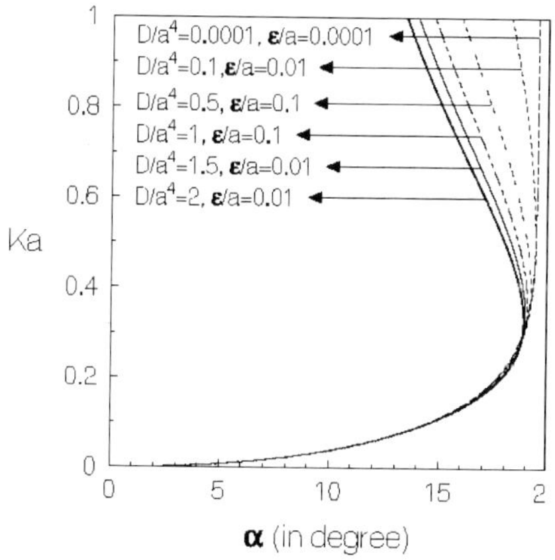

Fig. 8.7 Cut-off frequency $K_c a$ due to an incident wave of wavenumber λ_2 $\left(\frac{h}{a} = 2, \rho = 0.5\right)$

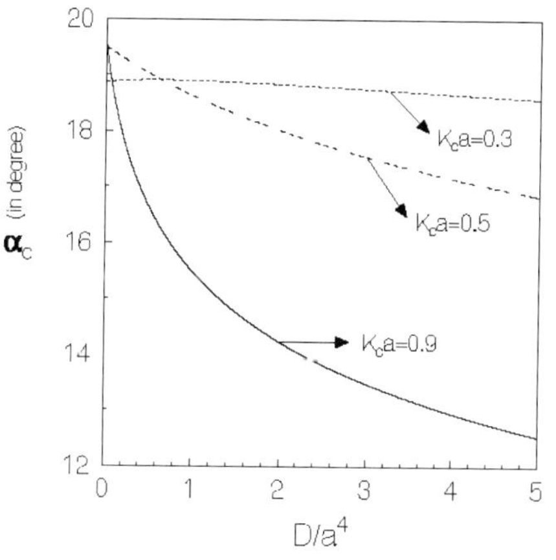

Fig. 8.8 Critical angle α_c for a fixed cut-off frequency $\left(\frac{\varepsilon}{a} = 0.01, \frac{h}{a} = 2, \rho = 0.5\right)$

When waves of wave number λ_1 propagates, the angle α_{λ_1} of the scattered waves of wave number λ_1 is given by

$$\tan \alpha_{\lambda_1} = \frac{\lambda_2 \sin \alpha}{\left(\lambda_1^2 - \lambda_2^2 \sin^2 \alpha\right)^{\frac{1}{2}}}. \tag{2.24}$$

For a two-layer fluid in which the upper layer has an ice-cover instead of a free surface, energy identities are derived in Chapter 7. These identities are used here as partial numerical checks for all the data points in obtaining the various curves for the reflection and transmission coefficients.

A. Cylinder in the lower layer

Let a horizontal circular cylinder of radius a have its axis at $y = f(<0)$ and its generator runs parallel to z-axis. Polar co-ordinates (r,θ) are defined in the (x, y)-plane by

$$x = r\sin\theta \quad \text{and} \quad y = f - r\cos\theta. \tag{2A.1}$$

Let the symmetric and antisymmetric multipoles be defined by $\phi_n^s\,(n \ge 0)$ and $\phi_n^a\,(n \ge 1)$ respectively. The multipoles are defined in the notation of Linton and Cadby (2002) by

$$\phi_n^{Is} = (-1)^n \int_0^\infty \cosh nk \cos(\gamma x \sinh k)\left(A(k)e^{\nu y} + B(k)e^{-\nu y}\right)dk, \tag{2A.2}$$

$$\phi_n^{IIs} = K_n(\gamma r)\cos n\theta + (-1)^n \int_0^\infty \cosh nk \cos(\gamma x \sinh k)C(k)e^{\nu y}\,dk, \tag{2A.3}$$

$$\phi_n^{Ia} = (-1)^{n+1} \int_0^\infty \sinh nk \sin(\gamma x \sinh k)\left(A(k)e^{\nu y} + B(k)e^{-\nu y}\right)dk, \tag{2A.4}$$

$$\phi_n^{IIa} = K_n(\gamma r)\sin n\theta + (-1)^{n+1} \int_0^\infty \sinh nk \sin(\gamma x \sinh k)C(k)e^{\nu y}\,dk, \tag{2A.5}$$

where $\nu = \gamma\cosh k$ and $A(k), B(k), C(k)$ are functions of k to be found such that the integrals exist in some sense. $K_n(z)$ is the modified Bessel function of second kind.

The functions ϕ_n^s and ϕ_n^a are singular solutions of the modified Helmholtz equation and satisfy the ice-cover condition (2.5) and the interface conditions (2.3) and (2.4) and are of outgoing nature at infinity. Then $A(k)$, $B(k)$ and $C(k)$ have the forms

$$A(k) = K\left\{\nu\left(D\nu^4 + 1 - \varepsilon K\right) + K\right\}e^{\nu(f-h)}/H(\nu), \tag{2A.6}$$

$$B(k) = K\left\{\nu\left(D\nu^4 + 1 - \varepsilon K\right) - K\right\}e^{\nu(f+h)}/H(\nu), \tag{2A.7}$$

$$\begin{aligned} C(k) = \big[K\rho\left\{\nu\left(D\nu^4 + 1 - \varepsilon K\right)\cosh\nu h - K\sinh\nu h\right\} \\ - \left\{(1-\rho)\nu + K\right\}\left\{\nu\left(D\nu^4 + 1 - \varepsilon K\right)\sinh\nu h - K\cosh\nu h\right\}\big]e^{\nu f}/H(\nu) \end{aligned} \tag{2A.8}$$

where $H(\nu)$ is given by (2.9) (with k replaced by ν).

The path of the integration in the integrals in (2A.2) to (2A.5) is indented below the poles at $k = \mu_1$ and $k = \mu_2$, where

$$\gamma\cosh\mu_j = \lambda_j,\ j = 1,2. \tag{2A.9}$$

The far-field forms of the multipoles, in the lower fluid, is given by

$$\phi_n^{IIs} : (-1)^n \pi i \left(C^{\mu_1} \cosh n\mu_1 e^{\pm i\beta_1 x + \lambda_1 y} + C^{\mu_2} \cosh n\mu_2 e^{\pm i\beta_2 x + \lambda_2 y} \right), \tag{2A.10}$$

$$\phi_n^{IIa} : \mp(-1)^n \pi \left(C^{\mu_1} \sinh n\mu_1 e^{\pm i\beta_1 x + \lambda_1 y} + C^{\mu_2} \sinh n\mu_2 e^{\pm i\beta_2 x + \lambda_2 y} \right), \tag{2A.11}$$

as $x \to \pm\infty$, where C^{μ_1} and C^{μ_2} are the residues of $C(k)$ at $k = \mu_1$ and $k = \mu_2$, and these are given by

$$C^{\mu_j} = \left[K\rho \left\{ \lambda_j \left(D\lambda_j^4 + 1 - \varepsilon K \right) \cosh \lambda_j h - K \sinh \lambda_j h \right\} - \left\{ (1-\rho)\lambda_j + K \right\} \right.$$

$$\left. \left\{ \lambda_j \left(D\lambda_j^4 + 1 - \varepsilon K \right) \sinh \lambda_j h - K \cosh \lambda_j h \right\} \right] e^{\lambda_j f} / \beta_j H'(\lambda_j), \; j = 1,2. \tag{2A.12}$$

Using the expansion

$$e^{\frac{1}{2}X\left(P+P^{-1}\right)} = \sum_{m=0}^{\infty} \frac{1}{2} \varepsilon_m \left(P^m + P^{-m} \right) I_m(X), \tag{2A.13}$$

where

$$\varepsilon_0 = 1, \; \varepsilon_m = 2, m \geq 1 \tag{2A.14}$$

$I_m(X)$ is the modified Bessel function of first kind, (2A.3) and (2A.5) can be expanded in terms of polar co-ordinates as

$$\phi_n^{IIs} = K_n(\gamma r) \cos n\theta + \sum_{m=0}^{\infty} A_{nm}^{(s)} I_m(\gamma r) \cos m\theta, \tag{2A.15}$$

$$\phi_n^{IIa} = K_n(\gamma r) \sin n\theta + \sum_{m=1}^{\infty} A_{nm}^{(a)} I_m(\gamma r) \sin m\theta, \tag{2A.16}$$

where

$$A_{nm}^{(s)} = \varepsilon_m (-1)^{m+n} \int_0^{\infty} e^{\gamma f} \cosh mk \cosh nk C(k) dk, \tag{2A.17}$$

$$A_{nm}^{(a)} = 2(-1)^{m+n} \int_0^{\infty} e^{\gamma f} \sinh mk \sinh nk C(k) dk. \tag{2A.18}$$

Obliquely incident wave train of wavenumber λ_1

Let us consider the case of an obliquely wave train of wave number λ_1 making an angle α with the positive x-axis, so that $\gamma = \lambda_1 \sin\alpha$. The incident wave potential (2.17) has the form

$$\phi_{inc}^{II} = e^{i\beta_1 x + \lambda_1 y} = e^{\lambda_1 f} \sum_{m=0}^{\infty} \varepsilon_m (-1)^m I_m(\gamma r) \left(\cosh m\nu \cos m\theta - i \sinh m\nu \sin m\theta \right) \tag{2A.19}$$

where

$$\cosh\nu = \frac{\lambda_1}{\gamma} = \frac{1}{\sin\alpha}.$$

To solve the scattering problem we write the potential function describing the fluid motion as

$$\phi_{\lambda_1} = \phi_{inc} + \sum_{n=0}^{\infty} \left(a_n \phi_n^a + b_n \phi_n^s \right), \tag{2A.20}$$

where a_n and b_n are unknown constants to be determined.

To find a_n and b_n the polar expansions of the multipoles (2A.3), (2A.5) and the incident wave (2A.19) are substituted into (2A.20). Applying the body boundary condition $\frac{\partial \phi_{\lambda_1}^{II}}{\partial r} = 0$ on $r = a$, and using the orthogonal properties of the trigonometric functions, we obtain two infinite systems of linear equations for the unknowns a_n and b_n as given by,

$$\frac{a_m}{Z_m} + \sum_{n=1}^{\infty} A_{nm}^{(a)} a_n = 2i(-1)^m e^{\lambda_1 f} \sinh m\nu, \; m = 1, 2, \cdots \tag{2A.21}$$

$$\frac{b_m}{Z_m} + \sum_{n=0}^{\infty} A_{nm}^{(s)} b_n = (-1)^{m+1} \varepsilon_m e^{\lambda_1 f} \cosh m\nu, \; m = 0, 1, \cdots \tag{2A.22}$$

where

$$Z_m = \frac{I_m'(\gamma r)}{K_m'(\gamma r)},$$

dash denoting derivative with respect to the arguments.

These two systems can be solved by truncation. Here 5×5 systems were used for numerical computations.

The far-field form for $\phi_{\lambda_1}^{II}$, in the lower layer, can be written as ~

$$\phi_{\lambda_1}^{II} \sim \begin{cases} e^{i\beta_1 x + \lambda_1 y} + R_{\lambda_1} e^{-i\beta_1 x + \lambda_1 y} + r_{\lambda_1} e^{-i\beta_2 x + \lambda_2 y} & \text{as } x \to -\infty, \\ T_{\lambda_1} e^{i\beta_1 x + \lambda_1 y} + t_{\lambda_1} e^{i\beta_2 x + \lambda_2 y} & \text{as } x \to \infty. \end{cases} \tag{2A.23}$$

Using (3.20), (3.10) and (3.11) we obtain the reflection and transmission coefficients as follows:

$$R_{\lambda_1} = \pi C^{\mu_1} \sum_{m=0}^{\infty} (-1)^m \{ i b_m \cosh m\mu_1 + a_m \sinh m\mu_1 \}, \tag{2A.24}$$

$$r_{\lambda_1} = \pi C^{\mu_2} \sum_{m=0}^{\infty} (-1)^m \{ i b_m \cosh m\mu_2 + a_m \sinh m\mu_2 \}, \tag{2A.25}$$

$$T_{\lambda_1} = 1 + \pi C^{\mu_1} \sum_{m=0}^{\infty} (-1)^m \{ i b_m \cosh m\mu_1 - a_m \sinh m\mu_1 \}, \tag{2A.26}$$

$$t_{\lambda_1} = \pi C^{\mu_2} \sum_{m=0}^{\infty} (-1)^m \{ i b_m \cosh m\mu_2 - a_m \sinh m\mu_2 \}. \tag{2A.27}$$

Obliquely incident wave train of wavenumber λ_2

We consider the case of an obliquely incident plane wave of wave number λ_2 making an angle α with the positive x-axis, so that $\gamma = \lambda_2 \sin\alpha$. The expansion of incident wave potential is the same as (2A.19), except that λ_1 is to be replaced by λ_2. The velocity potential ϕ_{λ_2} for this problem can again be expanded in terms of multipoles similar to (2A.20) and the equations for a_n and b_n are similar to (2A.21) and (2A.22) with λ_1 is to be replaced by λ_2.

The far-field forms of $\phi_{\lambda_2}^{II}$, in the lower layer, can be written as

$$\phi_{\lambda_2}^{II} \sim \begin{cases} e^{i\beta_2 x+\lambda_2 y} + R_{\lambda_2} e^{-i\beta_1 x+\lambda_1 y} + r_{\lambda_2} e^{-i\beta_2 x+\lambda_2 y} & \text{as } x \to -\infty, \\ T_{\lambda_2} e^{i\beta_1 x+\lambda_1 y} + t_{\lambda_2} e^{i\beta_2 x+\lambda_2 y} & \text{as } x \to \infty. \end{cases} \tag{2A.28}$$

Using the far-field forms of the multipoles given by (2A.10) and (2A.11) in ϕ_{λ_2} we find that the expressions for reflection coefficients R_{λ_2} and r_{λ_2} are similar to (2A.24) and (2A.25) with appropriate changes, and the transmission coefficients are given by

$$T_{\lambda_2} = \pi C^{\mu_1} \sum_{m=0}^{\infty} (-1)^m \{ib_m \cosh m\mu_1 - a_m \sinh m\mu_1\}, \tag{2A.29}$$

$$t_{\lambda_2} = 1 + \pi C^{\mu_2} \sum_{m=0}^{\infty} (-1)^m \{ib_m \cosh m\mu_2 - a_m \sinh m\mu_2\}. \tag{2A.30}$$

Normally incident wave train

Now for the case of normal incidence, $\alpha = 0$, the modified Helmholtz equation reduces to the Laplace's equation and solutions of Laplace's equation singular at $y = f < 0$ are $r^{-n}\cos n\theta$ and $r^{-n}\sin n\theta, n \geq 1$, and these have the integral representations

$$\frac{\cos n\theta}{r^n} = \frac{(-1)^n}{(n-1)!} \int_0^\infty k^{n-1} e^{-k(y-f)} \cos kx dk,$$

$$\frac{\sin n\theta}{r^n} = \frac{(-1)^{n+1}}{(n-1)!} \int_0^\infty k^{n-1} e^{-k(y-f)} \sin kx dk.$$

It is straightforward to add suitable solutions of Laplace's equation to the symmetric and anti-symmetric multipoles so that the boundary conditions are satisfied. We obtain

$$\phi_n^{Is} = \frac{(-1)^n}{(n-1)!} \int_0^\infty k^{n-1} \left(A(k)e^{ky} + B(k)e^{-ky}\right) \cos kx dk, \tag{2A.31}$$

$$\phi_n^{IIs} = \frac{\cos n\theta}{r^n} + \frac{(-1)^n}{(n-1)!} \int_0^\infty k^{n-1} C(k) e^{ky} \cos kx dk, \tag{2A.32}$$

$$\phi_n^{Ia} = \frac{(-1)^{n+1}}{(n-1)!} \int_0^\infty k^{n-1} \left(A(k)e^{ky} + B(k)e^{-ky}\right) \sin kx dk, \tag{2A.33}$$

$$\phi_n^{IIa} = \frac{\sin n\theta}{r^n} + \frac{(-1)^{n+1}}{(n-1)!}\int_0^\infty k^{n-1}C(k)e^{ky}\sin kxdk, \tag{2A.34}$$

where now

$$A(k) = K\{k(Dk^4 + 1 - \varepsilon K) + K\}e^{k(f-h)}/H(k), \tag{2A.35}$$

$$B(k) = K\{k(Dk^4 + 1 - \varepsilon K) - K\}e^{k(f+h)}/H(k), \tag{2A.36}$$

$$C(k) = [K\rho\{k(Dk^4 + 1 - \varepsilon K)\cosh kh - K\sinh kh\} - \{(1-\rho)k + K\}\{k(Dk^4 + 1 - \varepsilon K)\sinh kh - K\cosh kh\}]e^{kf}/H(k) \tag{2A.37}$$

and the path of integration is indented to pass beneath the poles of the above four integrands at $k = \lambda_1$ and $k = \lambda_2$. Here we have used the same notation without any confusion.

The multipoles (3.32) and (3.34) can be expanded about $r = 0$. Thus we obtain

$$\phi_n^{IIs} = \frac{\cos n\theta}{r^n} + \sum_{m=0}^{\infty} A_{nm} r^m \cos m\theta, \tag{2A.38}$$

$$\phi_n^{IIa} = \frac{\sin n\theta}{r^n} + \sum_{m=0}^{\infty} A_{nm} r^m \sin m\theta, \tag{2A.39}$$

where

$$A_{nm} = \frac{(-1)^{n+m}}{(n-1)!m!}\int_0^\infty k^{n+m-1}C(k)e^{ky}dk. \tag{2A.40}$$

Note that A_{nm} is the same for ϕ_n^{IIs} and ϕ_n^{IIa}.

The far-field form of the multipoles, in the lower layer, is given by ~

$$\phi_n^{IIs} \sim \frac{(-1)^n}{(n-1)!}\pi i\left(\lambda_1^{n-1}C^{\lambda_1}e^{\pm i\lambda_1 x + \lambda_1 y} + \lambda_2^{n-1}C^{\lambda_2}e^{\pm i\lambda_2 x + \lambda_2 y}\right), \tag{2A.41}$$

$$\phi_n^{IIa} \sim \mp\frac{(-1)^n}{(n-1)!}\pi\left(\lambda_1^{n-1}C^{\lambda_1}e^{\pm i\lambda_1 x + \lambda_1 y} + \lambda_2^{n-1}C^{\lambda_2}e^{\pm i\lambda_2 x + \lambda_2 y}\right), \tag{2A.42}$$

as $x \to \pm\infty$. Here C^{λ_1} C^{λ_2} are the residues of $C(k)$ at $k = \lambda_1$ and $k = \lambda_2$, given by

$$C^{\lambda_j} = [K\rho\{\lambda_j(D\lambda_j^4 + 1 - \varepsilon K)\cosh\lambda_j h - K\sinh\lambda_j h - \{(1-\rho)\lambda_j + K\}\} \{\lambda_j(D\lambda_j^4 + 1 - \varepsilon K)\sinh\lambda_j h - K\cosh\lambda_j h\}]e^{\lambda_j f}/H'(\lambda_j),\ j = 1,2. \tag{2A.43}$$

Normally incident wave train of wavenumber λ_1

The incident wave potential

$$\phi_{inc}^{II} = e^{i\lambda_1 x + \lambda_1 y}, \tag{2A.44}$$

when expanded about $r = 0$, has the form

$$\phi_{inc}^{II} = \sum_{m=0}^{\infty} \frac{(-1)^m}{m!} \lambda_1^m r^m (\cos m\theta - i \sin m\theta) e^{\lambda_1 f}. \tag{2A.45}$$

To solve this scattering problem we write

$$\phi_{\lambda_1}^{II} = \phi_{inc}^{II} + \sum_{n=1}^{\infty} a^n (a_n \phi_n^a + b_n \phi_n^s), \tag{2A.46}$$

where a_n and b_n are unknown constants to be determined.

To solve for a_n and b_n the polar expansions of the multipoles (2A.32), (2A.34) and the incident wave (2A.45) are substituted into (2A.46) and applying the body boundary condition $\frac{\partial \phi_{\lambda_1}^{II}}{\partial r} = 0$ on $r = a$ and using the orthogonal properties of the trigonometric functions, we obtain two infinite systems of linear equations for unknowns a_n and b_n which are

$$a_m - \sum_{n=1}^{\infty} a^{n+m} A_{nm} a_n = -i \frac{(-\lambda_1 a)^m}{m!} e^{\lambda_1 f}, \quad m = 1, 2, \cdots \tag{2A.47}$$

$$b_m - \sum_{n=1}^{\infty} a^{n+m} A_{nm} b_n = \frac{(-\lambda_1 a)^m}{m!} e^{\lambda_1 f}, \quad m = 1, 2, \cdots. \tag{2A.48}$$

Since left-hand sides of the systems of equations are of the same nature and the right-hand sides of the systems differ by a factor $-i$, we find that

$$a_n = -i b_n. \tag{2A.49}$$

Thus $\phi_{\lambda_1}^{II}$ is obtained as

$$\phi_{\lambda_1}^{II} = \phi_{inc}^{II} + \sum_{n=1}^{\infty} a^n b_n (\phi_n^s - i \phi_n^a). \tag{2A.50}$$

It follows immediately from (2A.41) and (2A.42) that as $x \to -\infty$

$$\phi_{\lambda_1}^{II} \sim \phi_{inc}^{II}. \tag{2A.51}$$

The far-field form for $\phi_{\lambda_1}^{II}$ in the lower fluid, can be written as

$$\phi_{\lambda_1}^{II} \sim \begin{cases} e^{i\lambda_1 x + \lambda_1 y} + R_{\lambda_1} e^{-i\lambda_1 x + \lambda_1 y} + r_{\lambda_1} e^{-i\lambda_2 x + \lambda_2 y} & \text{as } x \to -\infty, \\ T_{\lambda_1} e^{i\lambda_1 x + \lambda_1 y} + t_{\lambda_1} e^{i\lambda_2 x + \lambda_2 y} & \text{as } x \to \infty. \end{cases} \tag{2A.52}$$

Using (2A.50) we can obtain the reflection and transmission coefficients:

$$R_{\lambda_1} = r_{\lambda_1} \equiv 0, \tag{2A.53}$$

$$T_{\lambda_1} = 1 + 2\pi i \sum_{n=1}^{\infty} \frac{(-1)^n}{(n-1)!} a^n \lambda_1^{n-1} C^{\lambda_1} b_n, \tag{2A.54}$$

$$t_{\lambda_1} = 2\pi i \sum_{n=1}^{\infty} \frac{(-1)^n}{(n-1)!} a^n \lambda_2^{n-1} C^{\lambda_2} b_n. \tag{2A.55}$$

Normally incident wave train of wavenumber λ_2

For an incident wave of wave number λ_2 the mathematical analysis is the same except that λ_1 is to be replaced by λ_2 in the above equations. Also the far-field forms of $\phi_{\lambda_2}^{II}$, in the lower layer, can be written as

$$\phi_{\lambda_2}^{II} \sim \begin{cases} e^{i\lambda_2 x + \lambda_2 y} + R_{\lambda_2} e^{-i\lambda_1 x + \lambda_1 y} + r_{\lambda_2} e^{-i\lambda_2 x + \lambda_2 y} & \text{as } x \to -\infty, \\ T_{\lambda_2} e^{i\lambda_1 x + \lambda_1 y} + t_{\lambda_2} e^{i\lambda_2 x + \lambda_2 y} & \text{as } x \to \infty. \end{cases} \tag{2A.56}$$

Here also we find that the reflection coefficients R_{λ_2} and r_{λ_2} are identically zero. For the transmission coefficients we obtain

$$T_{\lambda_2} = 2\pi i \sum_{n=1}^{\infty} \frac{(-1)^n}{(n-1)!} a^n \lambda_1^{n-1} C^{\lambda_1} b_n, \tag{2A.57}$$

$$t_{\lambda_2} = 1 + 2\pi i \sum_{n=1}^{\infty} \frac{(-1)^n}{(n-1)!} a^n \lambda_2^{n-1} C^{\lambda_2} b_n. \tag{2A.58}$$

Numerical results

In the Figs. 8.9–8.12 the reflection and transmission coefficients are shown for the case of a wave train of wave number λ_1 obliquely incident on the cylinder submerged in the lower fluid. In all the plots immersion depth $-f/a$ is 2, the depth of the upper fluid layer h/a is 2 and ρ (density ratio) is 0.5, $\varepsilon/a = 0.01$ and $D/a^4 = 1.5$. The different curves correspond to different incident wave angle α, which are $15^0, 75^0, 80^0, 85^0, 89^0$. From

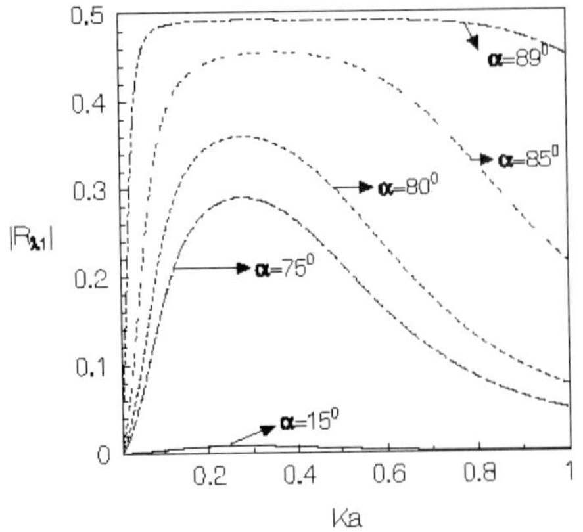

Fig. 8.9 Reflection coefficient due to wavenumber λ_1 incident on a cylinder in the lower layer ($\frac{D}{a^4} = 1.5, \frac{\varepsilon}{a} = 0.01, \frac{h}{a} = 2, \rho = .5, \frac{f}{a} = -2$)

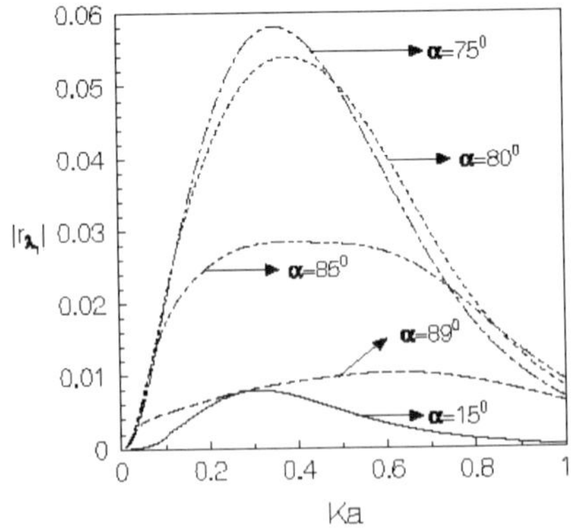

Fig. 8.10 Reflection coefficient due to wavenumber λ_1 incident on a cylinder in the lower layer ($\frac{D}{a^4} = 1.5, \frac{\varepsilon}{a} = .01, \frac{h}{a} = 2, \rho = .5, \frac{f}{a} = -2$)

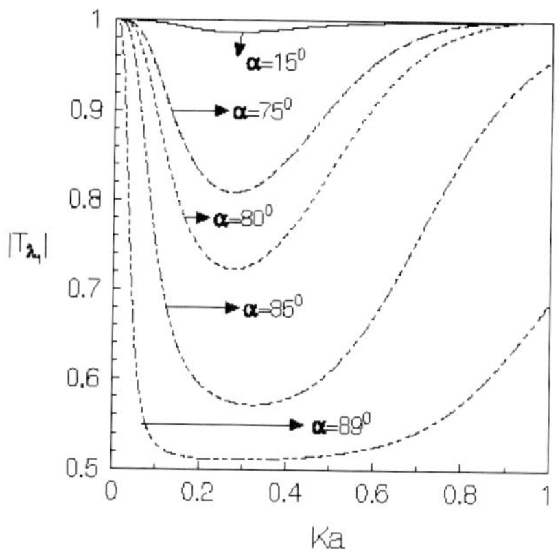

Fig. 8.11 Transmission coefficient due to a wave of wavenumber λ_1 incident on a cylinder in the lower layer $(\frac{D}{a^4} = 1.5, \frac{\varepsilon}{a} = .01, \frac{h}{a} = 2, \rho = .5, \frac{f}{a} = -2)$

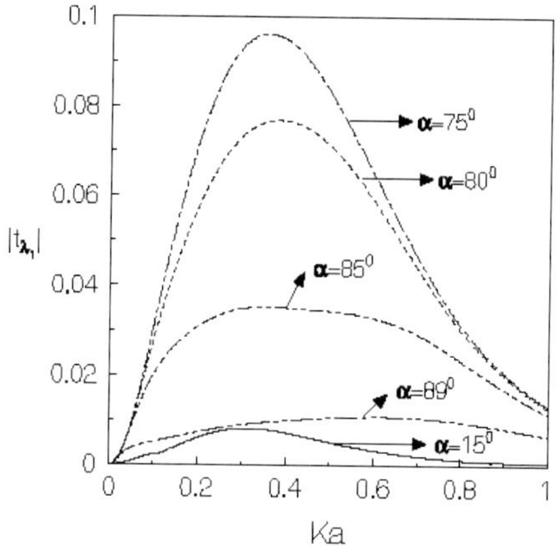

Fig. 8.12 Transmission coefficient due to wave of wavenumber λ_1 incident on a cylinder in lower layer $(\frac{D}{a^4} = 1.5, \frac{\varepsilon}{a} = .01, \frac{h}{a} = 2, \rho = .5, \frac{f}{a} = -2)$

Figs. 8.9 and 8.11 it is observed that as the angle of incidence increases, $|R_{\lambda_1}|$ increases while $|T_{\lambda_1}|$ decreases. Also $|R_{\lambda_1}|$ is somewhat smaller in comparison to that of Linton and Cadby (2002) and $|T_{\lambda_1}|$ is somewhat larger in comparison to that of Linton and Cadby (2002). This is due to the presence of the ice-cover. For $\alpha = 15°$, the reflection coefficient $|R_{\lambda_1}|$ is seen to be quite small. In fact for small α, $|R_{\lambda_1}|$ becomes negligible.

The reflection coefficient $|r_{\lambda_1}|$ and transmission coefficient $|t_{\lambda_1}|$ of waves of wave number λ_2 for an incident wave of wave number λ_1, shown in Figs. 8.10 and 8.12 respectively, are smaller in comparison to those for wave of wave number λ_1 but their non zero values show that there is some conversion of energy from one wave number to the other.

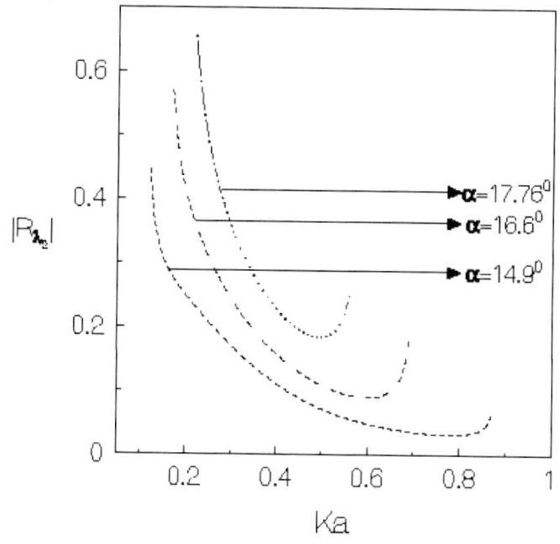

Fig. 8.13 Reflection coefficient due to a wave of wavenumber λ_2 incident on a cylinder in lower layer $(\frac{D}{a^4} = 1.5, \frac{\varepsilon}{a} = .01, \frac{h}{a} = 2, \rho = .5, \frac{f}{a} = -2)$

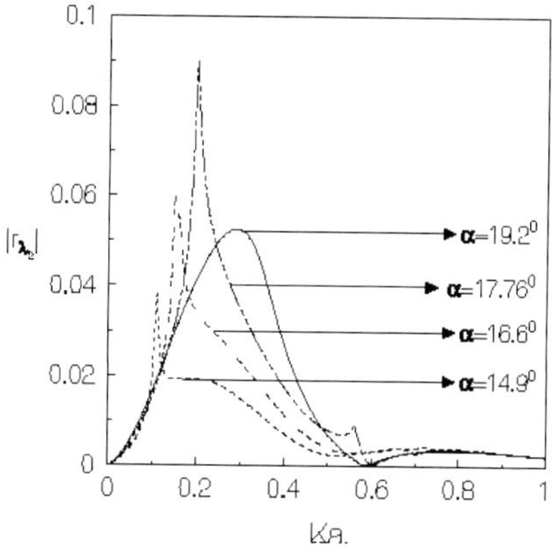

Fig. 8.14 Reflection coefficient due to a wave of wavenumber λ_2 incident on a cylinder in lower layer $(\frac{D}{a^4} = 1.5, \frac{\varepsilon}{a} = .01, \frac{h}{a} = 2, \rho = .5, \frac{f}{a} = -2)$

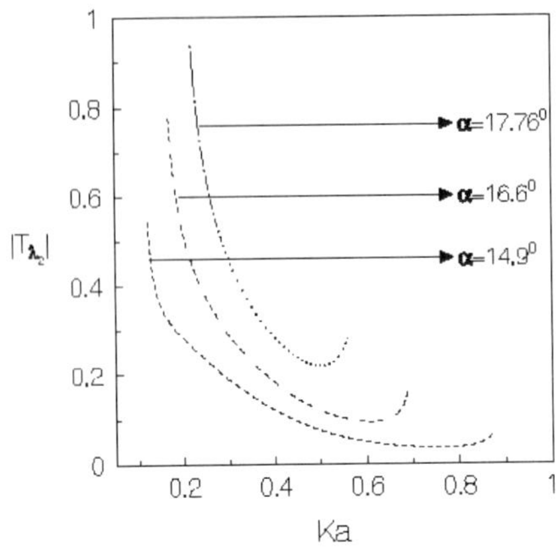

Fig. 8.15 Transmission coefficient due to a wave of wavenumber λ_2 incident on a cylinder in lower layer $(\frac{D}{a^4} = 1.5, \frac{\varepsilon}{a} = .01, \frac{h}{a} = 2, \rho = .5, \frac{f}{a} = -2)$

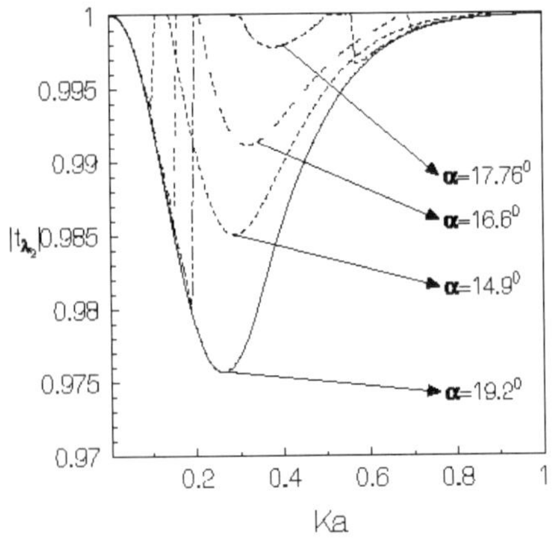

Fig. 8.16 Transmission coefficient due to a wave of wavenumber λ_2 incident on a cylinder in lower layer $(\frac{D}{a^4} = 1.5, \frac{\varepsilon}{a} = .01, \frac{h}{a} = 2, \rho = .5, \frac{f}{a} = -2)$

For the case of normal incidence, Figs. 8.17 and 8.18 show the transmission coefficients for the case of a incident wave of wave number λ_1 incident on a circular cylinder in the lower fluid for $\varepsilon/a = 0.01, h/a = 2, \rho = 0.5, f/a = -2$. The different curves correspond to different values of D/a^4, $D/a^4 = 0.1, 0.5, 1, 1.5, 2$. Figure 8.17 shows that $|T_{\lambda_1}|$ first decreases as Ka increases for low to moderate values of Ka but it increases as Ka further increases for any D/a^4. Figure 8.18 describes the behavior of $|t_{\lambda_1}|$ which is complimentary to the behavior of $|T_{\lambda_1}|$. Also Fig. 8.19 and Fig. 8.20 show the transmission coefficients due to a wave of wave number λ_2 incident on a cylinder in the lower layer. Figure 8.19 (8.20) describes the behavior of $|T_{\lambda_2}|$ which is similar to the behavior of $|t_{\lambda_2}|(|T_{\lambda_2}|)$.

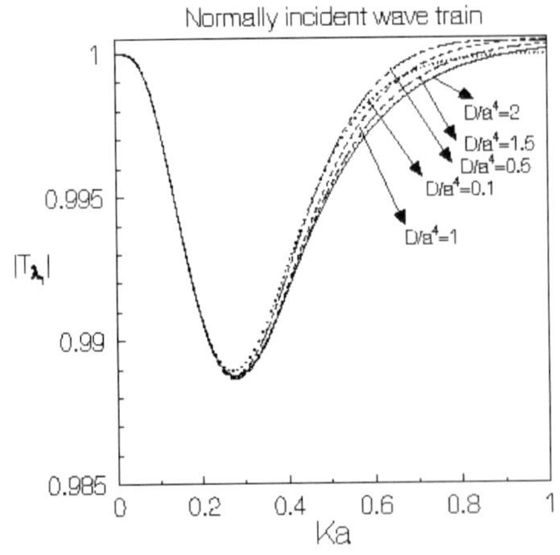

Fig. 8.17 Transmission coefficient due to wave of wavenumber λ_1 incident on a cylinder in lower layer $(\frac{D}{a^4} = 1.5, \frac{\varepsilon}{a} = .01, \frac{h}{a} = 2, \rho = .5, \frac{f}{a} = -2)$

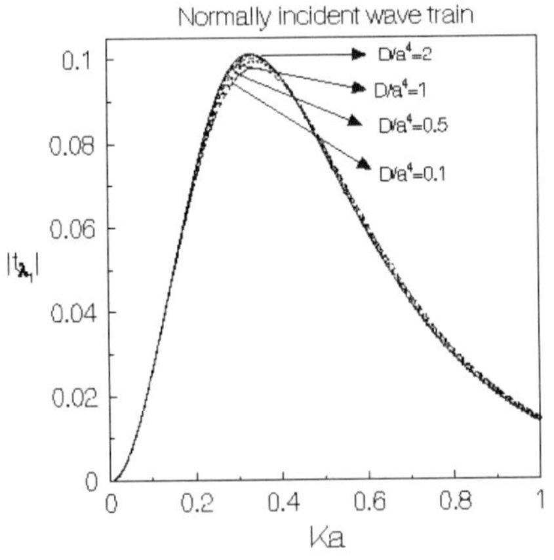

Fig. 8.18 Transmission coefficient due to a wave of wavenumber λ_1 incident on a cylinder in lower layer $(\frac{D}{a^4} = 1.5, \frac{\varepsilon}{a} = .01, \frac{h}{a} = 2, \rho = .5, \frac{f}{a} = -2)$

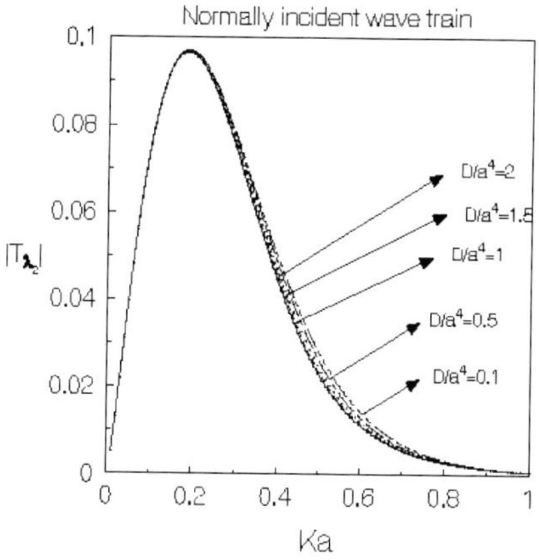

Fig. 8.19 Transmission coefficient due to a wave of wavenumber λ_2 incident on a cylinder in lower layer ($\frac{\varepsilon}{a} = .01, \frac{h}{a} = 2, \rho = .5, \frac{f}{a} = -2$)

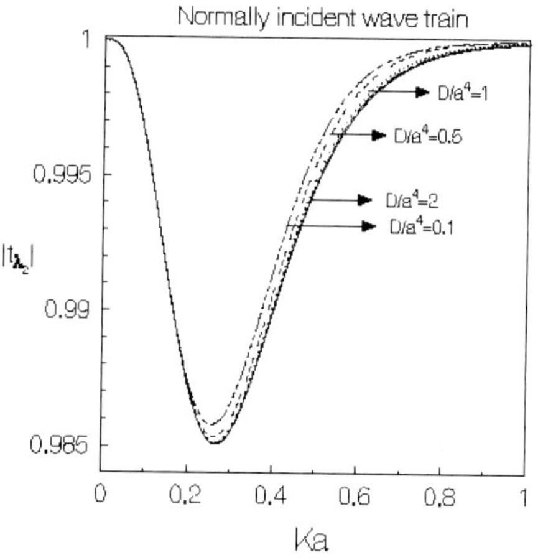

Fig. 8.20 Transmission coefficient due to a wave of wavenumber λ_2 incident on a cylinder in lower layer ($\frac{D}{a^4} = 1.5, \frac{\varepsilon}{a} = .01, \frac{h}{a} = 2, \rho = .5, \frac{f}{a} = -2$)

B. Cylinder in the upper layer

A horizontal circular cylinder of radius a has its axis at $y = f(>0)$ and its generator runs parallel to the z-axis ($f/a > 1$). Polar co-ordinates are again defined via (2A.1) and suitable multipoles have the forms

$$\phi_n^{Is} = K_n(\gamma r)\cos n\theta + \int_0^\infty \cosh nk \cos(\gamma x \sinh k)\left(A_n^{(0)}(k)e^{vy} + B_n^{(0)}(k)e^{-vy}\right)dk, \quad (2B.1)$$

$$\phi_n^{IIs} = \int_0^\infty \cosh nk \cos(\gamma x \sinh k) C_n^{(0)}(k)e^{vy}\,dk, \quad (2B.2)$$

$$\phi_n^{Ia} = K_n(\gamma r)\sin n\theta + \int_0^\infty \sinh nk \sin(\gamma x \sinh k)\left(A_n^{(1)}(k)e^{vy} + B_n^{(1)}(k)e^{-vy}\right)dk, \quad (2B.3)$$

$$\phi_n^{IIa} = \int_0^\infty \sinh nk \sin(\gamma x \sinh k) C_n^{(1)}(k)e^{vy}\,dk, \quad (2B.4)$$

where

$$A_n^{(j)}(k) = \frac{1}{2}\Big[\left\{v\left(Dv^4 + 1 - \varepsilon K\right) + K\right\}e^{-vh}\left\{(-1)^{n+j+1}\left((1-\rho)v - (1+\rho)K\right)e^{vf} - (1-\rho)(v-K)e^{-vf}\right\}\Big]/H(v), \quad (2B.5)$$

$$B_n^{(j)}(k) = \frac{1}{2}\Big[(-1)^{n+j+1}\left\{v\left(Dv^4 + 1 - \varepsilon K\right) + K\right\}(1-\rho)(v-K)e^{v(f-h)} - (1-\rho)(v-K)\left\{v\left(Dv^4 + 1 - \varepsilon K\right) - K\right\}e^{-v(f-h)}\Big]/H(v), \quad (2B.6)$$

$$C_n^{(j)}(k) = -\rho K\Big[(-1)^{n+j+1}\left\{v\left(Dv^4 + 1 - \varepsilon K\right) + K\right\}e^{v(f-h)} - \left\{v\left(Dv^4 + 1 - \varepsilon K\right) - K\right\}e^{-v(f-h)}\Big]/H(v), \; j = 0,1 \quad (2B.7)$$

where the contour is indented below the poles $k=\mu_1$ and $k=\mu_2$ in the complex k-plane.

The far-field form of these multipoles, in the lower fluid layer, is given by ~

$$\phi_n^{IIs} \sim \pi i\left(C_n^{(0)\mu_1}\cosh n\mu_1 e^{\pm i\beta_1 x+\lambda_1 y} + C_n^{(0)\mu_2}\cosh n\mu_2 e^{\pm i\beta_2 x+\lambda_2 y}\right), \qquad (2B.8)$$

$$\phi_n^{IIa} \sim \pm\pi\left(C_n^{(1)\mu_1}\sinh n\mu_1 e^{\pm i\beta_1 x+\lambda_1 y} + C_n^{(1)\mu_2}\sinh n\mu_2 e^{\pm i\beta_2 x+\lambda_2 y}\right), \qquad (2B.9)$$

as $x\to\pm\infty$, where $C_n^{(j)\mu_1}$ and $C_n^{(j)\mu_2}$ $(j=0,1)$ are the residues of $C_n^{(j)}(k)$ at $k=\mu_1$ and $k=\mu_2$, given by

$$C_n^{(j)\mu_i} = -\rho K\left[(-1)^{n+j+1}\left\{\lambda_i\left(D\lambda_i^4+1-\varepsilon K\right)+K\right\}e^{\lambda_i(f-h)}\right.$$
$$\left.-\left\{\lambda_i\left(D\lambda_i^4+1-\varepsilon K\right)-K\right\}e^{-\lambda_i(f-h)}\right]/\beta_i H'(\lambda_i),\ i=1,2,\ j=0,1. \qquad (2B.10)$$

The polar expansions of the multipoles, similar to the case when cylinder is in the lower fluid, are

$$\phi_n^{Is} = K_n(\gamma r)\cos n\theta + \sum_{m=0}^{\infty} B_{nm}^{(s)} I_m(\gamma r)\cos m\theta, \qquad (2B.11)$$

$$\phi_n^{Ia} = K_n(\gamma r)\sin n\theta + \sum_{m=1}^{\infty} B_{nm}^{(a)} I_m(\gamma r)\sin m\theta, \qquad (2B.12)$$

where

$$B_{nm}^{(s)} = \varepsilon_m \int_0^{\infty}\cosh mk\cosh nk\left\{(-1)^m A_n^{(0)}(k)e^{\gamma f} + B_n^{(0)}(k)e^{-\gamma f}\right\}dk, \qquad (2B.13)$$

$$B_{nm}^{(a)} = 2\int_0^{\infty}\sinh mk\sinh nk\left\{(-1)^{m+1} A_n^{(1)}(k)e^{\gamma f} + B_n^{(1)}(k)e^{-\gamma f}\right\}dk. \qquad (2B.14)$$

B.1 Obliquely incident wave train of wavenumber λ_1

For this problem ϕ_{inc}^I is given, in the upper fluid, by $e^{i\beta_1 x}g_1(y)$, where $g_1(y)$ is defined in (2.12). The polar expansion of ϕ_{inc}^I is given by

$$\phi_{inc}^I = \sum_{m=0}^{\infty}\varepsilon_m I_m(\gamma r)\left[(-1)^m M_1 e^{-\lambda_1(h-f)} + M_2 e^{\lambda_1(h-f)}\right]\cosh m\nu\cos m\theta$$
$$+\,i\sum_{m=0}^{\infty}\varepsilon_m I_m(\gamma r)\left[(-1)^{m+1} M_1 e^{-\lambda_1(h-f)} + M_2 e^{\lambda_1(h-f)}\right]\sin m\nu\sin m\theta, \qquad (2B.15)$$

where

$$\cosh\nu = \frac{\lambda_1}{\gamma} = \frac{1}{\sin\alpha},$$

and

$$M_{1,2}=\frac{\{(1-\rho)\lambda_1-K\}\{\lambda_1(D\lambda_1^4+1-\varepsilon K)\pm K\}}{K\rho\{\lambda_1(D\lambda_1^4+1-\varepsilon K)\cosh\lambda_1 h-K\sinh\lambda_1 h\}}. \tag{2B.16}$$

The velocity potential $\phi_{\lambda_1}^I$ is expanded similar as (3.20), where ϕ_n^s and ϕ_n^a are the symmetric and antisymmetric multipoles developed for the upper fluid respectively. After applying the body boundary condition, $\frac{\partial\phi_{\lambda_1}^I}{\partial r}=0$ on $r=a$ and also using the orthogonal properties of trigonometric functions, we obtain the two infinite system of linear equations

$$\frac{a_m}{Z_m}+\sum_{n=1}^{\infty}B_{nm}^{(a)}a_n=2i\left[(-1)^m M_1e^{-\lambda_1(h-f)}-M_2e^{\lambda_1(h-f)}\right]\sinh m\nu,\ m=1,2,\cdots \tag{2B.17}$$

$$\frac{b_m}{Z_m}+\sum_{n=0}^{\infty}B_{nm}^{(s)}b_n=\varepsilon_m\left[(-1)^{m+1}M_1e^{-\lambda_1(h-f)}-M_2e^{\lambda_1(h-f)}\right]\cosh m\nu,\ m=0,1,\cdots. \tag{2B.18}$$

These equations were solved by truncations to 5×5 systems to produce the numerical results. The reflection and transmission coefficients can be extracted from the far-field form of the potential $\phi_{\lambda_1}^I$, using (2A.20), (2B.8) and (2A.9) with (2A.23), and are given by

$$R_{\lambda_1}=\pi\sum_{m=0}^{\infty}\left\{ib_mC_n^{(0)\mu_1}\cosh m\mu_1-a_mC_n^{(1)\mu_1}\sinh m\mu_1\right\}, \tag{2B.19}$$

$$r_{\lambda_1}=\pi\sum_{m=0}^{\infty}\left\{ib_mC_n^{(0)\mu_2}\cosh m\mu_2-a_mC_n^{(1)\mu_2}\sinh m\mu_2\right\}, \tag{2B.20}$$

$$T_{\lambda_1}=1+\pi\sum_{m=0}^{\infty}\left\{ib_mC_n^{(0)\mu_1}\cosh m\mu_1+a_mC_n^{(1)\mu_1}\sinh m\mu_1\right\}, \tag{2B.21}$$

$$t_{\lambda_1}=\pi\sum_{m=0}^{\infty}\left\{ib_mC_n^{(0)\mu_2}\cosh m\mu_2+a_mC_n^{(1)\mu_2}\sinh m\mu_2\right\}. \tag{2B.22}$$

B.2 Obliquely incident wave train of wavenumber λ_2

For this problem ϕ_{inc}^I is given, in the upper fluid, by $e^{i\beta_2x}g_2(y)$, where $g_2(y)$ is defined in (2.12). The polar expansion of ϕ_{inc}^I is same as (2B.15), except that λ_1 is replaced by λ_2. The velocity potential $\phi_{\lambda_2}^I$ for this scattering problem can again be expanded in multipoles similar to (2A.20) and the equations for a_n and b_n are similar to (2B.17) and (2B.18) with λ_1 replaced by λ_2.

The reflection and transmission coefficients can be extracted from the far-field form of the potential $\phi_{\lambda_2}^I$ using (2A.20), (2B.8) and (2B.9) with (2A.28). The expressions for R_{λ_2} and r_{λ_2} are similar to (2B.19) and (2B.20) with appropriate changes, and the transmission coefficients are given by

$$T_{\lambda_2} = \pi\sum_{m=0}^{\infty}\left\{ib_m C_n^{(0)\mu_1}\cosh m\mu_1 + a_m C_n^{(1)\mu_1}\sinh m\mu_1\right\}, \tag{2B.23}$$

$$t_{\lambda_2} = 1+\pi\sum_{m=0}^{\infty}\left\{ib_m C_n^{(0)\mu_2}\cosh m\mu_2 + a_m C_n^{(1)\mu_2}\sinh m\mu_2\right\}. \tag{2B.24}$$

B.3 Normally incident wave train

Now for the case of normal incidence, $\alpha = 0$, the modified Helmholtz equation reduces to the Laplace's equation and solutions of Laplace's equation singular at $y = f > 0$ are $r^{-n}\cos n\theta$ and $r^{-n}\sin n\theta, n \geq 1$, and these have the integral representations

$$\frac{\cos n\theta}{r^n} = \frac{1}{(n-1)!}\int_0^\infty k^{n-1}e^{k(y-f)}\cos kx dk,$$

$$\frac{\sin n\theta}{r^n} = \frac{1}{(n-1)!}\int_0^\infty k^{n-1}e^{k(y-f)}\sin kx dk.$$

Here the appropriate multipoles have the forms

$$\phi_n^{Is} = \frac{\cos n\theta}{r^n} + \frac{1}{(n-1)!}\int_0^\infty k^{n-1}\left(A_n^{(0)}(k)e^{ky} + B_n^{(0)}(k)e^{-ky}\right)\cos kx dk, \tag{2B.25}$$

$$\phi_n^{IIs} = \frac{1}{(n-1)!}\int_0^\infty k^{n-1}C_n^{(0)}(k)e^{ky}\cos kx dk, \tag{2B.26}$$

$$\phi_n^{Ia} = \frac{\sin n\theta}{r^n} + \frac{1}{(n-1)!}\int_0^\infty k^{n-1}\left(A_n^{(1)}(k)e^{ky} + B_n^{(1)}(k)e^{-ky}\right)\sin kx dk, \tag{2B.27}$$

$$\phi_n^{IIa} = \frac{1}{(n-1)!}\int_0^\infty k^{n-1}C_n^{(1)}(k)e^{ky}\sin kx dk, \tag{2B.28}$$

where

$$A_n^{(j)}(k) = \frac{1}{2}\Big[\left\{k\left(Dk^4+1-\varepsilon K\right)+K\right\}e^{-kh}\left\{(-1)^{n+j+1}\left((1-\rho)k-(1+\rho)K\right)e^{kf} - (1-\rho)(k-K)e^{-kf}\right\}\Big]/H(k), \tag{2B.29}$$

$$B_n^{(j)}(k) = \frac{1}{2}\Big[(-1)^{n+j+1}\left\{k\left(Dk^4+1-\varepsilon K\right)+K\right\}(1-\rho)(k-K)e^{k(f-h)} - (1-\rho)(k-K)\left\{k\left(Dk^4+1-\varepsilon K\right)-K\right\}e^{-k(f-h)}\Big]/H(k), \tag{2B.30}$$

$$C_n^{(j)}(k) = -\rho K\Big[(-1)^{n+j+1}\left\{k\left(Dk^4+1-\varepsilon K\right)+K\right\}e^{k(f-h)} - \left\{k\left(Dk^4+1-\varepsilon K\right)-K\right\}e^{-k(f-h)}\Big]/H(k),\quad j=0,1, \tag{2B.31}$$

the contour being indented below the poles $k = \lambda_1$ and $k = \lambda_2$ in the complex k-plane.

The polar expansions of these multipoles, in the upper layer, valid for $r < f$, are

$$\phi_n^{Is} = \frac{\cos n\theta}{r^n} + \sum_{m=0}^{\infty} E_{nm}^{(s)} r^m \cos m\theta, \tag{2B.32}$$

$$\phi_n^{Ia} = \frac{\sin n\theta}{r^n} + \sum_{m=1}^{\infty} E_{nm}^{(a)} r^m \sin m\theta, \tag{2B.33}$$

where

$$E_{nm}^{(s)} = \frac{1}{(n-1)!m!} \int_0^\infty k^{m+n-1} \left\{ (-1)^m A_n^{(0)}(k) e^{kf} + B_n^{(0)}(k) e^{-kf} \right\} dk, \tag{2B.34}$$

$$E_{nm}^{(a)} = \frac{1}{(n-1)!m!} \int_0^\infty k^{m+n-1} \left\{ (-1)^{m+1} A_n^{(1)}(k) e^{kf} + B_n^{(1)}(k) e^{-kf} \right\} dk. \tag{2B.35}$$

We note that unlike the case of multipoles singular in the lower layer, the coefficients in the polar expansions of ϕ_n^{Is} and ϕ_n^{Ia} are not the same.

The far-field behavior of these multipoles, in the lower layer fluid, is given by

$$\phi_n^{IIs} \sim \frac{\pi i}{(n-1)!} \left(\lambda_1^{n-1} C_n^{(0)\lambda_1} e^{\pm i\lambda_1 x + \lambda_1 y} + \lambda_2^{n-1} C_n^{(0)\lambda_2} e^{\pm i\lambda_2 x + \lambda_2 y} \right), \tag{2B.36}$$

$$\phi_n^{IIa} \sim \pm \frac{\pi}{(n-1)!} \left(\lambda_1^{n-1} C_n^{(1)\lambda_1} e^{\pm i\lambda_1 x + \lambda_1 y} + \lambda_2^{n-1} C_n^{(1)\lambda_2} e^{\pm i\lambda_2 x + \lambda_2 y} \right), \tag{2B.37}$$

as $x \to \pm\infty$, where $C_n^{(0)\lambda_1}$ and $C_n^{(1)\lambda_1}$ are the residues of $C_n^{(0)}(k)$ and $C_n^{(0)}(k)$ at $k = \lambda_1$ and $k = \lambda_2$ respectively and are given by

$$C_n^{(j)\lambda_i} = -\rho K \Big[(-1)^{n+j+1} \left\{ \lambda_i \left(D\lambda_i^4 + 1 - \varepsilon K \right) + K \right\} e^{\lambda_i (f-h)}$$
$$- \left\{ \lambda_i \left(D\lambda_i^4 + 1 - \varepsilon K \right) - K \right\} e^{-\lambda_i (f-h)} \Big] / H'(\lambda_i), \; i = 1,2, \; j = 0,1. \tag{2B.38}$$

Normally incident wave train of wavenumber λ_1

For this case ϕ_{inc}^I is given, in the upper fluid, by $e^{i\lambda_1 x} g_1(y)(\alpha = 0)$, where $g_1(y)$ is defined in (2.12). The polar expansion of ϕ_{inc}^I is given by

$$\phi_{inc}^I = \sum_{m=0}^{\infty} \frac{(\lambda_1 r)^m}{m!} \Big[\left\{ (-1)^m M_1 e^{-\lambda_1 (h-f)} + M_2 e^{\lambda_1 (h-f)} \right\} \cosh m\theta$$
$$+ i \left\{ (-1)^{m+1} M_1 e^{-\lambda_1 (h-f)} + M_2 e^{\lambda_1 (h-f)} \right\} \sin m\theta \Big], \tag{2B.39}$$

where

$$M_{1,2} = \frac{\{(1-\rho)\lambda_1 - K\} \left\{ \lambda_1 \left(D\lambda_1^4 + 1 - \varepsilon K \right) \pm K \right\}}{K\rho \left\{ \lambda_1 \left(D\lambda_1^4 + 1 - \varepsilon K \right) \cosh \lambda_1 h - K \sinh \lambda_1 h \right\}}. \tag{2B.40}$$

To solve this scattering problem, the velocity potential $\phi_{\lambda_1}^I$ is expanded as in (3.20), where ϕ_n^s and ϕ_n^a are symmetric and antisymmetric multipoles obtained for the upper fluid. After applying the body boundary condition, $\frac{\partial \phi_{\lambda_1}^I}{\partial r} = 0$ on $r = a$, and using the orthogonal properties of trigonometric functions, we obtain the two infinite system of linear equations for unknown a_n and b_n given by

$$a_m - \sum_{n=1}^{\infty} a^{n+m} E_{nm}^{(a)} a_n = i \frac{(\lambda_1 a)^m}{m!} \left[(-1)^{m+1} M_1 e^{-\lambda_1 (h-f)} + M_2 e^{\lambda_1 (h-f)} \right], \; m = 1, 2; \cdots \quad \text{(2B.41)}$$

$$b_m - \sum_{n=1}^{\infty} a^{n+m} E_{nm}^{(s)} b_n = \frac{(\lambda_1 a)^m}{m!} \left[(-1)^m M_1 e^{-\lambda_1 (h-f)} + M_2 e^{\lambda_1 (h-f)} \right], \; m = 1, 2; \cdots. \quad \text{(2B.42)}$$

These equations were solved by truncating to 4×4 system to produce the result presented below.

The reflection and transmission coefficients can be obtained from the far-field form of the potential $\phi_{\lambda_1}^I$, using (2A.20), (2B.36) and (2A.37) with (2B.52), and are given by

$$R_{\lambda_1} = \pi \sum_{n=1}^{\infty} \frac{a^n}{(n-1)!} \lambda_1^{n-1} \left[-a_n C_n^{(1)\lambda_1} + i b_n C_n^{(0)\lambda_1} \right], \quad \text{(2B.43)}$$

$$r_{\lambda_1} = \pi \sum_{n=1}^{\infty} \frac{a^n}{(n-1)!} \lambda_2^{n-1} \left[-a_n C_n^{(1)\lambda_2} + i b_n C_n^{(0)\lambda_2} \right], \quad \text{(2B.44)}$$

$$T_{\lambda_1} = 1 + \pi \sum_{n=1}^{\infty} \frac{a^n}{(n-1)!} \lambda_1^{n-1} \left[a_n C_n^{(1)\lambda_1} + i b_n C_n^{(0)\lambda_1} \right], \quad \text{(2B.45)}$$

$$t_{\lambda_1} = \pi \sum_{n=1}^{\infty} \frac{a^n}{(n-1)!} \lambda_2^{n-1} \left[a_n C_n^{(1)\lambda_2} + i b_n C_n^{(0)\lambda_2} \right]. \quad \text{(2B.46)}$$

B.4 Normally incident wave train of wavenumber λ_2

In this case ϕ_{inc}^I is given by $e^{i\lambda_2 x} g_2(y)(\alpha = 0)$, where $g_2(y)$ is defined in (2.12). Here we find that the expressions for reflection coefficients R_{λ_2} and r_{λ_2} are similar to (2B.43) and (2B.44) with appropriate changes, and the transmission coefficients T_{λ_2} and t_{λ_2} are given by

$$T_{\lambda_2} = \pi \sum_{n=1}^{\infty} \frac{a^n}{(n-1)!} \lambda_1^{n-1} \left[a_n C_n^{(1)\lambda_1} + i b_n C_n^{(0)\lambda_1} \right], \quad \text{(2B.47)}$$

$$t_{\lambda_2} = 1 + \pi \sum_{n=1}^{\infty} \frac{a^n}{(n-1)!} \lambda_2^{n-1} \left[a_n C_n^{(1)\lambda_2} + i b_n C_n^{(0)\lambda_2} \right]. \quad \text{(2B.48)}$$

Numerical results

Figures 8.21–8.24 show the reflection and transmission coefficients for an incident wave of wave number λ_1 on a cylinder submerged in the upper fluid layer for $D/a^4 = 1.5, \varepsilon/a = 0.01, h/a = 2.5, \quad \rho = 0.5, f/a = 1.25$. The different curves correspond to $\alpha = 75^0, 80^0, 85^0, 89^0$. The curves are somewhat similar to those for scattering of an incident wave of wave number λ_1 by a circular cylinder in the lower fluid layer and display the same characteristics.

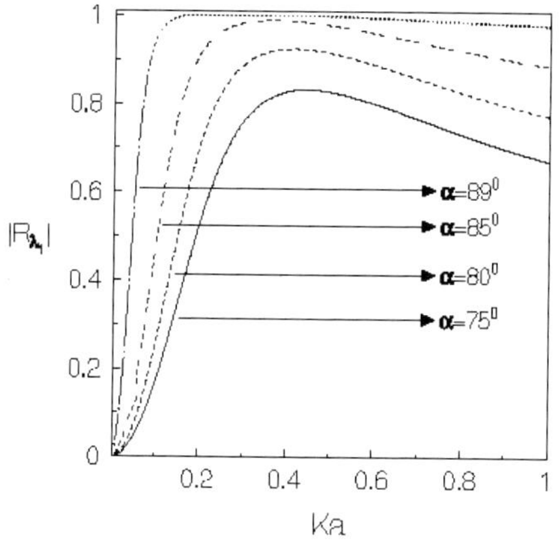

Fig. 8.21 Reflection coefficient due to a wave of wavenumber λ_1 incident on a cylinder in upper layer $(\frac{D}{a^4} = 1.5, \frac{\varepsilon}{a} = .01, \frac{h}{a} = 2.5, \rho = .5, \frac{f}{a} = 1.25)$

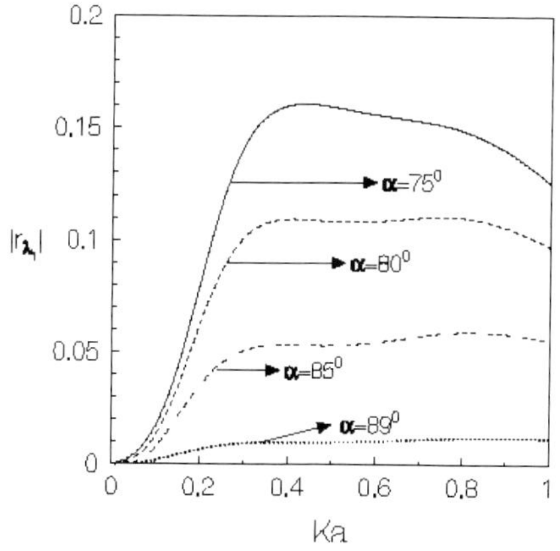

Fig. 8.22 Reflection coefficient due to a wave of wavenumber λ_1 incident on a cylinder in upper layer $(\frac{D}{a^4} = 1.5, \frac{\varepsilon}{a} = .01, \frac{h}{a} = 2.5, \rho = .5, \frac{f}{a} = 1.25)$

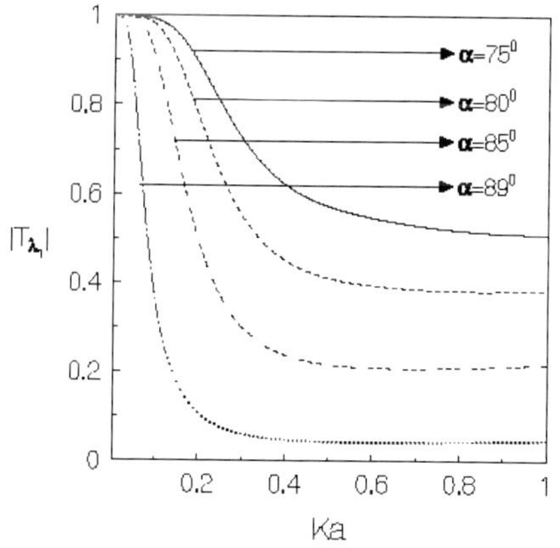

Fig. 8.23 Transmission coefficient due to a wave of wavenumber λ_1 incident on a cylinder in upper layer $(\frac{D}{a^4} = 1.5, \frac{\varepsilon}{a} = .01, \frac{h}{a} = 2.5, \rho = .5, \frac{f}{a} = 1.25)$

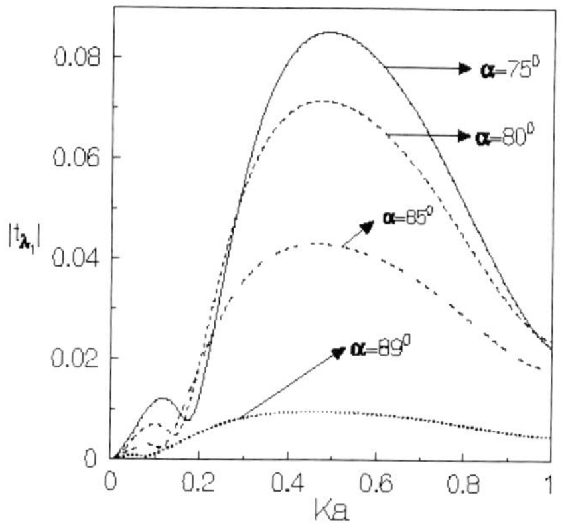

Fig. 8.24 Transmission coefficient due to a wave of wavenumber λ_1 incident on a cylinder in upper layer $(\frac{D}{a^4} = 1.5, \frac{\varepsilon}{a} = .01, \frac{h}{a} = 2.5, \rho = .5, \frac{f}{a} = 1.25)$

The case of an incident wave of wave number λ_2 is more interesting due to the presence of cut-off frequencies. Figures 8.25–8.28 show reflection coefficients $|R_{\lambda_2}|$, $|r_{\lambda_2}|$ and transmission coefficients $|T_{\lambda_2}|$, $|t_{\lambda_2}|$ against Ka. The different parameters are taken to be the same as in the previous set of figures and the different curves correspond to different values of α, viz. $\alpha = 15.07^0, 16.84^0, 17.93^0, 19.5^0$. When $\alpha = 19.5^0$, which is greater than the critical angle $\alpha_c = 19.13^0$, for the given value of different parameters, there are no waves of wave number λ_1 propagating in the fluid. Here we have the following cut-off frequencies: $K_c a = (0.1, 0.88); (0.15, 0.7); (0.2, 0.57)$ corresponding to the incident angles $15.07^0, 16.84^0, 17.93^0$ respectively. For these angles and for

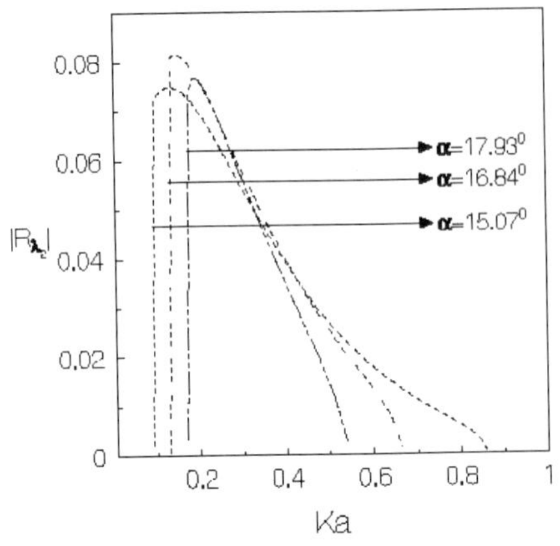

Fig. 8.25 Reflection coefficient due to a wave of wavenumber λ_1 incident on a cylinder in upper layer $(\frac{D}{a^4} = 1.5, \frac{\varepsilon}{a} = .01, \frac{h}{a} = 2.5, \rho = .5, \frac{f}{a} = 1.25)$

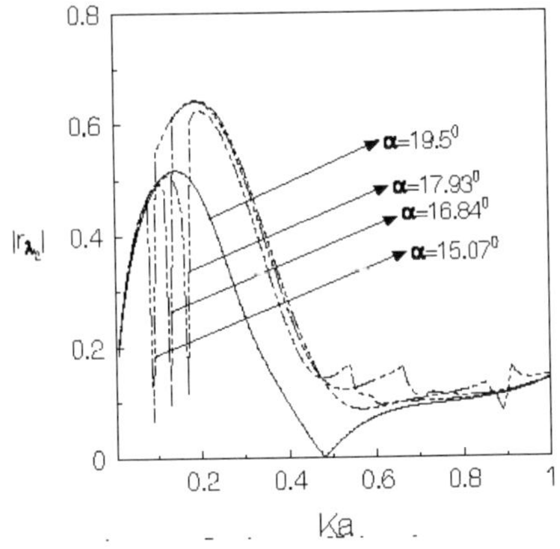

Fig. 8.26 Reflection coefficient due to a wave of wavenumber λ_1 incident on a cylinder in lower layer $(\frac{D}{a^4} = 1.5, \frac{\varepsilon}{a} = .01, \frac{h}{a} = 2.5, \rho = .5, \frac{f}{a} = 1.25)$

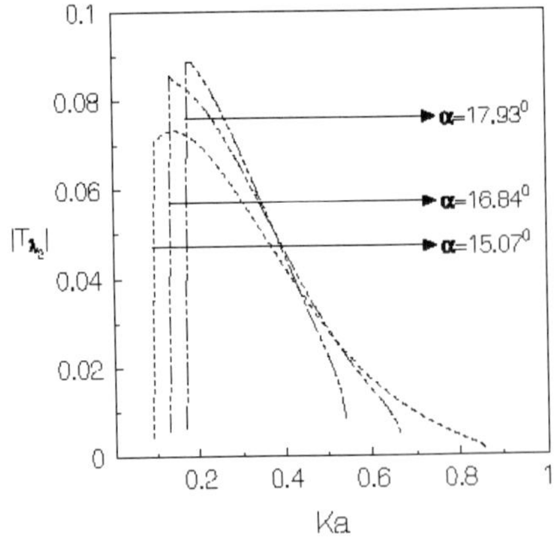

Fig. 8.27 Transmission coefficient due to a wave of wavenumber λ_2 incident on a cylinder in upper layer $(\frac{D}{a^4} = 1.5, \frac{\varepsilon}{a} = .01, \frac{h}{a} = 2.5, \rho = .5, \frac{f}{a} = 1.25)$

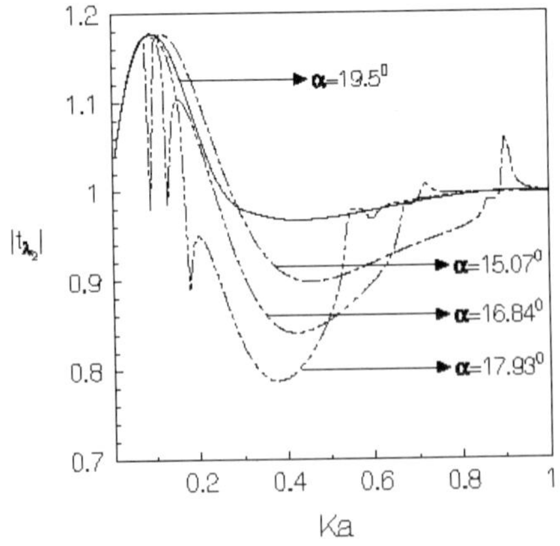

Fig. 8.28 Transmission coefficient due to a wave of wavenumber λ_2 incident on a cylinder in upper layer $(\frac{D}{a^4} = 1.5, \frac{\varepsilon}{a} = .01, \frac{h}{a} = 2.5, \rho = .5, \frac{f}{a} = 1.25)$

frequencies lying between two appropriate cut-off frequencies will there be conversion of energy from one mode to the other. These figures are shown in Fig. 8.25 and Fig. 8.27. The reflection coefficient $|r_{\lambda_2}|$ and transmission coefficient $|t_{\lambda_2}|$ for the wave of wave number λ_2 are shown in Fig. 8.26 and Fig. 8.28.

We observe from the curves of reflection and transmission coefficients for wave of wave number λ_2 that two spikes in each curve occur at the cut-off frequencies (cf. Fig. 8.26 and Fig. 8.28). For $\alpha = 19.5^0$ which is greater than the critical angle $\alpha_c = 19.13^0$, there is no spike in the curves of reflection and transmission coefficients in Fig. 8.26 and Fig. 8.28.

For the normally incident wave train, we choose $\varepsilon/a = 0.01$, $h/a = 2.5$, $\rho = 0.5$, $f/a = 1.25$ for which the reflection and transmission coefficients due to an incident wave of wave number λ_1 are depicted in Figs. 8.29–8.32. The different curves correspond to

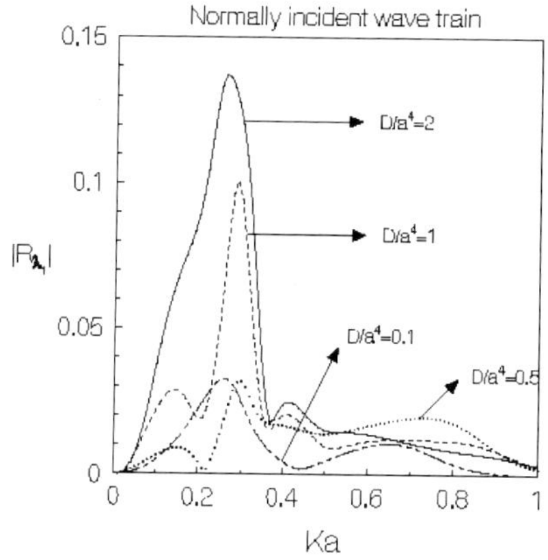

Fig. 8.29 Reflection coefficient due to a wave of wavenumber λ_1 incident on a cylinder in upper layer ($\frac{D}{a^4} = 1.5, \frac{\varepsilon}{a} = .01, \frac{h}{a} = 2.5, \rho = .5, \frac{f}{a} = 1.25$)

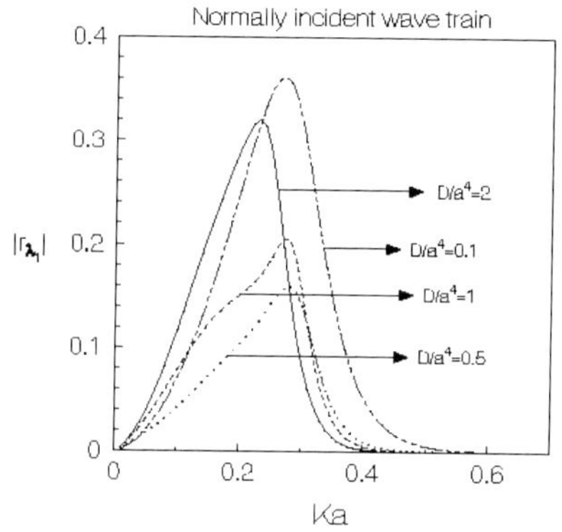

Fig. 8.30 Reflection coefficient due to a wave of wavenumber λ_1 incident on a cylinder in lower layer ($\frac{D}{a^4} = 1.5, \frac{\varepsilon}{a} = .01, \frac{h}{a} = 2.5, \rho = .5, \frac{f}{a} = 1.25$)

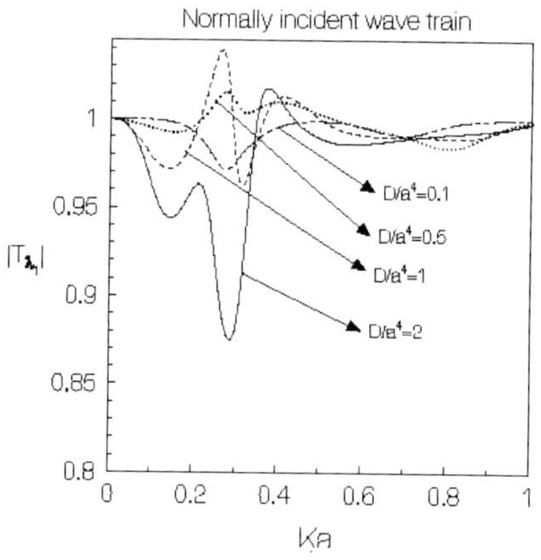

Fig. 8.31 Transmission coefficient due to a wave of wavenumber λ_1 incident on a cylinder in upper layer ($\frac{\varepsilon}{a} = .01, \frac{h}{a} = 2.5, \rho = .5, \frac{f}{a} = 1.25$)

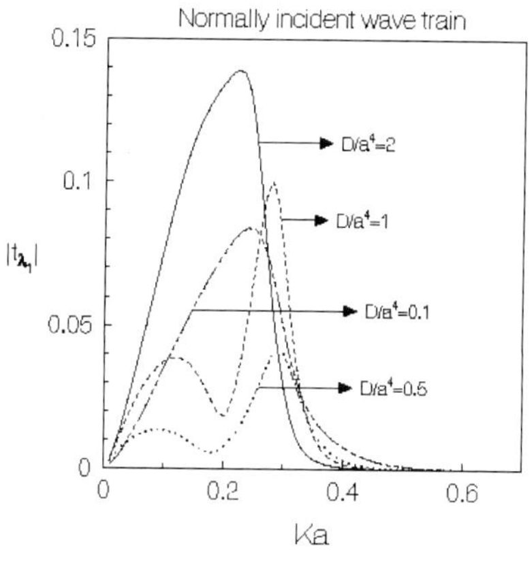

Fig. 8.32 Transmission coefficient due to a wave of wavenumber λ_1 incident on a cylinder in upper layer ($\frac{\varepsilon}{a} = .01, \frac{h}{a} = 2.5, \rho = .5, \frac{f}{a} = 1.25$)

$D/a^4 = 0.1, 0.5, 1, 2$. Figures 8.29, 8.30 and 8.32 show that the reflection coefficients $|R_{\lambda_1}|$, $|r_{\lambda_1}|$ and the transmission coefficient $|t_{\lambda_1}|$ first increase as Ka increases, each attains a maximum value and then decrease as Ka further increases. Figure 8.31 shows that transmission coefficients $|T_{\lambda_1}|$ first decreases as Ka increases, attains a minimum value and then increase as Ka further increases.

The reflection and transmission coefficients due to an incident wave of wave number λ_2 are shown in Figs. 8.33–8.36. The different curves correspond to $D/a^4 = 0.1, 0.5, 1, 2$. $|R_{\lambda_2}|$, $|r_{\lambda_2}|$ are shown in Fig. 8.33 and Fig. 8.34 and $|T_{\lambda_2}|$, $|t_{\lambda_2}|$ are shown in Fig. 8.35 and Fig. 8.36.

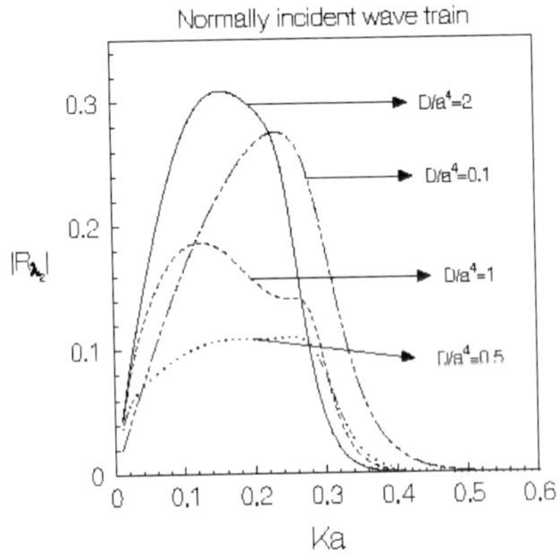

Fig. 8.33 Refl;ection coefficient due to a wave of wavenumber λ_2 incident on a cylinder in upper layer ($\frac{\varepsilon}{a} = .01, \frac{h}{a} = 2.5, \rho = .5, \frac{f}{a} = 1.25$)

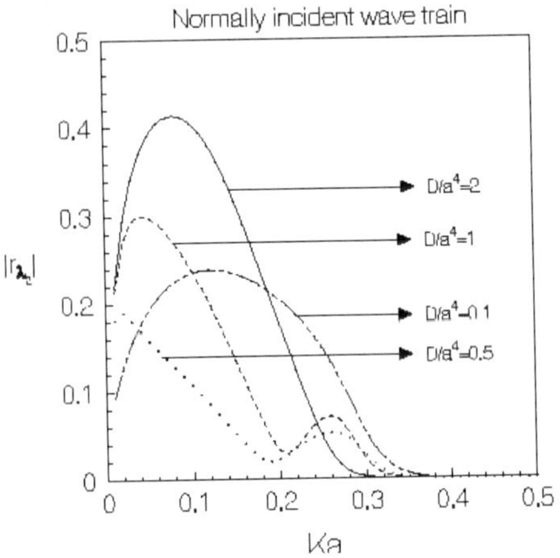

Fig. 8.34 Reflection coefficient due to a wave of wavenumber λ_2 incident on a cylinder in upper layer ($\frac{\varepsilon}{a} = .01, \frac{h}{a} = 2.5, \rho = .5, \frac{f}{a} = 1.25$)

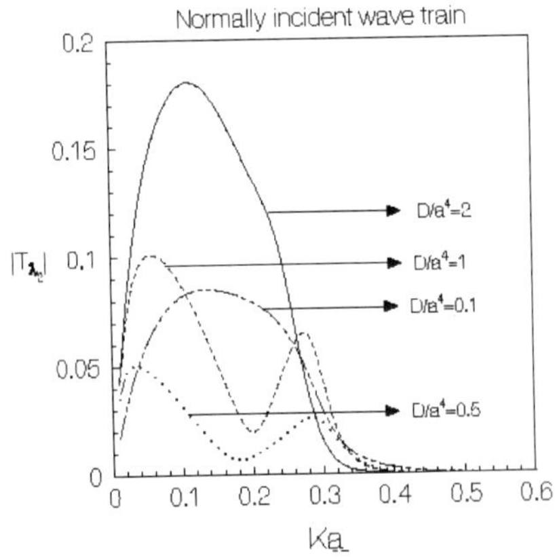

Fig. 8.35 Transmission coefficient due to a wave of wavenumber λ_2 incident on a cylinder in upper layer ($\frac{\varepsilon}{a} = .01, \frac{h}{a} = 2.5, \rho = .5, \frac{f}{a} = 1.25$)

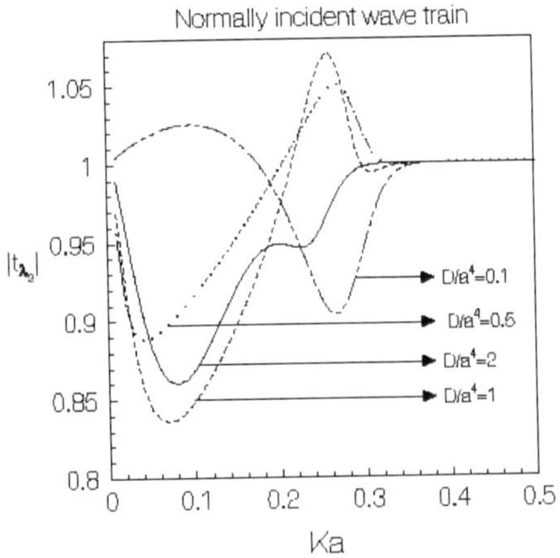

Fig. 8.36 Transmission coefficient due to a wave of wavenumber λ_2 incident on a cylinder in lower layer ($\frac{\varepsilon}{a} = .01, \frac{h}{a} = 2.5, \rho = .5, \frac{f}{a} = 1.25$)

Due to the presence of ice-cover, the figures for normally incident wave train are somewhat different from the figures given in the case of upper fluid with a free surface in two-layer fluid by Linton and McIver (1995). Reflection and transmission coefficients due to an incident wave of wave number λ_1 and λ_2 are oscillatory in nature and reflection coefficients for both the incident wave numbers, transmission coefficients $|t_{\lambda_1}|$ and $|T_{\lambda_2}|$ tend ultimately to zero for large Ka and also $|T_{\lambda_1}|$ and $|t_{\lambda_2}|$ tend to unity for large Ka. This may be attributed to interactions of the incident wave trains between the boundary of the circular cylinder, ice-cover surface and interface between two-layer. Also we observe that the peak values of the curves decrease as D/a^4 decreases in Figs. 8.29, 8.30, 8.32, 8.33, 8.34 and 8.35. But in Figs. 8.31 and 8.36 the peak value of the curves increase with decreasing D/a^4.

CHAPTER IX

Variable Bottom Topography

A. SINGLE-LAYER FLUID

9A.1 Scattering by small bottom undulations

When a train of surface waves propagating from infinity is incident on an obstacles present in water, it experiences partial reflection by and transmission over or below the obstacle. Determination of these coefficients is of some mathematical and physical importance and the corresponding problem is termed as the wave scattering problem. When an obstacle is present at the bottom of an otherwise laterally bounded ocean of uniform finite depth, the incident wave train is partially reflected by and transmitted over the obstacle. If the obstacle is of arbitrary geometrical shape, the problem of determining the reflection and transmission coefficients is in general a difficult task. An undulating bottom can be regarded as a bottom obstacle. Problems of wave scattering by an undulating bottom is of considerable interest in the literature on linearized theory of water waves due to their importance in finding the effects of naturally occurring bottom obstacles such as sand ripples on the wave propagation. One aim to study this class of problems is to investigate the mechanism of wave-induced mass transport that forms sand ripples of some wavelength. If this wavelength is half that of the incident wave, then these ripples produce strong reflected waves thus providing a model of breakwater to protect the offshore areas. This phenomenon of somewhat strong reflection was studied in the theory of water waves by a number of researchers like Mei (1985), Kirby(1986), Mei et al. (1988) and others.

There exists only one explicit solution for the two-dimensional problem of wave propagation over a particular bottom topography considered by Roseau (1976). For general bottom topography, a variety of approximate numerical methods have been devised in the literature. One such method uses conformal mapping by which the undisturbed fluid region with variable bottom is transformed into a uniform strip, e.g., Kreisel (1949), Fitz-Gerald (1976), Hamilton (1977). For an obstacle in the form of a small cylindrical deformation at the bottom, Miles (1981) employed a perturbation method followed by the finite cosine transform technique in the mathematical analysis to obtain the reflection and transmission coefficients up to first order. Davies (1982) considered wave scattering by a patch of sinusoidal undulations on an otherwise flat bottom using perturbation theory followed by an application of Fourier transform after introducing an artificial frictional term in the free surface condition (to ensure the existence of Fourier transform in the mathematical analysis). He obtained the reflection and transmission coefficients from the behavior of the velocity field at either infinity and observed that when the wavelength of the sinusoidal undulations is half that of the incident wave, a significant amount of wave reflection occurs. Davies and Heathershaw (1984) confirmed this theoretical result by conducting experiments in a wave tank.

A brief account of various problems of wave scattering by bottom undulations and their method of solution for normal as well as oblique incidence of a wave train is discussed in this section.

For an ocean of variable bottom topography given by $h = h(x, z)$, the governing equations are

$$\nabla^2 \phi = 0, \; -\infty < x, z < \infty, 0 \le y \le h(x, z) \tag{A1.1}$$

$$K\phi + \frac{\partial \phi}{\partial y} = 0 \text{ on } y = 0, \tag{A1.2}$$

$$\frac{\partial \phi}{\partial n} = 0 \text{ on } y = h(x, z), \tag{A1.3}$$

where $\frac{\partial}{\partial n}$ denotes the normal derivative at the bottom. For any general $h(x, z)$, it is not possible to write the solution for ϕ even in two dimensions $(h = h(x))$. In fact when $h(x)$ (for 2D) varies slowly, that is for bottom with mild variation, obtaining the solution for $\phi(x, y)$ is a topic of intense research work, since the late eighties of the last century, see Chamberlain and Porter (2005).

Henceforth we consider undulating bottom having local small cylindrical deformation so that the ocean bottom is represented by

$$y = h + \varepsilon\, c(x) \tag{A1.4}$$

where ε is a measure of smallness of the deformation, $c(x)$ is a function with compact support describing the shape of the bottom and $c(x) \to 0$ as $|x| \to \infty$, so that far away from the deformation, the ocean is of uniform finite depth h. Thus the incident surface wave train can be represented by

$$\phi_0(x, y) = \cosh k_0 (h - y) e^{ik_0 x}. \tag{A1.5}$$

where k_0 is the unique positive real root of the dispersion equation

$$k \tanh kh = K.$$

This incident wave train will be partially reflected by and transmitted over the bottom undulations (local). Thus the problem is to solve

$$\frac{\partial^2 \phi}{\partial x^2} + \frac{\partial^2 \phi}{\partial y^2} = 0, -\infty < x < \infty, \; 0 \le y \le \varepsilon\, c(x) \tag{A1.6}$$

with boundary condition

$$K\phi + \frac{\partial \phi}{\partial y} = 0 \text{ on } y = 0, \quad \text{(A1.7)}$$

the bottom condition

$$\frac{\partial \phi}{\partial n} = 0 \text{ on } y = h + \varepsilon\, c(x), \quad \text{(A1.8)}$$

and the infinity conditions

$$\phi(x,y) \sim \begin{cases} \phi_0(x,y) + R\phi_0(-x,y) \text{ as } x \to -\infty, \\ T\phi_0(x,y) \text{ as } x \to \infty, \end{cases} \quad \text{(A1.9)}$$

where R and T are the unknown reflection and transmission coefficients and the problem here is to determine them by some approximate method.

It is important to observe that if the bottom is of uniform finite depth (i.e., $\varepsilon = 0$), then $R \equiv 0$ and $T \equiv 1$ since the incident wave train is fully transmitted and there is no obstacle present in the fluid region to reflect the incident wave train.

As mentioned earlier, Kreisel (1949) employed conformal transformation to solve the problem for small undulation. However, here a simplified perturbation expansion will be employed to solve the problem described by (A1.6) to (A1.9) approximately. For this purpose, the bottom condition (A1.8) is expressed in an approximate form as given by

$$\frac{\partial \phi}{\partial y} - \varepsilon \frac{d}{dx}\left(c(x)\frac{\partial \phi}{\partial x}(x,h)\right) + O(\varepsilon^2) = 0 \text{ on } y = h. \quad \text{(A1.10)}$$

In view of the approximate bottom condition (A1.10) and the observation made above that a surface wave train propagating in an ocean of uniform finite depth does not experience any reflection, ϕ, R, T can be expanded in terms of the small parameter ε as

$$\phi = \phi_0 + \varepsilon\phi_1 + O(\varepsilon^2),\ R = \varepsilon R_1 + O(\varepsilon^2),\ T = 1 + \varepsilon T_1 + O(\varepsilon^2). \quad \text{(A1.11)}$$

Using this expansion in (A1.6), (A1.7), (A1.8) and (A1.9), we find that $\phi_1(x,y)$ satisfies the boundary value problem (BVP) described by

$$\frac{\partial^2 \phi_1}{\partial x^2} + \frac{\partial^2 \phi_1}{\partial y^2} = 0, -\infty < x < \infty, 0 \le y \le \varepsilon c(x), \quad \text{(A1.12)}$$

$$K\phi_1 + \frac{\partial \phi_1}{\partial y} = 0 \text{ on } y = 0, \quad \text{(A1.13)}$$

$$\frac{\partial \phi_1}{\partial x} = ik_0 \frac{d}{dx} c(x) e^{ik_0 x} \equiv q(x) \text{ on } y = h, \quad \text{(A1.14)}$$

$$\phi_1(x,y) \sim \begin{cases} R_1\phi_0(-x,y) \text{ as } x \to -\infty, \\ T_1\phi_0(x,y) \text{ as } x \to \infty. \end{cases} \quad \text{(A1.15)}$$

It may be noted that $\phi_1(x,y)$ behaves as outgoing waves as $|x| \to \infty$.

The BVP described (A1.12) to (A1.14) is now solved by employing Fourier integral transform and also by using Green's integral theorem.

Solution using Fourier transform

Since the conditions (A1.9) show that $\phi_1(x,.) \to c_1 e^{ik_0 x}$ as $|x| \to \infty$ where c_1 is independent of x, we assume k_0 to have a small positive imaginary part. This ensures that the Fourier transform of $\phi_1(x,.)$ defined by

$$\Phi_1(\xi,.) = \int_{-\infty}^{\infty} \phi_1(x,.) e^{i\xi x} dx \tag{A1.16}$$

exists. In the final result we make this positive imaginary part of k_0 to tend to zero. Thus $\Phi_1(\xi, y)$ satisfies

$$\frac{d^2\Phi_1}{dy^2} - \xi^2 \Phi_1 = 0, 0 \le y \le h, \tag{A1.17}$$

$$K\Phi_1 + \frac{d\Phi_1}{dy} = 0 \text{ on } y = 0, \tag{A1.18}$$

$$\frac{d\Phi_1}{dy} = \int_{-\infty}^{\infty} q(x) e^{i\xi x} dx \equiv Q(\xi) \text{ on } y = h. \tag{A1.19}$$

Solution of (A1.17) to (A1.19) is given by

$$\Phi_1(\xi, y) = Q(\xi) \frac{K \sinh \xi y - \xi \cosh \xi y}{\xi \Delta(\xi)} \tag{A1.21}$$

where

$$\Delta(\xi) = K \cosh \xi h - \xi \sinh \xi h. \tag{A1.22}$$

The zeroes of $\Delta(\xi)$ are the same as the roots of the dispersion equation $k \tanh kh = K$. Using Fourier inversion we find

$$\phi_1(x, y) = \frac{1}{2\pi i} \int_{-\infty}^{\infty} \frac{Q(\xi)(K \sinh \xi y - \xi \cosh \xi y)}{\xi \Delta(\xi)} e^{-i\xi x} d\xi \tag{A1.23}$$

where the path of integration is along the real axis in the complex ξ-plane from $\xi = -\infty$ to $\xi = \infty$ indented below the point $\xi = k_0$ and above the point $\xi = -k_0$ to ensure that $\phi_1(x, y)$ behaves as outgoing waves as $|x| \to \infty$. For $x < 0$, the path of integration of the integral in (A1.23) is closed by a semi-circle of large radius in the upper half of the complex ξ-plane as there is no contribution from the semi-circle. The only contribution is from the pole at $\xi = k_0$. Thus as $x \to -\infty$, we find

$$\phi_1(x, y) \to -\frac{iQ(k_0) \cosh k_0(h - y)}{k_0 h \sinh k_0 h \cosh k_0 h} e^{-ik_0 x}. \tag{A1.23}$$

It is easy to see that

$$Q(\xi) = k_0 \xi \int_{-\infty}^{\infty} c(x) e^{i(k_0 + \xi)x} dx \tag{A1.24}$$

Comparing with (A1.15) as $x \to -\infty$, we find from (A1.23) that

$$R_1 = -\frac{iQ(k_0)}{k_0 h + \sinh k_0 h \cosh k_0 h}.$$

Using (A1.24), this produces finally

$$R_1 = -\frac{ik_0}{h + \dfrac{\sinh^2 k_0 h}{K}} \int_{-\infty}^{\infty} c(x) e^{2ik_0 x} dx. \tag{A1.25}$$

Again, for $x > 0$, the path of integration of the integral in (A1.22) is closed by a semi-circle of large radius in the lower half of the complex ξ-plane as there is no contribution from the semi-circle. The only contribution is from the pole at $\xi = -k_0$. Thus we find that as $x \to \infty$

$$\phi_1(x, y) \to \frac{-iQ(-k_0)\cosh k_0(h-y)}{k_0 h + \sinh k_0 h \cosh k_0 h} e^{ik_0 x}. \tag{A1.26}$$

Computing with (A1.15) as $x \to \infty$, we find from (A1.26) that

$$T_1 = \frac{ik_0}{h + \dfrac{\sinh^2 k_0 h}{K}} \int_{-\infty}^{\infty} c(x) dx. \tag{A1.27}$$

The first order reflection and transmission coefficients $|R_1|$ and $|T_1|$ can be computed numerically once the form of the shape function $c(x)$ is known. It may be noted that the integrals in (A1.25) and (A1.27) are convergent since $c(x)$ is of compact support.

Solution using Green's function

We first construct Green's function $G(x, y; \xi, \eta)$ which satisfies

$$\nabla^2 G = 0, -\infty < x < \infty,\ 0 \le y \le h$$

except at (ξ, η),

$$G \sim \ln r \text{ as } r = \{(x-\xi)^2 + (y-\eta)^2\}^{1/2} \to 0,$$

$$KG + \frac{\partial G}{\partial y} = 0 \text{ on } y = 0,$$

$$\frac{\partial G}{\partial y} = 0 \text{ on } y = h$$

and G behaves as outgoing waves as $|x-\xi| \to \infty$, i.e., $G \sim$ a multiple of $\cosh k_0(h-y) e^{ik_0|x-\xi|}$ as $|x-\xi| \to \infty$. $G(x, y; \xi, \eta)$ is given by Thorne (1953)

$$G(x, y; \xi, \eta) = -4\pi i \frac{\cosh k_0(h-y)\cosh k_0(h-\eta)}{2k_0 h + \sinh 2k_0 h} e^{ik_0|x-\xi|}$$

$$-4\pi \sum_{n=1}^{\infty} \frac{\cos k_n(h-y)\cos k_n(h-\eta)}{2k_n h + \sin 2k_n h} e^{-k_n|x-\xi|}. \tag{A1.28}$$

We now apply the Green's integral theorem to $\phi_1(x,y)$ and $G(x,y;\xi,\eta)$ in the form

$$\int (\phi_1 \frac{\partial G}{\partial n} - G\frac{\partial \phi_1}{\partial n})ds = 0 \tag{A1.29}$$

over the region bounded externally by the lines $y = 0(-X \le x \le X), x = X(0 \le y \le h), y = h(-X \le x \le X), x = -X(0 \le y \le h)$ and internally by a circle C_0 with center at (ξ,η) and radius δ, and ultimately make $X \to \infty$ and $\delta \to 0$. It may be noted that there is no contribution to the integral in (A1.29) from the line $y = 0$ due to the free surface condition, no contribution from the line $x = \pm X$ because of the outgoing nature of both ϕ_1 and G as $|x| \to \infty$. The contribution from the circle C_0 as $\delta \to 0$ is $2\pi\phi_1(\xi,\eta)$ while the contribution from the line $y = h$ is $-\int_{-\infty}^{\infty} G(x,h;\xi,\eta)q(x)dx$ as $X \to \infty$. Thus we find

$$\phi_1(\xi,\eta) = \frac{1}{2\pi}\int_{-\infty}^{\infty} G(x,h;\xi,\eta)q(x)dx. \tag{A1.30}$$

We note from (A1.15) that

$$\phi_1(\xi,\eta) \to R_1\phi_0(-\xi,\eta) \text{ as } \xi \to -\infty \tag{A1.31}$$

and from (A1.28) that

$$G(x,h;\xi,\eta) \to -4\pi i \frac{\phi_0(-\xi,\eta)e^{ik_0x}}{2k_0h + \sinh 2k_0h} \text{ as } \xi \to -\infty. \tag{A1.32}$$

Using (A1.31) and (A1.32) in (A1.30), we find

$$R_1 = -\frac{2i}{2k_0h + \sinh 2k_0h}\int_{-\infty}^{\infty} e^{ik_0x}q(x)dx = -\frac{ik_0}{h + \dfrac{\sinh^2 k_0h}{K}}\int_{-\infty}^{\infty} e^{2ik_0x}c(x)dx \tag{A1.33}$$

and this coincides with (A1.25) obtained by using the Fourier transform.

Similarly, noting from (A1.15) that

$$\phi_1(\xi,\eta) \to T_1\phi_0(\xi,\eta) \text{ as } \xi \to \infty \tag{A1.34}$$

and from (A1.28) that

$$G(x,h;\xi,\eta) \to -4\pi i \frac{\phi_0(\xi,\eta)e^{-ik_0x}}{2k_0h + \sinh 2k_0h} \text{ as } \xi \to -\infty \tag{A1.35}$$

and using (A1.34) and (A1.35) in (A1.20), we find

$$T_1 = \frac{2ik_0^2}{2k_0h + \sinh 2k_0h}\int_{-\infty}^{\infty} c(x)dx = \frac{ik_0}{k_0 + \dfrac{\sinh^2 k_0h}{K}}\int_{-\infty}^{\infty} c(x)dx, \tag{A1.36}$$

and this coincides with (A1.27) obtained by using Fourier transform.

A patch of sinusoidal ripples at the bottom

Davies (1982) considered a patch of sinusoidal ripples at the bottom in the form

$$c(x) = \begin{cases} c \sin \lambda x, & -\dfrac{\pi m}{\lambda} \le x \le \dfrac{\pi m}{\lambda}, \\ 0 & \text{otherwise}, \end{cases} \tag{A1.37}$$

where m is a positive integer. For this case, it is obvious that $T_1 = 0$ as $c(x)$ is an odd function, and

$$R_1 = \frac{ck_0}{h + \dfrac{\sinh^2 k_0 h}{K}} \frac{(-1)^m \lambda}{\lambda^2 - 4k_0^2} \sin\left(\frac{2k_0}{\lambda}\pi m\right) = \frac{c}{2\left(h + \dfrac{\sinh^2 k_0 h}{K}\right)} H_m\left(\frac{2k_0}{\lambda}\right) \tag{A1.38}$$

where

$$H_m(z) = (-1)^m z \frac{\sin m\pi z}{z^2 - 1}. \tag{A1.39}$$

We note that $H_m(1) = m\pi/2$. This shows that the amplitude of the reflected wave increases in proportional to the number of ripples in the patch. Davies (1982) has given a graph for $H_m(z)$ for different values of m against z. It is obvious that $H_m(z)$ has a maximum value near $z = 1$ for different values of m. This shows that when a train of surface waves is incident on a patch of ripples of wave number λ, significant reflections of wave energy occurs when $2k_0 = \lambda$, i.e., when the ripple wave length is half the wave length of the incident wave field. As mentioned earlier, this phenomenon has practical application in the protection of coastal areas from the effect of a rough sea during storms. Instead of a curve for $H_m(z)$ depicted against z for different values of m, here we depict $|R_1|$ against Kh for different values of m (and c/h) in Fig. 9.1. A similar type of conclusion is arrived at from these figures.

Oblique scattering by a cylindrical deformation of the bottom

For oblique incidence of a train of surface water waves, Miles (1981) obtained the first order reflection and transmission coefficients using a perturbation method assuming the bottom deformation to be small and to have the form a long cylinder in the lateral direction. He used the finite Fourier cosine transform technique in the mathematical analysis. The mathematical description of the problem is as follows:

Solve the partial differential equation

$$\nabla^2 \psi = 0, -\infty < x, z < \infty, 0 \le y \le h + \varepsilon c(x) \tag{A1.40}$$

with boundary condition

$$K\psi + \frac{\partial \psi}{\partial y} = 0 \text{ on } y = 0, \tag{A1.41}$$

the bottom condition

$$\frac{\partial \psi}{\partial n} = 0 \text{ on } y = h + \varepsilon c(x). \tag{A1.42}$$

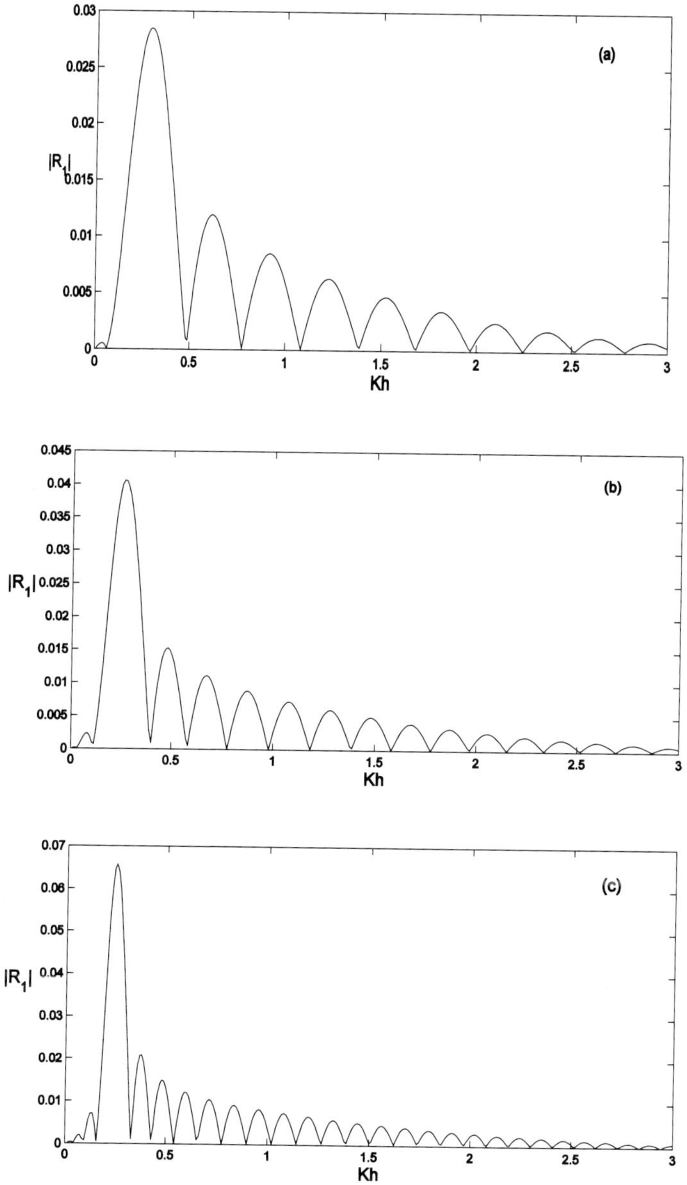

Fig. 9.1 Plot of $|R_1|$ against Kh for different values of m ((*a*) $m = 2$, (*b*) $m = 3$ (*c*) $m = 5$) with $c/h = 0.1$, and $\lambda/h = 1$.

Here ∇^2 is the three dimensional Laplacean operator, $c(x)$ is the shape function described earlier. We assume that a surface wave train represented by the velocity potential

$$\psi_0(x, y, z) = \cos hk_0(h - y)e^{ik_0(x\cos\theta + y\sin\theta)} \tag{A1.43}$$

is obliquely incident upon the bottom deformation from the direction of negative infinity, θ being the angle of incidence. It may be noted that $\theta = 0$ corresponds to the case of normal incidence considered above. Thus ψ has the asymptotic behavior given by

$$\psi(x,y,z) \sim \begin{cases} \psi_0(x,y,z) + R\psi_0(-x,y,z) \ \ as \ \ x \to -\infty, \\ T\psi_0(x,y,z) \text{ as } x \to \infty, \end{cases} \tag{A1.44}$$

where R and T, as before, denotes the reflection and transmission coefficients respectively.

Assuming ε to be small, the bottom condition (A1.45) can be expressed as

$$\frac{\partial \psi}{\partial y} - \varepsilon\{c(x)\frac{\partial^2 \psi}{\partial y^2} - c'(x)\frac{\partial \psi}{\partial x}\} + O(\varepsilon^2) = 0 \ \text{ on } \ y = h \tag{A1.45}$$

In view of the geometry of the problem, we can write

$$\psi(x,y,z) = \Phi(x,y)e^{ik_0 \sin\theta}, \tag{A1.46}$$

then $\Phi(x,y)$ satisfies

$$(\frac{\partial^2}{\partial x^2} + \frac{\partial^2}{\partial y^2} - \nu^2)\Phi = 0, \ -\infty < x < \infty, \ \ 0 \le y \le h + \varepsilon c(x) \tag{A1.47}$$

where $\nu = k_0 \sin\theta$,

$$K\Phi + \frac{\partial \Phi}{\partial y} = 0 \ \text{ on } \ y = 0, \tag{A1.48}$$

$$\frac{\partial \Phi}{\partial y} - \varepsilon\{\frac{\partial}{\partial x}(c(x)\frac{\partial \Phi}{\partial x}) - \nu^2 c(x)\Phi\} + O(\varepsilon^2) = 0 \ \text{ on } \ y = h \tag{A1.49}$$

and

$$\Phi(x,y) \sim \begin{cases} \Phi_0(x,y) + R\Phi_0(-x,y) \ \text{ as } \ x \to -\infty, \\ T\Phi_0(x,y) \text{ as } x \to \infty, \end{cases} \tag{A1.50}$$

where now

$$\Phi_0(x,y) = \cosh k_0(h-y)e^{i\mu x} \tag{A1.51}$$

with $\mu = k_0 \cos\theta$.

Method of solution

The form of the approximate boundary condition (A1.49) and the fact that a surface wave train propagating in an ocean of uniform finite depth experiences no reflection, suggest that, as before Φ, R, T has the perturbation expansion in terms of the small parameter ε

$$\Phi = \Phi_0 + \varepsilon\Phi_1 + O(\varepsilon^2), \ R = \varepsilon R_1 + O(\varepsilon^2), \ T = 1 + \varepsilon T_1 + O(\varepsilon^2). \tag{A1.52}$$

Using (A1.52) in (A1.47) to (A1.50) we find that $\Phi_1(x, y)$ satisfies the BVP described by

$$(\frac{\partial^2}{\partial x^2}+\frac{\partial^2}{\partial y^2}-\nu^2)\Phi_1 = 0, \ -\infty < x < \infty, \ 0 \le x \le h, \tag{A1.53}$$

$$K\Phi_1+\frac{\partial \Phi_1}{\partial y}=0 \ \text{ on } \ y=0, \tag{A1.54}$$

$$\frac{\partial \Phi_1}{\partial y}=i\mu\frac{d}{dx}(c(x)e^{i\mu x})-\nu^2 c(x)e^{i\mu x}\equiv p(x) \ \text{ on } \ y=h, \tag{A1.55}$$

$$\Phi_1(x,y)\sim\begin{cases} R_1\Phi_0(-x,y) \ \text{ as } \ x\to-\infty, \\ T_1\Phi_0(x,y) \text{ as } x\to\infty \end{cases} \tag{A1.56}$$

where now $\Phi_0(x, y)$ is given by (A1.51). Miles (1981) employed finite cosine transform to solve the BVP. Here we use the Green's function technique as above to solve it.

By using Green's integral theorem as used for the case of normal incidence, we obtain

$$\Phi_1(\xi,\eta)=\frac{1}{2\pi}\int_{-\infty}^{\infty} g(x,h;\xi,\eta)p(x)dx \tag{A1.57}$$

where $p(x)$ is given in (A1.55) and $g(x, y;\xi,\eta)$ is the Green's function associated with BVP and is given by (cf. Levine (1965))

$$g(x,y;\xi,\eta)=-4\pi i\sec\theta\frac{\cosh k_0(h-\eta)\cosh k_0(h-y)}{2k_0h+\sinh 2k_0h}e^{i\mu|x-\xi|}$$
$$-4\pi\sum_{n=1}^{\infty}\frac{k_n\cos k_n(h-\eta)\cos k_n(h-y)}{(2k_nh+\sin 2k_nh)(k_n^2+\nu^2)^{1/2}}e^{-(k_n^2+\nu^2)^{1/2}|x-\xi|}. \tag{A1.58}$$

Noting the behaviors of $\Phi(\xi,\eta)$ and $g(x,h;\xi,\eta)$ as $|\xi|\to\infty$, R_1 and T_1 can be evaluated as above. R_1 is found to be

$$R_1=-\frac{ik_0\sec\theta\cos 2\theta}{h+\frac{\sinh^2 k_0h}{K}}\int_{-\infty}^{\infty} e^{2i\mu x}c(x)dx \tag{A1.59}$$

while T_1 is found as

$$T_1=\frac{ik_0\sec\theta}{h+\frac{\sinh^2 k_0h}{K}}\int_{-\infty}^{\infty} c(x)dx. \tag{A1.60}$$

The results for normal incidence are recovered by putting $\theta = 0$ (i.e., for normal incidence) in (A1.59) and (A1.60). It is important to note that for $\theta=\frac{\pi}{4}$, R_1 vanishes independently of the shape of the barrier. These results were also obtained by Miles (1981) by a different method.

Effect of surface tension at the free surface

The effect of surface tension at the free surface can be considered. For the case of oblique incidence the mathematical description of the problem is as follows.

Solve the differential equation

$$\left(\frac{\partial^2}{\partial x^2}+\frac{\partial^2}{\partial y^2}-\nu^2\right)\Psi=0,\ -\infty<x<\infty,\ 0\le y\le h+\varepsilon c(x) \tag{A1.61}$$

with the boundary conditions

$$K\Psi+\frac{\partial\Psi}{\partial y}+M\frac{\partial^3\Psi}{\partial y^3}=0 \text{ on } y=0, \tag{A1.62}$$

$$\frac{\partial\Psi}{\partial n}=0 \text{ on } y=h+\varepsilon c(x) \tag{A1.63}$$

and the conditions as $x\to\pm\infty$ as

$$\Psi(x,y)\sim\begin{cases}\Psi_0(x,y)+R\Psi_0(-x,y) \text{ as } x\to-\infty,\\ T\Psi_0(x,y) \text{ as } x\to\infty,\end{cases} \tag{A1.64}$$

where

$$\Psi_0(x,y)=\cosh\kappa_0(h-y)e^{i\kappa_0 x\cos\theta}. \tag{A1.65}$$

In (A1.62), M arises due to the presence of surface tension at the free surface, and in (A1.60) κ_0 is the unique positive root of the transcendental equation

$$k(1+Mk^2\tanh kh)=K. \tag{A1.66}$$

Its other roots are $-\kappa_0, \pm i\kappa_n (n=1,2,...)$. Mandal and Basu (1990) solved this problem employing a perturbation analysis to obtain the first-order reflection and transmission coefficients as

$$R_1=-\frac{i\kappa_0\sec\theta\cos 2\theta}{h+\frac{(1+3M\kappa_0^2)\sinh^2\kappa_0 h}{K}}\int_{-\infty}^{\infty}e^{2i\kappa_0\cos\theta x}c(x)dx, \tag{A1.67}$$

$$T_1=\frac{i\kappa_0\sec\theta}{h+\frac{(1+3M\kappa_0^2)\sinh^2\kappa_0 h}{K}}\int_{-\infty}^{\infty}c(x)dx. \tag{A1.68}$$

In the absence of surface tension, these coincide with (A1.54) and (A1.60) respectively.

9A.2 Scattering by small bottom undulations of an ocean with an ice-cover

In polar regions, the ocean is covered with a thin sheet of ice, and as such there is a considerable interest in the investigation of various types of wave propagation problems in water with an ice-cover in recent times. The ice-cover is modeled as a thin sheet of an elastic plate of infinite extent having a very small thickness h_0 of which still a smaller part is immersed in water. Assuming linear theory and irrotational motion, the velocity potential function can be represented by $\mathrm{Re}\{\psi e^{-i\sigma t}\}$ for time-harmonic waves with angular frequency σ, where ψ satisfies

$$\nabla^2\psi = 0 \quad \text{in the fluid region,} \tag{A2.1}$$

the linearized ice-cover condition has the form

$$K\psi + \{D\nabla^4_{x,y} + 1\}\psi_y = 0 \text{ on } y = 0, \tag{A2.2}$$

and the sea-bed condition is

$$\psi_y = 0 \quad \text{on } y = h, \tag{A2.3}$$

for water basins of uniform finite depth h. Here the y-axis is chosen vertically downwards into the water and the (x, z)-plane is the rest position of the lower part of the ice-cover. In condition (A2.2) $\nabla^4_{x.z}$ denotes the two-dimensional biharmonic operator in the (x, z)-plane, $K = \sigma^2/g$ is the wave number where g is the gravity acceleration, $D = Eh_0^3/12(1-\nu^2)\rho g$ is the ice-thickness parameter where E is the Young's modulus, ν is the Poisson's ratio of the material of the ice-cover and ρ is the density of water.

For two-dimensional motion (i.e., independent of z), a possible solution for ψ, representing a train of water waves propagating just below the ice-cover along the positive x-direction, is given by

$$\phi_0(x,y) = \cosh k_0(h-y)\, e^{ik_0x}, \tag{A2.4}$$

where k_0 is the unique positive root of the transcendental equation

$$\Delta(k) \equiv (1+Dk^4)k\sinh kh - K\cosh kh = 0 \tag{A2.5}$$

It can be shown (Chung and Fox, 2002) that equation (A2.5) has two real roots $\pm k_0 (k_0 > 0)$, two pairs of complex conjugate roots $\pm\mu, \pm\bar{\mu}$ $(\mu = \alpha + i\beta, \alpha > 0, \beta > 0)$ and an infinite number of purely imaginary roots $\pm ik_n$ $(n = 1,2,\dots)$, where k_n $(n = 1,2,\dots)$ are real and positive and satisfy

$$k_n(1+Dk_n^4)\sin k_nh + K\cos k_nh = 0 \tag{A2.6}$$

so that $k_n \to n\pi/h$ as $n \to \infty$.

Here we consider wave scattering by a small cylindrical deformation of the bottom of a laterally unbounded ocean which is covered by a thin sheet of ice instead of having a free surface. The sea bed is described by

$$y = h + c(x), \tag{A2.7}$$

where $c(x)$ is a continuous bounded function and $c(x) \to 0$ as $|x| \to \infty$. The small positive quantity ϵ gives a measure of smallness of the elevation. Thus, the ocean is of uniform finite depth h on either far side of the elevation. As in section (9A.1), Mandal and Basu (2004) employed a simplified perturbation analysis involving the small parameter ϵ in the analysis to obtain the reflection and transmission coefficients upto first order of ϵ in terms of integrals involving the shape function $c(x)$ when a train of progressive waves propagating under the ice-cover is incident on the sea-bed deformation. This is now described below.

Formulation of the problem

Let a progressive wave train represented by the velocity potential $\mathrm{Re}\{\phi_0(x, y)e^{-i\sigma t}\}$, where $\phi_0(x, y)$ is given by (A2.4) and corresponds to the solution for an ocean of uniform finite depth h be incident from the direction of $x = -\infty$ upon the bottom deformation of an ice-covered ocean occupying the region $0 \le y \le h + \epsilon c(x)$. It will be partially reflected by and partially transmitted over the elevation of the ocean bottom. If $\phi(x, y)$ denotes the complex velocity potential describing the resulting motion in the fluid, then it satisfies the partial differential equation

$$\nabla^2 \phi = 0 \ \text{ for } \ 0 \le y \le h + \epsilon\, c(x), \qquad \text{(A2.8)}$$

where ∇^2 is the two-dimensional Laplacian in (x, y); the ice-covered condition is

$$K\phi + \left(D\frac{\partial^4}{\partial x^4} + 1\right)\phi_y = 0 \ \text{ on } \ y = 0, \qquad \text{(A2.9)}$$

The bottom condition is

$$\frac{\partial \phi}{\partial n} = 0 \ \text{ on } y = h + \epsilon c(x), \qquad \text{(A2.10)}$$

where $\frac{\partial}{\partial n}$ denotes the normal derivative at a point (x, y) on the bottom, and has the asymptotic behavior

$$\phi(x,y) \to \begin{cases} T\phi_0(x,y) & \text{as } x \to \infty, \\ \phi_0(x,y) + R\phi_0(x,y) & \text{as } x \to -\infty \end{cases} \qquad \text{(A2.11)}$$

where T and R denote, respectively, the transmission and reflection coefficients, which will have to be determined.

Solution of the problem

The bottom condition (A2.10) can be approximated as

$$\phi_y - \epsilon\frac{d}{dx}[c(x)\phi_x(x,h)] + O(\epsilon^2) = 0 \ \text{ on } \ y = h \qquad \text{(A2.12)}$$

In view of this approximation, and the fact that the incident field ϕ_0 is totally transmitted if there is no bottom deformation, we can expand ϕ, T and R in terms of ϵ as given by

$$\phi = \phi_0 + \epsilon\phi_1 + O(\epsilon^2), T = 1 + \epsilon T_1 + O(\epsilon^2), R = \epsilon R_1 + O(\epsilon^2) \quad \text{(A2.13)}$$

Using the expansions (A2.13) in equation (A2.8) and conditions (A2.9), (A2.11) and (A2.12), we find that ϕ_1 satisfies the boundary value problem described by

$$\nabla^2 \phi_1 = 0 \quad \text{for } 0 \le y \le h, \tag{A2.14}$$

$$K\phi_1 + \left(D\frac{\partial^4}{\partial x^4} + 1\right)\phi_{1y} = 0 \quad \text{on } y = 0, \tag{A2.15}$$

$$\phi_{1y} = q(x) \quad \text{on } y = h \tag{A2.16}$$

and

$$\phi_1(x, y) \to \begin{cases} T_1\phi_0(x, y) & \text{as } x \to \infty, \\ R_1\phi_0(x, y) & \text{as } x \to -\infty \end{cases} \tag{A2.17}$$

where

$$q(x) = ik_0\frac{d}{dx}(c(x)e^{ik_0x}). \tag{A2.18}$$

To solve this boundary value problem, we need to construct the Green's function $G(x, y; \xi, \eta)$, which satisfies

$$\nabla^2 G = 0 \text{ for } 0 \le y \le h \text{ except at } (x, y) = (\xi, \eta), \tag{A2.19}$$

where $0 < \eta < h$,

$$G \to \ln r \text{ as } r = \{(x - \xi)^2 + (y - \eta)^2\}^{1/2} \to 0, \tag{A2.20}$$

$$KG + \left(D\frac{\partial^4}{\partial x^4} + 1\right)G_y = 0 \quad \text{on } y = 0, \tag{A2.21}$$

$$G_y = 0 \text{ on } y = h, \tag{A2.22}$$

$$G \to \text{multiple of } e^{ik_0|x-\xi|} \text{ as } |x - \xi| \to \infty. \tag{A2.23}$$

The condition (A2.23) means that G represents an outgoing waves as $|x - \xi| \to \infty$. The construction of G is given in Mandal and Basu (2004) and is given by

$$\begin{aligned} G(x, y; \xi, \eta) = &-4\pi\sum_{n=1}^{\infty}\frac{(1 + Dk_n^4)\cos k_n(h - y)\cos k_n(h - \eta)}{2k_nh(1 + Dk_n^4) + (1 + 5Dk_n^4)\sin 2k_nh}e^{-k_n|x-\xi|} \\ &-4\pi i\frac{(1 + Dk_0^4)\cosh k_0(h - y)\cosh k_0(h - \eta)}{2k_0h(1 + Dk_0^4) + (1 + 5Dk_0^4)\sin 2k_0h}e^{-ik_0|x-\xi|} \\ &-4\pi i\frac{(1 + D\mu^4)\cosh \mu(h - y)\cosh \mu(h - \eta)}{2\mu h(1 + D\mu^4) + (1 + 5D\mu^4)\sin 2\mu h}e^{-i\mu|x-\xi|} \\ &-4\pi i\frac{(1 + D\bar{\mu}^4)\cosh \bar{\mu}(h - y)\cosh \bar{\mu}(h - \eta)}{2\bar{\mu}h(1 + D\bar{\mu}^4) + (1 + 5D\bar{\mu}^4)\sin 2\mu h}e^{-i\bar{\mu}|x-\xi|}. \end{aligned} \tag{A2.24}$$

Since Im $\mu > 0$, we find from (A2.24) that

$$G(x, y: \xi, \mu) \to -4\pi i \frac{(1 + Dk_0^4)\cosh k_0(h-y)\cosh k_0(h-\eta)}{2k_0 h(1 + Dk_0^4) + (1 + 5Dk_0^4)\sin 2k_0 h} e^{-ik_0|x-\xi|} \text{as } |x-\xi| \to \infty \tag{A2.25}$$

which shows that G satisfies the condition (A2.23).

An appropriate use of Green's integral theorem to $\phi_1(x, y)$ and $G(x, y; \xi, \eta)$ produces

$$\phi_1(\xi, \eta) = \frac{1}{2\pi}\int_{-\infty}^{\infty} G(x, h; \xi, \eta) q(x)\, dx. \tag{A2.26}$$

To obtain T_1, we note from (A2.17) and (A2.25) that as $\xi \to \infty$,

$$\begin{aligned} \phi_1(\xi, \eta) &\to T_1 \phi_0(\xi, \eta), \\ G(x, h; \xi, \eta) &\to -4\pi i A_0 e^{-ik_0 x}\phi_0(\xi, \eta), \end{aligned} \tag{A2.27}$$

where

$$A_0 = \frac{1 + Dk_0^4}{2k_0 h(1 + Dk_0^4) + (1 + 5Dk_0^4)\sinh 2k_0 h}. \tag{A2.28}$$

Using (A2.27) in (A2.26) (after making $\xi \to \infty$), we find that

$$T_1 = -2iA_0 \int_{-\infty}^{\infty} e^{-ik_0 x} q(x) dx = iA \int_{-\infty}^{\infty} c(x) dx, \tag{A2.29}$$

where

$$A = \frac{k_0}{h + K^{-1}(1 + 5Dk_0^4)\sinh^2 k_0 h}. \tag{A2.30}$$

Again, to find R_1, we use in (A2.26) the asymptotic results

$$\begin{aligned} \phi_1(\xi, \eta) &\to R_1 \phi_0(\xi, \eta) \text{ as } \xi \to -\infty, \\ G(x, y; \xi, \eta) &\to -4\pi i A_0 e^{ik_0 x}\phi_0(-\xi, \eta) \text{ as } \xi \to -\infty \end{aligned}$$

to obtain

$$R_1 = -2iA_0 \int_{-\infty}^{\infty} e^{ik_0 x} q(x) dx = -iA \int_{-\infty}^{\infty} c(x) e^{2ik_0 x} dx. \tag{A2.31}$$

For $D = 0$, the result (A2.31) coincides with the approximate result obtained earlier in section 9A.1 for the reflection by a low gently sloping bottom of the form $y = h + \epsilon c(x)$, k_0 now being defined as the real zero of ($k \tanh kh - K$). Thus, the effect of small bottom deformation on an incident wave train in an ice-covered ocean is qualitatively similar to the case in which the ocean has a free surface.

The integrals in (A2.29) and (A2.31) can be evaluated once the shape function $c(x)$ is known. We consider three special functional forms for the function $c(x)$.

Special forms of the bottom deformation

(i) $$c(x) = ae^{-\lambda|x|}(\lambda > 0), -\infty < x < \infty. \tag{A2.32}$$

In this case, the top of the deformation lies at the point $(0, h)$, and on either sides it decreases exponentially. Coefficients T_1 and R_1 are obtained as

$$T_1 = \frac{2ia}{\lambda}A, \qquad R_1 = -\frac{2ia}{\lambda^2 + 4k_0^2}A. \tag{A2.33}$$

(ii) $$c(x) = ae^{-\lambda x^2}(\lambda > 0), -\infty < x < \infty. \tag{A2.34}$$

In this case, the elevation is of the Gaussian profile, the top occurring at the point $(0, h)$ as in the case (i). Here, T_1 and R_1 are obtained as

$$T_1 = i\left(\frac{\pi}{\lambda}\right)^{1/2} aA, \qquad R_1 = -i\left(\frac{\pi}{\lambda}\right)^{1/2} aAe^{-\frac{k_0}{\lambda}}. \tag{A2.35}$$

(iii) $$c(x) = \begin{cases} a\sin(\lambda x + \delta), & L_1 \le x \le L_2 \\ 0 & \text{Otherwise,} \end{cases} \tag{A2.36}$$

where

$$L_1 = -\frac{n\pi + \delta}{\lambda}, \qquad L_2 = \frac{m\pi - \delta}{\lambda},$$

δ being a constant phase angle and m,n being positive integers. This represents a patch of sinusoidal ripples with wave number λ on an otherwise flat bottom. This case has been investigated in some detail by Davies (1982) when the ocean has a free surface. We obtain for this case

$$T_1 = \frac{ia}{\lambda}A\{(-1)^n - 1)\}m, \tag{A2.37}$$

and

$$R_1 = -\frac{ia\lambda}{\lambda^2 - 4k_0^2}\{(-1)^n e^{2ik_0L_1} - (-1)^m e^{2ik_0L_2}\} \tag{A2.38}$$

In the situation in which there is an integer number of ripple wavelengths in the patch $L_1 \le x \le L_2$ such that $m = n$ and $\delta = 0$, which was also considered by Davies (1982), we find that

$$T_1 = 0, \tag{A2.39}$$

and

$$R_1 = B(-1)^m \frac{\alpha\sin(m\pi\alpha)}{\alpha^2 - 1}, \tag{A2.40}$$

where

$$\alpha = \frac{\lambda}{2k_0}, B = \frac{a}{h + k^{-1}(1 + 5Dk_0^4)\sinh^2 k_0 h}. \tag{A2.41}$$

For $D = 0$, these results coincide with those obtained by Davies (1982) after using a somewhat elaborate mathematical analysis. The main observation in this case is that the first-order reflection coefficient is an oscillatory function of α which is twice the ratio of the wave numbers k_0 and λ. If this ratio becomes 0.5 (i.e., $\alpha = 1$), we find from (A2.40) that

$$R_1 = \frac{\pi}{2} B_0 m, \tag{A2.42}$$

where now B_0 is given by

$$B_0 = \frac{a}{h + \left\{1 + 4\left(1 + \frac{16}{D\lambda^4}\right)^{-1}\right\}\frac{\sin \lambda h}{\lambda}}. \tag{A2.43}$$

Thus R_1 in this case becomes a constant multiple of m, the number of ripples in the patch. By increasing m, it is possible to have large amount of reflection of the incident wave energy in the vicinity of the wave number near $\lambda/2$, as was true for an ocean with a free surface.

Numerical results

Three special forms for the bottom deformation have been considered above. However, the particular case of a patch of sinusoidal ripples on the ocean bed is of considerable significance due to the "ability of an undulating bed to reflect incident wave energy which is important in respect of both coastal protection, and of possible ripple growth if the bed is erodible", Davies (1982). Thus, we fix our attention in the numerical results for the first-order reflection coefficient $|R_1|$ given by equation (A2.40) for a patch of sinusoidal ripples with wave number λ having m number of ripple wavelengths in the patch.

To visualize the influence of the parameter D of the ice-cover thickness on the reflection coefficient $|R_1|$ for the case of a patch of sinusoidal ripples, $|R_1|$ is depicted against the wave number Kh in Fig. 9.2 for a single ripple ($m = 1$), $\lambda h = 1$, $a/h = 0.1$ and $D/h^4 = 0; 0.5$. The zero value of D/h^4 corresponds to the ocean with a free surface. The figure shows that the presence of nonzero D reduces the overall values of R_1, and also decreases the number of zeros of $|R_1|$ as a function of Kh; $|R_1|$ is an oscillating function of Kh. When $D = 0$, its peak value is attained when the wave number of the bottom undulations (kh) becomes approximately twice as large as the surface wave number (k_0h). For nonzero D, this peak value is slightly lower. Figure 9.3 depicts $|R_1|$ against Kh for $a/h = 0.1$, $\lambda h = 1$, $m = 2$, $D/h^4 = 0; 0.3; 0.5$. In this case, the number of ripples is increased to two. The same general feature of $|R_1|$ as in Fig. 9.2 is observed in this figure with the modification that the overall value of $|R_1|$ is now increased compared to the case of a simple ripple $m = 1$; the oscillating nature of $|R_1|$ against Kh is more pronounced and the number of zeros of $|R_1|$ is increased. This phenomenon becomes more evident when m is increased to six in Fig. 9.4. Figure 9.5 depicts $|R_1|$ for $D/h^4 = 0.3$, $a/h = 0.1$, $\lambda h = 1$ and $m = 1; 3; 5$. As the number of ripples m increases, the peak value of $|R_1|$ increases and $|R_1|$ becomes more oscillatory. This may be attributed to the increase in multiple reflections between the ice-cover and the ripples in the patch with the increase in the number of ripples.

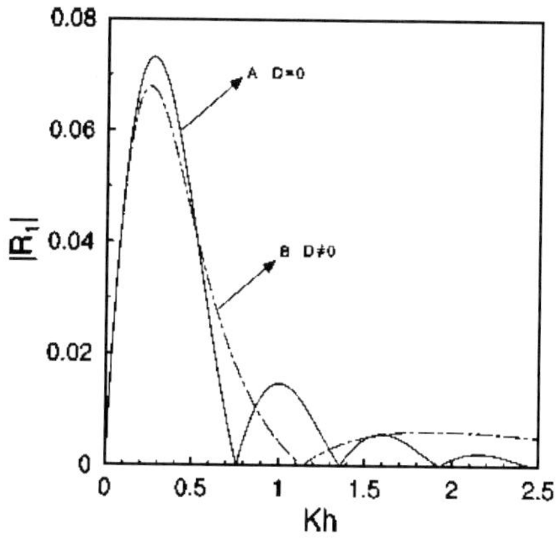

Fig. 9.2 Reflection coefficient in dependence of wave number for $\frac{a}{h} = 0.1; m = 1; \lambda h = 1$ and different $\frac{D}{h^4} = 0(A), 0.5(B)$

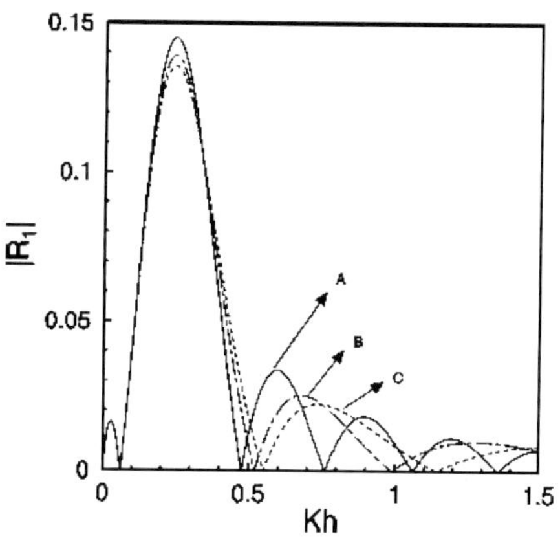

Fig. 9.3 Reflection coefficient in dependence of wave number for $\frac{a}{h} = 0.1; m = 2; \lambda h = 1$ and different $\frac{D}{h^4} = 0(A), 0.3(B), 0.5(c)$

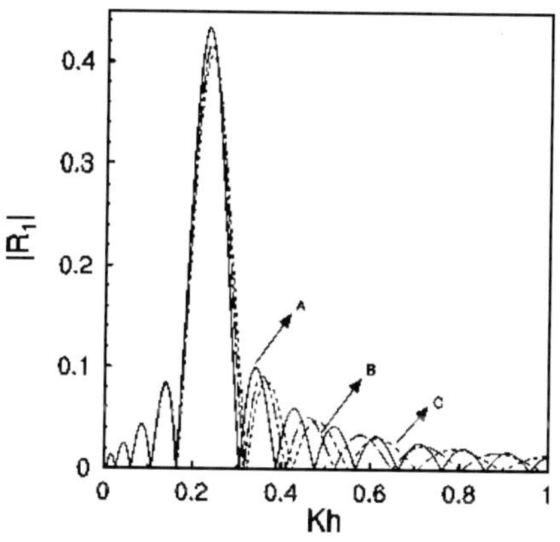

Fig. 9.4 Reflection coefficient in dependence of wave number for $\frac{a}{h} = 0.1; m = 6; \lambda h = 1$ and different $\frac{D}{h^4} = 0(A), 0.5(B)$

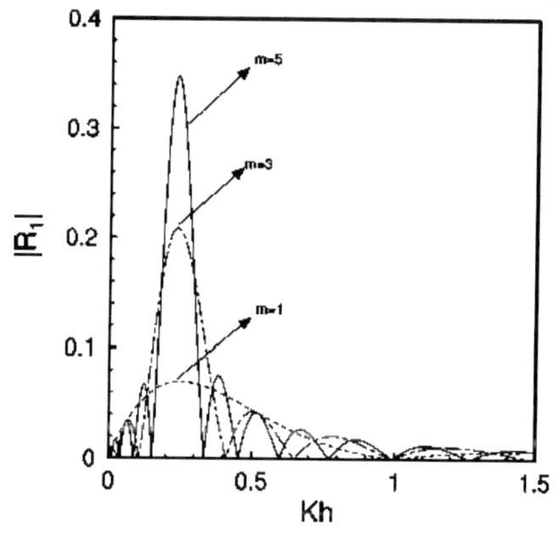

Fig. 9.5 Reflection coefficient in dependence of wave number for $\frac{a}{h} = 0.1; \lambda h = 1, \frac{D}{h^4} = 0.3$ and different $m = 1;3;5$

9A.3 Oblique scattering by undulations at the bed of an ice-covered ocean

To tackle the problem of oblique wave scattering by small cylindrical undulations of the bottom of an ocean with an *ice-cover*, Mandal and Maiti (2004) employed a similar perturbation technique directly to the governing partial differential equation and the boundary and infinity conditions for the potential function, after extracting out the z-dependence by exploiting the geometry of the problem, to obtain the first-order reflection and transmission coefficients in terms of integrals involving the shape function defining the undulations. The problem is described as follows

Solve the partial differential equation

$$\nabla^2 \phi = 0 \tag{A2.44}$$

in the region $0 \le y \le h + \varepsilon c(x), -\infty < x, z < \infty$, with the boundary conditions

$$K\phi + (D\nabla^4_{x,z} + 1)\phi_y = 0 \text{ on } y = 0, \tag{A2.45}$$

$$\phi_n = 0 \text{ on } y = h + \varepsilon c(x) \tag{A2.46}$$

together with suitable conditions as $x \to \pm\infty$ which will be stated shortly.

Here $c(x)$ is a continuous and bounded function describing the shape of the undulations of the ocean bed and $c(x) \to 0$ as $|x| \to \infty$, so that the ocean is of uniform finite depth far away from the undulations on either sides, and $\varepsilon (> 0)$ is a small parameter giving a measure of the smallness of the undulations.

We assume that a water wave train represented by the velocity potential $\phi_0(x, y, z)$, given by

$$\phi_0(x, y, z) = \cosh k_0(h - y)\, e^{ik_0(x \cos\theta + z \sin\theta)}, \tag{A2.47}$$

where k_0 is the unique real positive root of the transcendental equation

$$\Delta(k) \equiv k(Dk^4 + 1)\sinh kh - K\cosh kh = 0, \tag{A2.48}$$

is obliquely incident upon the undulations from a large distance in the direction of negative x-axis, then it undergoes partial transmission and reflection by the undulations. Thus the asymptotic behavior of $\phi(x, y, z)$ is given by

$$\phi \to \begin{cases} T\phi_0(x, y, z) & \text{as} \quad x \to \infty, \\ \phi_0(x, y, z) + R\phi_0(-x, y, z) & \text{as} \quad x \to -\infty \end{cases} \tag{A2.49}$$

where T and R are the transmission and reflection coefficients respectively and will have to be determined.

As ε is very small, we can approximate the bottom condition (A2.46) after neglecting $O(\varepsilon^2)$ terms as

$$-\phi_y + \varepsilon\{c'(x)\phi_x - c(x)\phi_{yy}\} = 0 \text{ on } y = h. \tag{A2.50}$$

In view of the geometry of the problem, we can assume that

$$\phi(x, y, z) = \psi(x, y)e^{i\nu z} \tag{A2.51}$$

where $\nu = k_0 \sin\theta$. Thus the z-dependence is extracted out, and the function $\psi(x, y)$ satisfies the BVP described by

$$\begin{aligned}
&\psi_{xx} + \psi_{yy} - \nu^2\psi = 0,\ 0 \le y \le h,\ -\infty < x < \infty,\\
&K\psi + \left\{ D\left(\frac{\partial^2}{\partial x^2} - \nu^2\right)^2 + 1 \right\}\psi_y = 0 \text{ on } y = 0,\\
&-\psi_y + \varepsilon\left\{ \frac{\partial}{\partial x}\left(c(x)\psi_x\right) - \nu^2 c(x) \right\} = 0 \text{ on } y = h, \qquad \text{(A2.52)}\\
&\psi(x,y) \to \begin{cases} T\psi_0(x,y) & \text{as} \quad x \to \infty,\\ \psi_0(x,y) + R\psi_0(-x,y) & \text{as} \quad x \to -\infty \end{cases}
\end{aligned}$$

where

$$\psi_0(x,y) = e^{ik_0 x\cos\theta}\cosh k_0(h - y). \qquad \text{(A2.53)}$$

This BVP is solved approximately upto first order of ψ by using a perturbation analysis applied to the governing PDE, the boundary conditions and infinity conditions.

Method of solution

Because of the approximate boundary condition $(A2.52)_3$ and the fact that a wave train propagating in an ocean of uniform finite depth h experiences no reflection, we may assume that ψ, T and R in (A2.52) can be expanded in terms of the small parameter ε as

$$\begin{aligned}
\psi(x,y) &= \psi_0(x,y) + \varepsilon\psi_1(x,y) + O(\varepsilon^2),\\
T &= 1 + \varepsilon T_1 + O(\varepsilon^2), \qquad \text{(A2.54)}\\
R &= \varepsilon R_1 + O(\varepsilon^2).
\end{aligned}$$

Using the expansions (A2.54) in Eq. (A2.52), we find that $\psi_1(x,y)$ satisfies the BVP described by

$$\begin{aligned}
&\psi_{1xx} + \psi_{1yy} - \nu^2\psi_1 = 0,\ 0 \le y \le h,\ -\infty < x < \infty,\\
&K\psi_1 + \left\{ D\left(\frac{\partial^2}{\partial x^2} - \nu^2\right)^2 + 1 \right\}\psi_{1y} = 0 \text{ on } y = 0,\\
&\psi_{1y} = ik_0\cos\theta\frac{\partial}{\partial x}\left(c(x)e^{ik_0 x\cos\theta}\right) - \nu^2 c(x) \qquad \text{(A2.55)}\\
&\quad \equiv q(x) \ on\ y = h,\\
&\psi_1(x,y) \to \begin{cases} T_1\psi_0(x,y) & \text{as} \quad x \to \infty,\\ R_1\psi_0(-x,y) & \text{as} \quad x \to -\infty. \end{cases}
\end{aligned}$$

We note that $\psi_1(x,y)$ behaves as an outgoing wave as $|x| \to \infty$.

By an appropriate use of Green's integral theorem, the solution of BVP is obtained as

$$\psi_1(\xi,\eta) = \frac{1}{2\pi}\int_{-\infty}^{\infty} G(x,h;\xi,\eta)q(x)dx.$$

where $G(x,h;\xi,\eta)$ is the corresponding Green's function and is given by Evans and Porter (2003)

$$G(x,h;\xi,\eta) = -4\pi\sum_{n=1}^{\infty}\frac{k_n(Dk_n^4+1)\cos k_n(h-y)\cos k_n(h-\eta)}{2k_nh(Dk_n^4+1)+(5Dk_n^4+1)\sin 2k_nh}\frac{e^{-(K_n^2+\upsilon^2)^{1/2}}\,|x-\xi|}{(K_n^2+v^2)^{1/2}}$$

$$-4\pi i\left[\frac{k_0(Dk_0^4+1)\cos k_0(h-y)\cos k_0(h-\eta)}{2k_0h(Dk_0^4+1)+(5Dk_0^4+1)\sin 2k_0h}\frac{e^{-(K_0^2+\upsilon^2)^{1/2}}\,|x-\xi|}{(K_0^2+v^2)^{1/2}}\right.$$

$$+\frac{\mu(D\mu^4+1)\cosh\mu(h-y)\cosh\mu(h-\eta)}{2\mu h(1+D\mu^4)+(5D\mu^4+1)\sinh 2\mu h}\frac{e^{i\mu'|x-\xi|}}{\mu'} \tag{A2.56}$$

$$\left.-\frac{\overline{\mu}(D\overline{\mu}^4+1)\cosh\overline{\mu}(h-y)\cosh\overline{\mu}(h-\eta)}{2\overline{\mu}h(1+D\overline{\mu}^4)+(5D\overline{\mu}^4+1)\sinh 2\overline{\mu}h}\frac{e^{i\overline{\mu}'|x-\xi|}}{\overline{\mu}'}\right]$$

where $\mu' = \left(\overline{\mu}^2 - \upsilon^2\right)^{1/2}$ and $-\overline{\mu}' = \{(-\overline{\mu})^2 - \upsilon^2\}^{1/2}$, and that of the square root has been chosen such that $\mu' = \mu, -\overline{\mu}' = -\overline{\mu}$ when $v = 0$.

Since μ' and $-\overline{\mu}'$ have positive imaginary parts, we find that, as $|x-\xi| \to \infty$,

$$G(x,h;\xi,\eta) \to -4\pi i\frac{k_0(Dk_0^4+1)\cos k_0(h-y)\cos k_0(h-\eta)}{2k_0h(Dk_0^4+1)+(5Dk_0^4+1)\sin 2k_0h}\frac{e^{-(K_0^2+v^2)^{1/2}}\,|x-\xi|}{(K_0^2+v^2)^{1/2}} \tag{A2.57}$$

so that G behaves as an outgoing wave for $|x-\xi| \to \infty$.

To obtain the first-order transmission and reflection coefficients T_1 and R_1 respectively, we note from $(A2.54)_4$ and (A2.57) that

$$\psi_1(\xi,\eta) \to \begin{cases} T_1\psi_0(\xi,\eta) & \text{as} \quad \xi \to \infty, \\ R_1\psi_0(-\xi,\eta) & \text{as} \quad \xi \to -\infty \end{cases} \tag{A2.58}$$

and

$$G(x,0;\xi,\eta) \to -4\pi i\frac{e^{\mp ik_0x\cos\theta}}{k_0\cos\theta}A\psi_0(\pm\xi,\eta) \text{ as } \xi \to \pm\infty \tag{A2.59}$$

where

$$A = \frac{1}{h + \dfrac{(1+5Dk_0^4)\sinh^2 k_0h}{K}}. \tag{A2.60}$$

Using the asymptotic results (A2.58) and (A2.59) in the representation (A2.55) we find that

$$\begin{aligned} T_1 &= -\frac{i}{k_0\cos\theta}A\int_{-\infty}^{\infty} e^{-ik_0x\cos\theta}q(x)dx \\ &= ik_0\sec\theta\,A\int_{-\infty}^{\infty} c(x)dx, \end{aligned} \tag{A2.61}$$

$$
\begin{aligned}
R_1 &= -\frac{i}{k_0 \cos\theta} A \int_{-\infty}^{\infty} e^{ik_0 x\cos\theta} q(x)dx \\
&= -ik_0 \sec\theta \cos 2\theta A \int_{-\infty}^{\infty} e^{2ik_0 x\cos\theta} q(x)dx.
\end{aligned} \tag{A2.62}
$$

The results for an ocean with a free surface are recovered by putting $D = 0$ in (A2.61) and (A2.62) where then, however, k_0 denotes the unique real positive zero of the transcendental equation

$$k \sinh kh - K \cosh kh = 0.$$

It is also interesting to note that R_1 vanishes identically for $\theta = \pi/4$, independently of the shape function $c(x)$. This was also observed in the case of an ocean with a free surface with or without surface tension.

We now consider three special types of undulations.

(i) $c(x) = ae^{-\lambda|x|} (\lambda > 0)$. Here the bottom undulation is maximum at $(0, h)$ and decreases exponentially on either side of $(0, h)$. In this case

$$
\begin{aligned}
T_1 &= \frac{2iak_0 A}{\lambda} \sec\theta, \\
R_1 &= -\frac{2iak_0 A\lambda}{\lambda^2 + 4k_0^2 \cos^2\theta} \sec\theta \cos 2\theta.
\end{aligned}
$$

(ii) $c(x) = ae^{-\lambda x^2} (\lambda > 0)$. Here the undulation is of Gaussian type having a maximum value at $(0, h)$. In this case

$$
\begin{aligned}
T_1 &= i\left(\frac{\pi}{\lambda}\right)^{1/2} k_0 aA \sec\theta, \\
R_1 &= i\left(\frac{\pi}{\lambda}\right)^{1/2} k_0 aA \sec\theta \cos 2\theta \, e^{-\frac{k_0^2 \cos^2\theta}{\lambda}}.
\end{aligned}
$$

(iii) $c(x) = \begin{cases} a \sin \lambda x, & -\dfrac{m\pi}{\lambda} \le x \le \dfrac{m\pi}{\lambda} \\ 0, & \text{otherwise}. \end{cases}$

This represents sinusoidal undulations of the bottom, having m number of patches and is of considerable physical interest. Davies (1982) earlier made somewhat elaborate study on the effect of sinusoidal undulations on the bottom of an ocean with a *free surface*, on an incident surface water wave train. In this case

$$T_1 \equiv 0,$$

$$R_1 = \sec^2\theta \cos 2\theta \, B(-1)^m \frac{\alpha}{\alpha^2 - 1} \sin(\alpha m\pi)$$

where $B = aA$, $\alpha = \dfrac{2k_0}{\lambda} \cos\theta$. It is interesting to note that when $\alpha \approx 1$, i.e., $\lambda \approx 2k_0 \cos\theta$,

$$R_1 \approx \frac{\pi}{2} \sec^2\theta \cos 2\theta \, Bm. \tag{A2.63}$$

The result (A2.63) has the implication that somewhat large reflection of the incident wave energy occurs when the bed wave number λ is twice the wave number component of the incident wave field along the x-direction, if the integer m denoting the number of patches is made large. This phenomenon has practical application in the construction of an efficient reflector of incident wave energy.

B. TWO-LAYER FLUID

9B.1 Oblique scattering by bottom undulations in a two-layer fluid

Scattering of waves obliquely incident on small cylindrical undulations at the bottom of a two-layer fluid wherein the upper layer has a free surface and the lower layer has an undulating bottom, is considered in this section. As mentioned earlier, there exists two modes of time-harmonic waves propagating at each of the free surface and the interface. Due to an oblique incident wave of a particular mode, reflected and transmitted waves of both the modes are created in general by the bottom undulations. For small undulations, a simplified perturbation analysis has been employed by Maiti and Mandal (2006) to obtain first-order reflection and transmission coefficients of both the modes due to oblique incidence of waves of again both modes, in terms of integrals involving the shape function describing the bottom. An account of this is described in this section.

We consider a two-layer fluid for which the upper layer has a free surface and the lower layer has small cylindrical undulations at the bottom. A Cartesian co-ordinate system is chosen in such a way that $y = -h$ denotes the undisturbed free surface while $y = 0$ denotes the undisturbed interface, the y-axis pointing vertically downwards. Then the bottom of the lower layer can be represented by $y = H + \varepsilon c(x)$ where $c(x)$ is bounded and continuous describing the shape of the bottom and $c(x) \to 0$ as $x \to \infty$, ε giving a measure of smallness of the bottom undulations. Thus the lower layer is of uniform finite depth H below the mean interface far away from the undulations on either side. To see the far-field behaviors of the potential functions under the usual assumptions of linear theory relevant to the present problem, we consider a two-layer fluid wherein the lower layer is of *uniform* finite depth H below the mean interface, and ρ_1 is the density of the lighter fluid occupying the upper layer while $\rho_2 (> \rho_1)$ is the density of the lower layer. As in Linton and Cadby (2002), the time-harmonic velocity potentials in the upper and lower layers can be described respectively by $Re\{\psi(x, y)e^{i\nu z}e^{-i\sigma t}\}$ and $Re\{\phi(x, y)e^{i\nu z}e^{-i\sigma t}\}$, σ denoting the angular frequency and t the time, where ψ, ϕ satisfy

$$(\nabla^2_{x,y} - \nu^2)\psi = 0 \quad \text{in the upper layer,} \tag{B1.1}$$

$$(\nabla^2_{x,y} - \nu^2)\phi = 0 \quad \text{in the lower layer.} \tag{B1.2}$$

The linearized free surface and interface conditions are

$$K\psi + \psi_y = 0 \quad \text{on} \quad y = -h, \tag{B1.3}$$

$$s(K\psi + \psi_y) = K\phi + \phi_y \quad \text{on} \quad y = 0, \tag{B1.4}$$

$$\psi_y = \phi_y \quad \text{on} \quad y = 0, \tag{B1.5}$$

where $s = \rho_1/\rho_2 (< 1)$ and $K = \sigma^2/g$, g being the acceleration due to gravity, and the bottom condition is

$$\phi_n = 0 \quad \text{on} \quad y = H + \varepsilon c(x), \tag{B1.6}$$

ϕ_n denoting the derivative normal to the bottom.

In such a two-layer fluid, the progressive gravity waves propagating at each of the free surface and the interface can be expressed by

$$f(k,y)e^{\pm i(k^2-\nu^2)^{1/2}x}(-h < y < 0), \cosh k(H-y)e^{\pm i(k^2-\nu^2)^{1/2}x}(0 < y < H) \tag{B1.7}$$

with

$$f(k,y) = \frac{\sinh kH\{k\cosh k(h+y) - K\sinh k(h+y)\}}{K\cosh kh - k\sinh kh} \tag{B1.8}$$

where k is real and positive and satisfies the dispersion relation

$$\Delta(k) \equiv (1-s)k^2 + K^2(s + \coth kh \coth kH) - kK(\coth kh + \coth kH) = 0. \tag{B1.9}$$

The equation (B1.9) has exactly two real and positive roots, m and M, say where $K < m < M$, so that, there exists two modes of waves propagating at each of the free surface and the interface.

Progressive waves of mode m are of the forms

$$f(m,y)e^{\pm(m^2-\nu^2)^{1/2}x}(-h < y < 0), \quad \cosh m(H-y)e^{\pm(m^2-\nu^2)^{1/2}x}(0 < y < H) \tag{B1.10}$$

where we must have $\nu < m$ for these waves to exists. Similarly progressive waves of mode M are of the forms

$$f(M,y)e^{\pm(M^2-\nu^2)^{1/2}x}(-h < y < 0), \quad \cosh M(H-y)e^{\pm(M^2-\nu^2)^{1/2}x}(0 < y < H) \tag{B1.11}$$

where we must have $\nu < M$ for these waves to exists.

An incident plane wave of mode m making an angle $\alpha(0 \le \alpha < \pi/2)$ with the positive x-axis has the forms

$$f(m,y)e^{imx\cos\alpha}(-h < y < 0), \quad \cosh m(H-y)e^{imx\cos\alpha}(0 < y < H). \tag{B1.12}$$

In this case

$$\nu = m\sin\alpha. \tag{B1.13}$$

Since $M > m$, we must have $(M^2 - \nu^2)^{1/2}$ to be real in this case, so that scattered waves of mode M exist for all values of m and all angles α. Thus if a wave train of mode m is obliquely incident at angle α with the positive x-axis on the cylindrical undulations at the bottom of a two-layer fluid, then reflected and transmitted waves of both the modes m and M for any angle of incidence α occurs and the far-field behaviors of ψ and ϕ are given by

$$\psi(x,y) \to f(m,y)\left(e^{imx\cos\alpha} + r^m e^{-imx\cos\alpha}\right) + R^m f(M,y)e^{-i(M^2-m^2\sin^2\alpha)^{1/2}x} \text{ as } x \to -\infty$$

$$\varphi(x,y) \to \cosh m(H-y)\left(e^{imx\cos\alpha} + r^m e^{-imx\cos\alpha}\right) + R^m \cosh M(H-y)e^{-i(M^2-m^2\sin^2\alpha)^{1/2}x} \text{ as } x \to -\infty \tag{B1.14}$$

and

$$\psi(x,y) \to t^m f(m,y)e^{imx\cos\alpha} + T^m f(M,y)e^{i(M^2-m^2\sin^2\alpha)^{1/2}x} \text{ as } x \to \infty,$$

$$\varphi(x,y) \to t^m \cosh m(H-y)e^{imx\cos\alpha} + T^m \cosh M(H-y)e^{i(M^2-m^2\sin^2\alpha)^{1/2}x} \text{ as } x \to \infty. \tag{B1.15}$$

In (B1.14) the constants r^m and R^m respectively denote the reflection coefficients associated with reflected waves of modes m and M respectively due to an obliquely incident wave of mode m. Similarly in (B1.15), t^m and T^m denote transmission coefficients associated with transmitted waves of modes m and M respectively due to an obliquely incident wave of mode m.

An incident wave of mode M making an angle α with the positive x-axis has the forms

$$f(M,y)e^{iMx\cos\alpha}(-h<y<0), \cosh M(H-y)e^{iMx\cos\alpha}(0<y<H). \tag{B1.16}$$

In this case

$$\nu = M\sin\alpha, \quad (m^2-\nu^2)^{1/2} = (m^2 - M^2\sin^2\alpha)^{1/2} \tag{B1.17}$$

For a given angle α, there may exists a value of m, i.e., K, for which $m = M\sin\alpha$ and thus $(m^2-\nu^2)^{1/2} = 0$. This may be termed as the cut-off frequency (in the terminology used by Linton and Cadby (2002)) and denoted by K_c. In Fig. 9.6, $K_c h$ is depicted against α for $s = 0.5$ and $H/h = 2,3,5,10,100$. For $H/h = 100$, the curves

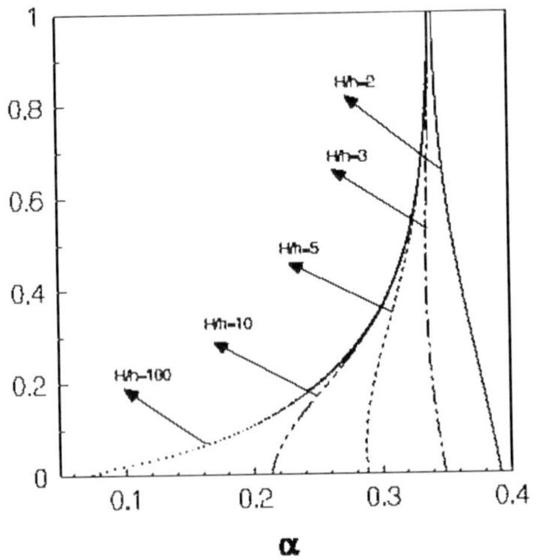

Fig. 9.6 Cut-off frequency $K_c h$ due to an incident wave of wavenumber M: $s = 0.5$

for $K_c h$ almost coincides with the curve given in Fig,1 of Linton and Cadby (2002) where the lower layer was taken to be infinitely deep. For an incident wave of mode M making an angle α with the positive x-axis for which the point (Kh, α) lies in the left side of the curve in Fig. 9.6, reflected and transmitted waves of mode m exists. However, if this point lies on the right side of this curve, then there do not exist reflected and transmitted waves of mode m.

Thus for an obliquely incident wave of mode M making an angle α with the positive x-axis where $\alpha < \sin^{-1}(\frac{m}{M})$, on the bottom undulations of the two-layer fluid, the far field behaviors of ψ and ϕ are given by

$$\psi(x,y) \to f(M,y)\left(e^{iMx\cos\alpha} + R^M e^{-iMx\cos\alpha}\right) + r^M f(m,y) e^{-i(m^2-M^2\sin^2\alpha)^{1/2}x} \text{ as } x \to -\infty,$$

$$\varphi(x,y) \to \cosh M(H-y)\left(e^{iMx\cos\alpha} + R^M e^{-iMx\cos\alpha}\right) + r^M \cosh m(H-y) e^{-i(m^2-M^2\sin^2\alpha)^{1/2}x} \text{ as } x \to -\infty \tag{B1.18}$$

and

$$\psi(x,y) \to T^M f(M,y) e^{iMx\cos\alpha} + t^M f(m,y) e^{i(m^2-M^2\sin^2\alpha)^{1/2}x} \text{ as } x \to \infty$$

$$\varphi(x,y) \to T^M \cosh M(H-y) e^{iMx\cos\alpha} + t^M \cosh m(H-y) e^{i(m^2-M^2\sin^2\alpha)^{1/2}x} \text{ as } x \to \infty. \tag{B1.19}$$

The constants r^M and R^M denote reflection coefficients associated with reflected waves of modes m and M respectively due to an obliquely incident wave of mode M. Similarly, the constants t^M and T^M denote transmission coefficients associated with transmitted waves of modes m and M respectively due to an obliquely incident wave of mode M. The angle of incidence α must satisfy $\alpha < \sin^{-1}(\frac{m}{M})$ for the coefficients r^M, t^M to exist.

If the incident wave has the mode m, then ψ, ϕ satisfy (B1.1), (B1.2) with $\nu = m\sin\alpha (0 \le \alpha < \pi/2)$, the conditions (B1.3) to (B1.5), the bottom condition

$$\frac{\partial \varphi}{\partial n} = 0 \text{ on } y = H + \varepsilon c(x) \tag{B1.20}$$

where $\frac{\partial}{\partial n}$ denotes the normal derivative to the bottom, and the infinity requirements (B1.14), (B1.15) involving the unknown coefficients r^m, R^m, t^m, T^m.

Similarly if the incident wave has the mode M, then ψ, ϕ satisfy (B1.1), (B1.2) with $\nu = M\sin\alpha(0 \le \alpha < \sin^{-1}(\frac{m}{M}))$, the conditions (B1.3) to (B1.5), (B1.20) and the infinity requirements (B1.18), (B1.19) involving the unknown coefficients r^M, R^M, t^M, T^M.

Determination of the reflection and transmission coefficients $r^{m,M}$, $R^{m,M}$, $t^{m,M}$, $T^{m,M}$ for a general type of bottom undulation is a difficult task. However, for small bottom undulations, an approximate method will be used here to obtain these upto first order. The clue for the approximate method is found by observing that the bottom condition (B1.20) can be approximated upto first order of ε as

$$-\varphi_y + \varepsilon \frac{\partial}{\partial x}\left\{c(x)\frac{\partial \varphi}{\partial x}(x,H)\right\} + O(\varepsilon^2) = 0 \quad \text{on} \quad y = H \tag{B1.21}$$

The lower layer $0 < y < H + \varepsilon c(x), -\infty < x < \infty$ reduces to the uniform strip $0 < y < H, -\infty < x < \infty$ in the following mathematical analysis using a perturbation technique.

The perturbation method

We consider a wave train of mode m to be obliquely incident at an angle α $(0 \le \alpha < \pi/2)$ on the bottom undulations. If there is no bottom undulation, then the incident wave train will propagate without any hindrance and there will be total transmission. This along with the approximate form (B1.21) of the bottom condition suggest that $\psi(x, y)$, $\phi(x, y)$, r^m, R^m, t^m and T^m can be expanded as

$$\begin{aligned}\psi = \psi_0 + \varepsilon\psi_1 + O(\varepsilon^2), \quad \phi = \phi_0 + \varepsilon\phi_1 + O(\varepsilon^2),\\ r^m = \varepsilon r_1^m + O(\varepsilon^2), \quad R^m = \varepsilon R_1^m + O(\varepsilon^2),\\ t^m = 1 + \varepsilon t_1^m + O(\varepsilon^2), \quad T^m = \varepsilon T_1^m + O(\varepsilon^2)\end{aligned} \tag{B1.22}$$

where

$$\psi_0(x, y) = f(m, y)e^{imx\cos\alpha}, \varphi_0(x, y) = \cosh m(H - y)e^{imx\cos\alpha}. \tag{B1.23}$$

Substituting the expansions (B1.22) in (B1.1), (B1.2), with $v = m \sin \alpha$, the conditions (B1.3), (B1.4), (B1.5), (B1.21) and (B1.14), (B1.15), and equating the coefficients of ε in both sides of the equations and the conditions, we find that the first-order functions $\psi_1(x, y)$, $\phi_1(x, y)$ satisfy the coupled boundary value problem described by

$$(\nabla^2_{x,y} - v^2)\psi_1 = 0, \quad -h < y < 0, \tag{B1.24}$$

$$(\nabla^2_{x,y} - v^2)\varphi_1 = 0, \quad 0 < y < H, \tag{B1.25}$$

with $v = m \sin \alpha$,

$$K\psi_1 + \psi_{1y} = 0 \quad \text{on} \quad y = -h, \tag{B1.26}$$

$$s(K\psi_1 + \psi_{1y}) = K\varphi_1 + \varphi_{1y} \quad \text{on} \quad y = 0, \tag{B1.27}$$

$$\psi_{1y} = \varphi_{1y} \quad \text{on} \quad y = 0, \tag{B1.28}$$

$$\varphi_{1y} = q(x) \quad \text{on} \quad y = H \tag{B1.29}$$

where

$$q(x) = im\cos\alpha \frac{d}{dx}\left(c(x)e^{imx\cos\alpha}\right) - m^2 \sin^2\alpha c(x)e^{imx\cos\alpha}, \tag{B1.30}$$

with the infinity requirements

$$\psi_1(x,y) \to r_1^m f(m,y)e^{-mx\cos\alpha} + R_1^m f(M,y)e^{-i(M^2-m^2\sin^2\alpha)^{1/2}x} \text{ as } \quad x \to -\infty,$$

$$\varphi_1(x,y) \to r_1^m \cosh m(H-y)e^{-imx\cos\alpha} + R_1^m \cosh M(H-y)e^{-i(M^2-m^2\sin^2\alpha)^{1/2}x} \text{as } x \to -\infty \tag{B1.31}$$

and

$$\psi_1(x,y) \to t_1^m f(m,y)e^{-imx\cos\alpha} + T_1^m f(M,y)e^{-i(M^2-m^2\sin^2\alpha)^{1/2}x} \text{ as } \quad x \to \infty,$$

$$\varphi_1(x,y) \to t_1^m \cosh m(H-y)e^{-imx\cos\alpha} + T_1^m \cosh M(H-y)e^{-i(M^2-m^2\sin^2\alpha)^{1/2}x} \text{ as } x \to \infty. \tag{B1.32}$$

A method based on the use of Green's integral theorem is now employed to solve this coupled boundary value problem and the first order coefficients r_1^m, R_1^m and t_1^m, T_1^m are obtained in terms of integrals involving the shape function.

To solve the coupled BVP described by (B1.24) to (B1.29), we construct Green's functions for the modified Helmholtz equation due to a source submerged in either of the two-layers. Let $G(x,y;\xi,\eta)$ and $G'(x,y;\xi,\eta)$ be the Green's functions in the upper and lower layers respectively due to a source submerged in the lower layer at $(\xi,\eta)(0<\eta<H)$, and $\mathcal{G}\,(x,y;\xi,\eta)$, $\mathcal{G}'\,(x,y;\xi,\eta)$ be the same due to a source submerged in the upper layer at $(\xi,\eta)(-h<\eta<0)$. Then *G*, *H* satisfy

$$(\nabla^2_{x,y} - \nu^2)G = 0 \quad \text{for} \quad -h<y<0,$$

$$(\nabla^2_{x,y} - \nu^2)G' = 0 \quad \text{except at } (\xi,\eta)(0<\eta<H) \quad \text{for} \quad 0<y<H,$$

$$G' \to K_0(\nu r) \quad \text{as} \quad r = \{(x-\xi)^2 + (y-\eta)^2\}^{1/2} \to 0,$$

where $K_0(z)$ denotes the modified Bessel function of second kind.

$$KG + G_y = 0 \quad \text{on} \quad y=-h,$$

$$s(KG+G_y) = KG' + G'y \quad \text{on} \quad y=0,$$

$$G_y = G'_y \quad \text{on} \quad y=0,$$

$$G'_y = 0 \quad \text{on} \quad y=H,$$

$$G, G' \quad \text{behave as outgoing waves as} \quad |x-\xi| \to \infty, \tag{B1.33}$$

while $\mathcal{G}$, $\mathcal{G}'$ satisfy

$$(\nabla^2_{x,y} - \nu^2)\mathcal{G} = 0 \quad \text{except at } (\xi,\eta)(-h<\eta<0) \text{ for } -h<y<0,$$

$$\mathcal{G} \to K_0(\nu r) \quad \text{as} \quad r = \{(x-\xi)^2 + (y-\eta)^2\}^{1/2} \to 0,$$

$$(\nabla^2_{x,y} - \nu^2)\mathcal{G}' = 0 \quad \text{for} \quad 0<y<H,$$

$$K\mathcal{G} + \mathcal{G}_y = 0 \quad \text{on} \quad y=-h,$$

$$s(K\mathcal{G} + \mathcal{G}_y) = K\mathcal{G}' + \mathcal{G}'_y \quad \text{on} \quad y=0,$$

$$\mathcal{G}_y = \mathcal{G}'_y \quad \text{on} \quad y=0,$$

$$\mathcal{G}'_y = 0 \quad \text{on} \quad y=H,$$

$$\mathcal{G}, \mathcal{G}' \quad \text{behave as outgoing waves as} \quad |x-\xi| \to \infty. \tag{B1.34}$$

The functions G, G' and $\mathcal{G}$, $\mathcal{G}'$ can be constructed following the method used by Mandal and Chakrabarti (1986). Expressions for G, G', $\mathcal{G}$, $\mathcal{G}'$ and their forms as $|x-\xi| \to \infty$ are given below.

Source submerged in the lower layer

Let (ξ,η) $(0<\eta<H)$ be the source, $G(x,y;\xi,\eta)$ and $G'(x,y;\xi,\eta)$ be the Green's functions for the modified Helmholtz equation in the upper and lower layers respectively satisfying (B1.33). These are obtained as

$$G(x,y;\xi,\eta) = \frac{K_0(\nu r) - K_0(\nu r')}{s} - 2\int_\nu^\infty \frac{F_{11}(k;y,\eta)\cos\{(k^2-\nu^2)^{1/2}|x-\xi|\}}{(k^2-\nu^2)^{1/2}\sinh kh \sinh kH\Delta(k)} dk,$$
$$-h<y<0, 0<\eta<H, \quad (B1.35)$$

$$G'(x,y;\xi,\eta) = K_0(\nu r) - K_0(\nu r') - 2\int_\nu^\infty \frac{F_{12}(k;y,\eta)\cos\{(k^2-\nu^2)^{1/2}|x-\xi|\}}{(k^2-\nu^2)^{1/2}\sinh kh \sinh kH\Delta(k)} dk,$$
$$0<y,\eta<H \qquad (B1.36)$$

where

$$r = \{(x-\xi)^2+(y-\eta)^2\}^{1/2}, r' = \{(x-\xi)^2+(y+2h+\eta)^2\}^{1/2},$$

$$\begin{aligned} F_{11}(k;y,\eta) = e^{-k(h+\eta)}\Big[&k\sinh kH(k\cosh ky - K\sinh ky)\\ &+\cosh kh(k\sinh kH - K\cosh kH)\{K\sinh k(h+y) - k\cosh k(h+y)\}\\ &+\frac{1}{s}(k\sinh kh - K\cosh kh)(k\sinh kH - K\cosh kH)\sinh k(h+y)\Big]\\ &+Ke^{-k(h+H)}\sinh k(h+\eta)\{k\cosh k(h+y) - K\sinh k(h+y)\}, -h<y<0, 0<\eta<H, \end{aligned}$$

$$\begin{aligned} F_{12}(k;y,\eta) \;=\; &\Big[e^{-k(h+\eta)}\{kK+(1-s)(K^2-k^2)\sinh kh\cosh kh\} +\\ &\sinh k(h+\eta)e^{-k(h+H)}\{(k\cosh kH - K\sinh kH)(K\cosh kh - k\sinh kh)\\ &-s(K^2-k^2)\sinh kh\cosh kH\}\Big]\cosh k(H-y)\\ &+e^{-k(h+H)}\Delta(k)\sinh kh\sinh kH\sinh k(h+\eta)\sinh k(H-y), \quad 0<y,\eta<H \end{aligned}$$

and the path of integration in the integrals in (B1.35) and (B1.36) are indented below the poles at $k=m, M$ on the real axis so as to ensure the outgoing nature of G, G' far away from the source $(\nu<m<M)$.

As $|x-\xi|\to\infty$, it can be shown that

$$G(x,y;\xi,\eta)\to -2\pi i\left[\frac{K\{m\cosh m(h+y)-K\sinh m(h+y)\}\cosh m(H-\eta)}{(m^2-\nu^2)^{1/2}\sinh mh\sinh mH\Delta'(m)}e^{i(m^2-\nu^2)^{1/2}|x-\xi|}\right.$$
$$\left.+\frac{K\{M\cosh M(h+y)-K\sinh M(h+y)\}\cosh M(H-\eta)}{(M^2-\nu^2)^{1/2}\sinh Mh\sinh MH\Delta'(M)}e^{i(M^2-\nu^2)^{1/2}|x-\xi|}\right],\ -h<y<0, 0<\eta<H$$

(B1.37)

and

$$G'(x,y;\xi,\eta)\to -2\pi iK\left[\frac{(K\cosh mh-m\sinh mh)\cosh m(H-\eta)\cosh m(H-y)}{(m^2-\nu^2)^{1/2}\sinh mh\sinh^2 mH\Delta'(m)}e^{i(m^2-\nu^2)^{1/2}|x-\xi|}\right.$$
$$+\frac{(K\cosh Mh-M\sinh Mh)\cosh M(H-\eta)\cosh M(H-y)}{(M^2-\nu^2)^{1/2}\sinh Mh\sinh MH\Delta'(M)}e^{i(M^2-\nu^2)^{1/2}|x-\xi|}, -h<y,\eta<H.$$

(B1.38)

Source submerged in the upper layer

Let $\mathcal{G}(x,y;\xi;\eta)$ and $\mathcal{G}'(x,y;\xi;\eta)$ denote the Green's functions for the modified Helmholtz equation in the upper and lower layers satisfying (4.2) due to a source at (ξ,η) submerged in the upper layer $(-h<\eta<0)$. These are obtained as

$$\mathcal{G}(x,y;\xi,\eta)=K_0(\nu r)-K_0(\nu r')-2\int_\nu^\infty\frac{F_{21}(k;y,\eta)\cos\{(k^2-\nu^2)^{1/2}|x-\xi|\}}{(k^2-\nu^2)^{1/2}\sinh kh\sinh kH\Delta(k)}dk,\ -h<y,\eta<0$$

(B1.39)

$$\mathcal{G}(x,y;\xi,\eta)=s\{K_0(\nu r)-K_0(\nu r')\}-2\int_\nu^\infty\frac{F_{22}(k;y,\eta)\cos\{(k^2-\nu^2)^{1/2}|x-\xi|\}}{(k^2-\nu^2)^{1/2}\sinh kh\sinh kH\Delta(k)}dk,\ 0<y<H, -h<\eta<0,$$

(B1.40)

where r, r' are the same as above,

$$\begin{aligned}F_{21}(k;y,\eta) &= ke^{-k(h+\eta)}\{(K\cosh kH-k\sinh kH)\cosh ky+\\ &\quad s\sinh kH(k\cosh ky-K\sinh ky)\}+e^{-kh}\{(1-s)\\ &\quad (k\sinh kH-K\cosh kH)-sKe^{-kH}\}\sinh k(y+\eta)\\ &\quad \{K\sinh k(h+y)-k\cosh k(h+y)\},\qquad -h<y,\eta<0\end{aligned}$$

$$\begin{aligned}F_{22}(k;y,\eta) &= s\Big[e^{-k(h+H)}\sinh k(h+\eta)\{(k\cosh kH-K\sinh kH)\\ &\quad (K\cosh kh-k\sinh kh)-s(K^2-k^2)\sinh kh\cosh kH\\ &\quad +(s-1)e^{kh}(K^2-k^2)\sinh kh\}+kKe^{-k(y+\eta)}\Big]\cosh k(H-y)\\ &+se^{-k(h+H)}\sinh kh\sinh kH\Delta(k)\sinh k(h+\eta)\sinh k(H-y),\ \ 0<y<H, -h<\eta<0,\end{aligned}$$

and the path of integration in the integrals in ($A4$) and ($A5$) are indented below the poles at $k=m, M$ on the real axis as before.

As $|x-\xi| \to 0$, it can be show that

$$\mathcal{G}(x,y;\xi,\eta) \to -2\pi isK\left[\frac{\{K\sinh m(h+\eta)-m\cosh m(h+\eta)\}}{(m^2-\nu^2)^{1/2}\sinh mh\sinh mH\Delta'(m)}\right.$$

$$\frac{\{K\sinh m(h+y)-m\cosh m(h+y)\}}{(K\cosh mh - m\sinh mh)}e^{i(m^2-\nu^2)^{1/2}|x-\xi|}$$

$$+\frac{\{K\sinh M(h+\eta)-M\cosh M(h+\eta)\}\{K\sinh M(h+y)-M\cosh M(h+y)\}}{(M^2-\nu^2)^{1/2}\sinh Mh\sinh MH(K\cosh Mh - M\sinh Mh)\Delta'(M)}$$

$$\left.e^{i(M^2-\nu^2)^{1/2}|x-\xi|}\right], -h<y,\eta<0 \quad \text{(B1.41)}$$

and

$$\mathcal{G}(x,y;\xi,\eta) \to -2\pi isK\left[\frac{\{m\cosh m(h+\eta)-K\sinh m(h+\eta)\}\cosh m(H-y)}{(m^2-\nu^2)^{1/2}\sinh mh\sinh mH\Delta'(m)}\right.$$

$$\left.e^{i(m^2-\nu^2)^{1/2}|x-\xi|}+\frac{\{M\cosh M(h+\eta)-K\sinh M(h+\eta)\}\cosh M(H-y)}{(M^2-\nu^2)^{1/2}\sinh Mh\sinh MH\Delta'(M)}e^{i(M^2-\nu^2)^{1/2}|x-\xi|}\right],$$

$$0<y<H, -h<\eta<0. \quad \text{(B1.42)}$$

We now apply the Green's integral theorem to $\psi_1(x,y)$ and $G(x,y;\xi,\eta)$ in the form

$$\int_c\left(\psi\frac{\partial G}{\partial n}-G\frac{\partial \psi_1}{\partial n}\right)ds = 0 \quad \text{(B1.43)}$$

where C is a contour in the (x,y)-plane formed by the lines $y=-h,0$ $(-X\le x\le X)$, $x=\pm X(-h\le y\le 0)$ and make $X=\infty$ ultimately. There will be no contribution to the integral from the line $y=-h$ due to the free surface conditions satisfied by ψ_1 and G. Thus we obtain

$$-\int_{-h}^{0}(\psi_1 G_x - G\psi_{1x})_{x=-\infty}dy+\int_{-\infty}^{\infty}(\psi_1 G_y - G\psi_{1y})_{y=0}dx+\int_{-h}^{0}(\psi_1 G_x - G\psi_{1x})_{x=\infty}dy = 0.$$

$$\text{(B1.44)}$$

Again we apply the Green's integral theorem to $\phi_1(x,y)$ and $G'(x,y;\xi,\eta)$ in the form

$$\int_{c'}\left(\varphi_1\frac{\partial G'}{\partial n}-G'\frac{\partial \varphi_1}{\partial n}\right)ds = 0 \quad \text{(B1.45)}$$

where C' is a contour formed by the lines $y=0,H$ $(-X\le x\le X)$, $x=\pm X(0\le y\le H)$ and a circle of small radius ε with centre at (ξ,η) $(0<\eta<H)$ and make $X\to\infty$ and $\varepsilon\to 0$ ultimately. We then obtain

$$-2\pi\phi(\xi,\eta)+\int_{-\infty}^{\infty}(\phi_1 G'_y - G'\phi_{1y})_{y=0}dx+\int_0^H(\phi_1 G'_x - G'\phi_{1x})_{x=-\infty}dy$$

$$-\int_{-\infty}^{\infty}(\phi_1 G'_y - G'\phi_{1y})_{y=H}dx-\int_0^H(\phi_1 G'_x - G'\phi_{1x})_{x=\infty}dy = 0. \quad \text{(B1.46)}$$

Multiplying (B1.44) by s and subtracting (B1.46) from this and using the interface conditions and the fact that

$$s\int_{-h}^{0}(\psi_1 G_x - G\psi_{1x})_{x=\pm X}\,dy + \int_{0}^{H}(\varphi_1 G'_x - G'\varphi_{1x})_{x=\pm X}\,dy \to 0 \text{ as } X \to \infty,$$

we obtain the representation for $\phi_1(\xi,\eta)$ $(0 < \eta < H)$, in the form, after using (B1.29)

$$\phi_1(\xi,\eta) = \frac{1}{2\pi}\int_{-\infty}^{\infty} G'(x,y;\xi,\eta)q(x)dx, \quad 0 < \eta < H. \tag{B1.47}$$

A somewhat similar procedure can be used to obtain a representation for $\psi_1(\xi,\eta)$ $(-h < \eta < 0)$, in the form

$$\psi_1(\xi,\eta) = \frac{1}{2\pi s}\int_{-\infty}^{\infty} \mathcal{G}'(x,y;\xi,\eta)q(x)dx, \quad -h < \eta < 0. \tag{B1.48}$$

The first-order reflection and transmission coefficients r_1^m, R_1^m and t_1^m, T_1^m are now obtained by making $\xi \to -\infty$ and $\xi \to \infty$ respectively in (B1.47) or (B1.48) and comparing with (B1.31) and (B1.32) (x, y being replaced by ξ, η). For this the forms of G' or $\mathcal{G}'$ as $\xi \to \pm\infty$, given in (B1.41) and (B1.42), have been used. Thus r_1^m, R_1^m are obtained as

$$r_1^m = -\frac{iKm(K\cosh mh - m\sinh mh)\cos 2\alpha}{\cos\alpha \sinh mh \sinh^2 mH \Delta'(m)}\int_{-\infty}^{\infty} e^{2imx\cos\alpha} c(x)dx, \tag{B1.49}$$

$$R_1^m = -\frac{iKm(K\cosh Mh - M\sinh Mh)\{\cos\alpha(M^2 - m^2\sin^2\alpha)^{1/2} - m\sin^2\alpha\}}{(M^2 - m^2\sin^2\alpha)^{1/2}\sinh Mh \sinh^2 MH\Delta'(M)}$$
$$\times \int_{-\infty}^{\infty} e^{i\{m\cos\alpha + (M^2 - m^2\sin^2\alpha)^{1/2}\}x} c(x)dx, \tag{B1.50}$$

while t_1^m, T_1^m are obtained as

$$t_1^m = \frac{iKm(K\cosh mh - m\sinh mh)}{\cos\alpha \sinh mh \sinh^2 mH\Delta'(m)}\int_{-\infty}^{\infty} c(x)dx, \tag{B1.51}$$

$$T_1^m = \frac{iKm(K\cosh Mh - M\sinh Mh)\{\cos\alpha(M^2 - m^2\sin^2\alpha)^{1/2} + m\sin^2\alpha\}}{(M^2 - m^2\sin^2\alpha)^{1/2}\sinh Mh \sinh^2 MH\Delta'(M)}$$
$$\times \int_{-\infty}^{\infty} e^{i\{m\cos\alpha - (M^2 - m^2\sin^2\alpha)^{1/2}\}x} c(x)dx. \tag{B1.52}$$

For a wave train of mode M obliquely incident at an angle $\alpha\left(0 \le \alpha < \sin^{-1}(\frac{m}{M})\right)$, the first-order coefficients r_1^M, R_1^M, t_1^M, T_1^M can be obtained by following a mathematical analysis described above. The results are as follows:

$$r_1^M = -\frac{iKM(K\cosh mh - m\sinh mh)\{\cos\alpha(m^2 - M^2\sin^2\alpha)^{1/2} - M\sin^2\alpha\}}{(m^2 - M^2\sin^2\alpha)^{1/2}\sinh mh \sinh^2 mH\Delta'(m)}$$
$$\times \int_{-\infty}^{\infty} e^{i\{(m^2 - M^2\sin^2\alpha)^{1/2} + M\cos\alpha\}x} c(x)dx, \tag{B1.53}$$

$$R_1^M = -\frac{iKM(K\cosh Mh - M\sinh Mh)\cos 2\alpha}{\cos\alpha \sinh Mh \sinh^2 MH\Delta'(M)}\int_{-\infty}^{\infty} e^{2iMx\cos\alpha} c(x)dx, \tag{B1.54}$$

$$t_1^M = \frac{iKM(K\cosh mh - m\sinh mh)\{\cos\alpha(m^2 - M^2\sin^2\alpha)^{1/2} + M\sin^2\alpha\}}{(m^2 - M^2\sin^2\alpha)^{1/2}\sinh mh \sinh^2 mH\Delta'(m)}$$

$$\times \int_{-\infty}^{\infty} e^{i\{M\cos\alpha - (m^2 - M^2\sin^2\alpha)^{1/2}\}x} c(x)dx, \quad \text{(B1.55)}$$

$$T_1^M = \frac{iKM(K\cosh Mh - M\sinh Mh)}{\cos\alpha\sinh Mh \sinh^2 MH\Delta'(M)} \int_{-\infty}^{\infty} c(x)dx. \quad \text{(B1.56)}$$

Numerical results

For sinusoidal undulations at the bottom of the two-layer fluid, the shape function $c(x)$ has the form

$$c(x) = \begin{cases} a\sin\gamma x & \text{for } -\dfrac{n\pi}{\gamma} \le x \le \dfrac{n\pi}{\gamma}, \\ 0 & \text{otherwise} \end{cases}$$

where n is a positive integer. In a single-layer ocean bed, sinusoidal undulations occur naturally. Davies (1982) studied this case somewhat elaborately and found that an undulating bed has the ability to reflect incident wave energy which has important implications in respect of coastal protection as well as possible ripple growth if the bed is erodible. For this reason, this particular bed form in a two-layer fluid is considered here and the first-order reflection and transmission coefficients are calculated explicitly and the numerical results are presented graphically against the wave number Kh.

Since $c(x)$ is an odd function, t_1^m, T_1^M vanish identically. The coefficients r_1^m, R_1^m, T_1^m, r_1^M, R_1^M, t_1^M are obtained as

$$r_1^m = (-1)^{n+1}\frac{Km(K\cosh mh - m\sinh mh)\cos 2\alpha}{\cos\alpha\sinh mh \sinh^2 mH\Delta'(m)} \frac{2a\gamma\sin\left(\dfrac{2nm\pi}{\gamma}\cos\alpha\right)}{\gamma^2 - 4m^2\cos^2\alpha}, \quad \text{(B1.57)}$$

$$R_1^m = (-1)^{n+1}\frac{Km(K\cosh Mh - M\sinh Mh)\{\cos\alpha(M^2 - m^2\sin^2\alpha)^{1/2} - m\sin^2\alpha\}}{(M^2 - m^2\sin^2\alpha)^{1/2}\sinh Mh \sinh^2 MH\Delta'(M)}$$

$$\times \frac{2a\gamma\sin\left\{\dfrac{n\pi}{\nu}(m\cos\alpha + (M^2 - m^2\sin^2\alpha)^{1/2})\right\}}{\gamma^2 - (m\cos\alpha + (M^2 - m^2\sin^2\alpha)^{1/2})^2}, \quad \text{(B1.58)}$$

$$T_1^m = (-1)^n\frac{Km(K\cosh Mh - M\sinh Mh)\{\cos\alpha(M^2 - m^2\sin^2\alpha)^{1/2} + m\sin^2\alpha\}}{(M^2 - m^2\sin^2\alpha)^{1/2}\sinh Mh \sinh^2 MH\Delta'(M)}$$

$$\times \frac{2a\gamma\sin\left\{\dfrac{n\pi}{\nu}(m\cos\alpha - (M^2 - m^2\sin^2\alpha)^{1/2})\right\}}{\gamma^2 - (m\cos\alpha - (M^2 - m^2\sin^2\alpha)^{1/2})^2}, \quad \text{(B1.59)}$$

$$r_1^M = (-1)^{n+1}\frac{KM(K\cosh mh - m\sinh mh)\{\cos\alpha(m^2 - M^2\sin^2\alpha)^{1/2} - M\sin^2\alpha\}}{(m^2 - M^2\sin^2\alpha)^{1/2}\sinh mh\sinh^2 mH\Delta'(m)}$$

$$\times\frac{2a\gamma\sin\left\{\dfrac{n\pi}{\nu}(M\cos\alpha + (m^2 - M^2\sin^2\alpha)^{1/2})\right\}}{\gamma^2 - (M\cos\alpha + (m^2 - M^2\sin^2\alpha)^{1/2})^2}, \tag{B1.60}$$

$$R_1^M = (-1)^{n+1}\frac{KM(K\cosh mh - m\sinh mh)\cos 2\alpha}{\cos\alpha\sinh Mh\sinh^2 MH\Delta'(M)}\frac{2a\gamma\sin\left(\dfrac{2nM\pi}{\gamma}\cos\alpha\right)}{\gamma^2 - 4M^2\cos^2\alpha}, \tag{B1.61}$$

$$t_1^M = (-1)^n\frac{KM(K\cosh mh - m\sinh mh)\{\cos\alpha(m^2 - M^2\sin^2\alpha)^{1/2} + M\sin^2\alpha\}}{(m^2 - M^2\sin^2\alpha)^{1/2}\sinh mh\sinh^2 mH\Delta'(m)}$$

$$\times\frac{2a\gamma\sin\left\{\dfrac{n\pi}{\nu}(M\cos\alpha - (m^2 - M^2\sin^2\alpha)^{1/2})\right\}}{\gamma^2 - (M\cos\alpha - (m^2 - M^2\sin^2\alpha)^{1/2})^2}. \tag{B1.62}$$

In the results (B1.60) to (B1.62), $0 \le \alpha < \sin^{-1}(\frac{m}{M})$.

The Figs. 9.7–9.9 depict $|r_1^m|$, $|R_1^m|$, $|T_1^m|$, the first order reflection and transmission coefficients due to a wave train of mode m obliquely incident at an angle α with the positive x-axis on the undulating bed in the lower layer for $a/h = 0.1$, $H/h = 2$, $s = 0.5$, $\gamma h = 1$, $n = 3$ and $\alpha = 0, .262, .524, 1.05, 1.31$ (measured in radians). The case $\alpha = 0$ represents the case of normal incidence. One feature that is common in all these figures is the oscillating nature of the absolute values of the first-order coefficients as

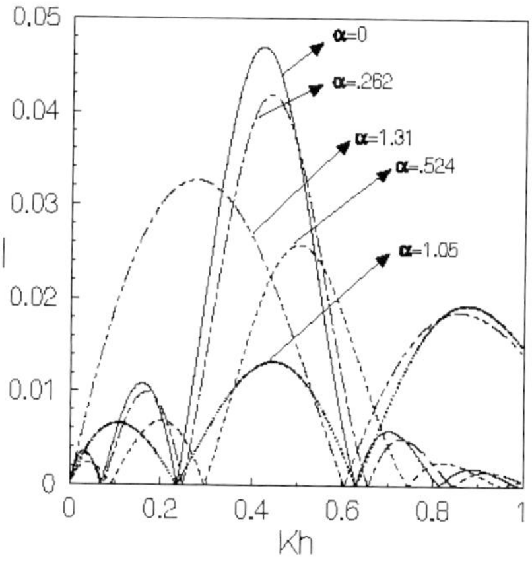

Fig. 9.7 First order reflection coefficient due to an incident wave of mode m: $a/h = 0.1$, $H/h = 2$, $s = 0.5$, $\gamma h = 1$ and $n = 3$

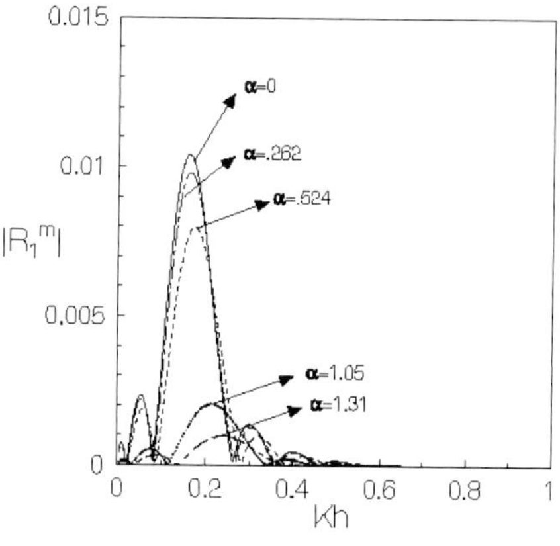

Fig. 9.8 First order reflection coefficient due to an incident wave of mode m: $a/h = 0.1$, $H/h = 2$, $s = 0.5$, $\gamma h = 1$ and $n = 3$

a function of the wave number Kh. For the case of normal incidence, the peak values of $|r_1^m|$, $|R_1^m|$ are the largest, while $|T_1^m|$ assumes the least peak value. The oscillating nature may be attributed due to multiple interaction of the incident wave train with the sinusoidal bottom, the free surface and the interface.

In the Figs. 9.10–9.12, plots of $|r_1^M|$, $|R_1^M|$, $|t_1^M|$ are shown against Kh for $a/h = 0.1$, $H/h = 2$, $s = 0.5$, $\gamma h = 1$, $n = 3$ and $\alpha = 0, .35, .36, .38, .39$. It may be noted that while for $\alpha = 0$ (the case of normal incidence), conversion of incident wave energy of mode M to reflected wave energy of mode m is always possible, for non-zero α (i.e., the case of oblique incidence), there will be no energy conversion if α exceeds a certain critical angle. The value of the critical angle obviously depends on Kh apart from other parameters. The Fig. 9.10 depicts $|r_1^M|$ for normal incidence ($\alpha = 0$) as well as for oblique incidence with α slightly less than the critical angle (viz. α =.35, .36,

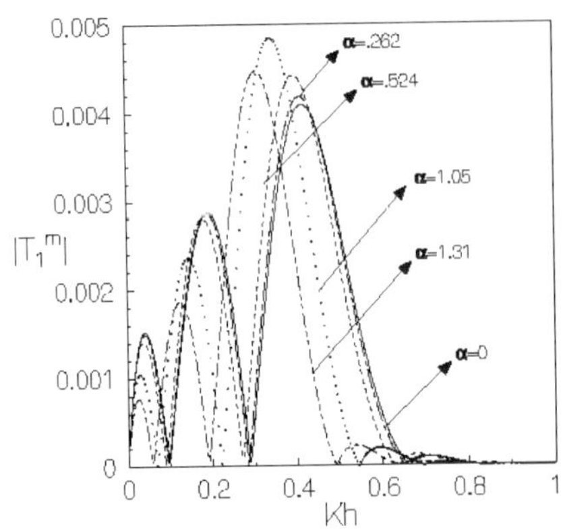

Fig. 9.9 First order transmission coefficient due to an incident wave of mode m: $a/h = 0.1$, $H/h = 2$, $s = 0.5$, $\gamma h = 1$ and $n = 3$

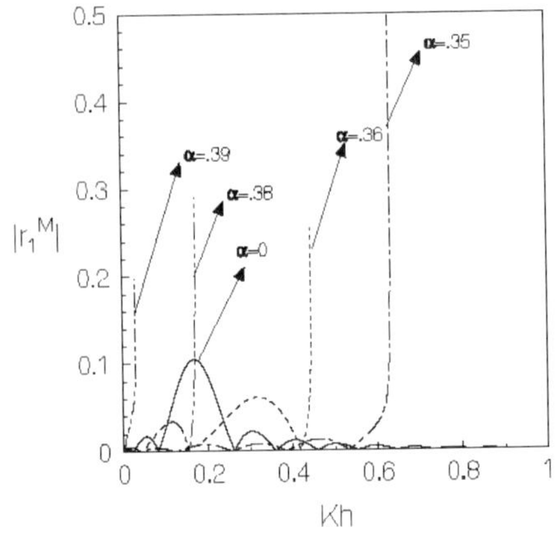

Fig. 9.10 First order reflection coefficient due to an incident wave of mode M: $a/h = 0.1$, $H/h = 2$, $s = 0.5$, $\gamma h = 1$ and $n = 3$

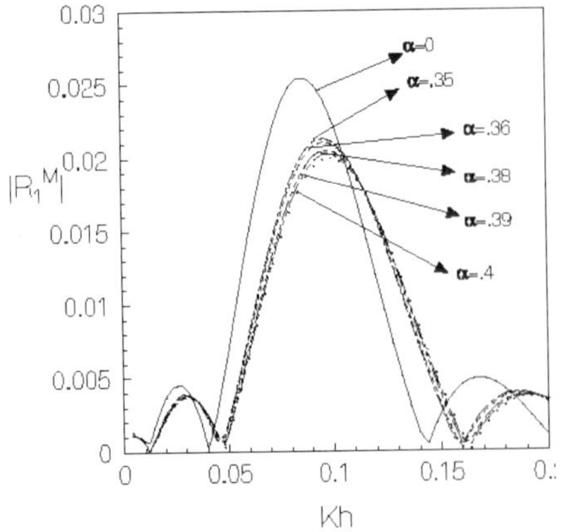

Fig. 9.11 First order reflection coefficient due to an incident wave of mode M: $a/h = 0.1$, $H/h = 2$, $s = 0.5$, $\gamma h = 1$ and $n = 3$

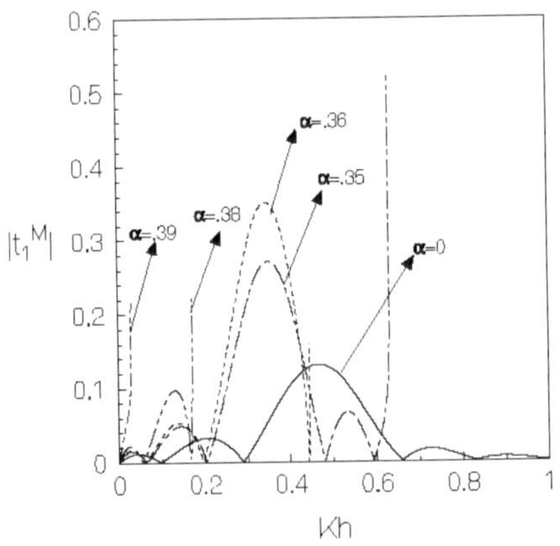

Fig. 9.12 First order transmission coefficient due to an incident wave of mode M: $a/h = 0.1$, $H/h = 2$, $s = 0.5$, $\gamma h = 1$ and $n = 3$

.38, .39). As α increases the range of Kh for which $|r_1^M|$ exists, decreases. The Fig. 9.12 depicting $|t_1^M|$ also displays the same characteristics. However, the Fig. 9.11, depicting $|R_1^M|$ does not show such behavior since reflection at mode M of incident wave energy of mode M always occurs whatever be the incident wave frequency.

The Figs. 9.13–9.15 show the curves for $|r_1^m|$, $|R_1^m|$, $|T_1^m|$ as a function of Kh for $a/h = 0.1$, $H/h = 2$, $\alpha = .35$, $\gamma h = 1$, $n = 3$ and different values of $s = 0.1, 0.4, 0.5, 0.56, 0.6$. As s increases, the curves for $|R_1^m|$ and $|T_1^m|$ become more oscillatory (cf. Figs. 9.14 and 9.15) while the curves for $|r_1^m|$ show that the peak values decrease (cf. Fig. 9.13).

The Figs. 9.16–9.18 depict the curves for $|r_1^M|$, $|R_1^M|$, and $|T_1^M|$ for the same values of different parameters as in Figs. 9.13–9.15.

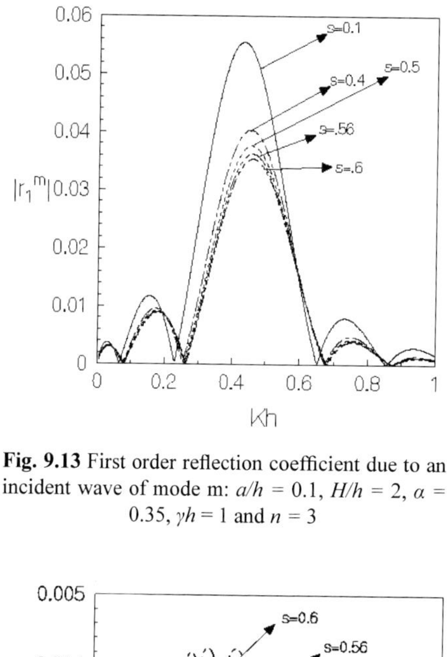

Fig. 9.13 First order reflection coefficient due to an incident wave of mode m: $a/h = 0.1$, $H/h = 2$, $\alpha = 0.35$, $\gamma h = 1$ and $n = 3$

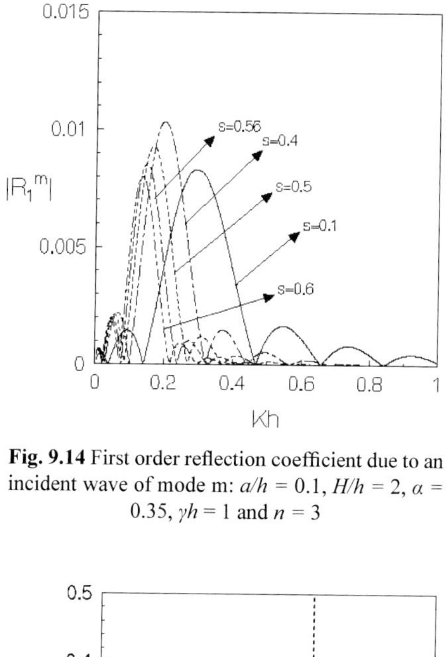

Fig. 9.14 First order reflection coefficient due to an incident wave of mode m: $a/h = 0.1$, $H/h = 2$, $\alpha = 0.35$, $\gamma h = 1$ and $n = 3$

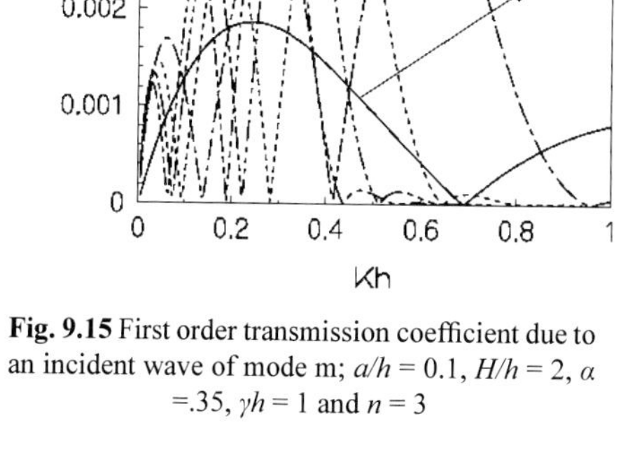

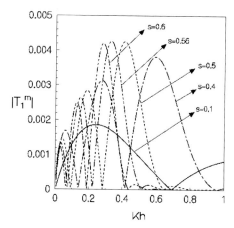

Fig. 9.15 First order transmission coefficient due to an incident wave of mode m; $a/h = 0.1$, $H/h = 2$, $\alpha = .35$, $\gamma h = 1$ and $n = 3$

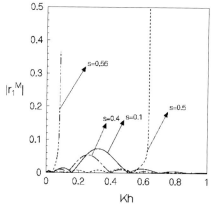

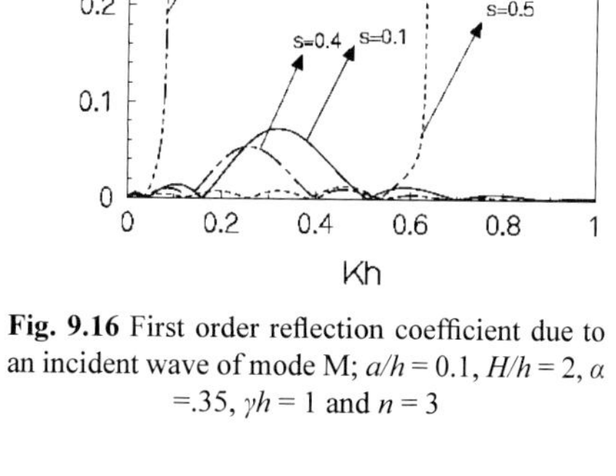

Fig. 9.16 First order reflection coefficient due to an incident wave of mode M; $a/h = 0.1$, $H/h = 2$, $\alpha = .35$, $\gamma h = 1$ and $n = 3$

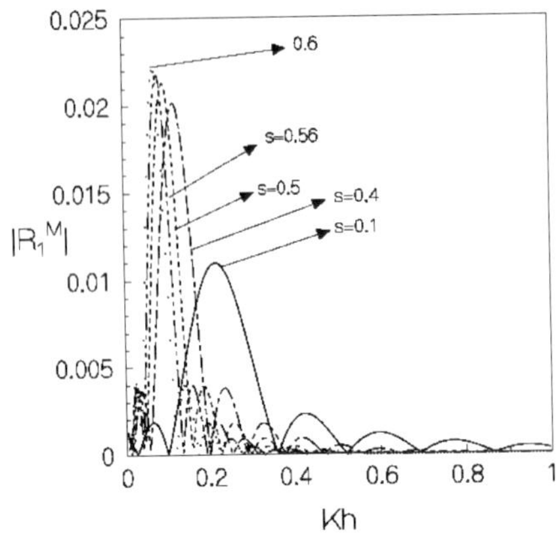

Fig. 9.17 First order reflection coefficient due to an incident wave of mode M; $a/h = 0.1$, $H/h = 2$, α =.35, $\gamma h = 1$ and $n = 3$

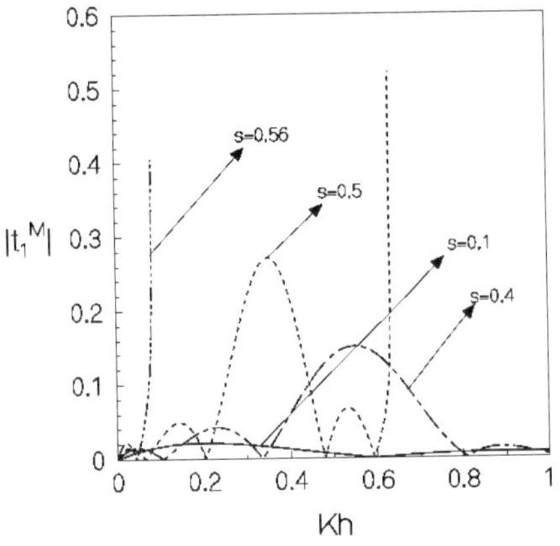

Fig. 9.18 First order transmission coefficient due to an incident wave of mode M; $a/h = 0.1$, $H/h = 2$, α =.35, $\gamma h = 1$ and n =3

The Figs. 9.16 and 9.18 show that conversion of the incident wave energy at mode M to reflected and transmitted wave energies at mode m is limited upto appropriate cut off values of Kh for different values of the density ratio s. The Fig. 9.15 depicting $|R_1^M|$ however does not show any cut-off values of Kh, which is obvious since there is no question of energy conversion from one mode to another here. However, increase of s results is more oscillatory nature in these coefficients.

9B.2 Scattering by bottom undulations in an ice-covered two-layer fluid

The scattering of plane surface waves by bottom undulations in an ice-covered ocean modelled as a two-layer fluid consisting of a layer of fresh water of lesser density above a deep layer of salt water has been investigated by Maiti and Mandal (2008). An account of this is described in this section.

Formulation of the problem

We consider an ideal and incompressible two-layer fluid of finite depth in which the upper layer is covered by an infinite but thin layer of ice sheet, while the lower layer has an undulating bottom topography. We use cartesian coordinates with the (x, z)-plane lying in the mean position of the interface and that of the y-axis directed vertically downwards and the plane $y = -h$ coinciding with the rest position of the ice-cover. The ice-cover is modeled as a thin elastic sheet having a uniform surface density $\varepsilon\rho$, where ε is a constant having the dimension of the length, ρ being the density of fresh water in the upper layer. The upper layer $-h < y < 0$, is referred to as region I, whilst the lower layer, $0 < y < H + \delta c(x)$ is region II. The potential in region I is ψ and that of region II is ϕ. The motion is assumed to be irrotational and two-dimensional and so that both ψ and ϕ are independent of z and satisfy Laplace's equation

$$\nabla^2\psi = 0 \qquad \text{in upper layer,}$$
$$\nabla^2\varphi = 0 \qquad \text{in lower layer} \tag{B2.1}$$

where $\nabla^2 = \frac{\partial^2}{\partial x^2} + \frac{\partial^2}{\partial y^2}$.

The linearized ice-cover condition is

$$K\psi + (1 - \varepsilon K + D\frac{\partial^4}{\partial x^4})\psi_y = 0 \;\; \text{on} \;\; y = -h, \tag{B2.2}$$

where D is the flexural rigidity of the ice-cover and is defined by

$$D = \frac{Eh_0^3}{12(1-\nu^2)\rho g}, \tag{B2.3}$$

E being Young's modulus and ν being Poisson's ratio, h_0 is the very small thickness of the ice-sheet.

The linearized boundary conditions at the interface are

$$\left.\begin{aligned} &\psi_y = \varphi_y, \\ &s(K\psi + \psi_y) = K\varphi + \varphi_y \end{aligned}\right\} \;\; \text{on} \;\; y = 0 \tag{B2.4}$$

where $K = \frac{\omega^2}{g}$, $s = \rho/\rho'(<1)$, ρ' being the density of the lower fluid, the time-dependence $e^{-i\omega t}$ always being suppressed. The bottom of the lower layer is represented by $y = H + \delta c(x)$ where $c(x)$ is a bounded and continuous function describing the shape of the undulations of the ocean bed and $c(x) \to 0$ as $|x| \to \infty$, so that the ocean is of uniform finite depth far away from the undulations on either side, and $\delta(>0)$ is a small parameter giving a measure of the smallness of the undulations.

The bottom condition is

$$\varphi_n = 0 \;\; on \;\; y = H + \delta c(x) \tag{B2.5}$$

where $\frac{\partial}{\partial n}$ denotes the normal derivative to the surface $y = H + \delta c(x)$. This can be approximated as

$$-\varphi_y + \delta\frac{\partial}{\partial x}\{c(x)\varphi_x(x,H)\} + O(\delta^2) = 0. \tag{B2.5$'$}$$

In a two-layer fluid with $-h < y < 0$ and $0 < y < H$ as the two layers, if a time-harmonic progressive wave with wavenumber u exists, then u satisfies the dispersion equation

$$\Delta(u) \equiv u^2(1-s)(Du^4 + 1 - \varepsilon K) - uK\{(Du^4 + 1 - \varepsilon K)(s\coth uh + \coth uH) + (1-s)\coth uh\}$$
$$+ K^2(s + \coth uh \coth uH) = 0. \tag{B2.6}$$

Thus dispersion equation can be shown to have only two real positive roots m and M, say, where $K < m < M$. Waves of wavenumber (or mode) m propagates at the ice-cover while that of wavenumber M propagates at the interface.

When a progressive plane wave of mode m is incident from the direction of $x = -\infty$ to the bottom undulations on the two-layer ocean, then $\psi(x, y)$ and $\phi(x, y)$ satisfy the infinity requirements described by

$$\psi(x,y) \to f(m,y)\left(e^{imx} + r^m e^{-imx}\right) + R^m f(M,y)e^{-iMx} \quad \text{as } x \to -\infty, \tag{B2.7}$$

$$\varphi(x,y) \to \cosh m(H-y)(e^{imx} + r^m e^{-imx}) + R^m \cosh M(h-y)e^{-iMx} \quad \text{as } x \to -\infty \tag{B2.8}$$

and

$$\psi(x,y) \to t^m f(m,y)e^{imx} + T^m f(M,y)e^{iMx} \text{ as } x \to \infty, \tag{B2.9}$$

$$\varphi(x,y) \to t^m \cosh m(H-y)e^{imx} + T^m \cosh M(H-y)e^{iMx} \text{ as } x \to \infty. \tag{B2.10}$$

where

$$f(u,y) = -\frac{\sinh uH}{K\cosh uh - u\sinh uh\{1-\varepsilon K + Du^4\}}$$
$$\times\{K\sinh u(h+y) - u(Du^4 + 1 - \varepsilon K)\cosh u(h+y)\}. \tag{B2.11}$$

In (B2.7) and (B2.8) the constants r^m, R^m denote respectively the reflection coefficients associated with reflected waves of modes m and M respectively and similarly in (2.9) and (2.10), t^m and T^m denote transmission coefficients associated with transmitted waves of modes m and M respectively due to an incident wave of mode m.

We also note that in the absence of any undulation at the bottom, the interface wave train propagates without any hindrance and there is total transmission. In view of this and of the approximate condition (B2.5') we can assume a perturbation expansion for ψ, ϕ, r^m, $R^m t^m$, T^m in terms of δ as

$$\begin{aligned}
\psi(x,y) &= \psi_0(x,y) + \delta\psi_1(x,y) + O(\delta^2),\\
\phi(x,y) &= \phi_0(x,y) + \delta\phi_1(x,y) + O(\delta^2),\\
t^m &= 1 + \delta t_1^m + O(\delta^2),\\
T^m &= \delta T_1^m + O(\delta^2),\\
r^m &= \delta r_1^m + O(\delta^2),\\
R^m &= \delta R_1^m + O(\delta^2).
\end{aligned} \tag{B2.12}$$

Substituting (B2.12) in equations (B2.1)–(B2.4) and (B2.5') we find that ψ_1, ϕ_1, satisfy the following coupled BVP described by following coupled BVP described by

$$\left.\begin{aligned}
&\nabla^2\psi_1 = 0 && -h < y < 0,\\
&\nabla^2\varphi_1 = 0 && 0 < y < H,\\
&K\psi_1 + (1-\varepsilon K + D\frac{\partial^4}{\partial x^4})\psi_{1y} = 0 && \text{on } y = -h,\\
&s(K\psi_1 + \psi_{1y}) = K\varphi_1 + \varphi_{1y} && \text{on } y = 0,\\
&\psi_{1y} = \varphi_{1y} && \text{on } y = 0,\\
&\varphi_{1y} = q(x) && \text{on } y = H
\end{aligned}\right\} \tag{B2.13}$$

where

$$q(x) = im\frac{d}{dx}\{c(x)e^{imx}\},$$

and the infinity requirements

$$\psi_1(x, y) \to r_1^m f(m, y)e^{-imx} + R_1^m f(M, y)e^{-iMx} \quad \text{as} \quad x \to -\infty,$$

$$\varphi_1(x, y) \to r_1^m \cosh m(H - y)e^{-imx} + R_1^m \cosh M(H - y)e^{-iMx} \quad \text{as} \quad x \to -\infty, \quad \text{(B2.14)}$$

and

$$\psi_1(x, y) \to t_1^m f(m, y)e^{imx} + T_1^m f(M, y)e^{iMx} \quad \text{as} \quad x \to \infty,$$

$$\varphi_1(x, y) \to t_1^m \cosh m(H - y)e^{imx} + T_1^m \cosh M(H - y)e^{iMx} \quad \text{as} \quad x \to \infty. \quad \text{(B2.15)}$$

Solution of the problem

The solution of the problem described by equations (B2.13) for the potentials ψ_1 and ϕ_1 are obtained by using Fourier transform technique.

To solve for ψ_1, ϕ_1 we decouple the above BVP by replacing the condition $(B2.13)_5$ with

$$\psi_{1y} = p(x) \qquad \text{on} \qquad y = 0, \qquad \text{(B2.16)}$$

and

$$\varphi_{1y} = p(x) \qquad \text{on} \qquad y = 0. \qquad \text{(B2.17)}$$

where $p(x)$ is an unknown function.

We now assume that m and M to have a small positive imaginary part so that ϕ_1, ψ_1 decrease exponentially as $|x| \to \infty$. This ensures the existence of Fourier transforms $\Psi_1(k, y)$ and $\Phi_1(k, y)$ of $\psi_1(x, y)$ and $\phi_1(x, y)$ respectively, defined by

$$\Psi_1(k, y) = \int_{-\infty}^{\infty} \psi_1(x, y)e^{-ikx}dx, \Phi_1(k, y) = \int_{-\infty}^{\infty} \varphi_1(x, y)e^{-ikx}dx. \qquad \text{(B2.18)}$$

Now $\Psi_1(k, y)$ satisfies the BVP

$$\left.\begin{aligned} &\Psi_{1yy} - k^2\Psi_1 = 0, && -h < y < 0, \\ &Dk^4 + (1 - \varepsilon K)\Psi_{1y} + K\Psi_1 = 0 && \text{on} \quad y = -h, \\ &\Psi_{1y} = \overline{p}(k) && \text{on} \quad y = 0 \end{aligned}\right\} \qquad \text{(B2.19)}$$

while $\Phi_1(k, y)$ satisfies

$$\left.\begin{aligned} &\Phi_{1yy} - k^2\Phi_1 = 0, && 0 < y < H, \\ &\Phi_{1y} = \overline{p}(k) && \text{on} \quad y = 0, \\ &\Phi_{1y} = \overline{q}(k) && \text{on} \quad y = H \end{aligned}\right\} \qquad \text{(B2.20)}$$

where $\overline{p}(k)$ and $\overline{q}(k)$ are the Fourier transforms of $p(x)$ and $q(x)$ respectively.

Then $\Psi_1(k, y)$ has the solution given by

$$\Psi_1(k,y) = -\frac{K\bar{q}(k)\{k\cosh k(y+h)(Dk^4+1-\varepsilon K) - K\sinh k(y+h)\}}{k\sinh kh \sinh kH\,\Delta(k)}, \quad -h<y<0, \tag{B2.21}$$

$$\Phi_1(k,y) = \frac{\bar{q}(k)}{k\sinh kH}\left[\cosh ky - \frac{K\{K\coth kh - k(Dk^4+1-\varepsilon K)\}}{\sinh kH\,\Delta(k)}\cosh k(H-y)\right], 0<y<H. \tag{B2.22}$$

We note that in the complex k-plane, Ψ_1 and Φ_1 have poles at the zeros of $\Delta(k)$. Using inverse Fourier transform

$$(\psi_1, \varphi_1) = \frac{1}{2\pi}\int_{-\infty}^{\infty}(\Psi_1, \Phi_1)e^{ikx}dk \tag{B2.23}$$

and noting the fact that $\Delta(k) = \Delta(-k)$, we obtain

$$\psi_1(x,y) = -\frac{K}{2\pi}\int_0^{\infty}\frac{k\cosh k(y+h)(Dk^4+1-\varepsilon K) - K\sinh k(y+h)}{k\sinh kh\sinh kH\,\Delta(k)}$$
$$\left\{\bar{q}(k)e^{ikx} + \bar{q}(-k)e^{-ikx}\right\}dk, \; -h<y<0, \tag{B2.24}$$

$$\varphi_1(x,y) = \frac{1}{2\pi}\int_0^{\infty}\frac{1}{k\sinh kH}\left[\cosh ky - \frac{K\{K\coth kh - k(Dk^4+1-\varepsilon K)\}\cosh k(H-y)}{\sinh kH\,\Delta(k)}\right]$$
$$\left\{\bar{q}(k)e^{ikx} + \bar{q}(-k)e^{-ikx}\right\}dk, 0<y<H, \tag{B2.25}$$

where the path in each integral is indented below the poles at $k = m, M$.

Now the first order reflection and transmission coefficients r_1^m, R_1^m and t_1^m, T_1^m can be found by comparing the behaviors of $\psi_1(x, y)$ or $\phi_1(x, y)$ as $x \to \mp\infty$ obtained from (B2.24) or (B2.25) using (B2.14) and (B2.15). To find the behaviors as $x \to \infty$, we rotate the contour in the integrals involving e^{ikx} in the first quadrant by an angle β $(0<\beta<\frac{\pi}{2})$ and the contour in the integrals involving e^{-ikx} in the fourth quadrant by the same angle β. As $x \to \infty$, the integral involving e^{ikx} will only contribute a term arising from the residues at $k = m, M$, while there will be no contribution from the integral involving e^{-ikx}. Thus we find

$$\psi_1(x,y) \to -iK\frac{m\cosh m(y+h)(Dm^4+1-\varepsilon K) - K\sinh m(y+h)}{m\sinh mh\sinh mH\Delta'(m)}\bar{q}(m)e^{imx}$$

$$-iK\frac{M\cosh M(y+h)(Dm^4+1-\varepsilon K) - K\sinh M(y+h)}{M\sinh Mh\sinh MH\,\Delta'(M)}\bar{q}(M)e^{iMx} \quad \text{as } x\to\infty, \tag{B2.26}$$

and

$$\phi_1(x,y) \to -iK\frac{\{K\coth mh - m(Dm^4+1-\varepsilon K)\}\bar{q}(m)}{m\sinh^2 mH\Delta'(m)}\cosh m(H-y)e^{imx}$$

$$-iK\frac{\{K\coth Mh - M(DM^4+1-\varepsilon K)\}\bar{q}(M)}{M\sinh^2 MH\Delta'(M)}\cosh M(H-y)e^{iMx} \quad \text{as } x\to\infty. \tag{B2.27}$$

Using

$$\bar{q}(m) = -m^2\int_{-\infty}^{\infty} c(x)dx, \quad \bar{q}(M) = -mM\int_{-\infty}^{\infty} c(x)e^{-i(M-m)x}dx \tag{B2.28}$$

we find that, after comparing (B2.26) (or(B2.27)) with (B2.15), t_1^m, T_1^m can be obtained as

$$t_1^m = \frac{iKm\{K\cosh mh - m\sinh mh(Dm^4+1-\varepsilon K)\}}{\sinh mh \sinh^2 mH\Delta'(m)}\int_{-\infty}^{\infty} c(x)dx, \tag{B2.29}$$

$$T_1^m = \frac{iKm\{K\cosh Mh - M\sinh Mh(DM^4+1-\varepsilon K)\}}{\sinh Mh \sinh^2 MH\Delta'(M)}\int_{-\infty}^{\infty} c(x)e^{-i(M-m)x}dx. \tag{B2.30}$$

In a similar manner r_1^m and R_1^m are obtained from the analysis of the behavior of ψ_1 or ϕ_1 in equations (B2.24) or (B2.25) as $x\to-\infty$ by rotating the path of the second integrals into a contour in the first quadrant, so that we must include the residue term at $k=m, M$ and then comparing with (B2.14), the expressions for r_1^m and R_1^m are obtained as

$$r_1^m = -\frac{iKm\{K\cosh mh - m\sinh mh(Dm^4+1-\varepsilon K)\}}{\sinh mh \sinh^2 mH\Delta'(m)}\int_{-\infty}^{\infty} e^{2imx}c(x)dx, \tag{B2.31}$$

$$R_1^m = -\frac{iKm\{K\cosh Mh - M\sinh Mh(DM^4+1-\varepsilon K)\}}{\sinh Mh \sinh^2 MH\Delta'(M)}\int_{-\infty}^{\infty} e^{i(m+M)x}c(x)dx. \tag{B2.32}$$

Similarly, when an incident wave of wave number M is incident on the bottom topography, the coefficients t_1^M, T_1^M and r_1^M, R_1^M are obtained as

$$t_1^M = \frac{iKM\{K\cosh mh - m\sinh mh(Dm^4+1-\varepsilon K)\}}{\sinh mh \sinh^2 mH\Delta'(m)}\int_{-\infty}^{\infty} e^{i(M-m)x}c(x)dx, \tag{B2.33}$$

$$T_1^M = \frac{iKM\{K\cosh Mh - M\sinh Mh(Dm^4+1-\varepsilon K)\}}{\sinh Mh \sinh^2 MH\Delta'(M)}\int_{-\infty}^{\infty} c(x)dx, \tag{B2.34}$$

$$r_1^M = -\frac{ikM\{K\cosh mh - m\sinh mh(Dm^4+1-\varepsilon K)\}}{\sinh mh \sinh^2 mH\Delta'(m)}\int_{-\infty}^{\infty} e^{i(m+M)x}c(x)dx, \tag{B2.35}$$

$$R_1^M = -\frac{iKM\{K\cosh Mh - M\sinh Mh(DM^4+1-\varepsilon K)\}}{\sinh Mh \sinh^2 MH\Delta'(M)}\int_{-\infty}^{\infty} e^{2iMx}c(x)dx. \tag{B2.36}$$

The BVP described by (B2.13) can be solved by employing Green's integral theorem after constructing the appropriate Green's functions in the two layers, and the same results for the first order reflection and transmission coefficients are obtained.

Particular case

We now consider the interaction of progressive waves of two different wavenumbers m and M with a patch of sinusoidal ripples on the bed. Davies (1982) earlier made a somewhat elaborate study in a single layer fluid, on the effect of sinusoidal undulations on the bed of an ocean with a free surface because of its importance in respect of coastal protection, and of possible ripple growth if the bed is erodible. We consider here the function $c(x)$ in the form

$$c(x) = \begin{cases} a \sin \gamma x & \text{for} -\dfrac{n\pi}{\gamma} \le x \le \dfrac{n\pi}{\gamma}, \\ 0 & \text{otherwise.} \end{cases} \tag{B2.37}$$

Since $c(x)$ is an odd function, t_1^m, T_1^M vanish identically. The other coefficients are given as

$$r_1^m = (-1)^{n+1} \frac{Km\{K \cosh mh - m \sinh mh(Dm^4 + 1 - \varepsilon K)\}}{\sinh mh \sinh^2 mH \Delta'(m)} \frac{2a\gamma \sin \dfrac{2nm\pi}{\gamma}}{\gamma^2 - 4m^2}, \tag{B2.38}$$

$$R_1^m = (-1)^{n+1} \frac{Km\{K \cosh Mh - M \sinh Mh(DM^4 + 1 - \varepsilon K)\}}{\sinh Mh \sinh^2 MH \Delta'(M)} \frac{2a\gamma \sin \dfrac{n(m+M)\pi}{\gamma}}{\gamma^2 - (m+M)^2}, \tag{B2.39}$$

$$T_1^m = (-1)^{n} \frac{Km\{K \cosh Mh - M \sinh Mh(DM^4 + 1 - \varepsilon K)\}}{\sinh Mh \sinh^2 MH \Delta'(M)} \frac{2a\gamma \sin \dfrac{n(M-m)\pi}{\gamma}}{\gamma^2 - (M-m)^2}, \tag{B2.40}$$

$$r_1^M = (-1)^{n+1} \frac{KM\{K \cosh mh - m \sinh mh(Dm^4 + 1 - \varepsilon K)\}}{\sinh mh \sinh^2 mH \Delta'(m)} \frac{2av \sin \dfrac{n(m+M)\pi}{\gamma}}{\gamma^2 - (m+M)^2}, \tag{B2.41}$$

$$R_1^M = (-1)^{n+1} \frac{KM\{K \cosh Mh - M \sinh Mh(DM^4 + 1 - \varepsilon K)\}}{\sinh Mh \sinh^2 MH \Delta'(M)} \frac{2a\gamma \sin \dfrac{2nM\pi}{\gamma}}{v^2 - 4M^2}, \tag{B2.42}$$

$$t_1^M = (-1)^{n} \frac{KM\{K \cosh mh - m \sinh mh(Dm^4 + 1 - \varepsilon K)\}}{\sinh mh \sinh^2 mH \Delta'(m)} \frac{2a\gamma \sin \dfrac{n(M-m)\pi}{\gamma}}{v^2 - (M-m)^2}. \tag{B2.43}$$

Numerical results

The expressions (B2.38) to (B2.43) are evaluated numerically for various values of the parameters. In Figs. 9.19 to 9.21 $|r_1^m|$, $|R_1^m|$, $|T_1^m|$ are plotted against Kh for three different values of the number of ripples n, and $a/h = 0.1$, $H/h = 2$, $s = 0.5$, $\gamma h = 1$, $D/h^4 = 1$ and $\varepsilon/h = .01$. In each of these figures the peak value increases with the number of ripples. Thus if the number of ripples increases indefinitely, the first order coefficients

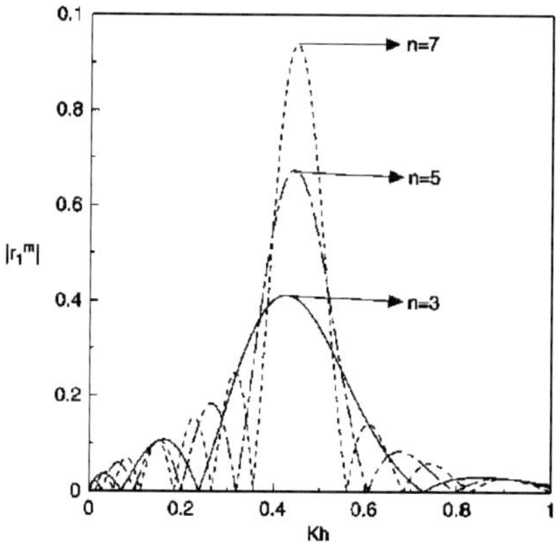

Fig. 9.19 First-order reflection coefficient at mode m due to an incident wave of mode m, $a/h = 0.1$, $H/h = 2$, $s = 0.5$, $\gamma h = 1$ and $D/h^4 = 1$, $\varepsilon/h = 0.01$

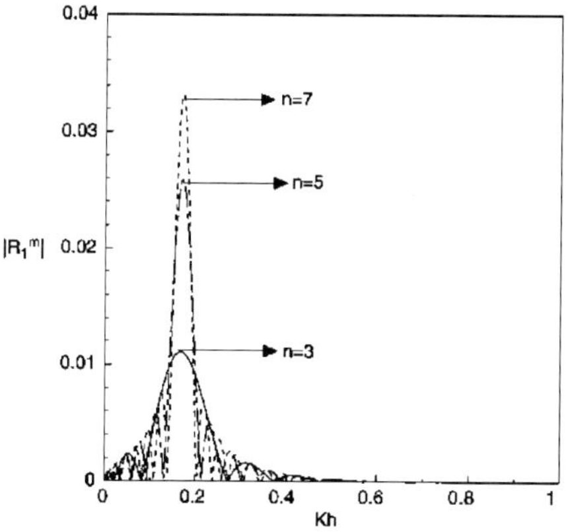

Fig. 9.20 First-order reflection coefficient at mode M due to an incident wave of mode m, $a/h = 0.1$, $H/h = 2$, $s = 0.5$, $\gamma h = 1$ and $D/h^4 = 1$, $\varepsilon/h = 0.01$

will become unbounded for certain values of Kh. This is what is known as Bragg resonance which occurs when the reflection coefficient for a single-layer fluid becomes much larger than the small parameter δ (cf. Mei (1985)). In this case the perturbation expansion (B2.12) is not valid. One feature that is common to all the three figures is the oscillating nature of the absolute values of the first order coefficients as a function of the wave number Kh. In Figs. 9.22 to 9.24 the first order reflection and transmission coefficients for incident wave train of wave number m are plotted against Kh for $a/h = 0.1$, $H/h = 2$, $s = 0.5$, $\gamma h = 1$, $n = 3$. In each of these figures, the results for different values of ice-cover parameter D/h^4 are plotted. In Figs. 9.23 and 9.24 we observe that peak values of the first-order reflection and transmission coefficient due to an incident wave of mode m increases with the increase of the flexural rigidity of the ice cover.

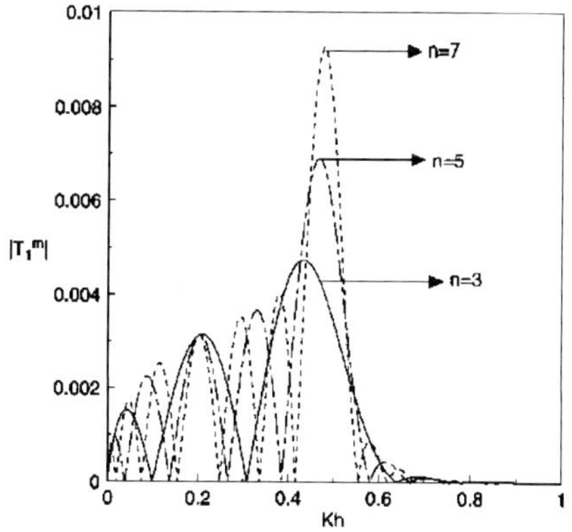

Fig. 9.21 First-order transmission coefficient at mode M due to an incident wave of mode m, $a/h = 0.1$, $H/h = 2$, $s = 0.5$, $\gamma h = 1$ and $D/h^4 = 1$, $\varepsilon/h = 0.01$

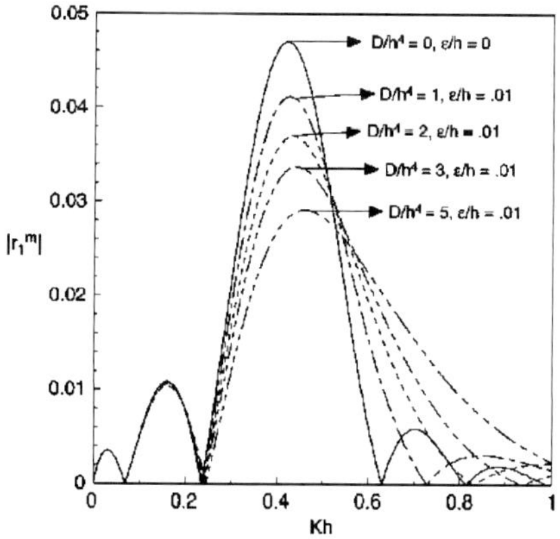

Fig. 9.22 First-order reflection coefficient at mode m due to an incident wave of mode m, $a/h = 0.1$, $H/h = 2$, $s = 0.5$, $\gamma h = 1$ and $n = 3$

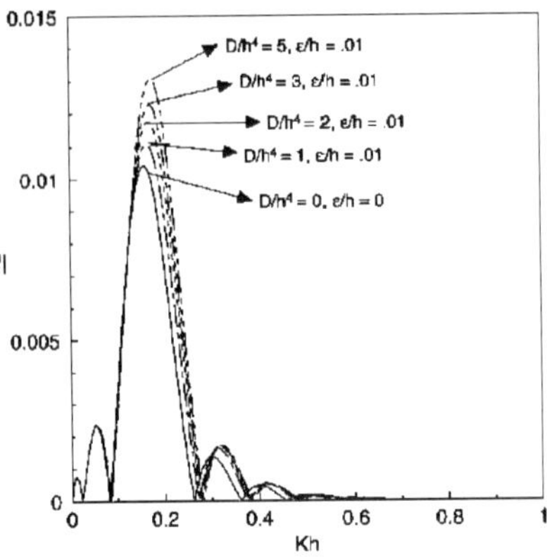

Fig. 9.23 First-order reflection coefficient at mode M due to an incident wave of mode m, $a/h = 0.1$, $H/h = 2$, $s = 0.5$, $\gamma h = 1$ and $n = 3$

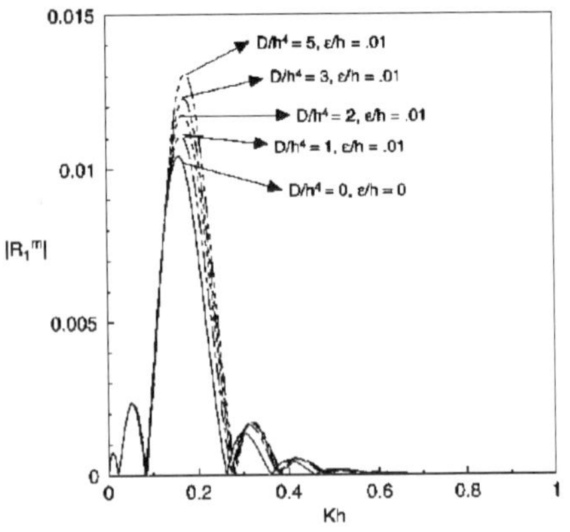

Fig. 9.24 First-order transmission coefficient at mode M due to an incident wave of mode m, $a/h = 0.1$, $H/h = 2$, $s = 0.5$, $\gamma h = 1$ and $n = 3$

In Figs. 9.25 to 9.27 the reflection and transmission coefficients $|r_1^m|$, $|R_1^m|$ and $|T_1^m|$ are depicted against Kh for different values of the density ratio s for $a/h = 0.1$, $H/h = 2$, $D/h^4 = 1$, $\varepsilon/h = .01$, $\gamma h = 1$, $n = 3$. In each of these three figures it is observed that as s increases the peak values of $|r_1^m|$, $|R_1^m|$ decrease, while peak value of $|T_1^m|$ increase, so that the first order coefficients are quite sensitive to the density ratio due to an incident wave at mode m.

In Figs. 9.28 to 9.30, $|r_1^M|$, $|R_1^M|$, $|t_1^M|$ are plotted against Kh for the same set of values of different dimensionless parameters. In these figures it is observed that the highest peak values increase with the increase of n, the number of patches of the sinusoidal undulation, as was observed for $|r_1^m|$, $|R_1^m|$, $|t_1^m|$. Here also, if the number of ripples increases indefinitely, then the first order coefficients due to an incident wave of mode M will become unbounded for certain values of Kh as was observed earlier for an incident wave of mode m. The same remark as made earlier also applies here.

In Figs. 9.31 to 9.32, it is observed that for different values of the parameter D/h^4, the curves for $|r_1^M|$, $|R_1^M|$, are almost same. This shows that these first-order corrections to the reflection coefficients are somewhat insensitive to the changes in the flexural rigidity of the ice-cover. This may be attributed to the fact that, since the ice-cover is somewhat above the interface, the interface wave is not much affected by the changes in the flexural rigidity of the ice-cover. Thus the ice-cover has not much effect on the first-order reflection coefficients due to a train of waves of mode M propagating at the interface. The Fig. 9.33 shows that the peak values of $|t_1^M|$ decrease with the increase of D/h^4. Thus the flexural rigidity of the ice cover affects the first-order transmission coefficient considerably.

In Figs. 9.34 to 9.36, $|r_1^M|$, $|R_1^M|$ and $|t_1^M|$ are depicted against Kh for different values of the density ratios s. These figures show that the density ratio is quite sensitive to the first order reflection and transmission coefficients due to an incident wave at mode M.

Choosing other forms of the shape function $c(x)$ such an exponentially decaying undulation or a Gaussion type undulation, the corresponding first-order corrections to the reflection and transmission coefficients can be obtained. The problem can also

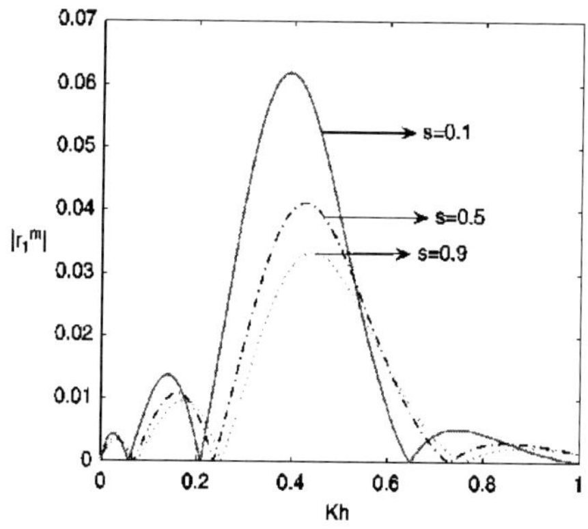

Fig. 9.25 First-order reflection coefficient at mode *m* due to an incident wave of mode *m*, $a/h = 0.1$, $H/h = 2$, $n = 3$, $\gamma h = 1$ and $D/h^4 = 1$, $\varepsilon/h = 0.01$

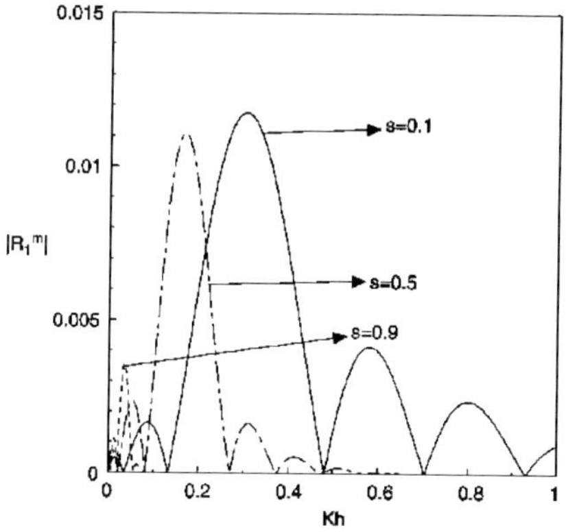

Fig. 9.26 First-order reflection coefficient at mode *M* due to an incident wave of mode *m*, $a/h = 0.1$, $H/h = 2$, $n = 3$, $\gamma h = 1$ and $D/h^4 = 1$, $\varepsilon/h = 0.01$

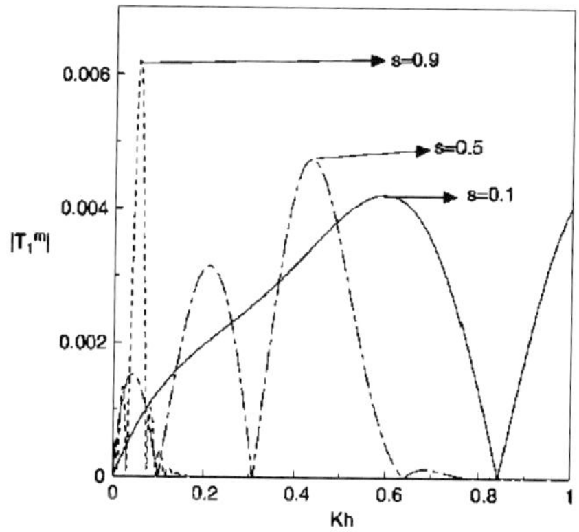

Fig. 9.27 First-order transmission coefficient at mode *M* due to an incident wave of mode *m*, $a/h = 0.1$, $H/h = 2$, $n = 3$, $\gamma h = 1$ and $D/h^4 = 1$, $\varepsilon/h = 0.01$

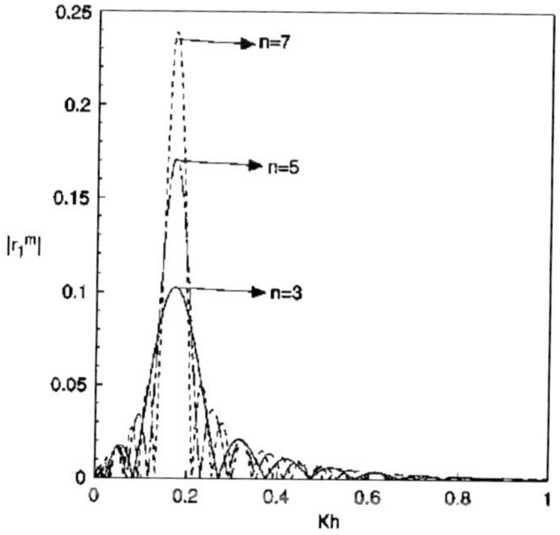

Fig. 9.28 First-order reflection coefficient at mode m due to an incident wave of mode M, $a/h = 0.1$, $H/h = 2$, $s = 0.5$, $\gamma h = 1$ and $D/h^4 = 1$, $\varepsilon/h = 0.01$

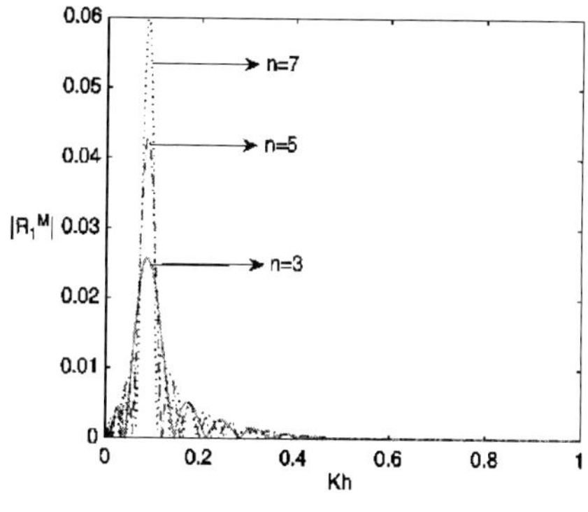

Fig. 9.29 First-order reflection coefficient at mode *M* due to an incident wave of mode M, $a/h = 0.1$, $H/h = 2$, $s = 0.5$, $\gamma h = 1$ and $D/h^4 = 1$, $\varepsilon/h = 0.01$

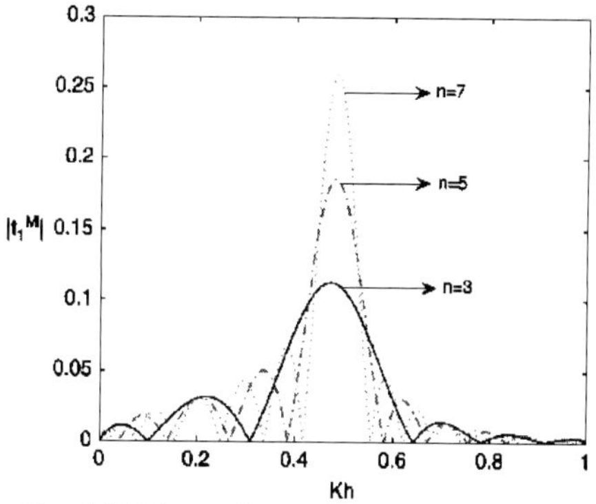

Fig. 9.30 First-order transmission coefficient at mode m due to an incident wave of mode M, $a/h = 0.1$, $H/h = 2$, $s = 0.5$, $\gamma h = 1$ and $D/h^4 = 1$, $\varepsilon/h = 0.01$

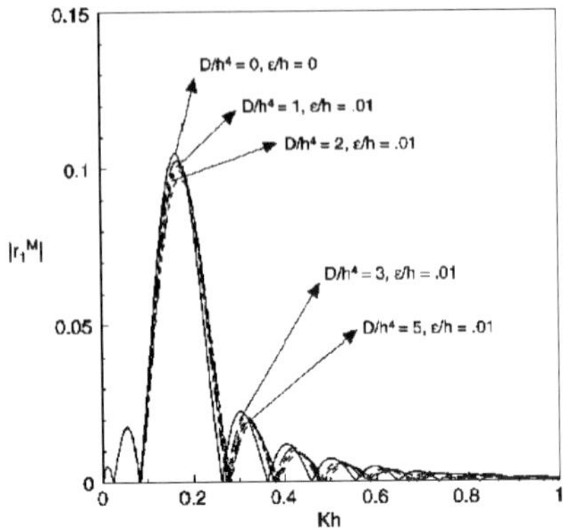

Fig. 9.31 First-order reflection coefficient at mode m due to an incident wave of mode M, $a/h = 0.1$, $H/h = 2$, $s = 0.5$, $\gamma h = 1$ and $n = 3$, $\varepsilon/h = 0.01$

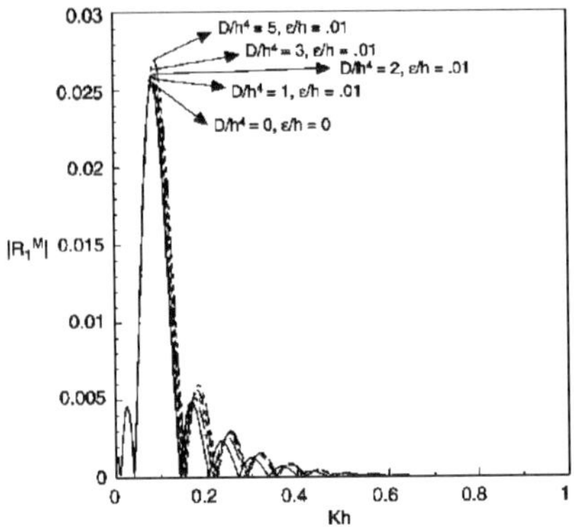

Fig. 9.32 First-order reflection coefficient at mode M due to an incident wave of mode M, $a/h = 0.1$, $H/h = 2$, $s = 0.5$, $\gamma h = 1$ and $n = 3$

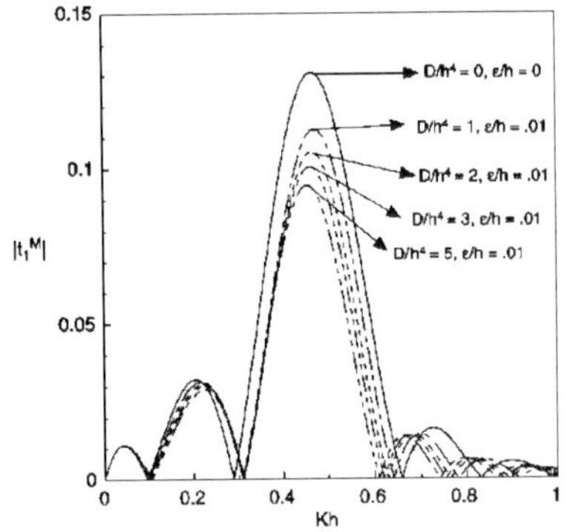

Fig. 9.33 First-order transmission coefficient at mode m due to an incident wave of mode M, $a/h = 0.1$, $H/h = 2$, $s = 0.5$, $\gamma h = 1$ and $n = 3$

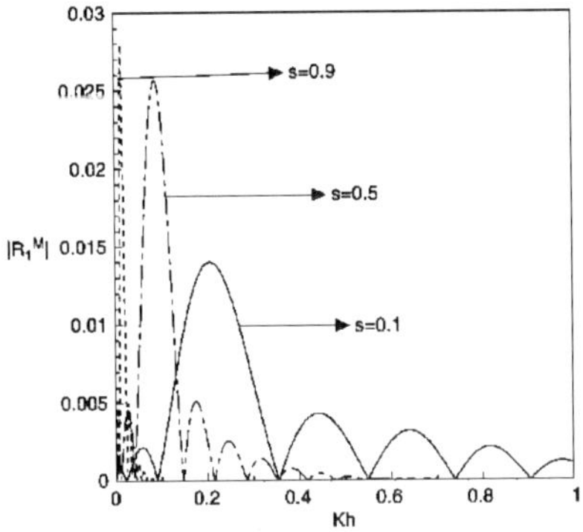

Fig. 9.34 First-order reflection coefficient at mode *m* due to an incident wave of mode *M*, $a/h = 0.1$, $H/h = 2$, $D/h^4 = 1$, $\gamma h = 1$ and $n = 3$, $\varepsilon/h = 0.01$

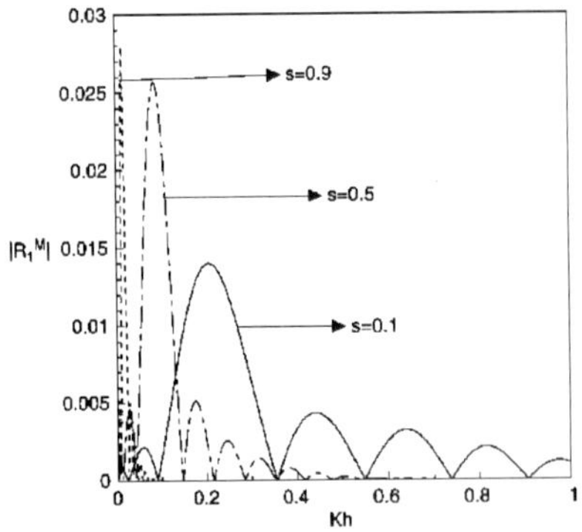

Fig. 9.35 First-order transmission coefficient at mode *M* due to an incident wave of mode *M*, $a/h = 0.1$, $H/h = 2$, $D/h^4 = 1$, $\gamma h = 1$ and $n = 3$, $\varepsilon/h = 0.01$

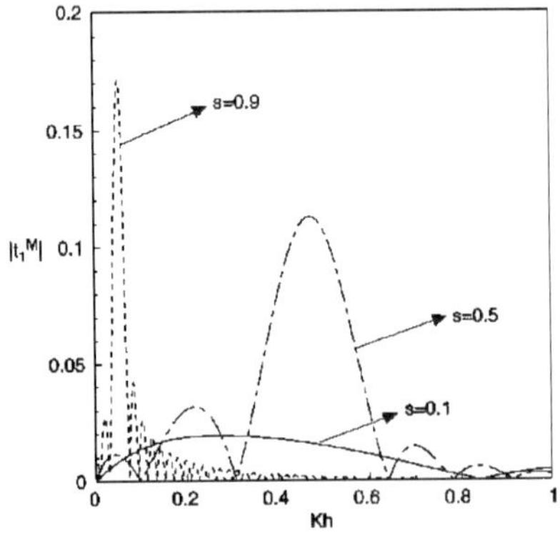

Fig. 9.36 First-order reflection coefficient at mode *m* due to an incident wave of mode M, $a/h = 0.1$, $H/h = 2$, $D/h^4 = 1$, $\gamma h = 1$ and $n = 3$, $\varepsilon/h = 0.01$

be treated by employing variational techniques such as the mild-slope approximation which has been used in a number of problems in a single-layer fluid with variable bottom topography. Also, as in Porter and Porter (2004), one can consider the ice-cover to be of variable thickness.

C. BOTTOM UNDULATIONS IN THE PRESENCE OF OBSTACLE

9C.1 Scattering by bottom undulations in the presence of a thin vertical barrier

In sections 9A and 9B we have considered bottom undulations as the only hindrance to the propagation of surface gravity waves. The additional effect of the presence of an obstacle in the form of a thin vertical rigid plate, either partially immersed or completely submerged, on the propagation of surface waves has been investigated by Mandal and Gayen (2006) and Mandal and De (2006) by using the perturbation analysis employed in the previous sections. An account of the problem considered by Mandal and Gayen (2006) is given in this section.

Formulation of the problem

We consider water of finite depth having small undulations at the bottom. A Cartesian co-ordinate system is chosen such that the xz-plane coincides with the undisturbed free surface, y-axis being taken vertically downwards into the fluid occupying the region described by $-\infty < x, z < \infty$, $0 < y < h + \varepsilon c(x)$. Here $c(x)$ is a continuous bounded function describing the shape of the bottom, $c(x) \to 0$ as $|x| \to \infty$ and ε is a very small dimensionless parameter giving a measure of smallness of the bottom undulations. Let a thin vertical plate be partially immersed upto a depth a below the mean free surface, and the plate whose position is described by $x = 0$, $0 < y < a$, be very long in the z direction so that the problem of ensuing motion due to a surface wave train incident normally on the plate, is two-dimensional and depends on x, y only. The incident wave train propagating from the direction of negative infinity is partially reflected by the plate and the bottom undulations and transmitted through the gap between the plate and the bottom. Assuming linear theory and irrotational motion, the velocity potential describing the fluid motion can be represented as $Re\{\phi^{inc}(x, y)e^{-i\sigma t}\}$ where σ is the circular frequency. Then $\phi(x, y)$ satisfies

$$\nabla^2\varphi = 0 \quad \text{in the fluid region,} \qquad \text{(C1.1)}$$

the free surface condition

$$K\varphi + \varphi_y = 0 \quad \text{on } y = 0, \qquad \text{(C1.2)}$$

where $K = \sigma^2/g$, g being the gravity, the plate condition

$$\varphi_x = 0 \quad \text{on } x = 0, 0 < y < a, \qquad \text{(C1.3)}$$

the bottom condition

$$\varphi_n = 0 \quad \text{on } y = h + \varepsilon c(x) \qquad \text{(C1.4)}$$

n denoting the normal derivative, the edge condition

$$r^{1/2}\nabla\varphi \quad \text{is bounded as } r = \{x^2 + (y-a)^2\}^{1/2} \to 0, \tag{C1.5}$$

and the infinity conditions

$$\varphi(x,y) \to \begin{cases} (e^{ik_0 x} + Re^{-ik_0 x})\psi_0(y) & \text{as } x \to -\infty, \\ Te^{ik_0 x}\psi_0(y) & \text{as } x \to \infty \end{cases} \tag{C1.6}$$

where

$$\psi_0(y) = N_0^{-1/2} \cosh k_0(h-y) \tag{C1.7}$$

with

$$N_0 = \frac{2k_0 h + \sinh 2k_0 h}{2k_0 h},$$

k_0 being the real positive root of the transcendental equation

$$k \tanh kh = K. \tag{C1.8}$$

In (C1.6), R and T denote respectively the unknown reflection and transmission coefficients. The main concern here is to find the coefficients approximately.

The bottom condition (C1.4) can be expressed approximately as

$$\varphi_y - \varepsilon \frac{d}{dx}\{c(x)\varphi_x\} + 0(\varepsilon^2) = 0 \quad \text{on} \quad y = h. \tag{C1.9}$$

This suggests that a perturbation technique can be employed to solve the BVP described by (C1.1) to (C1.6) approximately. This is described now.

Method of solution

The approximate boundary condition (C1.9) suggests that ϕ, R, T can be expanded in terms of ε as given by

$$\left.\begin{aligned} \varphi(x,y;\varepsilon) &= \varphi_0 + \varepsilon\varphi_1 + O(\varepsilon^2), \\ R(\varepsilon) &= R_0 + \varepsilon R_1 + O(\varepsilon^2), \\ T(\varepsilon) &= T_0 + \varepsilon T_1 + O(\varepsilon^2). \end{aligned}\right\} \tag{C1.10}$$

Substituting the expansions (C1.10) in (C1.1), (C1.2),(C1.3), (C1.5), (C1.6) and (C1.9) we find after equating the coefficients of ε^0 and ε from both sides, that the functions $\phi_0(x, y)$ and $\phi_1(x, y)$ satisfy the following BVPs:

BVP–I: The function $\phi_0(x, y)$ satisfies

$$\nabla^2\phi_0 = 0, 0 < y < h,$$

$$K\phi_0 + \phi_{0y} = 0,\ y = 0,$$

$$\phi_{0x} = 0, x = 0, 0 < y < a,$$

$$\phi_{0y} = 0, y = h,$$

$$r^{1/2}\nabla\phi_0 \text{ is bounded as } r \to 0,$$

$$\varphi_0(x, y) \to \begin{cases} (e^{ik_0x} + R_0 e^{-ik_0x})\psi_0(y) & \text{as } x \to -\infty \\ T_0 e^{ik_0x}\psi_0(y) & \text{as } x \to \infty. \end{cases} \tag{C1.11}$$

BVP–II: The function $\phi_1(x, y)$ satisfies

$$\nabla^2\phi_1 = 0, 0 < y < h,$$

$$K\phi_1 + \phi_{1y} = 0, y = 0,$$

$$\phi_{1x} = 0, x = 0, 0 < y < a,$$

$$\phi_{1y} = \frac{d}{dx}\{c(x)\phi_{0x}\}, y = h,$$

$$r^{1/2}\nabla\phi_1 \text{ is bounded as } r \to 0,$$

$$\varphi_1(x, y) \to \begin{cases} R_1 e^{-ik_0x}\psi_0(y) & \text{as } x \to -\infty, \\ T_1 e^{ik_0x}\psi_0(y) & \text{as } x \to \infty. \end{cases} \tag{C1.12}$$

It may be noted that the BVP-I corresponds to the problem of water wave scattering by a thin vertical barrier partially immersed in water of uniform finite depth h. This has been solved in the literature approximately in the sense that numerical estimates for R_0 and T_0 have been obtained by Losada et al. (1996), Mandal and Dolai (1994) and Porter and Evans (1995).

The BVP-II is a radiation problem in water of uniform finite depth h, in which, the bottom condition involves ϕ_0, the solution of BVP-I. Without solving $\phi_1(x, y)$ explicitly, R_1 and T_1 can be obtained in terms of integrals involving the shape function $c(x)$ and $\phi_{0x}(x, h)$. To show this, we apply Green's integral theorem to the functions $\phi_0(x, y)$ and $\phi_1(x, y)$ in the region bounded by the lines $y = 0$, $0 < x \le X$; $x = X$, $0 \le y \le h$; $y = h$, $-X \le x \le X$; $x = -X$, $0 \le y \le h$; $y = 0$, $-X < x < 0$; $x = 0+$, $0 \le y \le a$; $x = 0-$, $0 \le y \le a$ where X is large and positive, and ultimately make X to tend to infinity. This produces

$$2ik_0R_1 = \int_{-\infty}^{\infty} c(x)\varphi_{0x}^2(x, h)dx. \tag{C1.13}$$

Similarly, applying Green's integral theorem to $\chi_0(x, y) = \phi_0(-x, y)$ and $\phi_1(x, y)$ in the same region and making $X \to \infty$, we obtain

$$2ik_0T_1 = -\int_{-\infty}^{\infty} c(x)\varphi_{0x}(x, h)\varphi_{0x}(-x, h)dx. \tag{C1.14}$$

Thus both R_1 and T_1 are obtained in terms of integrals involving the shape function $c(x)$ and the zero-order potential function $\phi_0(x, y)$. Unfortunately $\phi_0(x, y)$ cannot be obtained analytically. However it can be expressed as

$$\varphi_0(x, y) = \begin{cases} (e^{ik_0x} + R_0 e^{-ik_0x})\psi_0(y) + \sum_{n=1}^{\infty} A_n e^{k_nx}\psi_n(y), x < 0, \\ T_0 e^{ik_0x}\psi_0(y) + \sum_{n=1}^{\infty} B_n e^{-k_nx}\psi_n(y), x > 0 \end{cases} \tag{C1.15}$$

where $\pm ik_n$ $(n = 1,2,...)$ are the purely imaginary roots of (C1.8), A_n, B_n $(n = 1,2,...)$ are unknown constants,

$$\psi_n(y) = N_n^{-1/2} \cos k_n(h-y)$$

with

$$N_n = \frac{2k_n h + \sin 2k_n h}{4k_n}. \tag{C1.16}$$

It can be shown that $R_0 = 1 - T_0$ and $A_n = -B_n$ $(n = 1,2,...)$ R_0 (and hence T_0) can be estimated numerically by using multi-term Galerkin approximations employed by Porter and Evans (1995). The same method can be used to estimate numerically the constants A_n $(n = 1,2,...)$. The details are given below.
Let

$$f(y) = \varphi_0(+0,y) - \varphi_0(-0,y) \text{ and } g(y) = \frac{\partial \varphi_0}{\partial x}(0,y), 0 < y < h. \tag{C1.17}$$

Then

$$f(y) = 0 \text{ for } a < y < h \text{ and } g(y) = 0 \text{ for } 0 < y < a.$$

Using (C1.15) in the definition of $g(y)$, we obtain

$$g(y) = ik_0(1-R_0)\psi_0(y) + \sum_{n=1}^{\infty} k_n A_n \psi_n(y), \quad 0 < y < h$$

and

$$g(y) = ik_0 T_0 \psi_0(y) + \sum_{n=1}^{\infty} k_n B_n \psi_n(y), \quad 0 < y < h.$$

Use of Havelock inversion theorem produces

$$\left.\begin{aligned} ik_0(1-R_0) = ik_0 T_0 &= \int_a^h g(y)\psi_0(y)dy, \\ k_n A_n = -k_n B_n &= \int_a^h g(y)\psi_n(y)dy. \end{aligned}\right\} \tag{C1.18}$$

Again using (C1.15) in the definition of $f(y)$, we similarly obtain

$$2R_0 = -\int_0^a f(y)\psi_0(y)dy, 2A_n = -\int_0^a f(y)\psi_n(y)dy. \tag{C1.19}$$

If we define

$$F(y) = \frac{f(y)}{2ik_0(1-R_0)}, 0 < y < a, \text{and } G(y) = -\frac{g(y)}{R_0}, a < y < h, \tag{C1.20}$$

then $F(y)$ and $G(y)$ satisfy the integral equations

$$\int_0^a F(t)\mathcal{K}_F(y,t)dt = \psi_0(y), \quad 0 < y < a \tag{C1.21}$$

and

$$\int_a^h G(t)\mathcal{K}_G(y,t)dt = \psi_0(y),\ \ a < y < h \tag{C1.22}$$

where

$$\mathcal{K}_F(y,t) = \sum_{n=1}^{\infty} k_n \psi_n(y)\psi_n(t),\ \ 0 < y,t < a \tag{C1.23}$$

and

$$\mathcal{K}_G(y,t) = \sum_{n=1}^{\infty} \frac{\psi_n(y)\psi_n(t)}{k_n}, a < y,\ t < h, \tag{C1.24}$$

together with

$$\int_0^a F(y)\psi_0(y)dy = \frac{1}{C} \text{ and } \int_a^h G(y)\psi_0(y)dy = C \tag{C1.25}$$

where

$$C = ik_0\left(1 - \frac{1}{R_0}\right). \tag{C1.26}$$

It may be noted that the functions $F(y)$, $G(y)$ and the constant C are all real. The integral equations (C1.21) and (C1.22) are solved by multi-term Galerkin approximations as in Porter and Evans (1995) given by

$$F(y) = \sum_{n=1}^{\infty} a_n f_n(y),\ 0 < y < a \tag{C1.27}$$

and

$$G(y) = \sum_{n=1}^{\infty} b_n g_n(y), a < y < h \tag{C1.28}$$

where

$$f_n(y) = \frac{d}{dy}\left[e^{-Ky}\int_y^a \overline{f}_n(u)e^{Ku}du\right], 0 < y < a$$

with

$$\overline{f}_n(y) = \frac{2(-1)^n}{\pi(2n+1)ah}(a^2 - y^2)^{1/2}U_{2n}\left(\frac{y}{a}\right) \tag{C1.29}$$

and

$$g_n(y) = \frac{2(-1)^n}{\pi\{(h-a)^2 - (h-y)^2\}^{1/2}}T_{2n}\left(\frac{h-y}{h-a}\right), a < y < h, \tag{C1.30}$$

U_{2n} and T_{2n} being the Chebyshev polynomials of second and first kinds respectively. The unknown coefficients a_n, b_n ($n = 0,1,\ldots, N$) are found by using the systems of linear equations

$$\sum_{n=0}^{\infty} a_n \mathcal{K}_{mn}^F = F_m, m = 0,1,\ldots, N, \tag{C1.31}$$

and

$$\sum_{n=0}^{\infty} b_n \mathcal{K}_{mn}^G = G_m, m = 0,1,\ldots, N \tag{C1.32}$$

where

$$\mathcal{K}_{mn}^F = \sum_{l=0}^{\infty} k_l \left(\int_0^a \psi_l(y) f_n(y) dy \right) \left(\int_0^a \psi_l(t) f_m(t) dt \right),$$

$$F_m = \int_0^a \psi_0(y) f_m(y) dy, \tag{C1.33}$$

and

$$\mathcal{K}_{mn}^G = \sum_{l=0}^{\infty} \frac{1}{k_l} \left(\int_a^h \psi_l(y) g_n(y) dy \right) \left(\int_a^h \psi_l(t) g_m(t) dt \right),$$

$$G_m = \int_a^h \psi_0(y) g_m(y) dy. \tag{C1.34}$$

Once a_n, b_n ($n = 0,1,\ldots, N$) are found, the real constant C can be obtained by using any one of the equations in (C1.25) after substituting from (C1.27) or (C1.28). R_0 then can be obtained by using (C1.26).

To find the constants A_n, we use either the second relation in (C1.15) or in (C1.19). Noting the relations in (C1.20) and the multi-term expansions (C1.27) or (C1.28), A_n is ultimately approximated as

$$A_n = -ik_0(1 - R_0) \sum_{l=0}^{N} a_l \int_0^a \psi_n(y) f_l(y) dy$$

or

$$A_n = -R_0 \sum_{l=0}^{N} b_l \int_a^h \psi_n(y) g_l(y) dy. \tag{C1.35}$$

In the numerical computations for R_1, both the sets of multi-term Galerkin approximations for $F(y)$ and $G(y)$ have been used. Almost the same numerical results for R_1 are obtained.

Thus R_1 and T_1 can be obtained numerically once the shape function $c(x)$ is known. Here we consider sinusoidal undulations at the bottom so that $c(x)$ can be taken in the form

$$c(x) = \begin{cases} c_0 \sin \lambda x, & -\dfrac{m\pi}{\lambda} \le x \le \dfrac{m\pi}{\lambda} \\ 0, & \text{otherwise} \end{cases} \tag{C1.36}$$

where m is a positive integer. Thus there exists m number of sinusoidal ripples at the bottom with wave number λ. In this case T_1 vanishes identically, and R_1 is given by

$$
\begin{aligned}
R_1 &= \frac{c_0 k_0 (R_0 - 1)}{2N_0} \left\{ \frac{\sin(\lambda - 2k_0)l}{\lambda - 2k_0} - \frac{\sin(\lambda + 2k_0)l}{\lambda + 2k_0} \right\} \\
&+ \frac{i c_0 k_0 R_0}{2N_0} \left\{ \frac{2(1 - \cos \lambda l)}{\lambda} - \frac{2\lambda}{\lambda^2 - 4k_0^2} + \frac{\cos(\lambda - 2k_0)l}{\lambda - 2k_0} + \frac{\cos(\lambda + 2k_0)l}{\lambda + 2k_0} \right\} \\
&+ \frac{i c_0}{N_0^{1/2}} \sum_{n=1}^{\infty} \left[\frac{k_n}{k_n^2 + (\lambda - k_0)^2} - \frac{k_n}{k_n^2 + (\lambda + k_0)^2} + \right. \\
&+ \left\{ \frac{(\lambda - k_0)\sin(\lambda - k_0)l - k_n \cos(\lambda - k_0)l}{k_n^2 + (\lambda - k_0)^2} \right. \\
&\left. \left. - \frac{(\lambda + k_0)\sin(\lambda + k_0)l - k_n \cos(\lambda + k_0)l}{k_n^2 + (\lambda + k_0)^2} \right\} e^{-k_n l} \right] \frac{k_n A_n}{N_n^{1/2}}.
\end{aligned}
\tag{C1.37}
$$

Numerical results

For numerical computation of R_1, we need to evaluate R_0 and the constants A_n (n = 1,2,...) associated with the solution $\phi_0(x, y)$ of the BVPI. As mentioned above, these are evaluated numerically by using multi-term Galerkin approximations. Only a few terms (at most three) in these approximations are sufficient to produce fairly accurate numerical estimates for R_1 (real and imaginary parts).

In our numerical computations the value of λh is chosen to be unity for Figs. 9.37 to 9.41. The Figs. 9.37 and 9.38 depict $|R_1|$ against the wave number Kh for different values of a/h and a single ripple (m = 1) and $c_0/h = 0.1$. From these two figures it is observed that $|R_1|$, regarded as a function of Kh is oscillatory in nature. The zeros of $|R_1|$ are shifted towards the left as the depth of the lower edge of the barrier increases. The crosses in Fig. 9.37 represent the data for $|R_1|$ in the absence of the barrier (for which case $R_0 \equiv 0$). These crosses almost lie on the curve for $|R_1|$ for a very small depth of the lower edge of the barrier ($a/h = 0.001$). This is obviously expected and also confirms the correctness of the numerical results.

The Fig. 9.39 depicts $|R_1|$ against Kh for different values of m, the number of ripples and fixed a/h and c_0/h. As m increases, $|R_1|$ increases, becomes more oscillatory and the number of zeros also increases. This is due to multiple interaction of the incident wave between the ripple tops, the barrier and the free surface.

The Figs. 9.40 and 9.41 show the effect of c_0/h (non-dimensional ripple amplitude) on $|R_1|$. As c_0/h increases, $|R_1|$ also increases, whatever be the plate length, and the zeros of $|R_1|$ remain unchanged with the change in c_0/h, but shift towards the left as a/h increases.

It is known that in the absence of the barrier, $|R_1|$ has peak value when $\alpha = \frac{2k_0}{\lambda} \approx 1$. A similar feature of $|R_1|$ is also evident in the Figs. 9.42 and 9.43 depicting $|R_1|$ against α for fixed c_0/h (=0.1), m(=2), Kh = .25 (in Fig. 9.42), Kh = 1.5 (in Fig. 9.43), in which the curve A represents the case of absence of barrier. However, the value of α for which

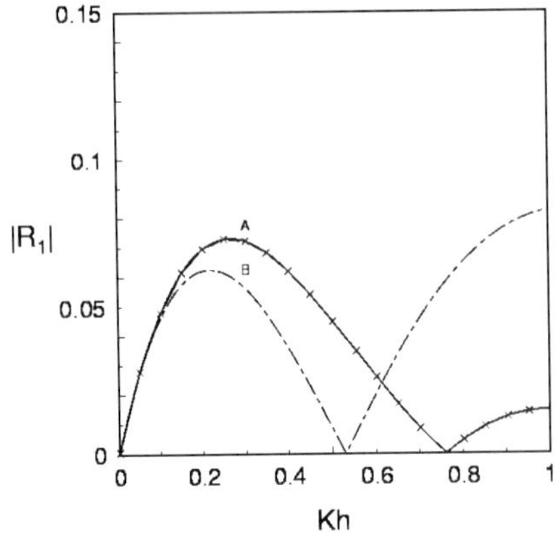

Fig. 9.37 $|R_1|$ for different plate-length with $c_0/h = 0.1$, $m = 1$: $a/h = 0.001(A)$, $0.5(B)$ crosses denote data for $|R_1|$ when there is no plate

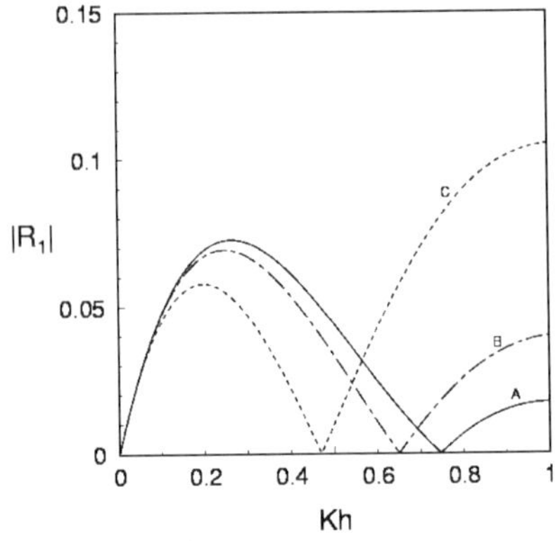

Fig. 9.38 $|R_1|$ for different plate-length with $c_0/h = 0.1$, $m = 1$: $a/h = 0.1(A)$, $0.3(B)$, $0.6(C)$

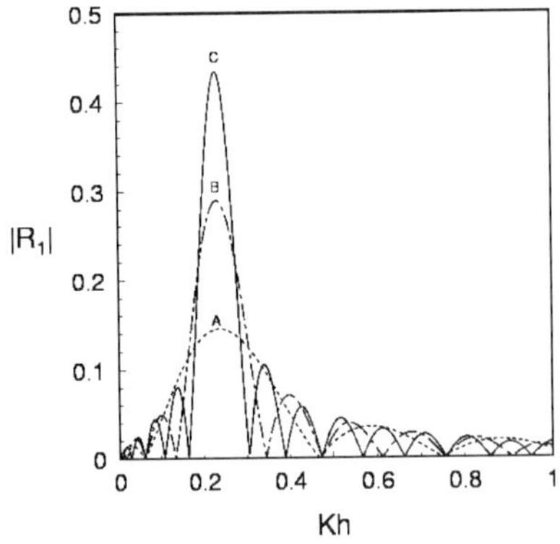

Fig. 9.39 $|R_1|$ for different of ripples with $c_0/h = 0.1$, $a/h = 0.1$, $m = 2(A)$, $4(B)$, $6(C)$

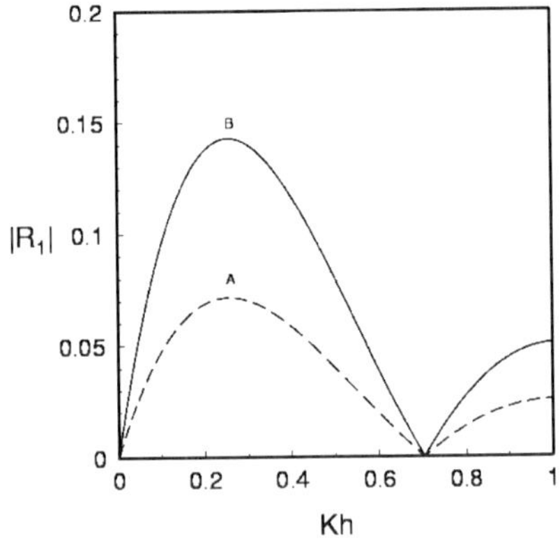

Fig. 9.40 $|R_1|$ for different ripple amplitude with $m = 1$: $a/h = 0.2$, $c_0/h = 0.1(A)$, $0.2(B)$

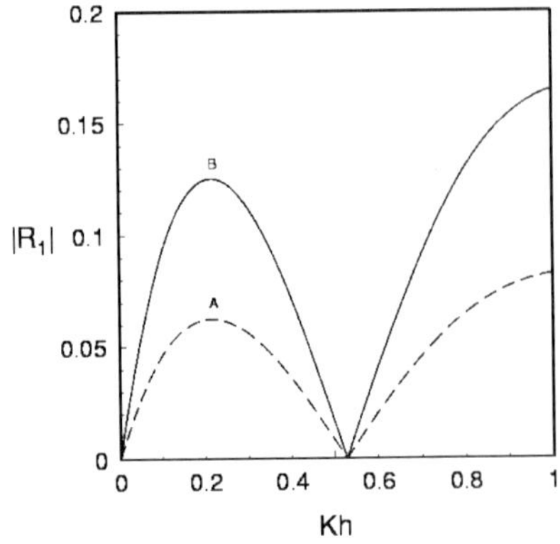

Fig. 9.41 $|R_1|$ for different ripple amplitude with $m = 1$: $a/h = 0.5$, $c_0/h = 0.1(A)$, $0.2(B)$

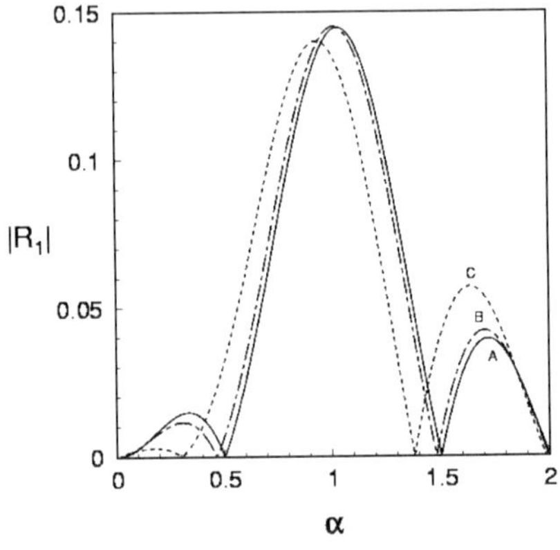

Fig. 9.42 $|R_1|$ against α with $c_0/h = 0.1$, $m = 2$, $Kh = 0.25$; $a/h = 0(A)$, $0.4(B)$, $0.8(C)$

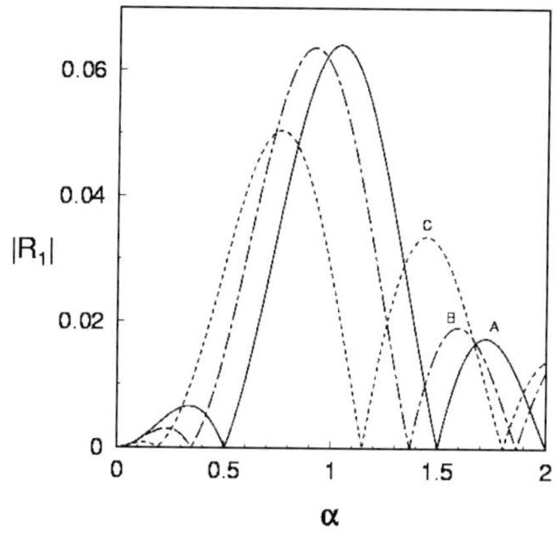

Fig. 9.43 $|R_1|$ against α with $c_0/h = 0.1$, $m = 2$, $Kh = 0.15$; $a/h = 0(A)$, $0.4(B)$, $0.8(C)$

$|R_1|$ attains its peak, is somewhat less than unity and the peak values are also reduced as a/h increases. These features are prominent for large wave numbers.

The case of a thin vertical plate submerged in water of finite depth with small undulations at the bottom has been studied by Mandal and De (2007) in a similar manner.

9C.2 Scattering by bottom undulations in the presence of a semi-infinite dock

In this section the problem of wave scattering by bottom undulations in the presence of a semi-infinite rigid dock is considered. This type of situation arises near the mouth of a wide river where the depth of the bottom becomes variable due to deposition of slits. The problem is formulated for the case of an obliquely incident train of surface waves. The dock is located on the upper surface of the channel ($x > 0$) and a train of progressive waves propagating from negative infinity is obliquely incident on the bottom having small undulations. A simplified perturbation method has been employed by Dhillon et al. (2013) to solvc the problem approximately. This is presented here.

A rectangular Cartesian co-ordinate system is chosen in which y-axis is taken vertically downwards and x-axis is along the mean free surface of the ocean. The position of the rigid dock is given by $y = 0, x > 0$. The bottom of the ocean with small undulations is described by $y = h + \varepsilon\, c(x)$. Here ε is a small non-dimensional positive number which gives a measure of smallness of the bottom undulations. The function $c(x)$ is a continuous bounded function which has a compact support so that far away from the undulations, the bottom of the ocean is of uniform finite depth h below the mean free surface. We consider a train of progressive waves represented by $Re\{\phi(x, y)e^{ivz-i\sigma t}\}$, is obliquely incident upon the dock from negative infinity making an angle θ with the mean free surface of ocean. Then $\phi(x, y)$ satisfies the following boundary value problem (BVP):

$$(\nabla^2 - v^2)\phi = 0 \quad \text{in the fluid region,} \tag{C2.1}$$

The free surface condition is given by

$$K\phi + \phi_y = 0 \quad \text{on } y = 0,\ x < 0, \tag{C2.2}$$

where $K = \sigma^2/g$, σ being the angular frequency of the incoming water-wave train and g being the acceleration due to gravity.

The condition of no motion of the gigid doc and bottom of ocean respectively are given by

$$\phi_y = 0 \quad \text{on } y = 0,\ x > 0, \tag{C2.3}$$

$$\phi_n = 0 \quad \text{on } y = h + \varepsilon\, c(x), \tag{C2.4}$$

$\frac{\partial}{\partial n}$ being the normal derivative.

The condition of boundedness of the velocity at the edge of the dock is given by

$$r^{1/2}\frac{\partial \phi}{\partial r} = 0 \quad \text{as } r = (x^2 + y^2)^{1/2} \to 0, \tag{C2.5}$$

The condition at infinity is given by

$$\phi(x, y) = \begin{cases} (e^{i\mu x} + Re^{-i\mu x})\psi_0(y) & \text{as } x \to -\infty, \\ 0 & \text{as } x \to \infty \end{cases} \tag{C2.6}$$

Here R is the reflection coefficient, i.e., the amplitude of the waves reflected by the dock and the bottom undulations. Also

$$\nu = k_0 \sin\theta, \quad \mu = k_0 \cos\theta,$$

$$\psi_0(y) = N_0 \cosh k_0(y - h) \tag{C2.7}$$

with $N_0 = \dfrac{2(k_0)^{\frac{1}{2}}}{(2k_0 h + \sinh 2k_0 h)^{\frac{1}{2}}}$.

and k_0 is the unique real positive root of the dispersion relation

$$k \tanh kh = K.$$

The bottom condition (C2.4) can be approximated up to the first order of the small parameter ε as

$$-\frac{\partial \phi}{\partial y} + \varepsilon\left[c'(x)\frac{\partial \phi}{\partial x} - c(x)\frac{\partial^2 \phi}{\partial x^2}\right] = 0 \quad \text{on } y = h. \tag{C2.8}$$

The approximate boundary condition (C2.8) suggests that ϕ and R can be expanded in terms of the perturbation parameter ε as

$$\phi(x,y;\varepsilon)=\phi_0(x,y)+\varepsilon\phi_1(x,y)+O(\varepsilon^2),$$
$$R(\varepsilon)=R_0+\varepsilon R_1+O(\varepsilon^2). \tag{C2.9}$$

Substituting the expansions (C2.9) in (C2.1)–(C2.3), (C2.5), (C2.6) and (C2.8) we find after equating the coefficients of ε^0 and ε^1 from both sides, that the functions $\phi_0(x, y)$ and $\phi_1(x, y)$ satisfy the following BVPs:

BVP-I: The function $\phi_0(x, y)$ satisfies

$$(\nabla^2-\nu^2)\phi_0(x,y)=0 \quad \text{in} \quad 0<y<h,-\infty<x<\infty,$$
$$K\phi_0(x,y)+\phi_{0y}(x,y)=0 \quad \text{on} \quad y=0, x<0,$$
$$\phi_{0y}(x,y)=0 \quad \text{on} \quad y=0, x>0,$$
$$r^{1/2}\frac{\partial\phi_0}{\partial r}=0 \quad \text{as} \quad r=(x^2+y^2)^{1/2}\to 0,$$
$$\phi_{0y}(x,y)=0 \quad \text{on} \quad y=h,$$
$$\phi_0(x,y)=\begin{cases}(e^{i\mu x}+R_0e^{-i\mu x})\psi_0(y) & \text{as } x\to-\infty,\\ 0 & \text{as } x\to\infty.\end{cases} \tag{C2.10}$$

BVP-II: The function $\phi_1(x, y)$ satisfies

$$(\nabla^2-\nu^2)\phi_1(x,y)=0 \quad \text{in} \quad 0<y<h,-\infty<x<\infty,$$
$$K\phi_1(x,y)+\phi_{1y}(x,y)=0 \quad \text{on} \quad y=0, x<0,$$
$$\phi_{1y}(x,y)=0 \quad \text{on} \quad y=0, x>0,$$
$$\phi_{1y}=\frac{d}{dx}\left(c(x)\frac{\partial\phi_0(x,h)}{\partial x}\right) \quad \text{on} \quad y=h,$$
$$r^{1/2}\frac{\partial\phi_1}{\partial r}=0 \quad \text{as} \quad r=(x^2+y^2)^{1/2}\to 0,$$
$$\phi_1(x,y)=\begin{cases}(R_1e^{-i\mu x})\psi_0(y) & \text{as } x\to-\infty,\\ 0 & \text{as } x\to\infty.\end{cases} \tag{C2.11}$$

BVP-I corresponds to the problem of water wave scattering by a rigid dock present in the surface of ocean with uniform constant depth h. Here the motion is described by zeroth order velocity potential $\phi_0(x, y)$ and the zeroth order reflection coefficient R_0. The BVP-II is a radiation problem in water of uniform finite depth h, in which, the bottom condition involves ϕ_0, the solution of BVP-I. Without solving $\phi_1(x, y)$ explicitly, R_1 can be determined in terms of integrals involving the shape function $c(x)$ and $\phi_0(x, h)$. To show this, we apply Green's integral theorem to the functions $\phi_0(x, y)$ and $\phi_1(x, y)$ in the regions bounded by the lines $y = 0, -X \le x \le X; x = \pm X, 0 \le y \le h; y = h, -X \le x \le X$ where X is large and positive, and ultimately make X to tend to infinity. This produces

$$2i\mu R_1=\int_{-\infty}^{\infty}c(x)\left[\phi_{0x}^2(x,h)+\nu\phi_0^2(x,h)\right]dx. \tag{C2.12}$$

Therefore, R_1 is derived in terms of integrals involving the shape function $c(x)$ and the zero-order potential function $\phi_0(x, y)$. We now briefly describe the method of solution of BVP-I.

Now, $\phi_0(x, y)$ can be expressed as

$$\phi_0(x,y)=\begin{cases}(e^{i\mu x}+R_0e^{-i\mu x})\psi_0(y)+\sum_{n=1}^{\infty}A_ne^{(k_n^2+\nu^2)^{1/2}x}\psi_n(y), & x<0,\\ \sum_{n=0}^{\infty}\frac{\varepsilon_n}{2}B_ne^{-(\frac{n^2\pi^2}{h^2}+\nu^2)^{1/2}x}\cos(\frac{n\pi}{h})y, & x>0.\end{cases} \tag{C2.13}$$

where $\varepsilon_0 = 1$ and $\varepsilon_n = 2$ for $n \geq 1$.

Here, $\pm ik_n$ $(n = 1,2,3,.....)$ are purely imaginary roots of the transcendental equation $k \tanh kh = K$ and

$$\psi_n(y) = N_n \cos k_n(y-h) \tag{C2.14}$$

with

$$N_n = \frac{2(k_n)^{\frac{1}{2}}}{(2k_nh+\sin 2k_nh)^{\frac{1}{2}}}, n \geq 1.$$

$A_n(n = 1,2,3,.....)$ *and* $B_n(n = 0,1,2,3,.....)$ are unknown constants to be determined along with the unknown reflection coefficient R_0.

We define

$$\alpha_0 = -i\mu = -i(k_0^2-\nu^2)^{\frac{1}{2}},\ \alpha_n = (k_n^2+\nu^2)^{\frac{1}{2}},\ n \geq 1, \gamma_n = ((\frac{n\pi}{h})^2+\nu^2)^{\frac{1}{2}}.$$

We now match ϕ_0 and $\frac{\partial \phi_0}{\partial x}$ at $x = 0$ and using the orthogonal properties of eigenfunctions $\psi_0(y)$ in $(0, h)$ we obtain

$$(1+R_0) = N_0 ik_0 \sin(ik_0h)\sum_{n=0}^{\infty}\frac{\varepsilon_n}{2}\frac{B_n}{-(k_0^2+(\frac{n\pi}{h})^2)},$$

$$\alpha_0(R_0-1) = N_0 ik_0 \sin(ik_0h)\sum_{n=0}^{\infty}\frac{\varepsilon_n}{2}\frac{B_n\gamma_n}{k_0^2+(\frac{n\pi}{h})^2}.$$

Using the above two equations we get

$$\sum_{n=0}^{\infty}\frac{\varepsilon_n}{2}\frac{B_n}{\gamma_n+\alpha_0} = \frac{-2\alpha_0R_0}{N_0K\cosh k_0h}, \tag{C2.15}$$

$$\sum_{n=0}^{\infty}\frac{\varepsilon_n}{2}\frac{B_n}{\gamma_n-\alpha_0} = \frac{2\alpha_0}{N_0K\cosh k_0h}. \tag{C2.16}$$

Now, using the orthogonal property of eigenfunctions $\psi_m(m = 1,2,....)$ and elimination of the constants A_m $(m = 1,2,....)$ we obtain,

$$\sum_{n=0}^{\infty} \frac{\varepsilon_n}{2} \frac{B_n}{\gamma_n - \alpha_m} = 0, \ m = 1,2,..... \tag{C2.17}$$

Combining the equations (C2.16) and (C2.17) we get,

$$\sum_{n=0}^{\infty} \frac{B_n}{\gamma_n - \alpha_m} = A\delta_{m0}, \ m = 0,1,2,.... \tag{C2.18}$$

where,

$$A = \frac{2\alpha_0}{N_0 K \cosh k_0 h}. \tag{C2.19}$$

Again, matching ϕ_0 and $\frac{\partial \phi_0}{\partial x}$ at $x = 0$ and using the orthogonal properties of eigenfunctions $\cos\frac{n\pi}{h}y$ $(n = 0,1,2,...)$ in $(0, h)$ we obtain

$$2c_{0m} + \sum_{n=0}^{\infty} A_n c_{nm} = \frac{h}{2} B_m, \ m \geq 0 \tag{C2.20}$$

$$\sum_{n=0}^{\infty} A_n \alpha_n c_{nm} = -\frac{h}{2} \gamma_m B_m, \ m \geq 0 \tag{C2.21}$$

where,

$$A_0 = R - 1$$

and

$$c_{nm} = \int_0^h \psi_n(y) \cos\frac{m\pi}{h} y \, dy$$

$$= \frac{N_n x_n \sin x_n h}{x_n^2 - \left(\frac{m\pi}{h}\right)^2}$$

with $x_0 = -ik_0$ and $x_n = k_n$ for $n \geq 1$.

Elimination of the constants $B_m (m = 0,1,2,...)$ from the equations (C2.20) and (C2.21) leads to an infinite system of equations

$$-2c_{0m}\gamma_m = \sum_{n=0}^{\infty} A_n(\alpha_n + \gamma_m)c_{nm} \tag{C2.22}$$

The zero-order reflection coefficient R_0 can be obtained by using the residue calculus as given in Linton and McIver (2001). For the sake of completeness we describe the method briefly, we consider the integral

$$\frac{1}{2\pi i} \int_{C_N} \frac{f(z)}{z - \alpha_m} dz, \tag{C2.23}$$

where $f(z)$ has simple poles at $z = \gamma_n (n = 0,1,2,...)$ and simple zeros at $z = \alpha_n (n = 1,2,...)$ and $f(z) = O(|z|^{-1})$ as $z \to \infty$ on C_N, C_N being a sequence of circles whose radius R_N increases without bound as $N \to \infty$ whilst avoiding the zeros of the integrand. The conditions on $f(z)$ are sufficient to ensure that the integral tends to zero as $N \to \infty$. If we further assume that $f(\alpha_0) = -1$, we obtain

$$\sum_{n=0}^{\infty} \frac{Res[f(z);\gamma_n]}{\gamma_n - \alpha_m} = \delta_{m0}, (m = 0,1,2,...). \tag{C2.24}$$

By comparing (C2.18) and (C2.24) we find

$$B_n = ARes[f(z);\gamma_n]. \tag{C2.25}$$

An appropriate form of the function $f(z)$ is given by

$$f(z) = \frac{\gamma_0 - \alpha_0}{z - \gamma_0} \prod_{n=1}^{\infty} \frac{(1 - \frac{z}{\alpha_n})(1 - \frac{\alpha_0}{\gamma_n})}{(1 - \frac{\alpha_0}{\alpha_n})(1 - \frac{z}{\gamma_n})}. \tag{C2.26}$$

We consider again another integral

$$\frac{1}{2\pi i} \int_{C_N} \frac{f(z)}{z + \alpha_m} dz, \tag{C2.27}$$

where $f(z)$ is the same as above. This integral also tends to zero as $N \to \infty$. Thus, for $m = 0$, this produces

$$\sum_{n=0}^{\infty} \frac{B_n}{\gamma_n + \alpha_0} = -Af(-\alpha_0). \tag{C2.28}$$

Comparison with (C2.15) then shows that

$$R_0 = f(-\alpha_0) = \frac{\alpha_0 - \gamma_0}{\alpha_0 + \gamma_0} \prod_{n=1}^{\infty} \frac{(1 + \frac{\alpha_0}{\alpha_n})(1 - \frac{\alpha_0}{\gamma_n})}{(1 - \frac{\alpha_0}{\alpha_n})(1 + \frac{\alpha_0}{\gamma_n})} = e^{-2i\theta} e^{2i\Theta} = e^{2i(\Theta - \theta)}, \tag{C2.29}$$

where

$$\Theta = \sum_{n=1}^{\infty} \tan^{-1}(\frac{\mu}{\gamma_n}) - \tan^{-1}(\frac{\mu}{\alpha_n}). \tag{C2.30}$$

It may be noted that $|R_0| = 1$, which is expected. Since $A_n (n = 1,2,....)$, $B_m (m = 1,2,....)$ and R_0 are all known, thus the function $\phi_0(x, y)$ is obtained in principle.

Hence, R_1 can be computed numerically from the equation (C2.12) once the shape function $c(x)$ is known. First we consider $c(x)$ to represent a sinusoidal variation for which $c(x)$ is chosen in the form

$$c(x) = \begin{cases} c_0 \sin \lambda(x - a) & a - \frac{m\pi}{\lambda} \le x \le a + \frac{m\pi}{\lambda}, \\ 0 & \text{otherwise,} \end{cases} \tag{C2.31}$$

where c_0 is the ripple amplitude and m is a positive integer denoting the number of sinusoidal ripples at the bottom with wave number λ. In this case, with the assumption that $|a| < \dfrac{m\pi}{\lambda}$, we obtain the first-order reflection coefficient as

$$
\begin{aligned}
R_1 = \frac{c_0}{2i\mu}\Bigg[& (N_0^2(\nu^2-\mu^2))\left(\frac{1}{\lambda^2-4\mu^2}\left[\begin{matrix}-2i\mu\sin\lambda a-\lambda\cos\lambda a\\ +\lambda(-1)^m e^{2i\mu(a-\frac{m\pi}{\lambda})}\end{matrix}\right]\right)\\
&-\frac{R_0^2}{\lambda^2-4\mu^2}\left[-2i\mu\sin\lambda a+\lambda\cos\lambda a-\lambda(-1)^m e^{-2i\mu(a-\frac{m\pi}{\lambda})}\right]\\
&+\frac{2R_0N_0^2(\mu^2-\nu^2)}{\lambda}[\cos\lambda a-(-1)^m]+\nu^2\int_{a-\frac{m\pi}{\lambda}}^{0}\Big(\sum_{n=1}^{\infty}A_n e^{\alpha_n x}N_n\Big)^2\sin\lambda(x-a)\,dx\\
&+\int_{a-\frac{m\pi}{\lambda}}^{0}\Big(\sum_{n=1}^{\infty}A_n\alpha_n e^{\alpha_n x}N_n\Big)^2\sin\lambda(x-a)\,dx \qquad\qquad (C2.32)\\
&-2N_0\sum_{n=1}^{\infty}\frac{A_n(\nu^2+i\mu\alpha_n)N_n}{\lambda^2+(\alpha_n+i\mu)^2}\left[(\alpha_n+i\mu)\sin\lambda a+\lambda\cos\lambda a-\lambda(-1)^m e^{(\alpha_n+i\mu)(a-\frac{m\pi}{\lambda})}\right]\\
&-2N_0R_0\sum_{n=1}^{\infty}\frac{A_n(\nu^2-i\mu\alpha_n)N_n}{\lambda^2+(\alpha_n-i\mu)^2}\left[(\alpha_n-i\mu)\sin\lambda a+\lambda\cos\lambda a-\lambda(-1)^m e^{(\alpha_n+i\mu)(a-\frac{m\pi}{\lambda})}\right]\\
&+(1+\nu^2)\int_0^{a+\frac{m\pi}{\lambda}}\sin\lambda(x-a)\left(\sum_{n=1}^{\infty}\frac{\varepsilon_n}{2}(-1)^n(-\gamma_n)B_n e^{-\gamma_n x}\right)^2 dx\Bigg].
\end{aligned}
$$

The second shape function is taken as

$$c(x) = b_0 e^{-\beta|x-b|} \quad (\beta > 0), \quad -\infty < x < \infty. \qquad (C2.33)$$

This corresponds to an exponentially damped undulation. The expression for the first-order correction to the reflection coefficient in this case is given by

$$
\begin{aligned}
R_1 = \frac{b_0}{2i\mu}\Bigg[& e^{-\beta b}\left(N_0^2(\nu^2-\mu^2)\left(\frac{1}{\xi+2i\mu}+\frac{R_0^2}{\xi-2i\mu}\right)+\frac{2R_0N_0^2(\nu^2-\mu^2)}{\xi}\right)\\
&+\int_{-\infty}^{0}\Big(\sum_{n=1}^{\infty}A_n\alpha_n e^{\alpha_n x}N_n\Big)^2 e^{\xi x}dx+\nu^2\int_{-\infty}^{0}\Big(\sum_{n=1}^{\infty}A_n e^{\alpha_n x}N_n\Big)^2 e^{\xi x}dx \qquad\qquad (C2.34)\\
&+2N_0\sum_{n=1}^{\infty}\frac{A_n(\nu^2+i\mu\alpha_n)N_n}{\beta+(\alpha_n+i\mu)}+2N_0R_0\sum_{n=1}^{\infty}\frac{A_n(\nu^2-i\mu\alpha_n)N_n}{\beta+(\alpha_n-i\mu)}\\
&+\int_0^b e^{-\xi(b-x)}\left(\sum_{n=1}^{\infty}\frac{\varepsilon_n}{2}(-1)^n(-\gamma_n)B_n e^{-\gamma_n x}\right)^2 dx+\nu^2\int_0^b e^{-\xi(b-x)}\left(\sum_{n=1}^{\infty}\frac{\varepsilon_n}{2}(-1)^n(-\gamma_n)B_n e^{-\gamma_n x}\right)^2 dx\\
&+\int_b^{\infty} e^{-\xi(b-x)}\left(\sum_{n=1}^{\infty}\frac{\varepsilon_n}{2}(-1)^n(-\gamma_n)B_n e^{-\gamma_n x}\right)^2 dx+\nu^2\int_b^{\infty} e^{-\xi(b-x)}\left(\sum_{n=1}^{\infty}\frac{\varepsilon_n}{2}(-1)^n(-\gamma_n)B_n e^{-\gamma_n x}\right)\Bigg].
\end{aligned}
$$

Unlike the sinusoidal bottom topography, in this case $|R_1|$ does not exhibit any resonating effect.

Numerical results

We have computed $|R_1|$ for different values of wave number Kh, for two types of shape functions $c(x)$ characterizing the uneveness of the bottom as mentioned earlier. The actual reflection R as given in equation (C2.6) is $|R| = |R_0 + \varepsilon R_1|$ upto first order where $|R_0| = 1$. For numerical computation the value of the non-dimensional parameter c_0/h is taken as 0.01 and λh as 1 for the Figs. 9.44 to 9.47 which depicts $|R_1|$ against Kh for $c(x)$ given by (C2.31).

In Figs. 9.44(a), 9.44(b) and 9.44(c), $|R_1|$ is plotted against Kh for $a/h = 3$ with number of ripples $m = 2,3,5$ and 7 and $\theta = \pi/9$ and $\pi/3$ respectively. In both figures it is observed that $|R_1|$ is oscillatory in nature and when the number of ripples increases, the general feature of $|R_1|$ remains the same but with the observation that the overall value of $|R_1|$ increases and the oscillatory nature of $|R_1|$ against Kh is more noticeable with the number of zeros of $|R_1|$ increasing. As the angle of incidence increases from

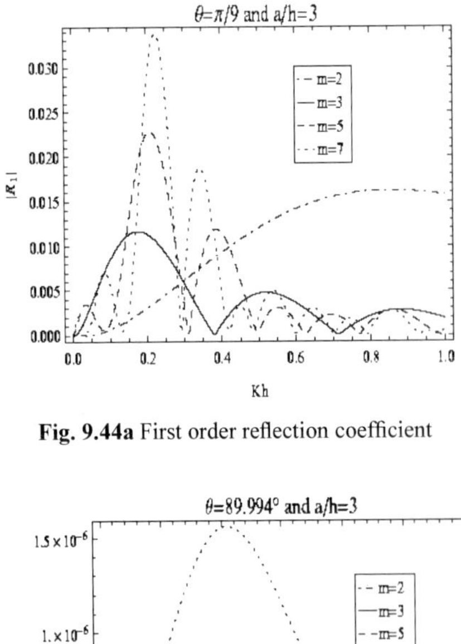

Fig. 9.44a First order reflection coefficient

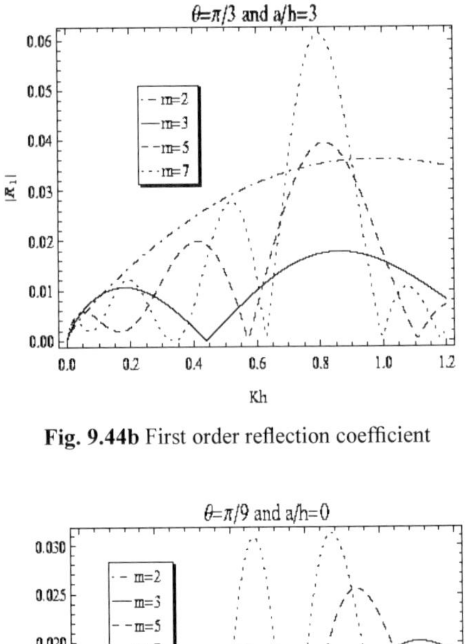

Fig. 9.44b First order reflection coefficient

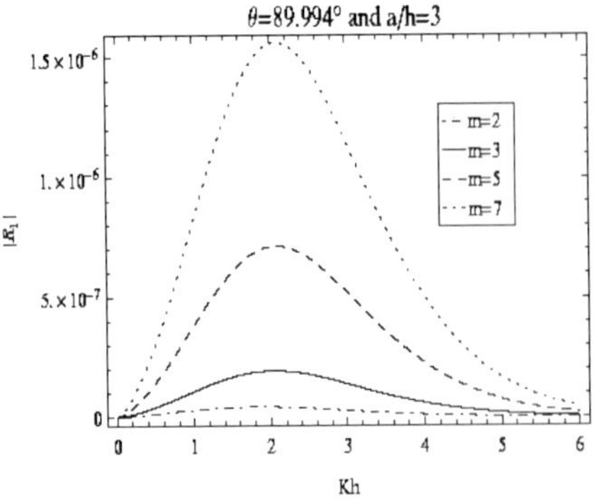

Fig. 9.44c First order reflection coefficient

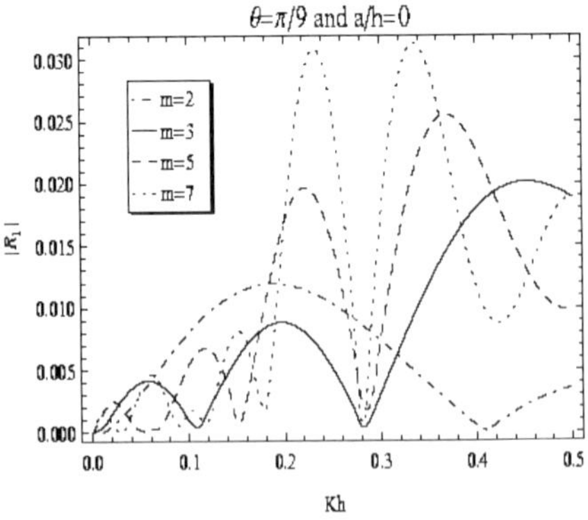

Fig. 9.45a First order reflection coefficient

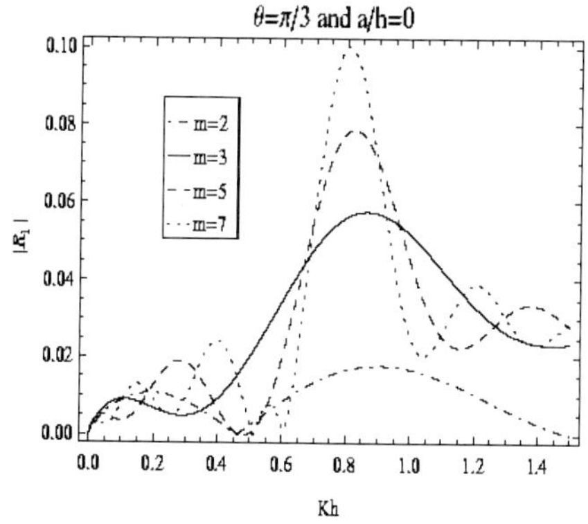

Fig. 9.45b First order reflection coefficient

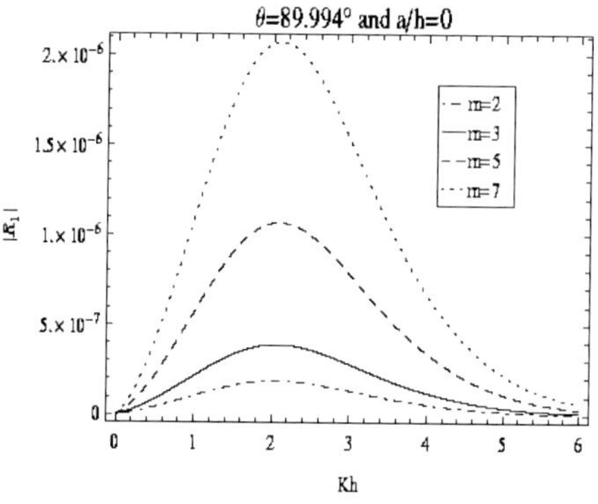

Fig. 9.45c First order reflection coefficient

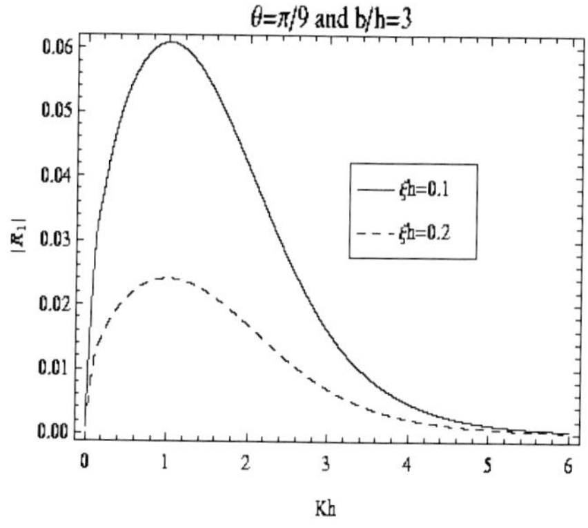

Fig. 9.46a First order reflection coefficient

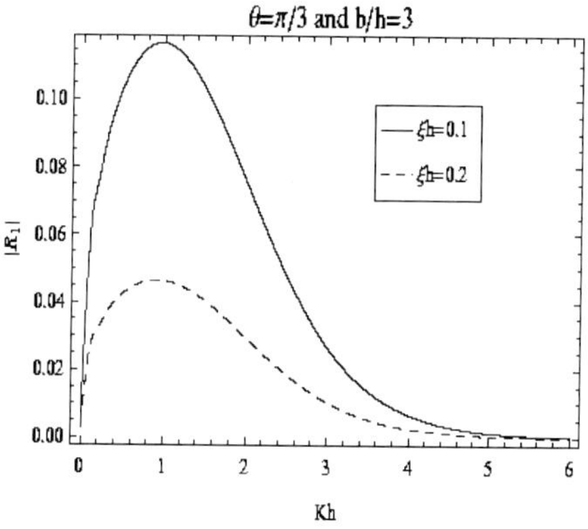

Fig. 9.46b First order reflection coefficient

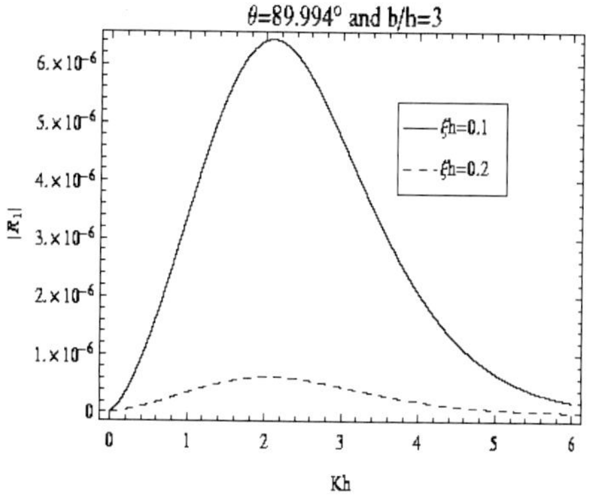

Fig. 9.46c First order reflection coefficient

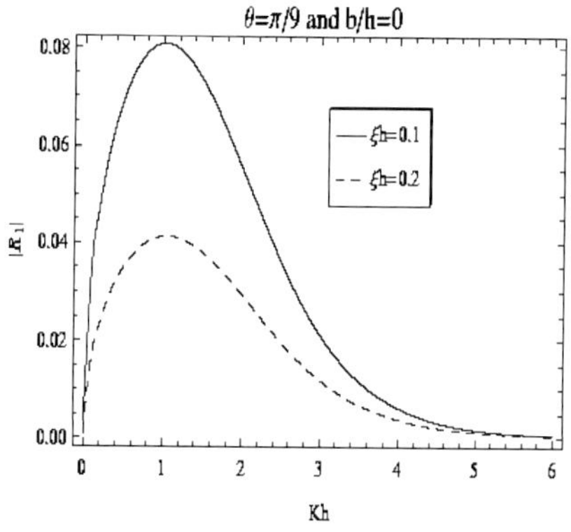

Fig. 9.47a First order reflection coefficient

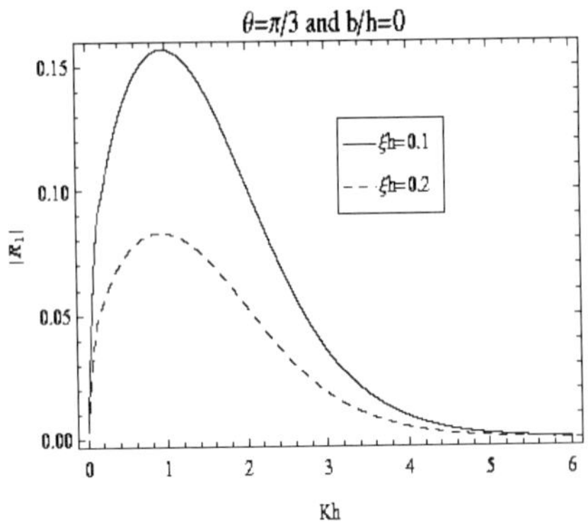

Fig. 9.47b First order reflection coefficient

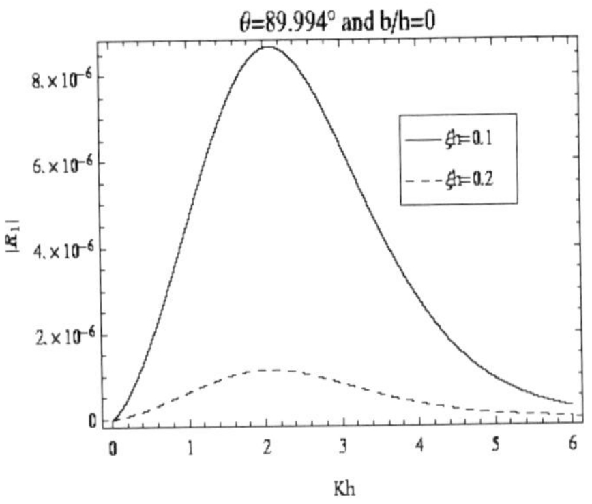

Fig. 9.47c First order reflection coefficient

$\pi/9$ to $\pi/3$, it is seen that the oscillatory nature of $|R_1|$ decreases and the peak value increases. In Fig. 9.44(c), θ has been taken as 89.994° (i.e., almost grazing incidence) with all other parameters fixed. We expect that $|R_1|$ should be very small which is indeed the case as is evident from the numerical computation. It is observed $|R_1|$ increase (but remains small), reaches a peak and then decreases to zero. The peak value increases as m increases from 2 to 7.

For Figs. 9.45(a), 9.45(b) and 9.45(c), a/h has been chosen to be zero, that is, the shape function is symmetric about $x = 0$, keeping other parameters fixed. Similar nature of $|R_1|$ is observed as seen in the above figures. Figure 9.45(c) again shows that the first-order reflection coefficient vanishes when the angle of incidence is taken very close to $\pi/2$. The value of $|R_1|$ is greater when $a/h = 0$ as compared $a/h = 3$. The oscillatory nature of $|R_1|$ in all the figures may be attributed due to multiple interactions of the incident wave train with sinusoidal bottom and the edge of the dock. Occurrence of zeros of $|R_1|$ for certain values of Kh implies that the sinusoidal bottom does not affect the incident waves at first order for certain frequencies. Also, a Bragg resonance interaction is observed to occur in all the cases.

Now, for the second shape function given by (C2.35) which represents an exponentially decaying bottom topography. $|R_1|$ is depicted against Kh for two different values of ξh (Figs. 9.46–9.47) and $b_0/h = 0.01$. It is seen that for each value of ξh, $|R_1|$ first increases with Kh, attains a maximum and then decreases as Kh is further increased. For all the figures it is observed that the peak value of $|R_1|$decreases as ξh increases.

The Figs. 9.46(a) and 9.46(b) show the variation of $|R_1|$ against Kh for $\theta = \pi/9$ and $\pi/3$ respectively with $b/h = 3$. It is observed that $|R_1|$ increases with the increase of the incident angle. Similar behavior is observed in the Figs. 9.47(a) and 9.47(b) where b/h is taken as zero, that is, symmetric about $x = 0$ and keeping all the other parameters fixed. Also, the value of $|R_1|$ is much greater when $b/h = 0$ compared to when $b/h = 3$.

For the Figs. 9.46(c) and 9.47(c), θ has been chosen as 89.994° with $b/h = 3$ and $b/h = 0$ respectively. The peak values of $|R_1|$ is observed to decrease with the increase of b/h and $|R_1|$ becomes negligible when θ is chosen close to $\pi/2$.

9C.3 Scattering by bottom undulations in the presence of surface discontinuity

In this section water wave scattering by bottom undulations in the presence of a discontinuity in the surface boundary condition is investigated by using a simplified perturbation method as has been done by Mandal and De (2009). This problem arises when wave propagation in the presence of two vast sheets of broken ice in cold areas such as Antarctic regions near the coast (with variable bottom topography) is considered.

We consider two-dimensional potential flow in an ocean of finite depth having small undulations at the bottom. A rectangular Cartesian co-ordinate system is chosen in which the y-axis is taken vertically downwards and $y = 0$ corresponds to the undisturbed upper surface of water. This upper surface is covered by two inertial surfaces of uniform area densities $\varepsilon_1\rho$ and $\varepsilon_2\rho$ (ρ being the density of water), the first occupies the region $y = 0$, $x < 0$ and the second occupies the region $y = 0$, $x > 0$, the direction of the positive x-axis being opposite to the direction of the incoming incident wave field. Thus there is a discontinuity in the surface boundary condition. The motion in water is assumed to be two-dimensional, irrotational and time-harmonic. If $\Phi(x, y, t) = \text{Re}\{\phi(x, y)e^{-i\omega t}\}$ describes the velocity potential for the two-dimensional motion in the fluid region, then the mathematical problem under consideration is to solve the following boundary value problem for ϕ satisfying the Laplace's equation

$$\nabla^2\phi = 0 \quad \text{in the fluid region}, \tag{C3.1}$$

the surface boundary conditions

$$K_1\phi + \phi_y = 0 \qquad \text{on} \quad y = 0,\ x > 0, \tag{C3.2}$$

$$K_2\phi + \phi_y = 0 \qquad \text{on} \quad y = 0,\ x < 0, \tag{C3.3}$$

producing a discontinuity in the surface boundary conditions at the point (0,0), where

$$K_1 = \frac{K}{1-\varepsilon_1 K}, \quad K_2 = \frac{K}{1-\varepsilon_2 K} \quad \text{with} \quad K = \frac{\omega^2}{g} \tag{C3.4}$$

and ε_1, $\varepsilon_2 < g/\omega^2$, so that time-harmonic progressive waves of circular frequency ω can propagate along the inertial surfaces, the bottom condition

$$\frac{\partial\phi}{\partial n} = 0 \quad \text{on} \quad y = h + \delta\, c(x), \tag{C3.5}$$

where $y = h + \delta c(x)$ denotes the bottom of an ocean of variable depth of small cylindrical undulations, δ is a small non-dimensional positive number which gives a measure of smallness of the bottom undulations, and $c(x)$ is a bounded continuous function and is such that $c(x) \to 0$ as $|x| \to \infty$ so that far away from the undulations the bottom is of uniform finite depth h below the mean free surface. The far field conditions are

$$\varphi(x,y) \sim \begin{cases} (\mathrm{e}^{\mathrm{i}k_0x} + R\mathrm{e}^{-\mathrm{i}k_0x})\psi_0^1(y) & \text{as } x \to -\infty, \\ T\mathrm{e}^{\mathrm{i}s_0x}\psi_0^2(y) & \text{as } x \to \infty, \end{cases} \tag{C3.6}$$

where,

$$\psi_0^1(y) = N_0^1 \cosh k_0(y-h), \qquad \psi_0^2(y) = N_0^2 \cosh s_0(y-h)$$

and

$$N_0^1 = \frac{2k_0^{1/2}}{(2k_0h + \sinh 2k_0h)^{1/2}}, \qquad N_0^2 = \frac{2s_0^{1/2}}{(2s_0h + \sinh 2s_0h)^{1/2}}.$$

Here k_0, s_0 are the real positive roots of transcendental equations $k \tanh kh = K_1$ and $s \tanh sh = K_2$ respectively, $e^{ik_0x}\,\psi_0^1(y)$ denotes the incident wave field, R and T denote respectively the reflection and transmission coefficients to be determined. In (C3.5), $\frac{\partial}{\partial n}$ denotes the derivative normal to the bottom.

The bottom condition (C3.5) can be expressed approximately as

$$\frac{\partial \phi}{\partial y} - \delta \frac{\mathrm{d}}{\mathrm{d}x}\left\{c(x)\frac{\partial \phi(x,h)}{\partial x}\right\} + \mathrm{O}(\delta^2) = 0 \quad \text{on} \quad y = h. \tag{C3.7}$$

The form of the approximate bottom condition (3.1) suggests that ϕ, R, T have the following perturbational expansions, in terms of the small parameter δ:

$$\left.\begin{aligned} \phi(x,y;\delta) &= \phi_0(x,y) + \delta\phi_1(x,y) + \mathrm{O}(\delta^2), \\ R(\delta) &= R_0 + \delta R_1 + \mathrm{O}(\delta^2), \\ T(\delta) &= T_0 + \delta T_1 + \mathrm{O}(\delta^2). \end{aligned}\right\} \tag{C3.8}$$

Substituting the expansions (C3.8) into the partial differential equation (C3.1), the surface boundary conditions (C3.2) and (C3.3), the approximate boundary condition (C3.7) and the far field boundary conditions (C3.6), we find after equating the coefficients of identical powers of δ^0 and δ^1 from both sides of the results, that $\phi_0(x, y)$ and $\phi_1(x, y)$ satisfy the following two boundary value problems, BVP-I and BVP-II, respectively.

BVP – I: The function $\phi_0(x, y)$ satisfies

$$\nabla^2\phi_0 = 0, \ 0 < y < h, \ -\infty < x < \infty,$$

$$K_1\phi_0 + \phi_{0y} = 0, \ y = 0, \ x < 0,$$

$$K_2\phi_0 + \phi_{0y} = 0, \ y = 0, \ x > 0,$$

$$\phi_{0y} = 0, \ y = h,$$

$$\phi_0(x,y) \to \begin{cases} (\mathrm{e}^{\mathrm{i}k_0x} + R_0\mathrm{e}^{-\mathrm{i}k_0x})\psi_0^1(y) & \text{as } x \to -\infty, \\ T_0\mathrm{e}^{\mathrm{i}s_0x}\psi_0^2(y) & \text{as } x \to \infty. \end{cases} \tag{C3.9}$$

BVP – II: The function $\phi_1(x, y)$ satisfies

$$\nabla^2\phi_1 = 0,\ 0 < y < h,\ -\infty < x < \infty,$$
$$K_1\phi_1 + \phi_{1y} = 0,\ y = 0,\ x < 0,$$
$$K_2\phi_1 + \phi_{1y} = 0,\ y = 0,\ x > 0,$$
$$\phi_{1y} = \frac{\mathrm{d}}{\mathrm{d}x}\{c(x)\phi_{0x}(x,h)\},\ y = h,$$

$$\phi_1(x,y) \to \begin{cases} R_1 \mathrm{e}^{-\mathrm{i}k_0 x}\psi_0^1(y) & \text{as } x \to -\infty, \\ T_1 \mathrm{e}^{\mathrm{i}s_0 x}\psi_0^2(y) & \text{as } x \to \infty. \end{cases} \tag{C3.10}$$

The BVP-I corresponds to the problem of water wave scattering by a discontinuity in the free-surface boundary condition in uniform finite depth water. Evans and Linton (1994) solved this problem using the residue calculus method and obtained the reflection and transmission coefficients explicitly.

The BVP-II is a radiation problem in uniform finite depth water having a discontinuity in the upper surface boundary condition and the bottom condition involves ϕ_0, the solution of BVP-I. Without solving for $\phi_1(x, y)$, R_1 and T_1 can be determined in terms of integrals involving the shape function $c(x)$ and $\phi_{0x}(x, h)$.

Now $\phi_0(x, y)$ has the expansion

$$\varphi_0(x,y) = \begin{cases} (\mathrm{e}^{\mathrm{i}k_0 x} + R_0 \mathrm{e}^{-\mathrm{i}k_0 x})\psi_0^1(y) + \sum_{n=1}^{\infty} A_n \mathrm{e}^{k_n x}\psi_n^1(y), x < 0, \\ T_0 \mathrm{e}^{\mathrm{i}s_0 x}\psi_0^2(y) + \sum_{n=1}^{\infty} B_n \mathrm{e}^{-s_n x}\psi_n^2(y), x > 0 \end{cases} \tag{C3.11}$$

where $\pm \mathrm{i}k_n$ and $\pm \mathrm{i}s_n (n = 1,2,...)$ are the purely imaginary roots of the transcendental equations $k \tanh kh = K_1$ and $s \tanh sh = K_2$ respectively. $A_n, B_n (n = 1,2,...)$ are unknown constants and

$$\psi_n^1(y) = N_n^1 \cos k_n(y-h), \qquad \psi_n^2(y) = N_n^2 \cos s_n(y-h)$$

where

$$N_n^1 = \frac{2k_n^{1/2}}{2k_n h + \sin 2k_n h}, \qquad N_n^2 = \frac{2s_n^{1/2}}{2s_n h + \sin 2s_n h}. \tag{C3.12}$$

Using the residue calculus method, the expressions for R_0 and T_0 were obtained analytically by Evans and Linton (1994) given by

$$R_0 = \frac{k_0 - s_0}{k_0 + s_0}\mathrm{e}^{2\mathrm{i}\alpha}, \qquad T_0 = \frac{2k_0}{k_0 + s_0}\mathrm{e}^{\mathrm{i}(\alpha+\beta)}, \tag{C3.13}$$

where

$$\alpha = \sum_{n=1}^{\infty}\left\{\tan^{-1}\left(\frac{k_0}{s_n}\right) - \tan^{-1}\left(\frac{k_0}{k_n}\right)\right\}, \beta = \sum_{n=1}^{\infty}\left\{\tan^{-1}\left(\frac{s_0}{k_n}\right) - \tan^{-1}\left(\frac{s_0}{s_n}\right)\right\}. \tag{C3.14}$$

The constants A_n, B_n were also be obtained analytically by the same method (In fact the analytical expressions for $B_n N_n^2 \cos s_n h$, ($r = 1,2,...$) were given by Evans and Linton (1994)). However, here we employ a matching procedure to obtain A_n, B_n numerically. For this purpose, we match ϕ_0 and ϕ_{0x} at $x = 0$ obtained from (C3.11) and use the orthogonality properties of the eigenfunctions $\psi_n^1(y)$ and $\psi_n^2(y)$ ($n = 1,2,...$). This produces the system of linear equations

$$\sum_{n=1}^{\infty} \frac{B_n N_n^2 \cos s_n h}{s_n - k_m} = \frac{T_0 N_0^2 \cosh s_0 h}{\mathrm{i}s_0 + k_m}, \tag{C3.15}$$

$$\sum_{n=1}^{\infty} \frac{A_n N_n^1 \cos k_n h}{k_n - s_m} = N_0^1 \cosh k_0 h \left\{ \frac{R_0}{ik_0 + s_m} - \frac{1}{ik_0 - s_m} \right\}, m = 1,2,... \tag{C3.16}$$

for the unknown constants A_n, B_n ($n = 1,2,...$). These are obtained numerically after truncation of the series (C3.15) and (C3.16).

To obtain R_1, we apply Green's integral theorem to the functions $\phi_0(x, y)$ and $\phi_1(x, y)$ in the region bounded by

$$y = 0, -X < x \le X; \; x = \pm X, 0 \le y \le h; \; y = h, -X < x \le X, (X > 0, Y > 0).$$

If L denotes the contour of this region then

$$\int_L \left(\varphi_0 \frac{\partial \varphi_1}{\partial n} - \varphi_1 \frac{\partial \varphi_1}{\partial n} \right) \mathrm{d}l = 0 \tag{C3.17}$$

where n is the outward normal to the line element dl. The free surface condition satisfied by $\phi_0(x, y)$ and $\phi_1(x, y)$ ensures that there is no contribution to the integral on the left side of the equation (C3.17) from the line $y = 0$, $-X \le x \le X$. As both ϕ_0 and ϕ_1 describe outgoing waves as $x \to \infty$, there is no contribution to the integral from the line $x = X$, $(0 \le y \le h)$ as $X \to \infty$. The only contributions arise from the integral along the line $x = -X$, $(0 \le y \le h)$ as $X \to \infty$ and the integral along the bottom.

Making $X \to \infty$ we obtain ultimately

$$2\mathrm{i}k_0 R_1 = \int_{-\infty}^{\infty} c(x) \varphi_{0x}^2(x, h) \mathrm{d}x. \tag{C3.18}$$

Thus R_1 given by (C3.18) can be obtained numerically once $c(x)$ is known.

To obtain T_1, we use the Green's integral theorem to the functions $\chi_0(x, y)(= \phi_0(-x, y))$ and $\phi_1(x, y)$ in the same region mentioned above and making $X \to \infty$, we find

$$2\mathrm{i}s_0 T_1 = -\int_{-\infty}^{\infty} c(x) \phi_{0x}(x, h) \phi_{0x}(-x, h) \mathrm{d}x, \tag{C3.19}$$

where, $\phi_0(x, h)$, $\phi_{0x}(-x, h)$ are obtained in terms of A_n, B_n ($n = 1,2,...$). T_1 given by (C3.19) can be obtained numerically once $c(x)$ is known. It may be noted that (C3.19) further produces

$$2\mathrm{i}s_0 T_1 = -\int_0^{\infty} \left(c(x) + c(-x) \right) \phi_{0x}(x, h) \phi_{0x}(-x, h) \mathrm{d}x. \tag{C3.20}$$

Thus if $c(x)$ is an odd function, then it is obvious that T_1 vanishes identically. Physically this means that if the surface discontinuity is centred above small antisymmetric bottom undulations, then at first order, there is no transmission of incident waves.

The expressions for R_1 and T_1 for two different shape functions such as sinusoidal and exponentially decaying forms, are given below.

Numerical results

The first order corrections to the reflection and transmission coefficients $|R_1|$ and $|T_1|$ calculated in the Appendix for two different shape functions, are depicted graphically against the wave number Kh for different values of various parameters.

The first shape function $c(x)$ is taken as

$$c(x) = \begin{cases} c_0 \sin \lambda x, & -\dfrac{m\pi}{\lambda} \le x \le \dfrac{m\pi}{\lambda}, \\ 0, & \text{otherwise} \end{cases} \tag{C3.21}$$

where m is a positive integer. This represents m sinusoidal ripples at the bottom with wave number λ. The expression for the first-order correction to the reflection coefficient is given by

$$R_1 = \frac{c_0}{2ik_0}\Bigg[-k_0^2 N_0^1 \Bigg\{ \frac{\lambda}{\lambda^2 - 4k_0^2}\left((-1)^m e^{-2ik_0 m\pi} - 1\right) + \frac{\lambda R_0^2}{\lambda^2 - 4k_0^2}\left((-1)^m e^{2ik_0 m\pi} - 1\right)$$

$$+ \frac{2R_0}{\lambda}\left((-1)^m - 1\right)\Bigg\} + \int_{-m\pi/\lambda}^{0}\left(\sum_{n=1}^{\infty} k_n A_n e^{k_n x} N_n^1\right)^2 \sin \lambda x \, dx + 2ik_0 N_0^1 \sum_{n=1}^{\infty} \frac{k_n A_n N_n^1 \lambda}{\lambda^2 + (k_n + ik_0)^2}$$

$$\left((-1)^m e^{\frac{m\pi(k_n + ik_0)}{\lambda}} - 1\right)$$

$$- 2ik_0 N_0^1 R_0 \sum_{n=1}^{\infty} \frac{k_n A_n N_n^1 \lambda}{\lambda^2 + (k_n + ik_0)^2}\left((-1)^m e^{\frac{m\pi(k_n - ik_0)}{\lambda}} - 1\right) + \frac{s_0^2 (N_0^2)^2 T_0 \lambda}{\lambda^2 - 4s_0^2}\left((-1)^m e^{\frac{2is_0 m\pi}{\lambda}} - 1\right)$$

$$+ \int_0^{\frac{m\pi}{\lambda}}\left(\sum_{n=1}^{\infty} s_n B_n e^{-s_n x} N_n^1\right)^2 \sin \lambda x \, dx + 2is_0 N_0^2 T_0 \sum_{n=1}^{\infty} \frac{s_n B_n N_n^2 \lambda}{\lambda^2 + (is_0 - s_n)^2}\left((-1)^m e^{\frac{m\pi(is_0 - s_n)}{\lambda}} - 1\right)\Bigg]. \tag{C3.22}$$

while the first-order correction to the transmission coefficients is

$$T_1 = 0. \tag{C3.23}$$

For this important case, $|R_1|$ is depicted against Kh in Figs. 9.48, 9.49 for $\lambda h = 1$. In figure 1, ε_1/h is taken as .02 while $\varepsilon_2/h = .01,.06$. $|R_1|$ is oscillatory in nature against

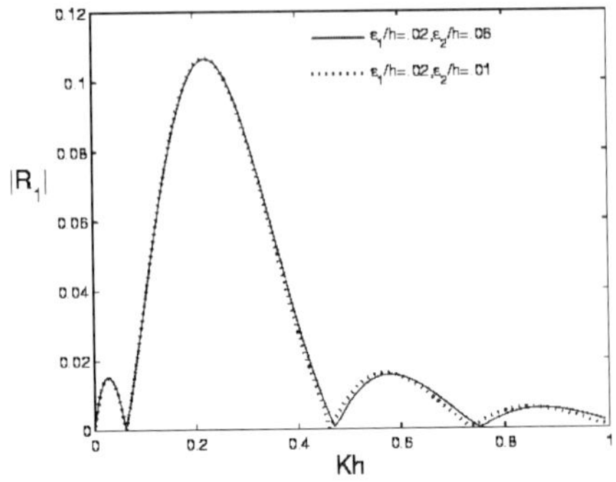

Fig. 9.48 $|R_1|$ for different ε_2/h

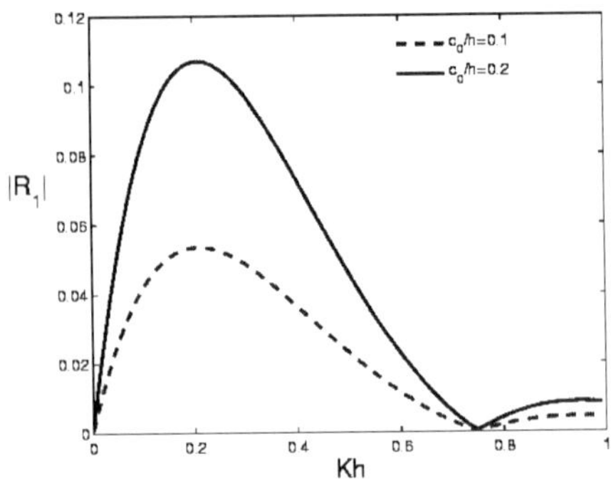

Fig 9.49 $|R_1|$ for different ripple amplitude

Kh and this may be attributed due to multiple interactions of the incident wave train with the sinusoidal bottom and the edge of the two inertial surfaces. Occurrence of zeros of $|R_1|$ for certain values of Kh implies that the sinusoidal bottom does not affect the incident waves at first order for certain frequencies. The effect of variation of ε_2/h while ε_1/h remains fixed is to shift the zeros of $|R_1|$ slightly.

In Fig. 9.49, $|R_1|$ as depicted against Kh for $c_0/h = 0.1,0.2$ and $\varepsilon_1/h = .02$, $\varepsilon_2/h = .01$. The peak values of $|R_1|$ are seen to increase with the increase c_0/h.

The second shape function is taken as

$$c(x) = c_0 e^{-\mu|x|}, -\infty < x < \infty, \mu > 0. \tag{C3.24}$$

This represents an exponentially decaying bottom topography. The expressions for the first-order correction to the reflection and transmission coefficients are given by

$$\begin{aligned} R_1 = \frac{c_0}{2ik_0}\Bigg[& -k_0^2 (N_0^1)^2 \left\{ \frac{1}{2ik_0 + \mu} + \frac{R_0^2}{\mu - 2ik_0} - \frac{2R_0}{\mu} \right\} + \int_{-\infty}^{0} \left(\sum_{n=1}^{\infty} k_n A_n e^{k_n x} N_n^1 \right)^2 e^{\mu x} dx \\ & + 2ik_0 N_0^1 \sum_{n=1}^{\infty} \frac{k_n A_n N_n^1}{\mu + (k_n + ik_0)} - 2ik_0 N_0^1 R_0 \sum_{n=1}^{\infty} \frac{k_n A_n N_n^1}{\mu + (k_n - ik_0)} + \frac{s_0^2 (N_0^2) T_0^2}{2is_0 - \mu} \\ & + \int_0^{\infty} \left(\sum_{n=1}^{\infty} s_n B_n e^{-s_n x} N_n^1 \right)^2 e^{-\mu x} dx + 2is_0 T_0 N_0^2 \sum_{n=1}^{\infty} \frac{s_n B_n N_n^2}{(is_0 - s_n) - \mu} \Bigg] \end{aligned}$$

and

$$\begin{aligned} T_1 = -\frac{c_0}{2is_0}\Bigg[& -s_0 k_0 T_0 N_0^1 N_0^2 \left\{ \frac{1}{\mu + ik_0} + \frac{R_0}{ik_0 - \mu} \right\} - is_0 T_0 N_0^2 \sum_{n=1}^{\infty} \frac{A_n k_n N_n^1}{is_0 - k_n - \mu} \\ & - ik_0 N_0^1 \left\{ \sum_{n=1}^{\infty} \frac{B_n k_n N_n^2}{ik_0 + s_n + \mu} + R_0 \sum_{n=1}^{\infty} \frac{B_n k_n N_n^2}{ik_0 - s_n - \mu} \right\} - \int_0^{\infty} \left(\sum_{n=1}^{\infty} A_n k_n e^{-(k_n+\mu)x} N_n^1 \right) \\ & \left(\sum_{n=1}^{\infty} B_n k_n e^{-s_n x} N_n^2 \right) dx \Bigg]. \end{aligned}$$

For this case, $|R_1|$ is depicted against Kh for different values of μh (Fig. 9.50) when $c_0/h = 0.1$, $\varepsilon_1/h = .02$ and $\varepsilon_1/h = .01$. It is seen that for each value of μh, $|R_1|$ first increases with Kh, attains a maximum and then it decreases with Kh. Also the peak value of $|R_1|$ decreases as μh increases. Figure 9.51 depicts $|R_1|$ against Kh when the left half surface is a free surface ($\varepsilon_2/h = 0$) and right one is an inertial surface for the exponentially decaying shape function. ε_1/h is taken as .01, .05, while $c_0/h = 0.1$ and $\mu h = 0.5$. As ε_1/h increases, overall values of $|R_1|$ decrease which is prominent for $Kh > 0.2$.

The Fig. 9.52 depicts $|T_1|$ for exponentially decaying bottom shape function for different values of μh when $\varepsilon_1/h = .02$, $\varepsilon_2/h = .01$ and $c_0/h = .1$. It is seen that maximum value of $|T_1|$ decreases as μh increases.

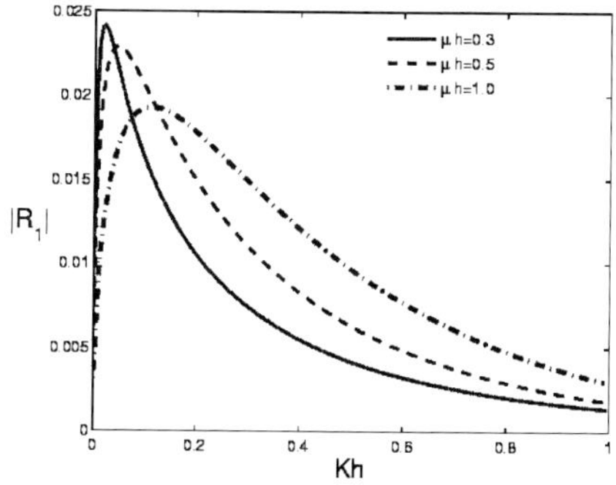

Fig. 9.50 $|R_1|$ for different μh

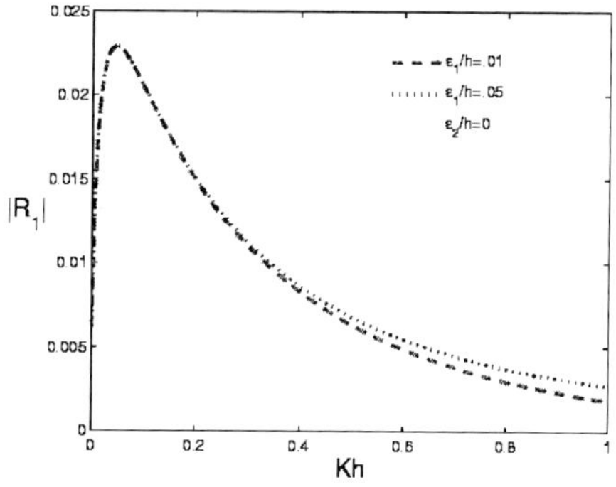

Fig. 9.51 $|R_1|$ for different ε_1/h

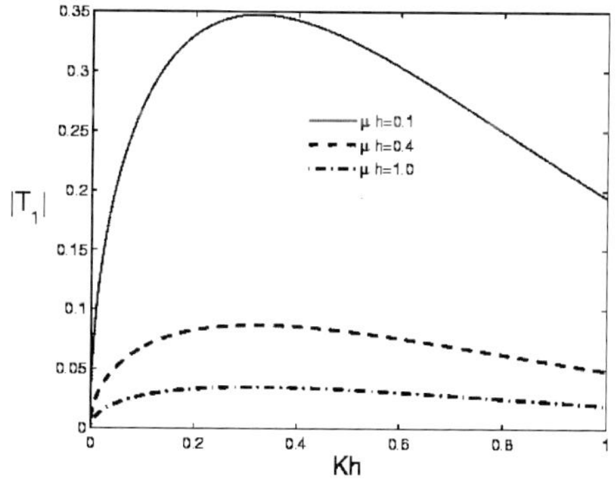

Fig. 9.52 $|T_1|$ for different μh.

References

Andrianov, A.I. and Hermans, A.J. 2003. The influence of water depth on the hydroelastic response of a Very Large Floating Platform, Marine Structures. 16: 355–371.

Balmforth, N.J. and Craster, R.V. 1999. Ocean waves and ice sheets. J. Fluid Mech. 395: 89–124.

Banerjea, S., Dolai, D.P. and Mandal, B.N. 1996. On waves due to rolling of a ship in water of finite depth. Arc. Appl. Mech. 67: 35–45.

Bender, C.J. and Dean, R.G. 2003. Wave transformation by two dimensional bathymetric anomalies with sloped transitions. Coastal Engineering. 56: 61–84.

Burke, J.E. 1964. Scattering of surface waves on an infinitely deep fluid. J. Math. Phys. 5: 805–819.

Chakrabarti, A., Ahluwalia, D.S. and Manam, S.R. 2003. Surface water waves involving a vertical barrier in the presence of an ice-cover. Int. J. Engng. Sci. 41: 1145–1162.

Cadby, J.R. and Linton, C.M. 2002. Three-dimensional water wave scattering in two-layer fluid. J. Fluid Mech. 423: 155–173.

Chakrabarti, A. 2000a. On the solution of the problem of scattering of surface water waves by a sharp discontinuity in the surface boundary condition. ANZIAM J. 42: 277–286.

Chakrabarti, A. 2000b. On the solution of the problem of scattering of surface water waves by the edge of an ice cover. Proc. R. Soc. Lond. A456: 1087–1099.

Chakrabarti, A., Mandal, B.N. and Gayen, R. 2005. The dock problem revisited. Inter. J. Math. & Math. Sci. 2005: 3459–3470.

Chakrabarti, A. and Martha, S.C. 2009. A note on energy-balance relations in surface water wave problems involving floating elastic plates. J. Adv. Res. Appl. Math. 12: 27–34.

Chakraborty, R. and Mandal, B.N. 2014a. Scattering of water waves by a submerged thin vertical elastic plate. Arch. Appl. Mech. 84: 207–217.

Chakraborty, R. and Mandal, B.N. 2014b. Wave scattering by rectangular trench. J. Engng. Math. 89: 101–112.

Chakraborty, R. and Mandal, B.N. 2015. Oblique wave scattering by a rectangular submarine trench. ANZIAM J. 56: 286–298.

Chakrabarti, R.N. and Mandal, B.N. 1983. Singularities in a two-fluid medium. International Journal of Mathematics and Mathematical Sciences. 6(4): 737–754.

Chamberlain, P.G. and Porter, D. 2005. Wave scattering in a two-layer fluid of varying depth. J. Fluid Mech. 524: 207–229.

Chung, H. and Fox, C. 2002. Calculation of wave-ice interaction using the Wiener-Hopf technique. New Zealand J. Math. 31: 1–18.

Chung, H. and Linton, C.M. 2005. Reflection and transmission of wave across a gap between two semi-infinite elastic plates on water. Q. Jl. Mech. Appt. Math. 58: 1–15.

Churchill, R.V., Brown, J.W. and Verhey, R.F. 1996. Complex Variables and Applications, McGraw-Hill (New York).

Das, D. and Mandal, B.N. 2005. A note on the solution of the dispersion equation for small-amplitude internal waves. Arch. Mech. 57: 449–457.

Das, D. and Mandal, B.N. 2006. Oblique wave scattering by a circular cylinder submerged beneath an ice-cover. Int. J. Engng. Sci. 44: 166–179.

Das, D. and Mandal, B.N. 2007. Wave scattering by a horizontal circular cylinder in a two layer fluid with an ice-cover. Int. J. Engng. Sci. 45: 842–812.

Das, D. and Mandal, B.N. 2009. Wave scattering by a circular cylinder half-immersed in water with an ice-cover, Int. J. Engng. Sci. 47: 463–474.

Das, D., Mandal, B.N. and Chakrabarti, A. 2008. Energy identities in water wave theory for free surface boundary condition with higher-order derivatives. Fluid Dyn. Res. 40: 253–272.

Das, P., Dolai, D.P. and Mandal, B.N. 1997. Oblique wave diffraction by two parallel thinbarriers with gaps. J. Wtry. Port Coast Ocean Engng. 123: 163–171.

Davies, A.G. 1982. The reflection of wave energy by undulations in the seabed, Dyn. Atmos. & Oceans. 8: 207–232.

Davies, A.G. and Heathershaw, A.D. 1984. Surface-wave propagation over sinusoidally varying topography. J. Fluid Mech. 144: 419–443.

De, S., Gayen, R. and Mandal, B.N. 2005. Water wave scattering by two partially immersed nearly vertical barriers, Wave Motion. 43: 167–175.

De, S., Mandal, B.N. and Chakrabarti, A. 2009. Water wave scattering by two submerged plane vertical barriers—Abel integral equations approach. J. Engng. Math. 65: 75–87.

De, S., Mandal, B.N. and Chakrabarti, A. 2010. Use of Abel integral equations in water wave scattering by-two surface-piercing barriers, Wave Motion. 47: 279–288.

De, S., Mandal, B.N. and Chakrabarti, A. 2013. Water wave scattering by two thin vertical barriers with apertures. IJAMES. 7(2): 161–175.

Dean, W.R. 1948. On the reflection of surface waves by a submerged circular cylinder. Proc. Camb. Phil. Soc. 44: 493–489.

Dean, R.G. and Ursell, F. 1959. Interaction of a fixed semi-immersed circular cylinder with a train of surface waves, M. I. T. Hydrodynamics Laboratory. Tech. Rep. no. 37.

Dhillon, H., Banerjea, S. and Mandal, B.N. 2014. Wave scattering by a thin vertical barrier in a two-layer fluid. Int. J. Engng. Sci. 78: 73–88.

Dhillon, H., Banerjea, S. and Mandal, B.N. 2013. Oblique wave scattering by a semi-infinite rigid dock in the presence of bottom undulations. Indian J. Pure Appl. Math. 44: 167–184.

England, A.H. 1971. Complex variable methods in elasticity. Interscience, New York.

Estrada, R. and Kanwal, R.P. 2000. Singular integral equations. Birkhauser, Boston.

Evans, D.V. 1970. Diffraction of water waves by a submerged vertical plate. J. Fluid Mech. 40: 433–451.

Evans, D.V. 1975. A note on the total reflection or transmission of surface waves in the presence of parallel obstacles. J. Fluid Mech. 67: 465–472.

Evans, D.V. 1976. A note on the waves produced by the small oscillations of a partially immersed vertical plate. J. Inst. Maths. Applics. 17: 135–140.

Evans, D.V. and Davies, T.V. 1968. Wave ice interaction, Rep. no. 1313, Davidson Laboratory, Stevents Inst. of Tech. New Jersey, USA.

Evans, D.V. and Fernyhough, M. 1995. Edge waves along periodic coastlines, Part 2. J. Fluid Mech. 297: 307–325.

Evans, D.V. and Linton, C.M. 1994. On step approximations for water wave problems. J. Fluid Mech. 278: 229–249.

Evans, D.V. and Morris, C.A.N. 1972. The effect of a fixed vertical barrier on obliquely incident surface waves in deep water. J. Inst. Maths. Applics. 9: 198–204.

Evans, D.V. and Porter, R. 2003. Wave scattering by narrow cracks in ice sheets floating on water of finite depth. J. Fluid Mech. 484: 143–165.

Fitz-Gerald, G.F. 1976. The reflection of plane gravity waves travelling in water of variable depth. Phil. Trans. Roy. Soc. Lond. 34: 49–89.

Fox, C. and Squire, V.A. 1990. Reflection and transmission characteristics at the edge of shore fast sea ice. J. Geophys. Res. 95: 11629–11639.

Fox, C. and Squire, V.A. 1994. On the oblique reflection and transmission of ocean waves at shore fast sea-ice. Phil. Trans. R. Soc. Lond. A347: 185–218.

Friedrichs, K.O. and Lewy, H. 1949. The dock problem. Comm. Pure Appl. Math. 2: 135–148.

Gabov, S.A., Sveshnikov, A.G. and Shatov, A.K. 1989. Dispersion of internal waves by an obstacle floating on the boundary separating two liquids. Prikl. Mat. Mech. 53: 727–730.

Gakhov, F.D. 1966. Boundary Value Problems. Pergamon Press, Oxford.

Garrison, C.J. 1969. On the interaction of an infinite shallow draft cylinder oscillating at the free surface with a train of oblique waves. J. Fluid Mech. 39: 227–255.

Gayen, R. and Mandal, B.N. 2009. Scattering of surface water waves by a floating elastic plate in two dimensions, SIAM J. Appl. Math. 69: 1520–1541.

Gayen, R., Mandal, B.N. and Chakrabarti, A. 2005. Water wave scattering by an ice-strip. J. Engng. Math. 53: 21–37.

Gayen, R., Mandal, B.N. and Chakrabarti, A. 2006. Water wave scattering by two sharp discontinuities in the surface boundary conditions. IMA J. Appl.-Math. 71: 811–831.
Gayen, R., Mandal, B.N. and Chakrabarti, A. 2007. Water wave diffraction by a surface strip. J. Fluid Mech. 571: 419–432.
Gol'dshtein, R.V. and Marchenko, A.V. 1989. The diffraction of plane gravitational waves by the edge of an ice-cover. PMM 53: 731–736.
Havelock, T.H. 1929. Forced surface waves on water. Phil. Mag. 8: 569–576.
Hamilton, J. 1977. Differential equations for long-period gravity waves on fluid of rapidly varying depth. J. Fluid Mech. 83: 289–310.
Hermans, A.J. 2004. Interaction of free-surface waves with floating flexible strips. J. Engng. Math. 49: 133–147.
Holford, R.L. 1964. Short surface waves in the presence of a finite dock I. Proc. Phil. Soc. 60: 957–983.
Holford, R.L. 1964. Short surface waves in the presence of a finite dock II. Proc. Phil. Soc. 60: 985–1011.
Jarvis, R.J. 1971. The scattering of surface waves by two vertical plane barriers. J. Inst. Maths. Applics. 7: 207–215.
John, F. 1948. Waves in the presence of an inclined barrier. Comm. Pure Appl. Math. 1: 149–200.
Jones, D.S. 1964. The Theory of Electromagnetism. Pergamon Press, New York.
Kanoria, M. and Mandal, B.N. 2002. Water wave scattering by a submerged circular arc shaped plate. Fluid Dyn. Res. 31: 317–331.
Kanoria, M., Mandal, B.N. and Chakrabarti, A. 1999. The wiener-Hopf solution of a class of mixed boundary value problems arising in surface water wave phenomena. Wave Motion. 29: 267–292.
Kashiwagi, M. 1998. A B-spline Galerkin scheme for calculation the hydroelastic response of a very large floating structure in waves. J. Mar. Sci. Tech. 3: 37–49.
Kashiwagi, M. 2000. Research on hydroelastic responses of VLFS: Recent progress and further work. Int. J. Offshore and Polar Engng. 10: 81–90.
Kashiwagi, M., Ten, I. and Yesunaga, M. 2006. Hydrodynamics of a body floating in a two-layer fluid of finite depth, part 2. Diffraction problem and wave incident motion. J. Mar. Sci. Tech. 11: 150–164.
Kagemoto, H., Masataka, F. and Motohika, M. 1998. Theoretical and experimental predictions of the hydroelastic response of a very large floating structure in waves. Appl. Ocean Res. 20: 135–144.
Khapasheva, T.I. and Korobkin, A.A. 2002. Hydroelastic behavior of compound floating plate in waves. J. Engng. Math. 44: 21–40.
Kirby, J.T. 1986. A general wave equation for waves over rippled beds. J. Fluid Mech. 162: 171–186.
Kirby, J.T. and Dalrymple, R.A. 1983. Propagation of obliquely incident water waves over a trench. J. Fluid Mech. 133: 47–63.
Kreisel, H. 1949. Surface waves, Quart. Appl. Math. 7: 21–44.
Kohout, A.L., Meylan, M.H., Sakai, S., Hanai, K., Leman, P. and Brossard, D. 2007. Linear water wave propagation through multiple elastic plates of variable properties. J. Fluids Struct. 23: 649–663.
Kuznetsov, N., Maz'ya, V. and Vainberg, B. 2002. Linear Water Waves. A Mathematical Approach, Cambridge University Press.
Lamb, H. 1932. Hydrodynamics, Dover. New York.
Landau, L.D. and Lifshitz, E.M. 1959. Theory of Elasticity, Pergamon Press. London.
Lassiter, J.B. 1972. The propagation of water waves over sediment pockets. Ph.D. thesis Massachusetts Institute of Technology.
Lawrie, J.B. and Abrahams, I.D. 1999. An orthogonality relation for a class of problems with higher-order boundary conditions; applications in sound-structure interaction. Q. J. Mech. Appl. Math. 52: 161–181.
Lee, J.J. and Ayer, R.M. 1981. Wave propagation over a rectangular trench. J. Fluid Mech. 110: 335–347.
Levine, H. 1965. Scattering of surface waves by a submerged circular cylinder. J. Math. Phys. 6: 1221–1243.
Levine, H. and Rodemich, E. 1958. Scattering of surface wives on an ideal fluid. Stanford Univ. Tech. Rep. No. 78, Math. Stat. Lab. 1958.
Linton, C.M. and Cadby, J.R. 2002. Scattering of oblique waves in a two-layer fluid. J. Fluid Mech. 461: 343–364.
Linton, C.M. and Chung, H. 2003. Reflection and transmission at the ocean/sea-ice boundary. Wave Motion 38: 43–52.
Linton, C.M. and McIver, M. 1995. The interaction of waves with horizontal cylinders in two-layer fluids. J. Fluid Mech. 304: 213–229.
Linton, C.M. and McIver, P. 2001. Handbook of Mathematical Techniques for Wave/Structure Interactions, Chapman & Hall/CRC.

Liu, H.W., Fu, D.J. and Sun, X.L. 2013. Analytical solution to the modified mild slope equation for reflection by rectangular break water with scour trenches. ASCE J. Engng. Mech. 139: 39–58.

Losada, I.J., Silva, R. and Losada, M.A. 1996. 3-D non breaking regular wave interaction with submerged breakwaters, Coastal Engng. 28: 229–248.

Maiti, P. and Mandal, B.N. 2004. Oblique wave scattering by undulations on the bed of an ice-covered ocean. Arch. Mech. 61: 485–493.

Maiti, P. and Mandal, B.N. 2006. Scattering of oblique waves by bottom undulations in a two-layer fluid, J. Appl. Math. & Computing. 22: 21–39.

Maiti, P. and Mandal, B.N. 2008. Water wave scattering by bottom undulations in an ice-covered two-layer fluid. Appl. Ocean Research. 30: 264–272.

Maiti, P. and Mandal, B.N. 2010. Wave scattering by a thin vertical barrier submerged beneath an ice-cover. Appl. Ocean Res. 32: 367–373.

Manam, S.R., Bhattacharjee, J. and Sahoo, T. 2006. Expansion formulae in wave structure interaction problems. Proc. R. Soc. Lond. A. 462: 1145–1162.

Mandal, B.N. and Basu, U. 1994. Oblique interface wave diffraction by a small bottom deformation in the presence of interfacial tension. Rev. Roum. Sci. Techn-M'ec. Appl. 39: 525–531.

Mandal, B.N. and Basu, U. 1990. A note on oblique water wave diffraction by a cylindrical deformation of the bottom in the presence of surface tension. Arch. Mech. 42: 723–727.

Mandal, B.N. and Basu, U. 2004. Wave diffraction by a small elevation of the bottom of an ocean with an ice-cover. Arch. Appl. Mech. 73: 812–822.

Mandal, B.N and Chakrabarti, A. 1989. A note on diffraction of water waves by a nearly vertical barrier. IMA J. Appl. Math. 43: 157–165.

Mandal, B.N. and Chakrabarti, A. 2000. Water Waves Scattering by Barriers. WIT press, Southampton. U.K.

Mandal, B.N. and Chakraborty, R. 2012. Water wave scattering by an undulating bottom, Bull. Cal. Math. Soc. 104(6): 533–546.

Mandal, B.N. and Chakrabarti, R.N. 1986. Two-dimensional source potential in two-fluid medium for the modified Helmholtz's equation, Int. J. Math. and Math. Sci. 9: 175–184.

Mandal, B.N. and De, S. 2006. Water wave scattering by two submerged nearly vertical barriers. ANZAM J. 48: 107–118.

Mandal, B.N. and De, S. 2007. Water wave scattering by bottom undulations in the presence of a thin submerged vertical plate. Int. J. Appl. Math & Engg Sci. 1: 193–205.

Mandal, B.N. and De, S. 2009. Surface wave propagation over small undulations at the bottom of an ocean with surface discontinuity. Geophys. & Astrophys. Fluid Dyn. 103: 19–30.

Mandal, B.N. and Dolai, D.P. 1994. Oblique water wave diffraction by thin vertical barriers in water of uniform finite depth. Appl. Ocean. Res. 16: 195–203.

Mandal, B.N. and Gayen, R. 2006. Water wave scattering by bottom undulations in the presence of a thin partially immersed barrier. Appl. Ocean Res. 28: 113–119.

Mandal, B.N. and Gayen, R. 2002. Water wave scattering by two symmetric circular arcs shaped thin plates. J. Engng. Math. 44: 297–303.

Mandal, B.N. and Goswami, S.K. 1984. Scattering of surface waves obliquely incident on fixed a half immersed circular cylinder. Math. Proc. Camb. phil. Soc. 96: 359–369.

Mandal, B.N. and Kanoria, M. 2004. Oblique wave scattering by thick barriers. J. Offshore Mechanics and Arctic Engng. 182: 100–108.

Mandal, B.N. and Maiti, P. 2000. Oblique wave scattering by undulations on the bed of an ice-covered ocean. Arch. Mech. 56: 495–493.

Martin, P.A. and Dalrymple, R.A. 1988. Scattering of long waves by cylindrical obstacles and gratings using matched asymptotic expansion. J. Fluid Mech. 108: 465–498.

McIver, M. and Urka, U. 1995. Wave scattering by circular arc shaped plates. J. Engng. Maths. 29: 575–589.

McIver, P. 1985. Scattering of water waves by two surface piercing vertical barriers. IMA J. Appl. Math. 34: 339–355.

McIver, P. 1994. Low frequency asymptotic of hydrodynamic forces on fixed and floating structures. Ocean Wave Engineering. M. Rahaman (ed.). Computational Mechanics Publications, Southampton, U.K.

Mei, C.C. 1985. Resonant reflection of surface water waves by periodic sandbars. J. Fluid Mech. 52: 315–335.

Mei, C.C. 1982. The applied Dynamics of Ocean surface waves. Wiley Inter science, U.S.A.

Mei, C.C. and Black, J.L. 1969. Scattering of surface waves by rectangular obstacles in waters of finite depth. J. Fluid Mech. 38: 499–511.

Mei, C.C., Hara, T. and Naciri, M. 1988. Note on Bragg scattering of water waves by parallel bars on the seabed. J. Fluid Mech. 186: 147–162.

Miles, J.W. 1981. Oblique surface wave diffraction by a cylindrical obstacle, Dyn. Atmos. & Oceans. 6: 121–123.

Miles, J.W. 1982. On surface wave diffraction by a trench. J. Fluid Mech. 115: 315–325.

Meylan, M.H. 1995. A flexible vertical sheet in waves. Int. J. Offshore & Polar Engng. 5: 105–110.

Meylan, M.H. and Squire, V.A. 1993. Finite floe reflection and transmission coefficients from a semi-infinite model. J. of Geophysical Research. 98(C7): 12537–12542.

Meylan, M.H. and Squire, V.A. 1994. The Response of Ice Floes to Ocean Waves. J. of Geophysical Research. 99(C1): 891–900.

Muskhelishvilli, N.I. 1953. Singular Integral Equations. Noordhoff, Gronigen.

Namba, Y. and Okhusu, M. 1999. Hydroelastic behavior of artificial islands in waves. Int. J. Offshore & Polar Engng. 9: 39–47.

Newman, J.N. 1976. The interaction of stationary vessel with regular waves. In: International Proceedings of the 11th Symposium on Naval Hydrodynamics, ONR.

Newman, J.N. 1994. Wave effect on deformable bodies. Appl. Ocean Res. 16: 47–59.

Noble, B. 1958. Methods Based on the Wiener-Hopf Technique. Pergamon Press, New York.

Okhusu, M. and Namba, Y. 2004. Hydroelastic analysis of a large floating structure. J. Fluids Struct. 19: 543–555.

Ogilive, T.F. 1963. First and second order forces on a cylinder submerged under a free surface. J. Fluid. Mech. 10: 451–472.

Parsons, N.F. and Martin, P.A. 1992. Scattering of water waves by submerged plates using hypersingular integral equations. Appl. Ocean Res. 14: 313–321.

Parsons, N.F. and Martin, P.A. 1994. Scattering of water waves by submerged curved and by surface-piercing flat plates. Appl. Ocean Res. 16: 129–139.

Peters, A.S. 1950. The effect of a floating mat on water waves. Commun. Pure Appl. Maths. 3: 319–354.

Porter, D. 1972. The transmission of surface waves through a gap in a vertical barrier. Proc. Camb. Phil. Soc. 71: 411–421.

Portet, R. and Eyans, D.V. 1995. Complementary approximations to wave scattering by vertical barriers. J. Fluid Mech. 294: 160–86.

Porter, R. 2002. Surface wave scattering by submerged cylinders of arbitrary cross-section. Proc. R. Soc. A. 458: 581–606.

Porter, D. and Porter, R. 2004. Approximations to wave scattering by an ice sheet of variable thickness over undulating topography. J. Fluid Mech. 509: 145–179.

Ralston, A. 1965. A First Course in Numerical Analysis. Mc Graw Hill Book Co., New York.

Rhodes-Robinson, P.F. 1971. On the forced surface waves due to a vertical wave maker in the presence of surface tension. Proc. Camb. Phil. Soc. 70: 323–337.

Rhodes-Robinson, P.F. 1984. On the generation of water waves at an inertial surface. J. Austral. Math. Soc. B. 25: 366–383.

Rhodes-Robinson, P.F. 1994. On wave motion a two-layered liquid of infinite depth in the presence of surface and interfacial tension. J. Austral. Math.l Soc. B. 35: 302–322.

Roseau, M. 1976. Asymptotic Wave Theory. North Holland American Elsevier. 311–347.

Shaw, D.C. 1985. Perturbational results for diffraction of water waves by nearly vertical barriers. IMA J. Appl. Math. 34: 99–117.

Sahoo, T., Yip, T.L. and Chwang, A.T. 2001. Scattering of surface waves by a semi-infinite floating elastic plate. Phys. Fluids. 13(11): 3215–3222.

Squire, V.A., Dugan, J.P., Wadahams, P., Rottier, P.J. and Liu, A.K. 1995. Of ocean waves and ice-sheets. Anun. Rev. Fluid Mech. 27: 115–168.

Sherief, H., Faltas, M.S. and Saad, E.I. 2003. Forced gravity waves in two layered fluids with the upper fluid having a free surface. Can. J. Phys. 81: 675–689.

Sherief, H.H., Faltas, M.S. and Saad, E.I. 2004. Axisymmetric gravity waves in two-ayered fluids with the upper fluid having a free surface. Wave Motion. 40: 143–161.

Stokes, G.G. 1847. On the theory of oscillatory waves, Trans. Camb. Phil. Soc. 8: 441–445, Reprinted in Mathematical and Physical Papers, London. 1: 314–326.

Ten, I. and Kashiwagi, M. 2004. Hydrodynamics of a body floating in a two-layer fluid of finite depth, part 1 Radiation problem. J. Mar. Sci. Technol. 9: 127–141.

Tkacheva, L.A. 2001. Scattering of surface waves by the edge of a floating elastic plate. J. Appl. Mech. & Tech. Phys. 42: 638–646.

Tkacheva, L.A. 2003. Plane problem of surface wave diffraction on a floating elastic plate. Fluid Dynamics. 38: 465–481.

Thorne, R.C. 1953. Multipole expansion in the theory of surface waves. Proc. Camb. Phil. Soc. 49: 707–716.

Ursell, F. 1947. The effect of a fixed vertical barrier on surface waves in deep water. Proc. Camb. Phil. Soc. 43: 374–382.

Ursell, F. 1950. Surface waves in deep water in the presence of a submerged circular cylinder, I, II. Proc. Camb. Phil. Soc. 46: 141–158.

Ursell, F. 1961. The transmission of surface waves under surface obstacles. Proc. Camb. Phil. Soc. 57: 638–668.

Ursell, F. 1968. The expansion of water-wave potentials at great distances. Proc. Camb. Phill. Soc. 64: 811–826.

Ursell, F., Dean, R.G. and Yu, Y.S. 1960. Forced Small Amplitude Water Waves: A Comparison of Theory and Experiment. J. Fluid Mechanics. 7: 33.

Varley, E. and Walker, J.D.A. 1989. A method for solving singular integro-differential equations. IMA J. Appl. Math. 43: 11–45.

Wehausen, J.V. and Laitone, E.V. 1960. Surface Waves, Handbuch Der Physik, Vol. 9, ed. Flugge S., Springer-Verlag, Berlin.

Weitz, M. and Keller, J.B. 1950. Reflection of water waves from ice in water of finite depth. Comm. Pure. Appl. Math. 3: 305–318.

Williams, T.D. and Squire, V.A. 2006. Scattering of flexural-gravity waves at the boundaries between three floating sheets with applications. J. Fluid Mech. 569: 113–140.

Wu, C., Watanabe, E. and Utsunamiya, T. 1995. An eigenfunction expansion matching method for analysing the wave-induced responses of an elastic floating plate. Appl. Ocean Res. 17: 301–310.

Xie, J.J., Liu, H.W. and Lin, P. 2011. Analytical solution for long wave reflection by a rectangular obstacle with two scours trenches. ASCE-Journal of Engng. Mech. 137: 39–58.

Yu, Y.S. and Ursell, F. 1961. Surface waves generated by an oscillating circular cylinder on water of finite depth: theory and experiments. J. Fluid Mech. 11: 529–551.

Yeung, R. and Nguyen, T. 2000. Radiation and diffraction of waves in a two-layer fluid. Int. Proc. 22nd Symp. on Noval Hydrodynamics, Washington.

Subject Index

T

U

V

W

Author Index